Pasteurellaceae

Biology, Genomics, and Molecular Aspects

Peter Kuhnert

Institute of Veterinary Bacteriology
Vetsuisse Faculty
University of Bern
Bern
Switzerland

and

Henrik Christensen

Department of Veterinary Pathobiology
Copenhagen University
Frederiksberg
Denmark

Caister Academic Press

Caister Academic Press
Norfolk, UK

www.caister.com

British Library Cataloguing-in-Publication Data
A catalogue record for this book is available from the British Library

ISBN: 978-1-904455-34-9

Printed and bound in Great Britain

Contents

Other Books of Interest

Mycobacterium: Genomics and Molecular Biology	2009
Real-Time PCR: Current Technology and Applications	2009
Clostridia: Molecular Biology in the Post-genomic Era	2009
Plant Pathogenic Bacteria: Genomics and Molecular Biology	2009
Microbial Production of Biopolymers and Polymer Precursors	2009
Plasmids: Current Research and Future Trends	2008
Vibrio cholerae: Genomics and Molecular Biology	2008
Pathogenic Fungi: Insights in Molecular Biology	2008
Corynebacteria: Genomics and Molecular Biology	2008
Leishmania: After The Genome	2008
Archaea: New Models for Prokaryotic Biology	2008
RNA and the Regulation of Gene Expression	2008
Legionella Molecular Microbiology	2008
Molecular Oral Microbiology	2008
Epigenetics	2008
Animal Viruses: Molecular Biology	2008
Segmented Double-Stranded RNA Viruses	2008
Acinetobacter Molecular Microbiology	2008
Pseudomonas: Genomics and Molecular Biology	2008
Microbial Biodegradation: Genomics and Molecular Biology	2008
The Cyanobacteria: Molecular Biology, Genomics and Evolution	2008
Coronaviruses: Molecular and Cellular Biology	2007
Real-Time PCR in Microbiology: From Diagnosis to Characterisation	2007
Bacteriophage: Genetics and Molecular Biology	2007
Candida: Comparative and Functional Genomics	2007
Bacillus: Cellular and Molecular Biology	2007
AIDS Vaccine Development: Challenges and Opportunities	2007
Alpha Herpesviruses: Molecular and Cellular Biology	2007
Pathogenic *Treponema*: Molecular and Cellular Biology	2007
PCR Troubleshooting: The Essential Guide	2006
Influenza Virology: Current Topics	2006
Microbial Subversion of Immunity: Current Topics	2006
Cytomegaloviruses: Molecular Biology and Immunology	2006
Papillomavirus Research: From Natural History To Vaccines and Beyond	2006
Epstein Barr Virus	2005
HIV Chemotherapy: A Critical Review	2005

Contributors

Ben Adler
Australian Research Council Centre of Excellence in Structural and Functional Microbial Genomics
Department of Microbiology
Monash University
Victoria
Australia

Ben.Adler@med.monash.edu.au

Nataliya V. Balashova
Department of Oral Biology
University of Medicine and Dentistry of New Jersey
Newark, NJ
USA

balashnv@umdnj.edu

Magne Bisgaard
Department of Veterinary Pathobiology
Copenhagen University
Frederiksberg
Denmark

mbi@life.ku.dk

Pat Blackall
Department of Primary Industries and Fisheries (Queensland)
Animal Research Institute
Yeerongpilly
Australia

Pat.Blackall@dpi.qld.gov.au

John D. Boyce
Australian Research Council Centre of Excellence in Structural and Functional Microbial Genomics
Department of Microbiology
Monash University
Victoria
Australia

John.Boyce@med.monash.edu.au

Jean F. Challacombe
DOE Joint Genome Institute and Bioscience Division
Los Alamos National Laboratory
Los Alamos, NM
USA

jchalla@lanl.gov

Henrik Christensen
Department of Veterinary Pathobiology
Copenhagen University
Frederiksberg
Denmark

hech@life.ku.dk

Jacqueline W. Chung
Department of Microbiology and Immunology
McGill University
Montreal
Canada

jacqueline.chung@mail.mcgill.ca

James W. Coulton
Department of Microbiology and Immunology
McGill University
Montreal
Canada

james.coulton@mcgill.ca

Joachim Frey
Institute of Veterinary Bacteriology
Vetsuisse Faculty
University of Bern
Bern
Switzerland

joachim.frey@vbi.unibe.ch

Thomas J. Inzana
Virginia-Maryland Regional College of Veterinary Medicine
Virginia Polytechnic Institute and State University
Blacksburg, VA
USA

tinzana@vt.edu

Mario Jacques
Département de pathologie et microbiologie
Faculté de médecine vétérinaire
Université de Montréal
St-Hyacinthe
Canada

mario.jacques@umontreal.ca

Scott C. Kachlany
Department of Oral Biology
University of Medicine and Dentistry of New Jersey
Newark, NJ
USA

kachlasc@umdnj.edu

Bożena M. Korczak
Institute of Veterinary Bacteriology
Vetsuisse Faculty
University of Bern
Bern
Switzerland

bozena.korczak@vbi.unibe.ch

Peter Kuhnert
Institute of Veterinary Bacteriology
Vetsuisse Faculty
University of Bern
Bern
Switzerland

peter.kuhnert@vbi.unibe.ch

Heather Maughan
Department of Zoology
University of British Columbia
Vancouver
Canada

maughan@zoology.ubc.ca

Niels Nørskov-Lauritsen
Aarhus University Hospital
Aarhus
Denmark

nnl@sks.aaa.dk

Rosemary Redfield
Department of Zoology
University of British Columbia
Vancouver
Canada

redfield@zoology.ubc.ca

Indra Sandal
Virginia Polytechnic Institute and State University
Blacksburg, VA
USA

isandal@vt.edu

Stefan Schwarz
Institute of Farm Animal Genetics
Friedrich-Loeffler-Institute (FLI)
Neustadt-Mariensee
Germany

stefan.schwarz@fli.bund.de

Shivakumara Siddaramappa
Virginia Polytechnic Institute and State University
Blacksburg, VA
USA

ssiddara@vt.edu

Sunita Sinha
Department of Zoology
University of British Columbia
Vancouver
Canada

sinha@zoology.ubc.ca

W. Edward Swords
Wake Forest University Health Sciences
Winston-Salem, NC
USA

wswords@wfubmc.edu

Lindsay Wilson
Department of Zoology
University of British Columbia
Vancouver
Canada

wilson@zoology.ubc.ca

Preface

The bacterial family *Pasteurellaceae*, honouring in its name Louis Pasteur, a pioneer of microbiology, comprises mainly commensal species but also includes important pathogens of man and mostly animals. Some of these bacteria represent real threats whereas others might impair health and welfare. In animals this leads to increased costs in livestock farming, reduced food production and significant economical loss. Since the last books published on the family in 1989 ('*Pasteurella* and Pasteurellosis') and 1995 ('*Haemophilus, Actinobacillus* and *Pasteurella*') we have entered the genomics era with new topics having arisen as well as new methods and approaches have been developed.

Exploring the biology, genomics and molecular aspects of *Pasteurellaceae* in a book can not be really comprehensive. The family expanded during the last 10 years from the classical 3 genera to a currently 13 and this number is certainly outdated by the time of publishing this book since it is expected to rise based on new information gained on many misclassified taxa but also due to new species that are going to be described. Therefore, it is essential to cover the current taxonomy and give an outlook on future developments in this field. Molecular methods especially based on DNA sequence comparison have enabled taxonomic clarification of many groups previously only known by phenotype. We therefore start with an overview of the recent changes and reclassifications with giving current names as well as a few hypotheses of how members of the family *Pasteurellaceae* and members of the vertebrates probably evolved in parallel. It is important to mention that we stick to the current and official nomenclature throughout the book and the reader is referred to this basic first chapter to clarify uncertainties in this respect. Closely related to it is the phylogeny of the family which helps a lot to resolve the often chaotic taxonomy that is still valid. This is as well of basic biological interest to elucidate the relationship and biodiversity of members of *Pasteurellaceae*. In a time of genome wide exploration of genetics it is tantamount to include this topic. A few genomes from species of the family have been sequenced and it is noteworthy to mention that the first ever sequenced bacterial genome in 1995 was a member of *Pasteurellaceae* thereby opening the genomic era. Staying at the DNA level there have nice basic studies been carried out in *Pasteurellaceae* on the mechanism of competence and DNA uptake which is presented in a separate chapter. A focus on the protein level is included with chapters dealing with proteome analysis, an aspect still marginally investigated but with promising initial results, as well as the topic of protein secretion, a quite multifaceted feature. Most research in *Pasteurellaceae* is done on those species which act as opportunistic or primary pathogens. Topics dealing with mechanisms involved in pathogenesis are included in a number of sections. Toxins are discussed in several parts of the book and additionally the RTX toxins are dealt in an own chapter. These types of toxins are quite often found in members of the family and are well studied. However, other mechanisms and structures have been determined to be involved in pathogenesis and are discussed in chapters on outer membrane proteins and iron uptake as well as on biofilm formation which includes the large topic of LPS and quorum sensing. Antibiotic resistance within *Pasteurellaceae* has for a long time been neglected but fortunately is now a field of research intensively followed and comprehensively summarized in one chapter. Finally, for every effort to study *Pasteurellaceae*, to treat and combat diseases as well as to determine new species, isolation and a solid diagnosis is paramount. How such a diagnosis should look like and which aspects have to be considered is elucidated in the final chapter of the book. Members of *Pasteurellaceae* are often difficult to identify since they are not easy to cultivate. A rather detailed description of isolation and identification of human as well as animal members of *Pasteurellaceae* is therefore given. Such description with specific reference to *Pasteurellaceae* has been missing for decades. A few small overlaps between chapters are there on purpose and are intertwining them.

The *Pasteurellaceae* are a big and diverse family given the many taxa they comprise and hosts they can be found in. The mainly commensal nature of *Pasteurellaceae* allows to investigate the role of the normal flora and its interactions with the host. On the other side given the fact that most infections caused by *Pasteurellaceae* affect animals we have natural models to study pathogenesis and host-pathogen-interactions which might also be representative for related human diseases. Certainly the family *Pasteurellaceae* will continue to provide an open field for future research activities for which this book might hopefully serve as a stimulating guideline.

Peter Kuhnert, University of Bern, Switzerland
Henrik Christensen, University of Copenhagen, Denmark

Taxonomy and Biodiversity of Members of *Pasteurellaceae*

1

Henrik Christensen and Magne Bisgaard

Abstract

The family *Pasteurellaceae* Pohl 1981 includes 38 properly classified species in addition to 24 misclassified species. The majority of taxa have been isolated from disease conditions in warm blooded animals and in particular in farm animals. These bacteria are obligate parasites or commensals of vertebrates, colonizing mainly the mucosal surfaces of the upper respiratory tract, oropharynx, and reproductive tracts and possibly also parts of the intestinal tract. Most taxa represent potential pathogens although mechanisms of virulence have remained unknown or doubtful until recent years. Both systemic and local infections have been reported for most taxa involved in diseases. However, pneumonia has been reported most frequently out of a number of other disease manifestations. Fossil remnants of members of *Pasteurellaceae* have never been reported and information on the diversification of taxa within the family can only be obtained by phylogenetic reconstruction. Most likely the ancestors of the current members of *Pasteurellaceae* might have been present in association with the ancestors of for example birds and dinosaurs. For marsupials, monotremes and reptiles information is very limited and further insight from these groups would be very helpful to test the hypothesis of co-evolution of host and parasite.

Introduction

Biodiversity is 'the number and distribution of species in the biosphere' (Bridgewater, 1988; Wilson and Fances, 1988). High biodiversity means many species demonstrating variability in biological properties. Biodiversity in this chapter refers to both the diversity of taxa of *Pasteurellaceae* Pohl 1981 (Anon, 1981) in specific habitats and to the diversity of members of *Pasteurellaceae* compared to other microorganisms.

Our present knowledge of the biodiversity of the family *Pasteurellaceae* mainly comes from investigation of taxa associated with disease in man and animals. Systematic investigations including all mucosal membranes within different phylogenetic groups of animals, however, remains to be carried out.

Presently, the family *Pasteurellaceae* is considered as a taxonomically relatively homogenous group of bacteria. The majority of taxa have been isolated from disease conditions in warm-blooded animals and in particular in farm animals. All taxa most probably have mucosal membranes as their primary habitat. They express a high diversity related to colonization of different animal habitats, hosts and ways of causing disease. Interestingly, most taxa represent potential pathogens and mechanisms of virulence have remained unknown or doubtful until recent years.

The aim of this chapter is to provide a guide to presently reported taxa of the family *Pasteurellaceae* and to provide an outline of how these taxa are distributed mainly in relation to habitats, hosts and diseases.

Classification

Overview

The family *Pasteurellaceae* includes 38 properly classified species in addition to 24 misclassified species (Table 1.1). The family *Pasteurellaceae* is classified as the only member of the order *Pasteurellales*. This order is included within the *Gamma-Proteobacteria* of the phylum *Proteobacteria* (Brenner *et al.*, 2005). Members of the bacterial family *Pasteurellaceae* have been characterized as Gram-negative, non-motile rods, coccoids or filaments with aerobic, microaerophilic or facultative anaerobic respiration and chemoorganotrophic metabolism and with the ability to reduce nitrate. In addition, 21 phenotypic characters have been found to be conserved for existing genera (Christensen *et al.*, 2007a).

The bacterial family *Pasteurellaceae* was originally proposed to include the genera *Pasteurella, Actinobacillus* and *Haemophilus* (Pohl, 1979). From the early 1980s a general agreement existed on the polyphyly within the genera *Actinobacillus, Pasteurella* and *Haemophilus* (Mutters *et al.*, 1989) and that these genera needed reclassification conserving the 'core' genera and formation of several new genera (Pohl, 1981; Deley *et al.*, 1990; Dewhirst *et al.*, 1992). The formation of *Mannheimia* was followed by proposals of *Gallibacterium, Histophilus, Avibacterium, Aggregatibacter* and most recently *Bibersteinia* (Angen *et al.*, 1999, 2003; Blackall *et al.*, 2005,

Table 1.1 Names of members of *Pasteurellaceae* with 'standing in nomenclature'

Species and subspecies*	Year**
Actinobacillus arthritidis	2002
A. capsulatus	1962
A. equuli subsp. *equuli*	1934
A. equuli subsp. *haemolyticus*	2002
A. hominis	1985
A. lignieresii	1910
A. pleuropneumoniae	1983
A. suis	1962
A. ureae	1986
[*A.*] *delphinicola*	1996
[*A.*] *indolicus*	1996
[*A.*] *minor*	1996
[*A.*] *muris*	1988
[*A.*] *porcinus*	1996
[*A.*] *rossii*	1990
[*A.*] *scotiae*	1998
[*A.*] *seminis*	1990
[*A.*] *succinogenes*	1999
Aggregatibacter actinomycetemcomitans	2006
Agg. aphrophilus	2006
Agg. segnis	2006
Avibacterium avium	2005
Av. endocarditidis	2007
Av. gallinarum	2005
Av. paragallinarum	2005
Av. volantium	2005
Bibersteinia trehalosi	2007

Table 1.1 *continued*

Species and subspecies*	Year**
Gallibacterium anatis	2003
Haemophilus aegyptius	1950
H. haemolyticus	1923
H influenzae	1917
H. parainfluenzae	1922
H. pittmaniae	2005
[*H.*] *ducreyi*	1923
[*H.*] *felis*	1999
[*H.*] *haemoglobinophilus*	1939
[*H.*] *paracuniculus*	1984
[*H.*] *parahaemolyticus*	1953
[*H.*] *paraphrohaemolyticus*	1971
[*H.*] *parasuis*	1969
Histophilus somni	2003
Lonepinella koalarum	1996
Mannheimia granulomatis	1999
M. glucosida	1999
M. haemolytica	1999
M. ruminalis	1999
M. varigena	1999
Nicoletella semolina	2005
Pasteurella canis	1985
P. dagmatis	1985
P. multocida subsp. *multocida*	1939
P. multocida subsp. *gallicida*	1985
P. multocida subsp. *septica*	1985
P. stomatis	1985
[*P.*] *aerogenes*	1974
[*P.*] *bettyae*	1990
[*P.*] *caballi*	1990
[*P.*] *langaaensis*	1985
[*P.*] *mairii*	1990
[*P.*] *pneumotropica*	1950
[*P.*] *skyensis*	2002
[*P.*] *testudinis*	1982
Phocoenobacter uteri	2000
Volucribacter psittacicida	2003
V. amazonae	2003

*Species unrelated to the three genera *Pasteurella*, *Actinobacillus* and *Haemophilus* are informally excluded by writing the genus name of these species in brackets leaving only *sensu stricto* taxa with these three genera (Mutters *et al.*, 1989).

**Year of valid publication for species on Approved Lists of Bacterial Names (Skerman *et al.*, 1980) and year of publication in International Journal of Systematic and Evolutionary Microbiology or Validation List of this journal. Only the most recent name for each species has been included. Further information can be found on List of Bacterial Names with Standing in Nomenclature (http://www.bacterio.cict.fr/index.html).

2007; Christensen *et al.*, 2003a; Nørskov-Lauritsen and Kilian, 2006). Further changes are needed to reclassify species excluded from *Pasteurella sensu stricto, Actinobacillus sensu stricto* and *Haemophilus sensu stricto* totalling 24 species with genus names in brackets (Table 1.1). The term '*sensu stricto*' is used to define the core group of closely related species meeting the criteria for proper classification. Reclassifications within *Pasteurellaceae* are shown in Table 1.2. The most recent name achieves 'Standing in nomenclature' even though older names have been included with the Approved Lists of Bacterial names (Skerman *et al.*, 1980) or after 1980 recognized by valid publication in *International Journal of Systematic and Evolutionary Microbiology* (*International Journal of Systematic Bacteriology*) or included with the Validation List of this journal.

The four new genera, *Lonepinella, Phocoenobacter, Nicoletella* and *Volucribacter,* have been formed based on isolates obtained from the specific hosts – koala, porpoise, horse and psittacine birds respectively (Osawa *et al.*, 1995; Foster *et al.*, 2000; Kuhnert *et al.*, 2004; Christensen *et al.*, 2004a). The taxonomy of these genera was initially based on a relatively low number of isolates (seven *Lonepinella,* one *Phocoenobacter,* 19 *Nicoletella,* 25 *Volucribacter*). The seven strains of *Lonepinella* all came from Australia. The single strain of *Phocoenobacter* used to name this genus with one species was isolated from a stranded harbour porpoise with unknown cause of death. The 19 isolates of *Nicoletella* came from Sweden and Switzerland and were all obtained from the airways of diseased horses (Kuhnert *et al.*, 2004). The 25 isolates of *Volucribacter* came from Europe and most isolates of *Volucribacter* were obtained from lesions although the pathogenic potential of members of this genus also remains to be fully investigated (Christensen *et al.*, 2004a). The genus *Volucribacter* is currently expanding based

Table 1.2 Reclassification within *Pasteurellaceae*

Named within or transferred to *Pasteurellaceae*		Reclassified within *Pasteurellaceae*	
Name	Year	Name	Year
Actinobacillus actinomycetemcomitans	1929	*Aggregatibacter actinomycetemcomitans*	2006
Haemophilus avium	1977	*Pasteurella avium*	1985
H. aphrophilus	1940	*Agg. aphrophilus*	2006
H. paraphrophilus	1968	*Agg. aphrophilus*	2006
H. paragallinarum	1969	*Avibacterium paragallinarum*	2005
H. pleuropneumoniae	1964	*A. pleuropneumoniae*	1983
H. segnis	1977	*Agg. segnis*	2006
P. anatis	1985	*Gallibacterium anatis*	2003
'*Pasteurella avicida*'* '*P. muriseptica/muricida*' '*P. cuniculicida*' '*P. suiseptica/suilla*' '*P. boviseptica/bollingeri*'	1923–1939	*P. multocida*	1939
P. avium	1985	*Av. avium*	2005
P. gallinarum	1955	*Av. gallinarum*	2005
P. granulomatis	1990	*M. granulomatis*	1999
P. haemolytica	1932	[*P.*] *trehalosi* or *Mannheimia haemolytica*	1990, 1999
P. trehalosi	1990	*Bibersteinia trehalosi*	2007
P. volantium	1985	*Av. volantium*	2005
P. ureae	1962	*A. ureae*	1986

*Quotation marks indicate that names were not included with Approved Lists of Bacterial Names (Skerman *et al.*, 1980).

on characterization of new isolates from mainly captive psittacine birds (Bertelsen *et al.*, 2007). The presence of these taxa in phylogenetically related animal species, their virulence and diversity of virulence expressed by members of *Phocoenobacter*, *Nicoletella* and *Volucribacter* also remain to be investigated. The tannin-degrading ability of members of *Lonepinella* should be further investigated, including the role of tannin degradation in mucosal surface ecology.

Further changes in classification and nomenclature might be necessary depending on access to whole genomic sequences from more species of the family *Pasteurellaceae* (Buckley and Roberts, 2006) which might lead to currently unknown constellations of species and genera. In addition, investigations of animals not previously sampled for *Pasteurellaceae* may add significantly to our present understanding of the diversity of *Pasteurellaceae* (Bisgaard, 1993).

The species concept of *Pasteurellaceae*

Species have to form monophyletic groups based on 16S rRNA gene sequence phylogenetic analysis (Chapter 2). However, some species are so closely related that they cannot be separated by 16S rRNA gene sequence comparison and more sensitive methods such as whole-genome comparison based on DNA–DNA hybridizations need to be applied (Christensen *et al.*, 2007a). Species of the *Pasteurellaceae* mostly have been defined by coherence of strains showing DNA reassociation values of more than 80–85% (Mutters *et al.*, 1989; Christensen *et al.*, 2007a). The comparison of housekeeping gene sequences has recently been used for genotypic definition of species (Nørskov-Lauritsen *et al.*, 2005). Kuhnert and Korczak (2006) proposed that sequence comparison of *recN* could replace DNA–DNA hybridizations. Good experiences with this method have subsequently been obtained from investigations of *Actinobacillus capsulatus* and *Avibacterium endocarditis* (Kuhnert *et al.*, 2007, Bisgaard *et al.*, 2007a). Further investigations are, however, needed before classical DNA–DNA hybridization can be replaced by other methods. A detailed paper on Minimal Standards for classification of *Pasteurellaceae* has been published by Christensen *et al.* (2007a).

The genomospecies category was originally suggested for species between which inter-bacterial transfer of genetic material and recombination was possible (genospecies) (Ravin, 1963). According to the present perception, an increased efficiency of homologous recombination controlled by mechanisms such as mismatch repair systems and the ability of RecA to promote DNA recombination is expected with an increased degree of sequence similarity (Feil and Spratt, 2001). For the same reason the biologic species concept meaning that species are interbreeding groups of organisms also has some relevance to bacteria.

The predominant mechanism for transfer of genetic material between members of *Pasteurellaceae* seems to be transformation (Tønjum *et al.*, 1990; Wang *et al.*, 2002; Fink and Geme, 2001; Bosse *et al.*, 2004; see also Chapter 4), and the ability of transformation of *Aggregatibacter aphrophilus* DNA into [*Haemophilus*] *paraphrophilus* was used as one of the arguments in favour of reclassification of [*H.*] *paraphrophilus* with *Agg. aphrophilus* (Nørskov-Lauritsen and Kilian, 2006). However, the ability of transformation is not always a reliable taxonomic criterion since transformation also can take place between distantly related species in different genera (Wang *et al.*, 2002). The ability of transformation has for example been found to depend on the sort and distribution of uptake signal sequences (Redfield *et al.*, 2006; see also Chapter 4). Furthermore, natural competence varies between strains and is in general higher in recent clinical isolates than in reference strains (Tønjum *et al.*, 1990; Fink and Geme, 2001). For this reason, transformation experiments are not generally used for species classification of *Pasteurellaceae*.

Unnamed taxa

Unnamed taxa are difficult to handle in the literature, in diagnostic laboratories, and in databases since they have no 'Standing in nomenclature'. This might lead to misidentification if ignored or recognized by different names posing the risk of failure as to treatment. For this reason all taxa should be named according the Prokaryotic Code (Lapage *et al.*, 1992). In practice, this means that names need to be approved either by publication in the official publication of the International Committee on Systematics of Prokaryotes (*International Journal of Systematic and Evolutionary Microbiology*) or if the names are published elsewhere, included on the Validation List of this journal. For taxa that cannot be separated phenotypically, the genomospecies category should be used as performed for members of *Actinobacillus* and *Gallibacterium* since such taxa might still be recognized by their genotype (Christensen *et al.*, 2002, 2003a).

Misclassification

Names in Latin used without 'Standing in nomenclature' and stated in quotation marks will only contribute to confusion since they represent 'dead ends' in relation to searches in public databases. Examples of taxa named in latin but not included with the List of Prokaryotic Names with Standing in Nomenclature are '*Mannheimia succiniciproducens*', '*Pasteurella multocida* subsp. *tigris*' and '*Actinobacillus porcitonsillarum*' (Capitini *et al.*, 2002; Lee *et al.*, 2002; Kuhnert *et al.*, 2005). The inclusion of these names with the NCBI Taxonomy Browser (http://www.ncbi.nlm.nih.gov/sites/entrez?db=taxonomy) is of no justification for a validated species name, it just indicates that it is 'represented in the genetic databases with at least one nucleotide or protein sequence' since 'NCBI taxonomy database is not a phylogenetic or taxonomic authority and should not be cited as such' (Wheeler *et al.*, 2000). Taxa excluded from *Pasteurellaceae* appear from Table 1.3. For these species, 'excluded' but not yet formally reclassified, the valid names must be used until a formal reclassification has been done.

Valid publication of unnamed taxa as new species is highly desirable; however, methodological limitations have often prevented this in the past. DNA sequence based classification has improved classification tremendously by providing systems for definitive typing and separation including global access to reference sequences through the internet.

Explanation of some difficult taxonomic cases

To start with a simple name then the epithet *haemolyticus* (e.g. in *parahaemolyticus*) refers to the ability to destroy red blood cells. The species

Table 1.3 Misclassified members of *Pasteurellaceae*

Original		Excluded from *Pasteurellaceae*	
Name	Year		Year
'*Actinobacillus mallei*'*	1933	Transferred to *Pseudomonas mallei*	1966
'*A. pseudomallei*'	1933	Transferred to *Pseudomonas pseudomallei*	1957
Haemophilus piscium	1950	Excluded from *Pasteurellaceae* (Probably belongs to *Aeromonas*)	1990
H. vaginalis	1955	Transferred to *Gardnerella vaginalis*	1980
H. equigenitalis	1983	Transferred to *Taylorella equigenitalis*	1984
'*Pasteurella anatipestifer*'	1953	Transferred to *Riemerella anatipestifer*	1993
P. lymphangitidis	1990	Excluded from *Pasteurellaceae* (Probably belongs to *Enterobacteriaceae*)	1991
'*P. pestis*'	1920	Transferred to *Yersinia pestis*	1944
'*P. piscicida*'	1968	Transferred to *Photobacterium damselae* subsp. *piscicida*	1995
'*P. pseudotuberculosis*'	1931	Transferred to *Yersinia pseudotuberculosis*	1965
'*P. tularensis*'	1923	Transferred to *Franciscella tularensis*	1947
'*P. gallicida*'	1980	Name rejected	1985

*Quotation marks indicate that names were not included with Approved Lists of Bacterial Names (Skerman *et al.*, 1980) or that they have been rejected.

Haemophilus haemolyticus thus 'Produces clear haemolytic zones on bovine or sheep blood agar' (Kilian, 2005) (see also Fig. 11.1). This species requires both X- (haemin) and V-factor (β-NAD) for growth. 'Classical' haemophili requiring both growth factors also include *Haemophilus influenzae* and *Haemophilus aegyptius*. For members of *Haemophilus* with only V-factor requirement, the prefix *para-* originally was included in the species epithet (Pittman, 1953; Zinnemann, 1981). As seen in Table 1.1, six species have the prefix *para*. The prefix *aphro* was introduced by Khirat (1940) for organisms requiring an elevated CO_2 atmosphere (capnophilic growth requirement). [*Haemophilus*] *paraphrohaemolyticus* therefore both needs V-factor and an elevated CO_2 atmosphere for optimal growth and it produces haemolytic zones. Unfortunately, the prefixes *para* and *aphro* have limited taxonomic importance since the V-factor requirement is not associated a particular group of *Pasteurellaceae* (Christensen *et al.*, 2007b) just as a CO_2 demand for growth can be expressed by many different members of the family *Pasteurellaceae* (Chapter 11). X-factor requirement is a genus specific characteristic of *Haemophilus sensu stricto* and *Histophilus*. However, all three growth factors (haemin, β-NAD, CO_2) as well as the ability to lyse red blood cells probably reflect the obligate parasitic lifestyle of these bacteria.

By the formation of the new genus *Aggregatibacter*, [*H.*] *paraphrophilus* was reclassified as *Agg. aphrophilus* mainly based on genotypic similarity and the limited value of V-factor requirement in classification. Consequently, the name [*H.*] *paraphrophilus* should no longer be used. The former type strain (NCTC 10557) has now just become a reference strain within *Agg. aphrophilus* since the most recent nomenclatural change has 'standing in nomenclature' and each species has one and only one type strain. Reclassification of other taxa within *Pasteurellaceae*, however, did not influence selection of type strains (Table 1.2).

The species [*Haemophilus*] *parahaemolyticus* associated with humans requires special attention since the name was previously used for the swine pathogen *Actinobacillus pleuropneumoniae* (Kilian *et al.*, 1978) and this might result in confusion (Nørskov-Lauritsen *et al.*, 2005) (see also Chapter 11). The only reliable, nearly full length, 16S rRNA gene sequence of the type strain of [*H.*] *parahaemolyticus* at present has GenBank/EMBL/DDBJ acc. no. AJ295746. This strain shows only 96% similarity to the 16S rRNA gene sequence with acc. no. M75076 of the type strain of [*H.*] *paraphrohaemolyticus* (see also Chapter 11 for reservations about the species *H. aegyptius*). The two species are somewhat distantly related to other species with the closest 16S rRNA similarity of around 95% to members of *Actinobacillus sensu stricto* and to *Mannheimia*.

Habitats and niches

Nearly all vertebrates investigated so far have allowed identification of one or more taxa of *Pasteurellaceae*. These bacteria are obligate parasites or commensals of vertebrates, colonizing mainly the mucosal surfaces of the upper respiratory tract, oropharynx, and reproductive tracts and possibly also parts of the intestinal tract (Olsen *et al.*, 2005) (Table 1.4 and Fig. 1.1). Out of 92 taxa of *Pasteurellaceae*, a majority of 67 have been isolated exclusively from the oropharynx and upper respiratory tract (Table 1.4 and Fig. 1.1). Six taxa have been isolated only from the urogenital tract and five from the intestinal tract including '*M. succiniciproducens*' from bovine rumen with potential for production of succinic acid (Lee *et al.*, 2002). Few taxa have been isolated from different ecological niches. These include *Histophilus somni*, *Haemophilus parainfluenzae*, taxa of *Gallibacterium* isolated from the upper respiratory and urogenital tract (Angen *et al.* 2003; Christensen *et al.*, 2003a; Bojesen *et al.*, 2007a; Kilian, 2005) in addition to *Mannheimia varigena*, *Mannheimia* sp. and taxon 6 of Bisgaard isolated from the upper respiratory and intestinal tract (Angen *et al.*, 1999; Bisgaard, 1993). Including taxa with more habitats, 79 out of 92 taxa have been isolated from the respiratory tract, 17 from the urogenital tract and ten from the intestine (Table 1.4 and Fig. 1.1).

Members of *Pasteurellaceae* identified by sequencing of 16S rRNA clone libraries

Isolation of many taxa of *Pasteurellaceae* often pose problems to diagnostic laboratories due to specific growth requirements (see also

Table 1.4 Habitats and hosts of members of *Pasteurellaceae*

Taxon[a]	Phylogenetic group[b]	Main habitat[d]	Hosts[d]
Actinobacillus arthritidis	*Actinobacillus sensu stricto*	NI[c]	Horses
A. capsulatus	*Actinobacillus sensu stricto*	Upper respiratory tract	Leporidae
A. equuli subsp. *equuli*	*Actinobacillus sensu stricto*	Upper respiratory and intestinal tract	Horses, pigs
A. equuli subsp. *haemolyticus*	*Actinobacillus sensu stricto*	Oropharynx and upper respiratory tract	Horses
A. hominis	*Actinobacillus sensu stricto*	Upper respiratory tract	Man
A. lignieresii	*Actinobacillus sensu stricto*	Oropharynx	Cattle, ruminants
A. pleuropneumoniae	*Actinobacillus sensu stricto*	Upper respiratory tract	Pigs
A. suis	*Actinobacillus sensu stricto*	Upper respiratory tract	Pigs
A. ureae	*Actinobacillus sensu stricto*	Upper respiratory tract	Man
Actinobacillus genomospecies 1 and 2	*Actinobacillus sensu stricto*	Oropharynx and upper respiratory tract	Horses
[*A.*] *delphinicola*	Delphinicola	NI	Sea mammals
[*A.*] *indolicus*	Parasuis	Upper respiratory tract	Pigs
[*A.*] *minor*	Minor	Upper respiratory tract	Pigs
[*A.*] *muris*	Rodent	Upper respiratory tract	Rodents
[*A.*] *porcinus*	Porcinus	Upper respiratory tract	Pigs
[*A.*] *rossii*	Rossii	Urogenital tract	Pigs
[*A.*] *scotiae*	Delphinicola	NI	Sea mammals
[*A.*] *seminis*	Seminis	Urogenital tract	Sheep
[*A.*] *succinogenes*	Succinogenes	Intestinal tract	Cattle
'A. porcitonsillarum'	Minor	Tonsils	Pigs
Aggregatibacter actinomycetemcomitans	*Aggregatibacter*	Oropharynx	Primates
Agg. aphrophilus	*Aggregatibacter*	Oropharynx	Man
Agg. segnis	*Aggregatibacter*	Oropharynx	Man
Avibacterium avium	Avian, *Avibacterium*	Upper respiratory tract	Birds
Av. endocarditidis	Avian, *Avibacterium*	NI	Galliformes
Av. gallinarum	Avian, *Avibacterium*	Upper respiratory tract	Galliformes
Av. paragallinarum	Avian, *Avibacterium*	Upper respiratory tract	Galliformes
Av. volantium	Avian, *Avibacterium*	Upper respiratory tract	Birds
Avibacterium (*P.*) sp. A	Avian, *Avibacterium*	Upper respiratory tract	Birds
Bibersteinia (*P.*) *trehalosi*	*Bibersteinia*	Upper respiratory tract	Ruminants

Table 1.4 *continued*

Taxon[a]	Phylogenetic group[b]	Main habitat[d]	Hosts[d]
Bisgaard taxon 5	Taxon 5	Oropharynx and upper respiratory tract	Guinea pigs
Bisgaard taxon 6	Succinogenes	Oropharynx and upper respiratory tract	Guinea pigs
Bisgaard taxon 7	Taxon 7	Oropharynx and upper respiratory tract	Guinea pigs
Bisgaard taxon 8	*Actinobacillus sensu stricto*	Oropharynx and upper respiratory tract	Guinea pigs
Bisgaard taxon 10	Succinogenes	Oropharynx and upper respiratory tract	Horses
Bisgaard taxon 14	Testudinis	Upper respiratory tract	Birds
Bisgaard taxon 16	ND	Oropharynx and upper respiratory tract	Dogs
Bisgaard taxon 17	Rodent	NI	Rodents
Bisgaard taxon 22	Rodent	Upper respiratory tract	Galliformes
Bisgaard taxon 26	*Actinobacillus sensu stricto*	Upper respiratory tract	Anseriformes
Bisgaard taxon 32	Testudinis	Upper respiratory tract	Hawks
Bisgaard taxon 34	NI	Upper respiratory tract	Psittaciformes
Bisgaard taxon 35	Haemoglobinophilus	NI	Dog
Bisgaard taxon 40	Testudinis	Upper respiratory tract	Gulls
Bisgaard taxon 45 and 46	*Pasteurella sensu stricto*	Oropharynx	Large cats (*Panthera*)
Gallibacterium anatis	Avian, *Gallibacterium*	Urogenital and upper respiratory tract	Birds
Gallibacterium genomospecies 1 and 2 ('*A. salpingitidis*')	Avian, *Gallibacterium*	Urogenital and upper respiratory tract	Galliformes
Gallibacterium (I-V) (Bisgaard taxon 2 and 3)	Avian, *Gallibacterium*	Upper respiratory and urogenital tract	Birds
Haemophilus aegyptius	*Haemophilus*	Oropharynx and upper respiratory tract	Man
H. haemolyticus	*Haemophilus*	Oropharynx and upper respiratory tract	Man
H. influenzae	*Haemophilus*	Oropharynx and upper respiratory tract	Man
H. parainfluenzae	*Haemophilus*	Oropharynx, respiratory and genital tracts	Man
H. pittmaniae	*Haemophilus*	Oropharynx and upper respiratory tract	Man
[*H.*] *ducreyi*	Ducreyi	Urogenital tract	Man
[*H.*] *felis*	Felis	Upper respiratory tract	Cats
[*H.*] *haemoglobinophilus*	Haemoglobinophilus	Urogenital tract	Dog
[*H.*] *paracuniculus*	Paracuniculus	Intestinal tract	Leporidae

Table 1.4 *continued*

Taxon[a]	Phylogenetic group[b]	Main habitat[d]	Hosts[d]
[*H.*] *parasuis*	Parasuis	Upper respiratory tract	Pigs
[*H.*] *parahaemolyticus*	Parahaemolyticus	Upper respiratory tract	Man
[*H.*] *paraphrohaemolyticus*	Parahaemolyticus	Upper respiratory tract	Man
'*H. influenzaemurium*'	Rodent	Upper respiratory tract	Rodents
Histophilus somni	*Histophilus*	Upper respiratory tract urogenital tract	Ruminants
Lonepinella koalarum	*Lonepinella*	Intestinal tract	Koala
Mannheimia glucosida	*Mannheimia*	Upper respiratory tract	Ruminants
M. haemolytica	*Mannheimia*	Upper respiratory tract	Ruminants
M. granulomatis	*Mannheimia*	Upper respiratory tract	Cattle, *Leporidae*
M. ruminalis	*Mannheimia*	Intestinal tract	Ruminants
M. varigena	*Mannheimia*	Upper respiratory and intestinal tract	Pigs, Ruminants
Mannheimia sp.	*Mannheimia*	Upper respiratory and intestinal tract	Ruminants
MBEL55E '*Mannheimia succiniciproducens*'	Succiniciproducens	Bovine rumen	Bovines
Nicoletella semolina	*Nicoletella*	Respiratory tract	Horses
Pasteurella canis	*Pasteurella sensu stricto*	Oropharynx	Carnivora
P. dagmatis	*Pasteurella sensu stricto*	Oropharynx	Carnivora
P. stomatis	*Pasteurella sensu stricto*	Oropharynx	Carnivora
P. multocida subsp. *gallicida*	*Pasteurella sensu stricto*	Oropharynx and upper respiratory tract	Birds, mammals
P. multocida subsp. *multocida*	*Pasteurella sensu stricto*	Oropharynx and upper respiratory tract	Birds, mammals
P. multocida subsp. *septica*	*Pasteurella sensu stricto*	Oropharynx and upper respiratory tract	Birds, mammals
Pasteurella sp. B	*Pasteurella sensu stricto*	Oropharynx	Probably carnivora
[*P.*] *bettyae*	Bettyae	Urogenital tract	Man
[*P.*] *caballi* including Bisgaard taxon 42	Caballi	Oropharynx and upper respiratory tract	Horses, pigs
[*P.*] *aerogenes*	Seminis	Intestinal and urogenital tract	Pigs
[*P.*] *langaaensis*	Langaa	Upper respiratory tract	Galliformes
[*P.*] *mairii*	Seminis	Intestinal and urogenital tract	Pigs
[*P.*] *pneumotropica*	Rodent	Upper respiratory tract	Rodents
[*P.*] *testudinis*	Testudinis	Oropharynx and upper respiratory tract	Tortoise

Table 1.4 *continued*

Taxon[a]	Phylogenetic group[b]	Main habitat[d]	Hosts[d]
Phocoenobacter uteri	*Phocoenobacter*	Urogenital tract	Sea-mammals
Volucribacter psittacicida	Avian, *Volucribacter*	Upper respiratory tract	Psittaciformes
V. amazonae (Bisgaard taxon 33)	Avian, *Volucribacter*	Upper respiratory tract	Psittaciformes
Volucribacter I (Bisgaard taxon 2 and 3)	Avian, *Volucribacter*	Upper respiratory tract	Psittaciformes
Volucribacter II (Bisgaard taxon 2 and 3)	Avian, *Volucribacter*	Upper respiratory tract	Psittaciformes

[a]References to publication of common names are provided at List of Prokaryotic names with Standing in Nomenclature (http://www.bacterio.cict.fr/) and at the ICSP Subcommittee on the Taxonomy of Pasteurellaceae homepage (http://www.the-icsp.org/taxa/Pasteurellaceaelist.htm). The species *P. skyensis*, *P. lymphangitidis* and *H. piscium* are not included since they have been found unrelated to the family (Christensen *et al*., 2007a).

[b]16S rRNA groups according to Olsen *et al*. (2005) and current taxonomy.

[c]NI, limited record,

[d]Selected references: Angen *et al.*, 1999, 2003; Bisgaard 1993, Bisgaard and Mutters, 1986; Bisgaard *et al.*, 2007a–c; Blackall *et al.*, 2001, 2005, 2007; Christensen *et al.*, 2003a,b, 2004a, 2005a–c, 2006; Frederiksen, 1993; Friis-Møller *et al.*, 2001; Kilian, 1976; Kilian *et al.*, 1978; Kuhnert *et al.*, 2005, 2007; Mutters *et al.*, 1989; Nørskov-Lauritsen *et al.*, 2005; Nørskov-Lauritsen and Kilian, 2006; Oliveira and Pijoan, 2004; Osawa *et al.*, 1995; Rapp-Gabrielsen, 1999; Ribeiro *et al.*, 1989; Tegtmeier *et al.*, 1999a,b; Van Ostaaijen *et al.*, 1997.

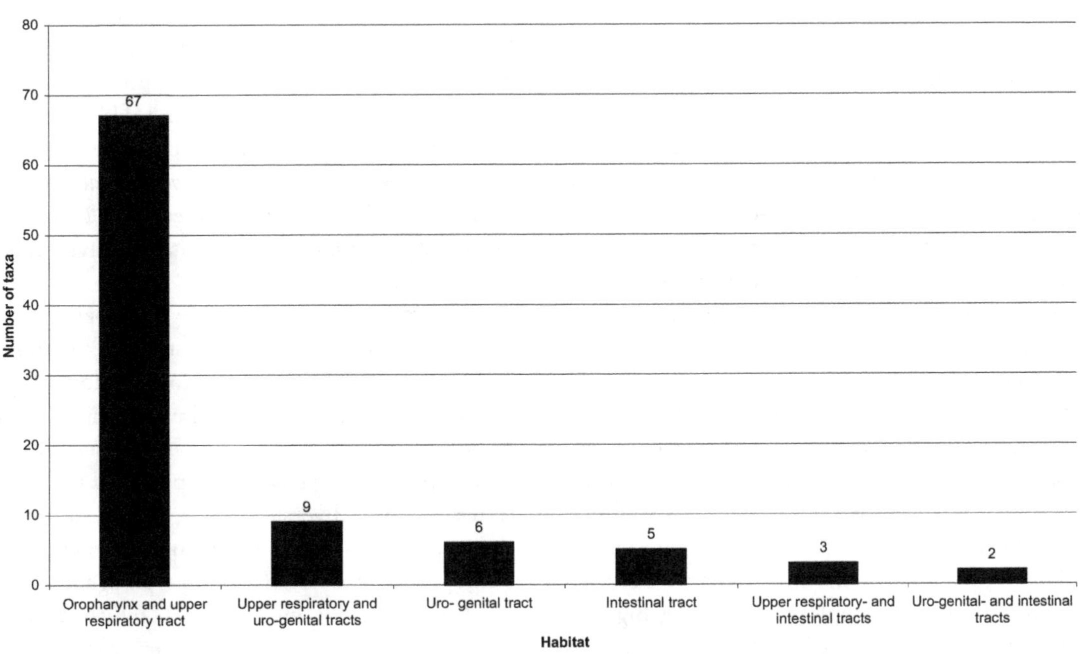

Figure 1.1 Distribution of taxa in relation to habitats.

Chapter 11). In addition, knowledge of the existence of non-culturable members of *Pasteurellaceae* remains to be generated. For the same reason our present knowledge of *Pasteurellaceae* might be biased toward taxa cultivated. Investigations based on direct sequencing of 16S rRNA genes directly from PCR amplified DNA extracts have answered some of these questions. Investigation of aspirates from the lungs of healthy people by direct sequencing confirmed that the respiratory organs are sterile (Flanagan *et al.*, 2007). A patient infected by *Pseudomonas aeruginosa* in the lungs, and investigated by endotracheal aspirates allowed sequences to be determined that showed similarities with some members of *Pasteurellaceae* (Flanagan *et al.*, 2007). Quite surprisingly, one patient on a single day allowed numerous different sequences with high similarity to *H. haemolyticus* to be determined. This species probably colonized the lungs from the oral or nasal region. Other sequences showed high similarity with the type strain of *H. parainfluenzae*.

In another investigation, biopsy samples from the human oesophagus were analysed and six sequences were found with high similarity to members of *Pasteurellaceae*. Closest relatedness was reported to *Aggregatibacter segnis*, *H. parainfluenzae* and *Agg. aphrophilus* (Pei et *al.*, 2004).

From human subgingival plaques, Paster *et al.* (2001) identified sequences from clone libraries with high similarity to *H. parainfluenzae*, *Aggregatibacter actinomycetemcomitans* and *H. influenzae*. For the same type of samples, Kroes *et al.* (1999) identified two sequences with high similarity to *H. parainfluenzae* and *Agg. segnis*. These observations confirmed the presence of human associated taxa already previously known.

Leser *et al.* (2002) showed the presence of sequences related to [*Actinobacillus*] *minor*, [*Actinobacillus*] *porcinus* and [*Pasteurella*] *aerogenes* in pig intestinal samples, however, their frequency was very low. A similar approach was used to document the presence of members of *Mannheimia* and *Bibersteinia trehalosi* in nasal and oropharyngeal samples from healthy bighorn sheep (Safaee *et al.*, 2006). These observations seem to confirm observations based on isolation of these bacteria from healthy North American domesticated and wild ruminants (Ward *et al.*, 1997, 1999).

Comparison of 16S rRNA clone libraries and information from metagenomics datasets with cultured members of *Pasteurellaceae* seems to represent one of the most promising ways to characterize the biodiversity of these bacteria. Further investigations within this area are highly needed.

Biodiversity at population level

Most species of *Pasteurellaceae* associated with diseases of economic importance have been grouped below the species level according to genotype and phenotype including serology and virulence. However, unique subgroups or populations of biological relevance remains to be identified.

A clonal population structure has been shown for the seven species including *H. influenzae*, *Agg. actinomycetemcomitans*, *A. pleuropneumoniae*, *P. multocida*, *B. trehalosi*, [*Haemophilus*] *parasuis* and *Mannheimia* (Musser *et al.*, 1987; 1988; Hampson *et al.*, 1993; Poulsen *et al.*, 1994; Angen *et al.*, 1997; Davies *et al.*, 1997a, b; Blackall *et al.*, 1997, 1998). These investigations have been based upon multilocus enzyme electrophoresis (MLEE) and subsequently been confirmed by multilocus sequence typing (MLST) for *H. influenzae* (Meats *et al.*, 2003). MLST documented the existence of 'cryptic species' of *H. influenzae*. On the other hand, [*H.*] *parasuis* was concluded to be without a 'clonal framework' when analysed by MLST (Olvera *et al.*, 2006). MLST has also been performed with *P. multocida* and *Agg. actinomycetemcomitans* (Davies et *al.*, 2004; Eriksen *et al.*, 2005).

Characterization of members of *Pasteurellaceae* by different typing methods have been used for classification, epidemiological tracing of specific types involved in disease or for setting up vaccination strategies. As to biodiversity, typing has mostly provided information about the diversity at the population level. Serotyping has been used routinely with *A. pleuropneumoniae*, *Avibacterium paragallinarum*, *M. haemolytica*, *P. multocida*, *H. influenzae* and [*H.*] *parasuis*, however, genotyping systems are now being more frequently used in the form of MLST providing a definitive type but PFGE or AFLP typing have also been used with more taxa. PCR based typing for capsular genes is one of the most frequent typing methods used with *P. multocida* and PCR based typing for the distribution of *apx* gene variants and presence of the *toxA* gene are frequently used with *A. pleuropneumoniae* and *P. multocida*, respectively. Further references can be found in Christensen *et al.* (2007a).

Host specificity

The host specificity of members of *Pasteurellaceae* has only been systematically investigated for major potential pathogens of the family such as *P. multocida, H. influenzae, A. pleuropneumoniae* and *Av. paragallinarum* (Christensen and Bisgaard, 2003; Blackall, 1999; Kilian, 2005; Olsen and Møller, 2005). Systematic samplings from healthy animals have only rarely been reported (e.g. Snipes *et al.*, 1988; Muhairwa *et al.*, 2001). Knowledge of host associations has mostly been assumed indirectly from systematic recordings of the origin of isolates such as described by Bisgaard (1993). As previously pointed out caution is very much required as to which taxa were in fact studied in papers not including genotypic characterization. With the advent of a more safe basis for correct diagnosis the way has been paved for improving our knowledge on the possible complexity of factors determining adaptation of these taxa.

Specific associations with vertebrate hosts have been documented with a dominance of taxa isolated from birds, followed by man, pigs, ruminants, dogs and cats, rodents and horses (Bisgaard, 1993) (Table 1.4 and Fig. 1.2). As stated previously, this ranking might be biased towards isolations from man and from diseased domesticated animals. The species [*Actinobacillus*] *succinogenes* and *Lonepinella koalarum* have been isolated from the alimentary track of a cow and koala, respectively, where they probably live in symbiosis with the host (Guettler *et al.*, 1999, Osawa *et al.*, 1995). Although the majority of taxa have been isolated from birds and mammals, [*Pasteurella*] *testudinis* was isolated from the desert tortoise (*Gopherus agassizii*) (Snipes and Biberstein, 1982) just as *P. multocida* has recently been found in the saliva of 'komodo dragons' (lizards) (Montgomery *et al.*, 2002). To what extend these observations might represent passive pass-through via feed remains to be investigated. Host associations have also been documented for some of the suggested virulence factors as described in the following. From an evolutionary point of view, host adaptation probably evolved primarily and virulence secondary (Falkow, 1997).

Host specific virulence factors of *Pasteurellaceae*

Specific adaptations between pathogen and host have been reported in relation to the RTX (repeat in toxin). This group of toxins is further described in Chapter 7. In *A. equuli* subsp. *haemolyticus*, a RTX operon has been reported and named *aqx* (Berthoud *et al.*, 2002). Stronger cytotoxicity against horse lymphocytes compared to pig lymphocytes of cell extract of *E. coli* harbouring the *aqxCA* genes was subsequently reported by Kuhnert *et al.* (2003). In *Mannheimia* and *B. trehalosi* the RTX homologue has been named leukotoxin (Lkt) (Confer *et al.*, 1990; Highlander *et al.*, 2000). The cytotoxic activity of *M. haemolytica* extract was higher towards ruminant leucocytic and platelet cells compared with similar cells from horse, pig or man (Shewen and Wilkie, 1982; Lo *et al.*, 1985; Davies and Baillie, 2003). However, the leukotoxin from *M. varigena* was active towards both pig and cattle cells (Chang *et al.*, 1993). The DNA sequence of the leukotoxin operon has been characterized in detail in the species *M. haemolytica, Mannheimia glucosida, M. varigena* and *B. trehalosi*. Certain regions of the *lktA* gene are probably adapted towards the host antibody response from different ruminants (Chang *et al.*, 1993; Larsen *et al.*, 2007).

The leukotoxin (Ltx) has also been found in *Agg. actinomycetemcomitans* (Kolodrubetz *et al.*, 1989). The polymorphonuclear leukocytes, monocytes and lymphocytes of primates closest related with man were most susceptible to the leukotoxin of *Agg. actinomycetemcomitans* and cytotoxicity was not found for other mammals or other cell types except man (Tsai *et al.*, 1979; Taichman *et al.*, 1987).

Another host specificity mechanism has been recognized with respect to proteins involved in iron uptake by the bacteria. Porcine transferrin binds selectively to *A. pleuropneumoniae* and a host-specific uptake occurred compared with transferrin from other animals (Morton and Williams, 1989; Gonzalez *et al.*, 1990; see also

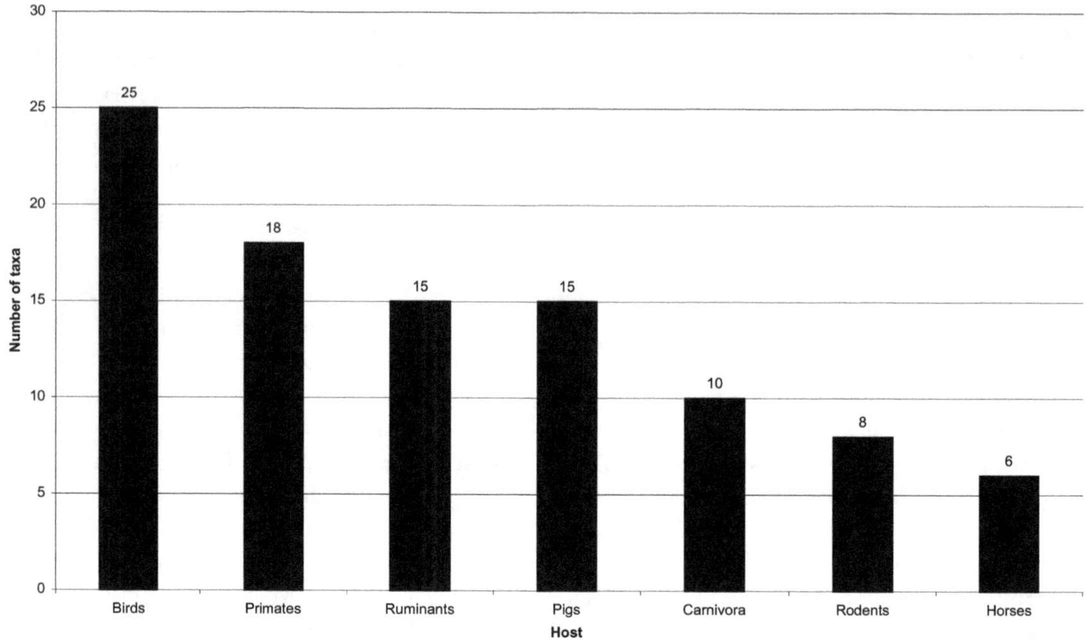

Figure 1.2 Distribution of taxa in relation to hosts.

further information in Chapter 8 about iron uptake). *P. multocida* and *M. haemolytica* isolated from cattle were both able to bind transferrin from goat, sheep and cattle (Ogunnariwo and Schryvers, 1990; Yu and Schryvers, 1994). However, for *P. multocida* serotype B:2,5 associated with haemorrhagic septicaemia the binding was highest towards bovine transferrin (the natural host), with a weak binding to sheep transferrin and no binding to rabbit or chicken transferrins (Veken *et al.*, 1994). Bovine transferrin was also bound selectively to *Hist. somni* while porcine transferrin bound selectively to *[H.] parasuis*. *H. influenzae* and *H. haemolyticus* take up iron only from human transferrin and not from lactoferrin, ovine or porcine transferrins (Morton and Williams, 1989; Fink and Geme, 2001). In conclusion, iron uptake systems among members of *Pasteurellaceae* have been found host specific in several cases.

For most species of *Pasteurellaceae* the observed host specificity, however, remains to be explained just as the biology behind the multitude of animal species from which other taxa have been isolated should be addressed. With the access to definitive sequence based typing methods, observed clonal adaptation for taxa with a broad host spectrum should be subjected to further investigation (Davies *et al.*, 2004). Furthermore investigation to explain the biology behind the observation that *M. haemolytica* serovar A1 affects cattle while serovar A2 affects sheep are also highly indicated.

Wild animals

Only few publications seem to exist on the occurrence of *Pasteurellaceae* in wild animals. Most often information has been obtained from diseased animals and the background for diseases observed in wild animals is often poorly documented. So far one might speculate only that a similar situation exists in wild animals as observed for farm animals. However, wild animals phylogenetically related to farm animals might also be colonized by special populations of already known species, as observed for *Mannheimia granulomatis*, a distinct population of which has affected roe deer while cattle seem to be affected by a different clone (Bojesen *et al.*, 2007b).

P. multocida is known to cause dramatic mortality in migrating web-footed birds (Christensen *et al.*, 1998) just as *M. haemolytica* and *B. trehalosi* have caused dramatic losses in bighorn sheep (Foreyt *et al.*, 1994; Ward *et al.*, 1997). In these cases epidemiological links were documented between outbreaks in wild and domestic animals. In other cases a similar relationship has not been observed (Villard *et al.*, 2006). Recent investigations of captured wild birds from zoos and animal parks have shown specific taxa to be associated with these birds (Bertelsen *et al.*, 2007). It is therefore important to investigate if endangered animal species are colonized by organisms related with taxa with a known disease potential to prevent eventual eradication as exemplified by demonstration of a new genomospecies of *Pasteurella* from tigers related to *P. multocida* (Christensen *et al.*, 2005a).

Geographic distribution

The geographic distribution of members of *Pasteurellaceae* seems to have been associated mostly with animal and human movements, and it is assumed that members of *Pasteurellaceae* follows the animal hosts since these bacteria are obligate parasites. *A. pleuropneumoniae* causing pleuropneumonia in pigs have been demonstrated in most pig-keeping countries (Sebunya and Saunders, 1983) just as a worldwide distribution of *P. multocida* has been known for a long time (Carter and Bain, 1960). Movement of wild animals and migrating birds may also contribute to spreading of associated members of *Pasteurellaceae* (Christensen *et al.*, 1998).

Disease

Members of *Pasteurellaceae* have been categorized as primary or potential pathogens or as commensals. These terms are very general and other factors may determine the outcome of infections including the immune status of the host and co-infections by other microorganisms. Diversity in the disease potential seems to exist for most species. For the same reasons, the pathogenic potential stated in Table 1.5 seems valid only if all factors are considered.

Consistent criteria have not been used to investigate the host spectrum and colonization of these organisms and challenge experiments have been reported frequently only for the best-known pathogens such as *A. pleuropneumoniae*, *P. multocida*, *Agg. actinomycetemcomitans*, *M. haemolytica*, *[Haemophilus] ducreyi*, *[H.] parasuis* and *Av. paragallinarum*.

Clinical signs and lesions may be suggestive of the taxa involved in diseases; however, diagnostic tests are needed to properly identify species involved in clinical cases and to detect healthy carriers. Healthy carriers are observed as part of the epidemiology of most infections with members of *Pasteurellaceae*. Pathognomonic lesions and symptoms are not observed and an unambiguous diagnosis depends on access to relevant diagnostic methods as stated in Chapter 11. The term pasteurellosis might be ambiguous and adds to confusion since similar lesions might be caused by other genera of *Pasteurellaceae* as well as *Yersinia pseudotuberculosis*, *Riemerella anatipestifer* and *Ornithobacterium rhinotracheale* (Bisgaard *et al.*, 2008).

Most common diseases

Both systemic and local infections have been reported for most taxa involved in diseases. However, pneumonia has been reported most frequently out of a number of other disease manifestations (Bisgaard, 1993; Olsen *et al.*, 2005). Nearly half the taxa isolated from the respiratory tract have been found associated with or documented as the primary cause of pneumonia and other respiratory tract infections (Tables 1.4 and 1.5). As a consequence of bacteraemia, other organs may be affected as well. This is exemplified by *H. influenzae* and *[H.] parasuis* which may result in meningitis in man and pigs, respectively although the primary habitat for both seems to be the upper respiratory tract (Table 1.4).

Among *Pasteurellaceae*, *H. influenzae* seems to represent the most important pathogen for humans resulting in meningitis and pneumonia, mainly in children (Fink and Geme, 2001). *Agg. actinomycetemcomitans* is responsible for localized juvenile periodontitis and subacute bacterial endocarditis, while *[H.] ducreyi* is associated with genital ulcers (chancroid) (Nørskov-Lauritsen *et al.*, 2006; Fink and Geme, 2001). In relation to clinical veterinary microbiology diseases due to the species *P. multocida* and *A. pleuropneumoniae* are to be considered among the most common and most costly infections of pigs, resulting in atrophic rhinitis and pneumonia, respectively (Jong, 1999). In poultry, *P. multocida* may cause fowl cholera resulting in major losses worldwide (Glisson *et al.*, 2003). In ruminants, *P. multocida* is responsible for pneumonia and haemorrhagic septicaemia (De Alwis, 1992). *P. multocida* is also associated with pneumonia in pigs but is not

Table 1.5 Pathogenic potential and major diseases reported for members of *Pasteurellaceae.* Only taxa with available information have been included (and even for them the information should be taken with caution)

Taxon[a]	Pathogenic potential[b]	Main disease(s)[b]
Actinobacillus capsulatus	NI[c]	Arthritis, septicaemia
A. equuli subsp. *equuli*	Opportunistic	Septicaemia 'sleepy foal disease'
A. equuli subsp. *haemolyticus*	Commensal	
A. hominis	NI	Respiratory tract infections
A. lignieresii	Opportunistic	Actinobacillosis, pyogranulomatous lesions, 'wooden tongue'
A. pleuropneumoniae	Primary pathogen	Fibrinous necrotizing pneumonia
A. suis	Opportunistic	Pneumonia, septicaemia
A. ureae	NI	Meningitis
[*A.*] *indolicus*	Commensal	
[*A.*] *minor*, *Haemophilus* gr. minor	Commensal	
[*A.*] *porcinus*	Commensal	
[*A.*] *rossii*	Opportunistic	Reproductive tract infections
[*A.*] *seminis*	Opportunistic	Epididymitis
'*A. porcitonsillarum*'	Commensal	
Aggregatibacter actinomycetemcomitans	Opportunistic	Juvenile periodontitis (localized aggressive)
Agg. segnis	Opportunistic	NI
Agg. aphrophilus	Opportunistic	Endocarditis
Avibacterium gallinarum	Opportunistic	Chronic fowl cholera-like lesions
Av. paragallinarum	Primary pathogen	Infectious coryza
Av. endocarditidis	Opportunistic	Valvular endocarditidis
Bibersteinia trehalosi	Opportunistic	Pneumonia, septicaemia
Gallibacterium anatis	Opportunistic	Salpingitis, peritonitis, systemic infections
Gallibacterium genomospecies 1 and 2 ('*A. salpingitidis*')	Opportunistic	Salpingitis
Gallibacterium (I-V) (Bisgaard taxon 2 and 3)	Opportunistic	Salpingitis, systemic infections
H. influenzae	Opportunistic	Meningitis, pneumonia
H. pittmania	Opportunistic	Various infections
[*H.*] *ducreyi*	Primary pathogen	Chancroid
[*H.*] *parasuis*	Opportunistic	'Glässer's disease' (polyserositis, meningitis, polyarthritis), bronchopneumonia
Histophilus somni	Opportunistic	Respiratory tract diseases, trombotic meningoencephalitis; urogenital disease
Lonepinella koalarum	Commensal	
Mannheimia haemolytica	Opportunistic	Pneumonia, septicaemia
M. granulomatis	Opportunistic	Panniculitis, bronchopneumonia
M. glucosida	Opportunistic	Various infections
M. varigena	Opportunistic	Pneumonia
M. ruminalis	Commensal	
P. canis, *P. dagmatis*, *P. stomatis*	NI	Bite wounds

Table 1.5 *continued*

Taxon[a]	Pathogenic potential[b]	Main disease(s)[b]
Pasteurella multocida subsp. *multocida*	Opportunistic	Pneumonia, fowl cholera, atrophic rhinitis, haemorrhagic septicaemia
P. multocida subsp. *gallicida*	Opportunistic	Fowl cholera
P. multocida subsp. *septica*	Opportunistic	Septicaemia, bite wounds
Bisgaard taxon 45 and 46	NI	Bite wounds
[*P.*] *aerogenes*, [*P.*] *mairii*	Opportunistic	Abortion
[*P.*] *bettyae*	Opportunistic	Ulcers, genitourinary infections
[*P.*] *caballi* and Bisgaard taxon 42	NI	Pneumonia, bite wounds
[*P.*] *pneumotropica*	NI	Pneumonia
[*P.*] *testudinis*	NI	Pneumonia
Bisgaard taxon 14	NI	Septicaemia
Volucribacter psittacicida, *V. amazonae* (taxon 33)	NI	Pneumonia
MBEL55E '*Mannheimia succiniciproducens*'	Commensal	

[a]References to publication of names are provided at List of Prokaryotic names with Standing in Nomenclature (http://www.bacterio.cict.fr/) and at the ICSP Subcommittee on the Taxonomy of Pasteurellaceae homepage (http://www.the-icsp.org/taxa/Pasteurellaceaelist.htm)

[b]For selected references se the footnote to Table 1.4.

[c]NI, no or limited record.

considered as the main cause of disease (Carter and Bain 1960, Pijoan, 1999). In ruminants *M. haemolytica* causes mainly pneumonia or 'shipping fever' also called 'stockyard-' or 'transit fever' in feedlot cattle (Frank, 1989). *Av. paragallinarum* causes 'infectious coryza' an important disease of fowl while *Hist. somni* mainly is associated with bovine respiratory tract disease and thrombotic meningoencephalitis (Blackall, 1999; Angen *et al.*, 2003). Serious disease conditions caused by [*H.*] *parasuis* include polyserositis and bronchopneumonia in pigs (Oliveira and Pijoan, 2004) whereas *B. trehalosi* mostly has been isolated from pneumonia and systemic infections of sheep (Blackall *et al.*, 2007). Analysis of human infections caused by dog and cat bites, showed *P. multocida* and *Pasteurella canis* to be most frequently isolated both from abscesses and non-purulent wounds with lymphangitis (Talan *et al.*, 1999).

Multiple infections

Most members of *Pasteurellaceae* typically behave as opportunistic pathogens, often contributing to disease as secondary invaders complicating mycoplasma and virus infections and resulting in serious disease complexes. Virus and/or mycoplasma are no doubt involved in the aetiology of most of the pneumonias of domesticated animals and in general, but the importance of *Pasteurellaceae* as secondary invaders should not be underestimated. Some of the most common examples of co-infections with *P. multocida* are *Bordetella bronchiseptica* in relation to atrophic rhinitis, *Mycoplasma hyopneumoniae* in relation to pig pneumonia (Jong, 1999; Ross, 1999) and virus in relation to wasting syndrome and pneumonia (Fuentes and Pijoan, 1987; Pallares *et al.*, 2002). Increased mortality in broilers have been associated with *Mycoplasma synoviae* and *Av. gallinarum* infections (Droual *et al.*, 1992). Systematic investigations on the aetiology of bovine pneumonia showed that bacterial isolations were dominated by members of *Pasteurellaceae* including *P. multocida*, *Hist. somni* and *M. haemolytica*. However, five other bacterial species as well as three viruses were also involved (Tegtmeier *et al.*, 1999a, b). Examination of bacteria involved in foal septicaemia showed mixed infection in more than half the cases, and members of *Actinobacillus* to be involved together with *Escherichia coli* and *Klebsiella pneumoniae* (Wilson and Madigan,

1989). Infection of layers with members of *Gallibacterium* was found to be associated with both different pathogenic bacteria as well as virus (Gerlach, 1977). Infections with other members of *Pasteurellaceae* have frequently been related to previous vaccination with live virus (see, for example, Droual *et al.*, 1992). Finally, it should be emphasized that factors other than infections including management, immunosuppression due to various reasons, imbalances related to mucosal membranes etc. might predispose for infections due to *Pasteurellaceae*.

A typical example of infection with a primary pathogen (*Av. paragallinarum*) is shown in Fig. 1.3, while Fig. 1.4 shows pathological material from examples of infections associated with opportunistic members *Pasteurella multocida* (Fig. 1.4A–E), *Mannheimia haemolytica* (Fig. 1.4F) and [*Haemophilus*] *parasuis* (Fig. 1.4G and H) demonstrating fowl cholera in turkeys and chickens (Fig. 1.4A–C), pneumonia in pigs (Fig. 1.4D and E), shipping fever in cattle (Fig. 1.4F), acute fibrinopurulent peritonitis in pig (Fig. 1.4G) and acute fibrinous arthritis (Fig. 1.4H) in pig.

Evolution

Information about the evolutionary relationships of members of *Pasteurellaceae* is of importance to classification and to understand the mechanisms behind host adaptation and virulence properties to improve our knowledge of development of these bacteria. Fossil remnants of members of *Pasteurellaceae* have never been reported, and information on the diversification of taxa within the family can be obtained only by phylogenetic reconstruction.

A common monophyletic origin of members of *Pasteurellaceae* including identification of a number of monophyletic subgroups has been inferred through DNA–DNA hybridizations, rRNA–DNA hybridizations and 16S rRNA sequence comparisons and recently also by housekeeping gene sequence comparisons (Pohl, 1981; DeLey *et al.*, 1990; Dewhirst *et al.*, 1992; Angen *et al.*, 2003; Christensen *et al.*, 2004b; Kuhnert *et al.*, 2004; Olsen *et al.*, 2005). The phylogeny is described in detail in Chapter 2. Estimation of the age of the monophyletic groups can be done by molecular comparison, assuming a molecular clock. For members of *Pasteurellaceae*, molecular time estimation is of particular interest since such information might be compared with the age of the host-associated groups known from fossil records and phylogenetic trees inferred from 16S rRNA gene sequences. A similar approximated age of host and parasite might be taken as evidence for parallel evolution.

Figure 1.3 Typical infection (infectious coryza) by primary pathogen (*Avibacterium paragallinarum*). With permission from and thanks to Professor Pat Blackall. A colour version of this figure is located in the plate section at the back of the book.

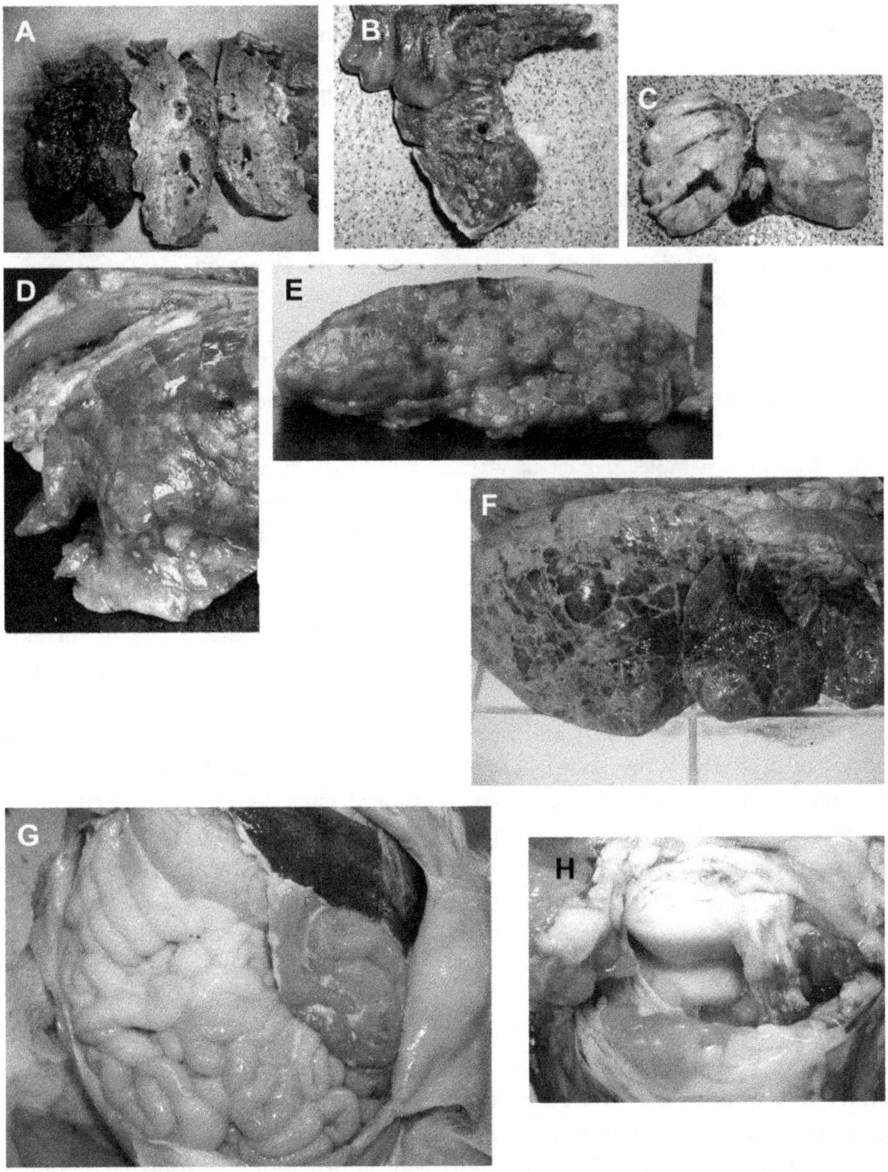

Figure 1.4 (A–E) Typical infection associated with the opportunistic pathogen *Pasteurella multocida*. (A) Pneumonia (fowl cholera) in turkey (cross-section of lung) and (B) pneumonia (fowl cholera) in chicken (cross-section of lung). (C) Normal chicken lung (normal left, affected right). (D and E) Necrotizing bronchopneumonia of pig (E, cross-section of lung). (F) Typical bovine infection (acute haemorrhagic fibrinonecrotizing pneumonia) associated with the opportunistic pathogen *Mannheimia haemolytica*. (G and H) Typical infection of pig associated with the opportunistic pathogen [*Haemophilus*] *parasuis*. (G) Acute fibrinopurulent peritonitis and (H) acute fibrinous arthritis. With permission from and thanks to Professor M. Bisgaard, Professor H. E. Jensen and PhD student S. Pors. A colour version of this figure is located in the plate section at the back of the book.

The Testudinis and Avian groups (Table 1.4) have been found to contain taxa mainly isolated from birds and the strongest tendency for more specific host associations has been observed between *Av. gallinarum* and *Av. paragallinarum* and galliform birds and between *Volucribacter* and taxon 34 and birds of the Psittacciformes order. At least 22 16S rRNA groups have so far been demonstrated to be associated with single animal hosts (Table 1.4), and at least seven have two or more hosts including *Pasteurella sensu stricto* isolated from both birds and mammals (Table 1.4) (Christensen, 2005). It is noteworthy that all monotypic taxa have been of mammalian origin although an explanation for this can not be given.

The analysis of species or species-like taxa characterized by 16S rRNA gene sequencing has indicated that members of the family *Pasteurellaceae* have evolved in close association with either birds or mammals reflecting the evolution of their hosts. Similar but not always identical relationships have been determined by comparison of housekeeping gene sequences (Christensen *et al.*, 2004b; Chapter 2).

To estimate the evolutionary time when *Pasteurellaceae* became associated with the different groups of animals a molecular clock was assumed which means that the divergence in 16S rRNA and amino acid sequences between taxa might be linearly related to the evolutionary time assuming these taxa diverged from a common ancestor. To get a rough approximation, the average time for each group was based on 1% divergence corresponding to averages of 66 and 60 million years (Myr) for 16S rRNA and protein sequences, respectively (conversion factors based on various references listed in Christensen, 2005). For the monophyletic groups, average ages ranged from 152 (*Pasteurella sensu stricto*) to 561 Myr (Avian) when calculations were based on 16S rRNA gene sequences, however, shorter intervals of 30 (*Gallibacterium*) to 276 Myr (*Actinobacillus sensu stricto*) were found for the housekeeping genes. The age of the family was estimated to be around 700 Myr based on 16S rRNA and around 900 Myr based on housekeeping genes. Both estimates correspond to formation of members of *Pasteurellaceae* in the most recent part of Precambrian, some time earlier than the vertebrates in Cambrian (510–550 Myr) and to the formation of the whole family at once (Christensen, 2005).

The ages of the monophyletic groups based on 16S rRNA gene sequences covers the rise of birds (Late Jurassic, 160 Myr) and mammals (Late Triassic, 220 Myr) (Benton, 1997). The majority of the mammals and the birds nearly went to extinction at the end of Cretaceous and the present forms probably evolved in Early to Mid-Eocene (57–47 Myr). These groups are thought to have been the ancestors of the present mammals evolving mainly during Palaeocene and Early Eocene (60–50 Myr) (Benton, 1997). This scenario for the evolution of the animals has probably influenced the evolution of members of *Pasteurellaceae*.

The strongest dilemma is that 16S rRNA phylogenetic analysis tends to reflect host groups. At first sight, an excellent correspondence seems to exist, but unfortunately the corresponding groups of *Pasteurellaceae* estimated from 16S rRNA gene sequence divergence are much older than the host animals. The housekeeping gene estimates might indicate that at least some of the monophyletic genera-like groups formed in parallel with the formation of modern mammals and birds during the first part of Tertiary. The quite divergent phylogenies and molecular time estimates obtained by 16S rRNA gene sequence comparison compared with housekeeping genes are difficult to explain (Fig. 1.5).

Two hypotheses might be formulated to explain the much older age of parasites compared with hosts. Firstly, the 16S rRNA gene sequence might evolve much faster than previously estimated. If this holds true, the evolution of many other bacterial groups will need reconsideration. Alternatively, it can be assumed that the parasites were present in the ancestors of the current host groups. This hypothesis seems the most likely and might, for example, imply that the current members of *Pasteurellaceae* might have been present as common ancestors of, for example, birds and dinosaurs. Better conversion factors are obviously required to investigate this problem further.

For marsupials, monotremes and reptiles no information is available about divergence between members of the *Pasteurellaceae* and further insight from these groups would be very helpful to test the hypothesis of co-evolution of host and parasite. For example, the monotremes (egg-laying mammals) are remnants of an ancient radiation dating back to the Lower Cretaceous period (130 Myr).

Conclusion

Investigation of the biodiversity of *Pasteurellaceae* is important to gain more knowledge about the *Pasteurellaceae* colonizing man and animals. These bacteria share a limited number of niches but express a high level of diversity. Proper classification and a stable nomenclature

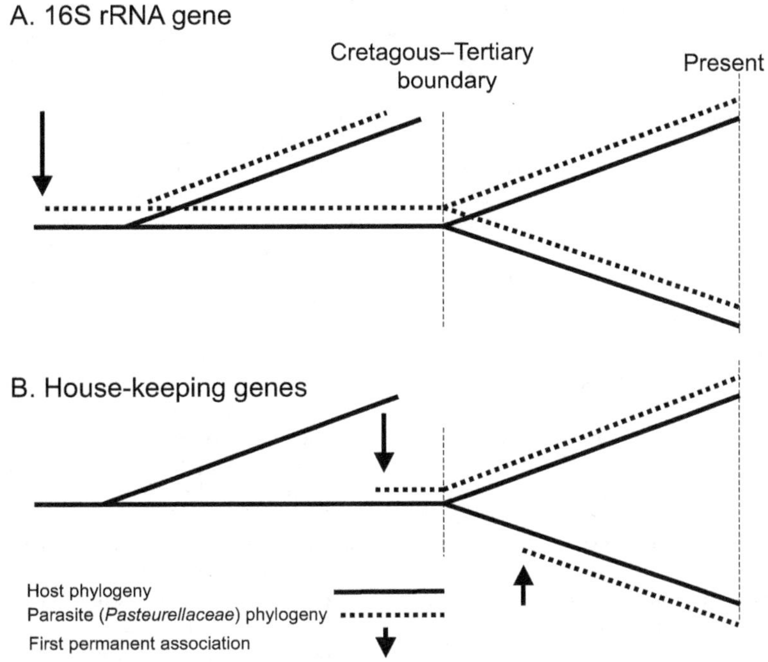

Figure 1.5 Evolutionary model based on phylogeny combined with molecular clock estimation indicates either (A) gradual evolution of host–parasite association and common ancestry if based on 16S rRNA gene sequences or (B) sudden association if based on housekeeping gene sequence comparison

is a prerequisite for investigation of members of *Pasteurellaceae*. More precise definition of the specific populations of bacteria involved in disease will enable more precise prevention and treatment. Relations behind host adaptation and colonization of mucus membranes combined with the ability to co-exist peacefully with the respective animals have remained unknown for decades, just as the role of virulence factors for the most is incompletely understood. Maintenance and evolution of these organisms seem to rely on a persisting colonization of mucous membranes, contact exposure and shedding. Precipitation of diseases seems to be associated with host factors rather than unique virulence factors. Fossil remnants of members of *Pasteurellaceae* have never been reported and information on the diversification of taxa within the family can only be obtained by phylogenetic reconstruction. Most likely the ancestors of the present members of *Pasteurellaceae* might have been present in the common ancestors of, for example, birds and dinosaurs. For marsupials, monotremes and reptiles no information is available about divergence between members of the *Pasteurellaceae* and further insight from these groups would be very important and also helpful to test the hypothesis of co-evolution of host and parasite.

Web resources

Homepage of the Subcommittee of the International Committee of Systematics of Prokaryotes briefly stating 'Names with Standing in Nomenclature':
http://www.the-icsp.org/taxa/Pasteurellaceaelist.htm)

Full record of the taxonomy of the family is found at the List of Prokaryotic Names edited by Professor Jean Euzeby: http://www.bacterio.cict.fr/

A full record of Bisgaard's taxa is found at: http://www.bisgaard.eu

References

Angen, Ø., Caugant, D.A., Olsen, J.E., and Bisgaard, M. (1997). Genotypic relationships among strains classified under the [*Pasteurella*] *haemolytica*-complex as indicated by ribotyping and multilocus enzyme electrophoresis. Zentralbl. Bakteriol. *286*, 333–354.

Angen, Ø., Mutters, R., Caugant, D.A., Olsen, J.E., and Bisgaard, M. (1999). Taxonomic relationships of the [*Pasteurella*] *haemolytica* complex as evaluated by DNA–DNA hybridizations and 16S rRNA sequencing with proposal of *Mannheimia haemolytica* gen. nov., comb. nov., *Mannheimia granulomatis* comb. nov., *Mannheimia glucosida* sp. nov., *Mannheimia ruminalis* sp. nov. and *Mannheimia varigena* sp. nov. Int. J. Syst. Bacteriol. *49*, 67–86.

Angen, Ø., Ahrens, P., Kuhnert, P., Christensen, H., and Mutters, R. (2003). Proposal of *Histophilus somni* gen. nov., sp. nov. for the three species *incertae sedis 'Haemophilus somnus', 'Haemophilus agni'* and *'Histophilus ovis'.* Int. J. Syst. Evol. Microbiol. *53*, 1449–1456.

Anon. (1981). Validation of the publication of new names and new combinations previously effectively published outside the IJSB. List no. 7. Int. J. Syst. Bacteriol. *31*, 382–383.

Benton, M.J. (1997). Vertebrate Palaeontology, 2nd edn. (London: Chapman & Hall).

Bertelsen, M.F., Bojesen, A.M., Christensen, H., and Bisgaard, M. (2007). Prevalence and genetic characterization of *Pasteurellaceae* isolated form captive psittacine birds. XV Congress of the World Veterinary Poultry Association, 10–15 September 2007, Beijing, China.

Berthoud, H., Frey, J. and Kuhnert, P. (2002). Characterization of Aqx and its operon: the hemolytic RTX determinant of *Actinobacillus equuli*. Vet. Microbiol. *87*, 159–174.

Bisgaard, M. (1993). Ecology and significance of *Pasteurellaceae* in animals. Zentralbl. Bakteriol. *279*, 7–26.

Bisgaard, M., and Mutters, R. (1986). Characterization of some previously unclassified *'Pasteurella'* spp. obtained from the oral cavity of dogs and cats and description of a new species tentatively classified with the family *Pasteurellaceae* Pohl 1981 and provisionally called taxon 16. Acta Pathol. Microbiol. Immunol. Scand. Sect. B, 94, 177–184.

Bisgaard., M., Christensen, J.P., Bojesen, A.M., and Christensen, H. (2007a). Classification of *Avibacterium endocarditidis* sp. nov. isolated from valvular endocarditis in chickens. Int. J. Syst. Evol. Microbiol. 57, 1729–1734.

Bisgaard, M., Günther, R., and Christensen, H. (2007 b). Further investigation on the involvement of taxon 14 in upper respiratory tract infections and blepharoconjunctivitis in turkeys. 4th International Symposium on Turkey Production, Berlin 21–23 June.

Bisgaard, M., Korczak, B.M., Busse, H.-J., Kuhnert, P., Bojesen, A.M., and Christensen, H. (2007c). Five new species of *Gallibacterum* – a diagnostic enigma. WVPC15.

Bisgaard, M., Bojesen, A.M., and Christensen, J.P. (2008). Infections caused by species of *Pasteurellaceae, Ornithobacterium* and *Riemerella*: an introduction. In Poultry Diseases, 6th edn. M. Pattison, P.F. McMullin and D. Alexander, eds. (London: Elsevier), pp. 146–148.

Blackall, P.J. (1999). Infectious coryza: overview of the disease and new diagnostic options. Clin. Microbiol. Rev. *12*, 627–632.

Blackall, P.J., Trott, D.J., Rapp-Gabrielson, V., and Hampson, D.J. (1997). Analysis of *Haemophilus parasuis* by multilocus enzyme electrophoresis. Vet. Microbiol. *56*, 125–134.

Blackall, P.J., Fegan, N., Chew, G.T.I., and Hampson, D.J. (1998). Population structure and diversity of avian isolates of *Pasteurella multocida* from Australia. Microbiology *144*, 279–289.

Blackall, P., Angen, Ø., Fegan, N., Blackall, L., Mutters, R., and Bisgaard, M. (2001). Characterisation of a novel *Mannheimia* sp. from Australian feedlot cattle. Aust. Vet. J. 79, 634–639.

Blackall, P.J., Christensen, H., Beckenham, T., Blackall, L.L., and Bisgaard, M. (2005). Reclassification of *Pasteurella gallinarum, [Haemophilus] paragallinarum, Pasteurella avium* and *Pasteurella volantium* as *Avibacterium gallinarum* gen. nov., comb. nov., *Avibacterium paragallinarum* comb. nov., *Avibacterium avium* comb. nov. and *Avibacterium volantium* comb. nov. Int. J. Syst. Evol. Microbiol. *55*, 353–362.

Blackall, P.J., Bojesen, A.M., Christensen, H., and Bisgaard, M. (2007). Reclassification of *[Pasteurella] trehalosi* as *Bibersteinia trehalosi* gen. nov., comb. nov. Int. J. Syst. Evol. Microbiol. *57*, 666–674.

Bojesen, A.M., Christensen, H., Nielsen, S.S., Bisgaard, M. (2007a). Host-specific bacterial lineages in the taxon 2 and 3 complex of *Pasteurellaceae*. Syst. Appl. Microbiol. *30*, 119–127.

Bojesen, A.M., Larsen, J., Pedersen, A.G., Morner, T., Mattson, R., Bisgaard, M. (2007b). Identification of a novel *Mannheimia granulomatis* lineage from lesions in roe deer (*Capreolus capreolus*). J. Wildl. Dis. 43, 345–352.

Bosse, J.T., Nash, J.H., Kroll, J.S. and Langford, P.R. (2004). Harnessing natural transformation in *Actinobacillus pleuropneumoniae*: a simple method for allelic replacements. FEMS Microbiol. Lett. *233*, 277–281.

Brenner, D.J., Krieg, N.R. and Staley, J.T. (2005). Bergey's Manual of Systematic Bacteriology Vol. 2. 2nd ed. Vol. 2. ed. (New York: Springer).

Bridgewater, P.B. (1988). Biodiversity and landscape. Earth-Science Reviews *25*, 486–491.

Buckley, M., and Roberts, R.J. (2006). Reconciling microbial systematics & genomics. American Academy of Microbiology. Colloquium report. (Washington DC: American Society for Microbiology).

Capitini, C.M., Herrero, I.A., Patel, R., Ishitani, M.B., and Boyce, T.G. (2002). Wound infection with *Neisseria weaveri* and a novel subspecies of *Pasteurella multocida* in a child who sustained a tiger bite. Clin. Infect. Dis. *34*, e74–e76.

Carter, G.R., and Bain, R.V.S. (1960). Pasteurellosis (*Pasteurella multocida*). A review stressing recent developments. Vet. Rev. Annot. *6*, 105–128.

Chang, Y.F., Ma, D.P., Shi, J. and Chengappa, M.M. (1993). Molecular characterization of a leukotoxin gene from a *Pasteurella haemolytica*-like organism, encoding a new member of the RTX toxin family. Infect. Immun. *61*, 2089–2095.

Christensen, H. (2005). Systematics of selected members of the bacterial family *Pasteurellaceae* associated with animals. Thesis. Doctoral of Veterinary Science. The Royal Veterinary and Agricultural University, Copenhagen.

Christensen, H. and Bisgaard, M. (2003). The Genus *Pasteurella* In: The Prokaryotes: an evolving electronic resource for the microbiological community. Ver. 3.13. Ed. Dworkin, M., and Lyons, C. (New York: Springer-Verlag).

Christensen, H., Bisgaard, M., Angen, Ø., and Olsen, J.E. (2002). Final classification of Bisgaard taxon 9 as *Actinobacillus arthritidis* sp. nov. and recognition of a novel genomospecies for equine strains of *A. lignieresii*. Int. J. System. Evol. Microbiol. I, 1239–1246.

Christensen, H., Bisgaard, M., Bojesen, M., Mutters, R., and Olsen, J.E. (2003a). Genetic relationships among strains of biovars of avian isolates classified as '*Pasteurella haemolytica*', '*Actinobacillus salpingitidis*' or *Pasteurella anatis* with proposal of *Gallibacterium anatis* gen. nov., comb. nov. and description of additional genomospecies within *Gallibacterium* gen. nov. Int. J. System. Evol. Microbiol. 53, 275–287.

Christensen, H., Foster, G., Christensen, J.P., Pennycott, T., Olsen, J.E., and Bisgaard, M. (2003b). Phylogenetic analysis by 16S rRNA sequence comparison of avian taxa of Bisgaard and characterization and description of two new taxa of *Pasteurellaceae* J. Appl. Microbiol. 95, 354–363.

Christensen, H., Bisgaard, M., Aalbæk, B., and Olsen, J.E. (2004a). Reclassification of Bisgaard taxon 33 with proposal of *Volucribacter psittacicida* gen.nov., sp. nov. and *Volucribacter amazonae* as new members of *Pasteurellaceae*. Int. J. Syst. Evol. Microbiol. 54, 813–818.

Christensen, H., Kuhnert, P., Olsen, J.E., and Bisgaard, M. (2004b). Comparative phylogeny between the housekeeping genes *atpD*, *infB*, *rpoB* and 16S rDNA within *Pasteurellaceae*. Int. J. Syst. Evol. Microbiol. 54, 1601 – 1609.

Christensen, H., Bisgaard, M., Angen, Ø., Frederiksen, W., and Olsen, J.E. (2005a). Characterization of sucrose negative variants of *Pasteurella multocida* including isolates from large cat bite-wounds. J. Clin. Microbiol. 43, 259–270.

Christensen, H., Bisgaard, M., Fink, M., and Olsen, J.E. (2005b). Classification and description of new taxa of *Pasteurellaceae* associated with *Muridae*. IUMS conference, 23–28 July, San Francisco. Poster abstract.

Christensen, H., Kuhnert, P., Bisgaard, M., Mutters, R., Dziva, F., and Olsen J.E. (2005 c). Emended description of porcine [*Pasteurella*] *aerogenes*, [*Pasteurella*] *mairii* and [*Actinobacillus*] *rossii*. Int. J. System. Evol. Microbiol. 55, 209–223.

Christensen, H., Hommez, J., Olsen, J.E., and Bisgaard, M. (2006). [*Pasteurella*] *caballi* infection not limited to horses – a closer look at Taxon 42 of Bisgaard. Lett. Appl. Microbiol. 43, 424–429.

Christensen, H., Kuhnert, P., Busse, H.-J., Frederiksen, W. and Bisgaard, M. (2007a). Proposed minimal standards for the description of genera, species and subspecies of the *Pasteurellaceae*. Int. J. Syst. Evol. Microbiol. 57, 166–178.

Christensen, H., P.J.Blackall & M. Bisgaard. (2007b). New V-factor requiering species/taxa of *Pasteurellaceae* obtained form lesions in different species of birds. Oral presentation. XV Congress of the World Veterinary Poultry Association. September 10–15, 2007. Beijing.

Christensen, J.P., Dietz, H.H. and Bisgaard, M. (1998). Phenotypic and genotypic characters of isolates of *Pasteurella multocida* obtained from back-yard poultry and from two outbreaks of avian cholera in avifauna in Denmark. Avian Pathol. *27*, 373–381.

Confer, A.W., Panciera, R.J., Clinkenbeard, K.D., and Mosier, D.A. (1990). Molecular aspects of virulence of *Pasteurella haemolytica*. Can. J. Vet. Res. 54, S48–52.

Davies, R.L., and Baillie, S. (2003). Cytotoxic activity of *Mannheimia haemolytica* strains in relation to diversity of the leukotoxin structural gene *lktA*. Vet. Microbiol. *92*, 263–279.

Davies, R.L., Arkinsaw, S., and Selander, R.K. (1997a). Evolutionary genetics of *Pasteurella haemolytica* isolates recovered from cattle and sheep. Infect. Immun. *65*, 3585–3593.

Davies, R.L., Arkinsaw, S. and Selander, R.K. (1997b). Genetic relationships among *Pasteurella trehalosi* isolates based on multilocus enzyme electrophoresis. Microbiology *143*, 2841–2849.

Davies, R.L., MacCorquodale, R., and Reilly, S. (2004). Characterisation of bovine strains of *Pasteurella multocida* and comparison with isolates of avian, ovine and porcine origin. Vet. Microbiol. *99*, 145–158.

De Alvis, M.C.L. (1992). Haemorrhagic septicaemia. A general review. Br. Vet. J. *148*, 99–112.

DeLey, J., Mannheim, W., Mutters, R., Piechulla, K., Tytgat, R., Segers, P., Bisgaard, M., Frederiksen, W., Hinz, K.-H., and Vanhoucke, M. (1990). Inter- and intrafamilial similarities of rRNA cistrons of the *Pasteurellaceae*. Int. J. System. Bacteriol. *40*, 126–137.

Dewhirst, F.E., Paster, B.J., Olsen, I., and Fraser, G.J. (1992). Phylogeny of 54 representative strains of species in the family *Pasteurellaceae* as determined by comparison of 16S rRNA sequences. J. Bacteriol. *174*, 2002–2013.

Droual, R., Shivaprasad, H.L., Meteyer, C.U., Shapiro, D.P., and Walker, R.L. (1992). Severe mortality in broiler chickens associated with *Mycoplasma synoviae* and *Pasteurella gallinarum*. Avian Dis. 36, 803–807.

Eriksen, K.T., Haubek, D. and Poulsen, K. (2005). Intragenomic recombination in the highly leukotoxic JP2 clone of *Actinobacillus actinomycetemcomitans*. Microbiology *151*, 3371–3379.

Falkow, S. (1997). What is a pathogen? ASM News *63*, 359–365.

Feil, E.J., and Spratt, B.G. (2001). Recombination and the population structures of bacterial pathogens. Ann. Rev. Microbiol. 55, 561–90.

Fink, D.L., and Geme, J.W.St. III. (2001). The genus *Haemophilus*. In The Prokaryotes: an evolving electronic resource for the microbiological community. Ver. 3.13, M. Dworkin and C. Lyons, C. eds. (New York: Springer-Verlag).

Flanagan, J.L., Brodie, E.L., Weng, L., Lynch, S.V., Garcia, O. Brown, R., Hugenholtz, P., DeSantis, T.Z., Andersen, G.L., Wiener-Kronish, J.P., and Bristow, J. (2007). Loss of bacterial diversity during antibiotic treatment of intubated patients colonized with *Pseudomonas aeruginosa*. J. Clin. Microbiol. 45, 1954–1962.

Foreyt, W.J., Snipes, K.P., and Kasten, R.W. (1994) Fatal pneumonia following inoculation of healthy bighorn sheep with *Pasteurella haemolytica* from healthy domestic sheep. J. Wildlife Dis. *30*, 137–145.

Foster, G., Ross, H.M., Malnick, H., Willems, A., Hutson, R.A., Reid, R.J., and Collins, M.D. (2000). *Phocoenobacter uteri* gen. nov., sp. nov., a new member of the family *Pasteurellaceae* Pohl (1979) 1981 isolated from a harbour porpoise (*Phocoena phocoena*). Int. J. System. Evol. Microbiol. *50*, 135–139.

Frank, G.H. (1989). Pasteurellosis of cattle. In: *Pasteurella* and Pasteurellosis. Adlam, C. and Rutter, J.M. (eds.). (London Academic Press: London), pp. 197–222.

Frederiksen, W. (1993). Ecology and significance of *Pasteurellaceae* in Man – an update. Zentralbl. Bakteriol. *279*, 27–34.

Friis-Møller, A., Christensen, J.J., Fussing, V., Hesselbjerg, A., Christiansen, J., and Bruun, B. (2001). Clinical significance and taxonomy of *Actinobacillus hominis*. J. Clin. Microbiol. *39*, 930–935.

Fuentes, M.C., and Pijoan, C. (1987). Pneumonia in pigs induced by intranasal challenge exposure with pseudorabies virus and *Pasteurella multocida*. Am. J. Vet. Res. *48*, 1446–1448.

Gerlach, H. (1977). Die bedeuting von *Pasteurella haemolytica* in Hühnerbeständen. Der praktische Tierarzt 5, 324–328.

Glisson, J.R., Hofacre, C.L., and Christensen, J.P. (2003). Fowl cholera. In: Diseases of Poultry 11th edition. Saif, Y.M., Barnes, H.J., Glisson, J.R., Fadly, A.M., McDougald, L.R., and Swayne, D.E. (Eds). (Ames Iowa State Press), pp. 658–676.

Gonzalez, G.C., Caamano, D.L., and Schryvers, A.B. (1990). Identification and characterization of a porcine-specific transferrin receptor in *Actinobacillus pleuropneumoniae*. Mol. Microbiol. *4*, 1173–1179.

Guettler, M.V., Rumler, D., and Jain, M.K. (1999). *Actinobacillus succinogenes* sp. nov., a novel succinic-acid-producing strain form the bovine rumen. Int. J. System. Bacteriol. *49*, 207–216.

Hampson, D.J., Blackall, P.J., Woodward, J.M., and Lymbery, A.J. (1993). Genetic analysis of *Actinobacillus pleuropneumoniae*, and comparison with *Haemophilus* spp. Taxon 'minor group' and Taxon C. Zentralbl. Bakteriol. *279*, 83–91.

Highlander, S.K., Fedorova, N.D., Dusek, D.M., Panciera, R., Alvarez, L.E., and Rinehart, C. (2000). Inactivation of *Pasteurella* (*Mannheimia*) *haemolytica* leukotoxin causes partial attenuation of virulence in a calf challenge model. Infect. Immun. *68*, 3916–3922.

Jong, de M.F. (1999). Progressive and nonprogressive atrophic rhinitis. In: Diseases of Swine. 8th ed. Straw, B.E., D'Allaire, S., Mengeling, W.L and Taylor D.J. (Eds.). (Ames: Iowa State Univ. Press), pp. 355–384.

Khirat, O. (1940). Endocarditis due to a new species of *Haemophilus*. J. Path. Bacteriol. 50, 497–505.

Kilian, M. (1976). A taxonomic study of the genus *Haemophilus*, with the proposal of a new species. J. Gen. Microbiol. *93*, 9–62.

Kilian, M. (2005). Genus *Haemophilus* Winslow, Broadhurst, Buchanan, Krumwiede, Rogers and Smith 1917, 561^{AL}. In *Bergey's Manual of Systematic Bacteriology*, 2nd edn, Vol. 2, ed. Brenner, D.J, Krieg, N.R. & Staley, J.T. (New York: Springer), pp. 883–904.

Kilian, M., Nicolet, J. and Biberstein, E.L. (1978). Biochemical and serological characterization of *Haemophilus pleuropneumoniae* (Matthews and Pattison 1961) Shope 1964 and proposal of a neotype strain. Int. J. System. Bacteriol. *28*, 20–26.

Kolodrubetz, D., Dailey, T., Ebersole, J., and Kraig, E. (1989). Cloning and expression of the leukotoxin gene from *Actinobacillus actinomycetemcomitans*. Infect. Immun. *57*, 1465–1469.

Kroes, I., Lepp, P.W., and Relman, D.A. (1999). Bacterial diversity within the human subgingival crevice. Proc. Natl. Acad. Sci. USA *96*, 14547–14552.

Kuhnert, P., and Korczak, B.M. (2006). Prediction of whole genome DNA–DNA similarity, determination of G+C content and phylogenetic analysis within the family *Pasteurellaceae* by multilocus sequence analysis (MLSA). Microbiology *152*, 2537–2548.

Kuhnert, P., Berthoud, H., Straub, R., and Frey, J. (2003). Host cell specific activity of RTX toxins from haemolytic *Actinobacillus equuli* and *Actinobacillus suis*. Vet. Microbiol. *92*, 161–167.

Kuhnert, P., Korczak, B., Falsen, E., Straub, R., Hoops, A., Boerlin, P., Frey, J., and Mutters, R. (2004). *Nicoletella semolina* gen. nov., sp. nov., a new member of *Pasteurellaceae* isolated from horses with airway disease. J. Clin. Microbiol. *42*, 5542–5548.

Kuhnert, P., Schlatter, Y., and Frey, J. (2005). Characterization of the type I secretion system of the RTX toxin ApxII in '*Actinobacillus porcitonsillarum*'. Vet. Microbiol. *107*, 225–232.

Kuhnert, P., Korczak, B., Christensen, H. and Bisgaard, M. (2007). Emended description of *Actinobacillus capsulatus* Arseculeratne 1962, 38AL. Int. J. Syst. Evol. Microbiol. *57*, 625–632.

Lapage, S.P., Sneath, P.H.A., Lessel, E.F., Skerman, V.B.D., Seeliger, H.P.R., and Clark, W.A. (1992). International Code of Nomenclature of Bacteria and Statutes of the International Committee on Systematic Bacteriology and Statutes of the Bacteriology and Applied Microbiology Section of the International Union of Microbiological Societies. Bacteriological Code. 1990 Revision. (Washington D. C.: American Society for Microbiology).

Larsen, J., Christensen, H., Bisgaard, M., Angen, Ø., Ahrens, P. and Olsen, J.E. (2007). Tracing the origin and evolution of the leukotoxin operon in genus *Mannheimia*. J. Mol. Evol. *64*, 423–437.

Lee, P.C., Lee, S.Y., Hong, S.H. and Chang, H.N. (2002). Isolation and characterization of a new succinic acid-producing bacterium, *Mannheimia succiniciproducens* MBEL55E, from bovine rumen. Appl. Microbiol. Biotechnol. 58, 663–668.

Leser, T.D., Amenuvor, J.Z., Jensen, T.K., Lindecrona, R.H., Boye, M. and Møller, K. (2002). Culture-independent analysis of gut bacteria: the pig gastrointestinal tract microbiota revisited. Appl. Environ. Microbiol. *68*, 673–690.

Lo, R.Y., Shewen, P.E., Strathdee, C.A. and Greer, C.N. (1985). Cloning and expression of the leukotoxin gene of *Pasteurella haemolytica* A1 in *Escherichia coli* K-12. Infect. Immun. *50*, 667–671.

Meats, E., Feil, E.J., Stringer, S., Cody, A.J., Goldstein, R., Kroll, J.S., Popovic, T. and Spratt, B.G. (2003). Characterization of encapsulated and noncapsulated *Haemophilus influenzae* and determination of phylogenetic relationships by multilocus sequence typing. J. Clin. Microbiol. 41, 1623–1636.

Montgomery, J.M., Gillespie, D., Sastrawan, P., Fredeking, T.M. and Stewart, G.L. (2002). Aerobic salivary bacteria in wild and captive Komodo dragons. J. Wildl. Dis. *38*, 545–551.

Morton, D.J., and Williams, P. (1989). Utilization of transferrin-bound iron by *Haemophilus* species of human and porcine origins. FEMS Microbiol. Lett. *53*, 123–127.

Muhairwa, A.P., Mtambo, M.M.A., Christensen, J.P., and Bisgaard, M. (2001). Occurrence of *Pasteurella multocida* and related species in village free ranging chickens and their animal contacts in Tanzania. Vet. Microbiol. *78*, 139–153.

Musser, J.M., Rapp, V.J. and Selander R.K. (1987). Clonal diversity in *Haemophilus pleuropneumoniae*. Infect. Immun. *55*, 1207–1215.

Musser, J.M., Kroll, J.S., Moxon, E.R., and Selander, R.K. (1988). Clonal population structure of encapsulated *Haemophilus influenzae*. Infect. Immun. *56*, 1837–1845.

Mutters, R., Mannheim, W., and Bisgaard, M. (1989). Taxonomy of the Group. In *Pasteurella* and Pasteurellosis. C. Adlam and J.M. Rutter, ed. (London: Acadademic Press), pp. 3–34.

Nørskov-Lauritsen, N., Bruun, B., and Kilian, M. (2005). Multilocus sequence phylogenetic study of the genus *Haemophilus* with description of *Haemophilus pittmaniae* sp. nov. Int. J. System. Evol. Microbiol. *55*, 449–456.

Nørskov-Lauritsen, N., and Kilian, M. (2006). Reclassification of *Actinobacillus actinomycetemcomitans*, *Haemophilus aphrophilus*, *Haemophilus paraphrophilus* and *Haemophilus segnis* as *Aggregatibacter actinomycetemcomitans* gen. nov., comb. nov., *Aggregatibacter aphrophilus* comb. nov. and *Aggregatibacter segnis* comb. nov., and emended description of *Aggregatibacter aphrophilus* to include V factor-dependent and V factor-independent isolates. Int. J. Syst. Evol. Microbiol. *56*, 2135–2146.

Ogunnariwo, J.A. and Schryvers, A.B. (1990). Iron acquisition in *Pasteurella haemolytica*: expression and identification of a bovine-specific transferrin receptor. Infect. Immun. *58*, 2091–2097.

Oliveira, S., and Pijoan, C. (2004). *Haemophilus parasuis*: new trends on diagnosis, epidemiology and control. Vet Microbiol. *99*, 1–12.

Olsen, I., and Møller, K. (2005). Genus II *Actinobacillus* Brumpt 1910, 849[AL]. In: Bergey's Manual of Systematic Bacteriology, 2nd ed., Vol. 2. Garrity, G.M. *et al.* (New York: Springer), pp. 866–883.

Olsen, I., Dewhirst, F.E., Paster, B.J. & Busse, H.-J. (2005). Family *Pasteurellaceae. In Bergey's Manual of Systematic Bacteriology*, 2nd edn, Vol. 2, ed. Brenner, D.J, Krieg, N.R. and Staley, J. (New York: Springer), pp. 851–856.

Olvera, A., Cerdà-Cuéllar, M., and Aragon, V. (2006). Study of the population structure of *Haemophilus parasuis* by multilocus sequence typing. Microbiology *152*, 3683–3690.

Osawa, R., Rainey, F., Fujisawa, T., Lang, E., Busse, H.J., Walsh, T.P., and Stackebrandt, E. (1995). *Lonepinella koalarum* gen. nov., sp. nov., a new tannin-protein complex degrading bacterium. Syst. Appl. Microbiol. *18*, 368–373.

Pallares, F.J., Halbur, P.G., Opriessnig, T., Sorden, S.D., Villar, D., Janke, B.H., Yaeger, M.J., Larson, D.J., Schwartz, K.J., Yoon, K.J., and Hoffman, L.J. (2002). Porcine circovirus type 2 (PCV-2) coinfections in US field cases of postweaning multisystemic wasting syndrome (PMWS). J. Vet. Diagn. Invest. *14*, 515–519.

Paster, B.J., Boches, S.K, Galvin, J.L., Ericson, R.E., Lau, C.N., Levanos, V.A., Sahasrabudhe, A., and Dewhirst, F.E. (2001). Bacterial diversity in human subgingival plaque. J. Bacteriol. *183*, 3770–3783.

Pei, Z., Bini, E.J., Yang, L., Zhou, M., Francois, F., and Blaser, M.J. (2004). Bacterial biota in the human distal esophagus. Proc. Natl. Acad. Sci. USA *101*, 4250–4255.

Pijoan, C. (1999). Pneumonic pasteurellosis. In Diseases of Swine, 8th edn. B.E. Straw, S. D'Allaire, W.L. Mengeling, and D.J. Taylor, eds. (Ames: Iowa State University Press), pp. 511–520.

Pittman, M. (1953). A classsification of the haemolytic bacteria of the genus *Haemoplilus*: *Haemophilus haemolyticus* Bergey *et al.* and *Haemophilus parahaemolyticus* nov. spec. J. Bacteriol. 65, 750–751.

Pohl, S. (1979). Reklassifizierung der Gattungen *Actinobacillus* Brumpt 1910, *Haemophilus* Winslow *et al.* 1917 und *Pasteurella* Trevisan 1887 anhand phänotypischer und molekularen Daten, insbesondere der DNS-Verwandschaften bei DNS:DNS Hybridisierung in vitro und Vorschlag einer neuen Familie, *Pasteurellaceae*. Inagural Dissertation, Fachbereich Biologie der Philipps-Universität Marburg-Lahn, (Marburg: Mauersberger Druck).

Pohl, S. (1981). DNA relatedness among members of *Haemophilus*, *Pasteurella* and *Actinobacillus*. In: *Haemophilus*, *Pasteurella* and *Actinobacillus*. M. Kilian, W. Frederiksen, E.L. Biberstein, eds. (London: Acadademic Press), pp. 245–253.

Poulsen, K., Theilade, E., Lally, E.T., Demuth, D.R. and Kilian, M. (1994). Population structure of *Actinobacillus actinomycetemcomitans*: a framework for studies of disease-associated properties. Microbiology *140*, 2049–2060.

Rapp-Gabrielson, V.J. (1999). *Haemophilus parasuis*. In Diseases of Swine, 8th edn. B.E. Straw, S. D'Allaire, W.L. Mengeling, and D.J. Taylor, eds. (Ames: Iowa State University Press), pp. 511–520.

Ravin, A.W. (1963). Experimental approaches to the study of bacterial phylogeny. The American Naturalist *97*, 307–318.

Redfield, R.J., Findlay, W.A., Bosse, J., Kroll, J.S., Cameron, A.D., Nash, J.H. (2006). Evolution of competence and DNA uptake specificity in the *Pasteurellaceae*. BMC Evol. Biol. *6*, 82.

Ribeiro, G.A., Carter, G.R., Frederiksen, W., and Riet-Correa, F. (1989). *Pasteurella haemolytica*-like bacterium from a progressive granuloma of cattle in Brazil. J. Clin. Microbiol. *27*, 1401–1402.

Ross, R.F. (1999). Mycoplasmal diseases. In: Diseases of Swine. 8th ed. Straw, B.E., D'Allaire, S., Mengeling, W.L., and Taylor D.J. (Eds.). (Ames: Iowa State Univ. Press), pp. 495–509.

Safaee, S., Weiser, G.C., Cassirer, E.F., Ramey, R.R., and Kelley, S.T. (2006). Molecular diversity in bighorn sheep revealed by culture-independent methods. J. Wildl. Dis. *42*, 545–555.

Sebunya, T.N., and Saunders, J.R. (1983). *Haemophilus pleuropneumoniae* infection in swine: A review. J. Am. Vet. Med. Assoc. *182*, 1331–1337.

Shewen, P.E., and Wilkie, B.N. (1982). Cytotoxin of *Pasteurella haemolytica* acting on bovine leukocytes. Infect. Immun. *35*, 91–94.

Skerman, V.B.D., McGowan, V., and Sneath, P.H.A. (1980). Approved Lists of Bacterial Names. Int. J. Syst. Bacteriol. *30*, 225–420.

Snipes, K.P., and Biberstien, E.L. (1982). *Pasteurella testudinis* sp. nov.: a parasite of desert tortoises (*Gopherum agassizi*). Int. J. Syst. Bact. *32*, 201–210.

Snipes, K.P., Carpenter, T.E., Corn, J.L., Kasten, R.W., Hirsh, D.C., Hird, D.W., and McCapes, R.H. (1988). *Pasteurella multocida* in wild mammals and birds in California: Prevalence and virulence for turkeys. Avian Dis. *32*, 9–15.

Taichman, N.S., Simpson, D.L., Sakurada, S., Cranfield, M., DiRienzo, J. and Slots, J. (1987). Comparative studies on the biology of *Actinobacillus actinomycetemcomitans* leukotoxin in primates. Oral Microbiol. Immunol. 2, 97–104.

Talan, D.A., Citron, D.M., Abrahamian, F.M., Moran, G.J. and Goldstein, E.J. (1999). Bacteriologic analysis of infected dog and cat bites. Emergency Medicine Animal Bite Infection Study Group. N. Engl. J. Med. *340*, 85–92.

Tegtmeier, C., Uttenthal, A., Friis, N.F., Jensen, N.E. and Jensen, H.E. (1999). Pathological and microbiological studies on pneumonic lungs from Danish calves. J. Vet. Med. B *46*, 693–700.

Tegtmeier, C, Bloch, B., Jensen, N.E. and Jensen, H.E. (1999b). Initial lung lesions in two calves experimentally infected with *Haemophilus somnus*. Zentralbl. Veterinarmed. B. *46*, 517–523.

Tsai, C.C., McArthur, W.P., Baehni, P.C., Hammond, B.F., and Taichman, N.S. (1979). Extraction and partial characterization of a leukotoxin from a plaque-derived Gram-negative microorganism. Infect. Immun. *25*, 427–439.

Tønjum, T., Bukholm, G., and Bøvre, K (1990). Identification of *Haemophilus aphrophilus* and *Actinobacillus actinomycetemcomitans* by DNA–DNA hybridization and genetic transformation. J. Clin. Microbiol. *28*, 1994–1998.

Van Ostaaijen, J., Frey, J., Rosendal, S., and MacInnes, J.I. (1997). *Actinobacillus suis* strains isolated from healthy and diseased swine are clonal and carry *apx*ICABD var. suis and *apx*IICA var. suis toxin genes. J. Clin. Microbiol. *35*, 1131–1137.

Veken, J.W., Oudega, B., Luirink, J., and Graaf, F.K.de (1994). Binding of bovine transferrin by *Pasteurella multocida* serotype B:2, 5, a strain which causes haemorrhagic septicaemia in buffalo and cattle. FEMS Microbiol. Lett. *115*, 253–257.

Villard, L., Gauthier, D., Lacheretz, A., Abadie, G., Game, Y., Maurin, F., Richard, Y., Borges, E., and Kodjo, A. (2006). Serological and molecular comparison of *Mannheimia haemolytica* and *Pasteurella trehalosi* strains isolated from wild and domestic ruminants in the French Alps. Vet. J. *171*, 545–550.

Wang, Y., Goodman, S.D., Redfield, R.J., and Chen, C. (2002). Natural transformation and DNA uptake signal sequences in *Actinobacillus actinomycetemcomitans*. J. Bacteriol. *184*, 3442–3449.

Ward, A.C., Hunter, D.L., Jaworski, M.D., Benolkin, P.J., Dobel, M.P., Jeffress, J.B., and Tanner, G.A. (1997). *Pasteurella* spp. in sympatric bighorn and domestic sheep. J. Wildlife Dis. *33*, 544–557.

Ward, A.C., Dyer, N.W., and Fenwick, B.W. (1999). *Pasteurellaceae* isolated from tonsillar samples of commercially-reared American bison (*Bison bison*). Can. J. Vet. Res. *63*, 161–165.

Wheeler, D.L., Chappey, C., Lash, A.E., Leipe, D.D., Madden, T.L., Schuler, G.D., Tatusova, T.A., and Rapp, B.A. (2000). Database resources of the National Center for Biotechnology Information. Nucleic Acids Res. 28, 10–14.

Wilson, E.O., and Fances, M.P. (1988). Biodiversity, National Academy Press, March 1988.

Wilson, W.D., and Madigan, J.E. (1989). Comparison of bacteriologic culture of blood and necropsy specimens for determining the cause of foal septicemia: 47 cases (1978–1987). J. Am. Vet. Med. Assoc. *195*, 1759–1763.

Yu, R., and Schryvers, A.B. (1994). Transferrin receptors on ruminant pathogens vary in their interaction with the C-lobe and N-lobe of ruminant transferrins. Can. J. Microbiol. *40*, 532–540.

Zinnemann, K. (1981). Some historical aspects of the genera *Haemophilus*, *Pasteurella* and *Actinobacillus*, and some of their species. In: *Pasteurella* and Pasteurellosis. Adlam, C. and Rutter, J.M. eds. (London: Academic Press), pp. 1–10.

Phylogeny of *Pasteurellaceae*

2

Bożena M. Korczak and Peter Kuhnert

Abstract

This chapter provides an overview on the DNA-based phylogeny of the family *Pasteurellaceae* and the genetic relatedness between taxa taking into account the various gene targets and approaches applied in the literature. The classical 16S rRNA gene based phylogeny as well as phylogenies based on housekeeping genes are described. Moreover, the strength and weakness of the different trees and their topology are discussed based on the phylogenetic groups resolved. The data should help to get a clearer picture on the recent, current and future classification and also provide information to genetic characterization of members of the family. The history of phylogeny applied to the family as well as the phylogenetic history of the family is thereby presented. In this way it is the story of the search for the optimal phylogenetic marker without giving a final conclusive suggestion but it is also a resource for choosing the appropriate gene target(s) for people investigating the phylogeny of groups of *Pasteurellaceae*.

Introduction

Phylogeny as a tool for delineating genealogic association or evolution has become increasingly useful for investigating the relationship of microorganisms. In bacteriology it is an indispensable approach for proper classification of taxa as well as for identification and diagnosis. DNA–DNA hybridization studies as a rough estimate of genetic similarities between species have been performed for a long time with *Pasteurellaceae* and helped establishing the family (Mutters *et al.*, 1985; Pohl, 1981). However, the first report investigating phylogeny within *Pasteurellaceae* based on nucleic acid sequences dates back 20 years. At that time direct sequencing of RNA was easier than sequencing DNA and the template was in this case already abundantly present, namely the 16S ribosomal RNA (16S rRNA). This molecule was introduced to phylogeny by the well known work of Woese (1987) allowing the delineation of phylogenetic relationships by sequence comparison of 16S rRNA or its genes, respectively. In the first publication applying this promising new approach, Chuba *et al.* (1988) used only seven species in their study. However, they included the type species of the three genera *Haemophilus*, *Actinobacillus* and *Pasteurella* (usually referred to as 'HAP') which for a long time were the only recognized genera within *Pasteurellaceae*. Moreover, by inclusion of at that time [*Actinobacillus*] *actinomycetemcomitans* (now *Aggregatibacter actinomycetemcomitans*) and [*Haemophilus*] *aphrophilus* (now *Aggregatibacter aphrophilus*) actually members of a fourth genus were also investigated. Furthermore, two additional representatives from the genus *Actinobacillus*, namely *Actinobacillus equuli* and *Actinobacillus hominis* were analysed. Results from this limited study, not only due to the small number of strains but also based on the short sequences of 450 basepairs (bp) generated, already indicated that [*A.*] *actinomycetemcomitans* was wrongly attributed to this genus. The three other actinobacilli were found closely related and *A. lignieresii*, *A. equuli* and *A. hominis* are nowadays recognized as true *Actinobacillus* species (Kuhnert *et al.*, 2007). Later rRNA–DNA hybridization showed the heterogeneity of the family with at least seven rRNA branches (De Ley *et al.*, 1990). This problematic within the 'HAP' group was then shown very impressively by the two publications of Dewhirst *et al.* (1992, 1993) analyzing a total of 70 strains of *Pasteurellaceae* based on 16S rRNA sequences thereby expanding the initial phylogenetic study of Chuba *et al.* (1988) significantly. The branching was extremely complex and there was no close relation between branching and taxonomic position of strains, what just represented the many misclassified species. Nevertheless, it took some time until this finding became common knowledge and revision of the family became accepted. For some years it seemed to be a taboo to break the old 'HAPpy' family apart and suggest new genera. With the exception of the species *Lonepinella koalarum* (Osawa *et al.*, 1995) representing a new genus (which however was hardly noticed) we almost missed the opportunity to do so in the last millennium. But then the paper of Angen and coworkers appeared in 1999 suggesting the genus *Mannheimia* including the species *M. haemolytica, M. glucosida, M. varigena, M. granulomatis* and *M. ruminalis* (Angen *et al.*, 1999). Fortunately enough, this investigation was not only based on phylogeny using 16S rRNA gene but was a thorough poylphasic study which leaved no doubt for the necessity to create a new genus, even though its acceptance in the scientific community was not unanimous.

Since then there has been a cascade of changes, including reclassification as well as description of many new taxa within *Pasteurellaceae* based on 16S rRNA gene sequence phylogeny but also in respect to phenotypic characterization. In addition to the 16S rRNA gene, a few housekeeping genes have been applied for phylogenetic purposes. The family has undergone important changes, in particular at the genus level, and in a very short time the number of genera has increased to 13. *Phocoenobacter uteri* representing also a new genus came to the scene in 2000 (Foster *et al.*, 2000). Then in 2003 strains previously referred to as [*Haemophilus*] *somnus*, [*Haemophilus*] *agni* or '*Histophilus ovis*' were analysed by 16S rRNA and *rpoB* gene sequencing. This, together with earlier investigations by DNA–DNA reassociation, showed that all these organisms are very closely related to each other and can be classified within the same species, *Histophilus somni* currently being the only member of the genus (Angen *et al.*, 2003). In the same year investigation of genetic relation of bacteria previously classified as the avian [*Pasteurella*] *haemolytica*-'*Actinobacillus salpingitidis*'-[*Pasteurella*] *anatis* complex ended in the description of the new genus *Gallibacterium* with *G. anatis* and at least one genomospecies (Christensen *et al.*, 2003a). In 2004, 16S rRNA gene sequence comparison of previously phenotypically characterized strains (Bisgaard *et al.*, 1999) resulted in assignment of the new genus *Volucribacter*, which comprises the two species *V. psittacicida* and *V. amazonae* (Christensen *et al.*, 2004a). Later that year organisms cultured from airway infection of horses were investigated in a complex phylogenetic analysis, including 16S rRNA, *rpoB*, and *infB* gene sequences, which in a polyphasic approach provided powerful evidence for the new genus *Nicoletella*, currently consisting of only one species *N. semolina* (Kuhnert *et al.*, 2004). Phylogeny based on 16S rRNA gene sequences was also useful for description of the genus *Avibacterium* in 2005, comprising *Av. paragallinarum, Av. gallinarum, Av. avium* and *Av. volantium* (Blackall *et al.*, 2005). In 2006, accurate investigation of genetic relatedness between [*Actinobacillus*] *actinomycetemcomitants*, [*Haemophilus*] *aphrophilus*, [*Haemophilus*] *paraphrophilus*, [*Haemophilus*] *segnis* and a few other members of the family *Pasteurellaceae* provided arguments that these species should be placed to the new genus *Aggregatibacter*, and they have been renamed to *Agg. actinomycetemcomitants, Agg. aphrophilus* (combined from former species [*H.*] *aphrophilus* and [*H.*] *paraphrophilus*) and *Agg. segnis* (Nørskov-Lauritsen and Kilian, 2006). Finally, as of the end of 2007 the last genus was *Bibersteinia*, resulting from the reclassification of [*Pasteurella*] *trehalosi*, which was supported by phylogenetic analysis of 16S rRNA gene sequences (Blackall *et al.*, 2007).

Bioinformatic tools for phylogeny

There is one topic worth mentioning before we move on to get a closer look at the phylogeny of the *Pasteurellaceae*, which is: how can phylogenetic relationship properly be reconstructed? In the style of the quote attributed to Winston Churchill that 'the only statistics you can trust are those you falsified yourself' one is tempted to say 'the only phylogenetic tree you can trust is the one you constructed yourself'. In a simplified view phylogenetic reconstruction is a mathematical approach using sophisticated algorithms to reflect as much as possible biological principles, in this case mutational changes representing the timespan between separation of two or more organisms. How far these models can in fact represent the historical flow and evolution as well as account for selection changes (if any) and other influences on mutational rate is, of course, a case for dispute. This is certainly one reason why there are various approaches and algorithms nowadays available and phylogeny has developed into a very complex if not complicated scientific field. The authors really do not feel competent to claim that they have accounted for all the unknowns and that they have chosen the most appropriate approach for the phylogenetic reconstructions shown in this chapter. Rather, we try to give a simple representation of the phylogenetic relationships within the family without going into too much details, e.g. about branching, distances and evolutionary aspects. The aim is more to show the big context. Own experience has taught us that the choice of algorithms, the numbers of species included and, of course, the quality of the 'raw data', i.e. the DNA sequence (including the question how diverse copies of 16S rRNA genes within a species should be handled), can all influence the tree structure. This should in no way be a pretext or argument that more appropriate analyses cannot be done, and in some cases very nice conclusions can indeed be drawn from intensive phylogenetic comparisons, but this was not our intention for this chapter.

Basically, there are three particular families of methods used for phylogenetic reconstruction which are distance matrix, maximum parsimony and maximum likelihood. Maximum parsimony is based on the minimum number of base changes which are needed to build the tree. Maximum likelihood considers various parameters that might influence phylogenetic reconstruction, e.g. identical sites between sequences based on multiple mutations or transition/transversion bias. This is less well handled by the distance-based trees but corrections like Jukes–Cantor can still to some extent count for this. A tree is then built from the distance matrix by, for example, the most common used method based on Neighbour-Joining (Saitou and Neil, 1987). Maximum likelihood is considered the most appropriate approach; however, parameters have to be set accordingly and the calculation takes a great deal of computational resources, especially with large trees.

Bootstrap analysis should be included with all methods as measures of robustness of nodes. Ideally, trees are built based on different methods and consistency between results indicates the trustfulness of the phylogenetic relationships.

In our hands Bionumerics from Applied Maths NV (Sint-Martens-Latem, Belgium) proved to be very useful for phylogenetic analyses since it is a comprehensive database program including several modules. Thereby, not only DNA sequences can be handled but also phenotypic traits or gel pictures (fragment analysis) as well as other markers. Moreover, it includes many algorithms for phylogenetic analyses and is therefore very flexible. However, the program is rather expensive and for that reason not always affordable.

Alternatively, there are several programs publicly available for phylogenetic analysis, e.g. MEGA4 (Tamura *et al.*, 2007), ClustalX (Thompson *et al.*, 1997) or PHYLIP. Comprehensive lists of available software can be found on the web, e.g. evolution.genetics.washington.edu/phylip/software.html.

Phylogeny based on 16S rRNA genes

The 16S rRNA gene has massively improved our understanding of microbial evolution. Since the 16S rRNA itself has a function in the ribosome and therefore in translation from DNA to protein it is vital and present in all so far known prokaryotes. The 16S rRNA is a linear molecule which, however, shows basepairing with itself resulting in linear, double-strand and hairpin regions (Gutell *et al.*, 1994). The structural and functional constraints results in highly conserved as well as more variable regions within the gene. There are several advantages to that and to the 16S rRNA gene in general: (i) the information gained from the sequence allows delineation of distant relations from more conserved regions as well as closer ones from the variable parts of the gene; (ii) it allows definition of universal primers from conserved regions for amplification and sequencing of the gene; (iii) the size of about 1.5 kb is reasonable for data generation and handling; and (iv) a large amount of data for 16S rRNA gene from a broad variety is available for phylogenetic comparison of numerous bacterial species from a large number of families. The 16S rRNA gene-based phylogenetic analysis developed to a gold standard for genetic comparison of bacteria and its analysis has become a prerequisite for the description of new species and is therefore one of the pillars in taxonomy (Stackebrandt *et al.*, 2002). One of the conclusions from 16S rRNA gene sequence comparisons is that strains which show more than 3% divergence in their 16S rRNA gene sequence should be regarded separate species as determined by comparison to the reference method DNA–DNA hybridization, whereas strains that are less than 3% divergent may or may not be members of different species and DNA–DNA hybridization or equivalent methods have to be applied (Stackebrandt and Goebel, 1994). Therefore, 3% 16S rRNA divergence is applied as a conservative criterion for demarcating species.

Currently, the 16S rRNA gene sequence-based phylogeny is recognized as most useful for investigation of relatedness also within the family *Pasteurellaceae*. Since the first descriptions of 16S rRNA sequences from *Pasteurellaceae* (Chuba *et al.*, 1988), many increasingly large 16S rRNA gene-based trees have been published (Dewhirst *et al.*, 1992, 1993; Olsen *et al.*, 2005; Christensen *et al.*, 2004b; Korczak *et al.*, 2004; Angen *et al.*, 2003). The 16S rRNA genes of all named and several unnamed species-like taxa have been determined and are available, e.g. on GenBank (Benson *et al.*, 2007). One important issue having some influence on phylogenetic analysis is the copy number of rRNA operons present in bacterial genomes. Exploring the genome sequences of some *Pasteurellaceae* results in three copies of 16S rRNA gene for *Actinobacillus pleuropneumoniae* L20 (NC_009053), five for *Hist. somni* 129Pt (NC_008309), and six for *Haemophilus influenzae* KW20 Rd (NC_000907), *[Haemophilus] ducreyi* 35000HP (NC_002940), *'Mannheimia succiniciproducens'* MBEL55E (NC_006300) and *Pasteurella multocida* Pm70 (NC_002663). The degree of allelic heterogeneity varies between taxa and in some cases 16S rRNA gene sequences could differ up to several percent. It has been shown that at least for gamma-proteobacteria, the selection of a single 16S rRNA allele is sufficient for inferring phylogenies, and that orthologous alleles from the same strain would have generated similar phylogenetic trees, however, presence of heterogeneous alleles which appear in the sequence as ambiguous bases makes the analysis difficult (Olivier *et al.*, 2005). This has become especially problematic since direct sequencing of PCR products amplified from genomic DNA became routinely used. Own experience using our protocol (Kuhnert *et al.*, 2002) have shown that in certain cases the number of ambiguous positions is very high. In the case of *[Haemophilus] felis* we were even not able to properly determine the 16S rRNA gene sequence in our lab, probably due to an experimental artefact, since the 16S rRNA gene sequence of the type strain is available in GenBank. The ambiguous positions mounted up to over 3% within the type strain of *[H.] felis* (P. Kuhnert, unpublished data). A similar situation was observed with the 16S rRNA gene sequence of the type strain of *L. koalarum* in our hands. Cloning of the corresponding 16S rRNA genes might be the only way to definitely answer the question if this is a PCR artefact or really representing the intra-strain variability (Acinas *et al.*, 2004). Divergence of rRNA operons has been found more often in *Archaea* but also in *Bacteria* and its implication for phylogeny has been discussed (Acinas *et al.*, 2005; Olivier *et al.*, 2005). The problem within the family *Pasteurellaceae* in this context should not be overestimated but it is worth pointing out in the view of new genera and species that certainly will be described in the future.

Taking the above-mentioned aspects into account, we constructed an updated 16S rRNA gene based phylogenetic tree of the *Pasteurellaceae* (Fig. 2.1). It is based on a Jukes–Cantor corrected distance matrix and on Neighbour-Joining tree formation using sequences of type strains representing nearly all currently described spe-

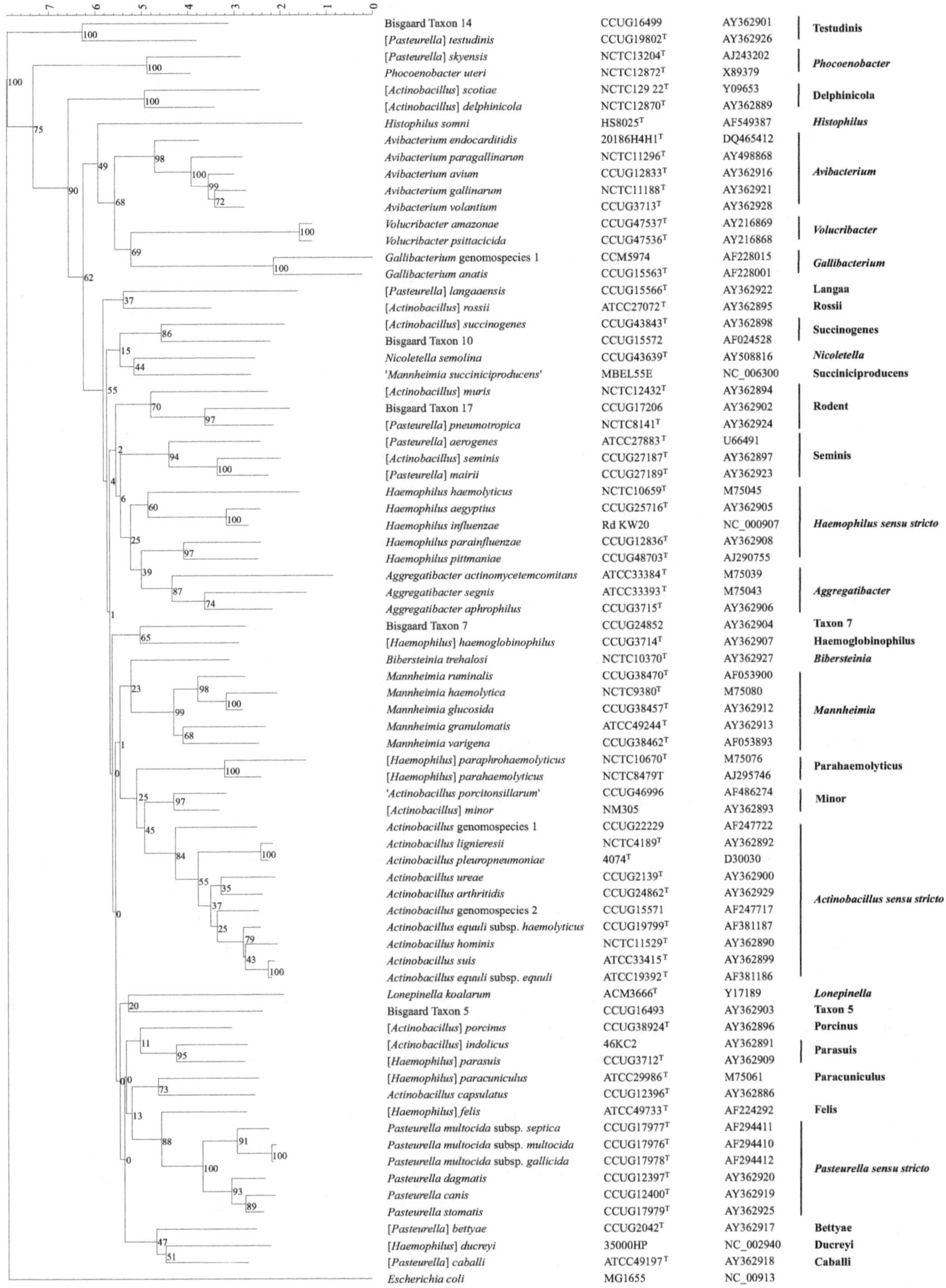

Figure 2.1 Phylogenetic tree based on the 16S rRNA gene from members of *Pasteurellaceae*. *E. coli* was included as an outgroup to root the tree. The Neighbour-Joining tree was built from a Jukes–Cantor corrected distance matrix in Bionumerics v.4.6. Bootstrap values of 500 trees are indicated as per cent confidence values for the branching. The distance bar on top represents per cent differences.

cies and a few candidates for new species as well as some genome sequences. The tree topology and branching shows that phylogenetically many members of *Pasteurellaceae* are unrelated to the taxa to which they have been previously classified mainly based on their phenotypic features. The groups and monotypic taxa correspond to ones formerly defined by others (Christensen *et al.*, 2004b; Korczak *et al.*, 2004; Olsen *et al.*, 2005). The tree shows about 30 groups, in general representing the current and putative new genera, including monophyletic branches. These groups and their members as well as the monotypic taxa are also listed in Table 1.4. Generally the tree shows clearly that there are still many misclassified species which need a final proper classification.

Similarities of 16S rRNA gene sequences within the family are more than 89% and within a genus it is generally more than 93%. However, these values are an approximation and cannot be taken as definitive. They may also change with more taxa of the family will be described.

It has been shown that in some cases the 16S rRNA-based phylogeny does not support results provided by DNA–DNA hybridization studies. Based on the study of Dewhirst (1992) *Actinobacillus capsulatus* was suggested to be excluded from the genus *Actinobacillus*, although DNA–DNA reassociation values showed that this taxon was correctly named (Escande *et al.*, 1984; Mutters *et al.*, 1989). Moreover, phylogenetic analysis based on *rpoB*, *infB* and *recN* unambiguously clarified that this species belongs to the genus *Actinobacillus* (Kuhnert *et al.*, 2007). This finding indicates that even such stable markers and phylogenetic gold standard as the 16S rRNA gene might be subjected to lateral transfer (Yap *et al.*, 1999). Furthermore, it shows that the 16S rRNA gene not only might in some cases be too conserved, and therefore unable to differentiate between some closely related but ecological distinct groups of bacteria (Fox *et al.*, 1992), but also has to be complemented by other phylogenetic approaches based on different genes. A few such 'phylogenes', i.e. genes suitable for phylogenetic analyses, have been applied for the *Pasteurellaceae* or at least for some groups of the family.

Phylogeny based on selected housekeeping genes

For a number of reasons housekeeping genes have proved to be very promising targets for clarifying phylogenetic relationships within the various bacteria. These genes encode proteins that are essential for the cell metabolism and physiology, and therefore should be present in all species and comparable between different organisms. This kind of genes normally evolve at a higher rate than the 16S rRNA gene, thereby providing a better resolution for differentiation of closely related species (Lawrence *et al.*, 1991). It has been a big challenge to find powerful approaches being suited for this purpose since not every gene is of the equal value in investigation of phylogenetic relationships. The ideal phylogene has to fulfil certain criteria: (i) changes in the sequence should occur randomly; (ii) rates of changes should be constant in order to properly represent evolutionary distances; and (iii) the size of the sequence has to be large enough to provide sufficient information (Woese, 1987). Whole-genome comparisons showed that there are 30–40 candidate genes for that purpose (Ludwig and Schleifer, 1999; Zeigler, 2003). Several such housekeeping genes have been used to investigate the phylogeny within certain genera or even the entire family of *Pasteurellaceae*.

Using amino acid sequences instead of nucleic acid sequences could be an advantage for comparison of distanly related groups, e.g. loosely related genera within the family. The nucleic acid level includes all sequence variations which could be too high to properly resolve these phylogenies whereas the amino acid sequence level reduces diversity by neglecting silent substitutions and third codon positions. In our analyses we concentrated on the DNA sequence-based phylogeny of *Pasteurellaceae*. Nevertheless, the phylogeny using amino acid sequences could provide more useful information for certain targets.

Phylogeny based on *infB*

The single-copy *infB* gene encodes the translation initiation factor 2 (IF2) that is involved in the initiation of protein synthesis in prokaryotes. The structure of the protein IF2 is highly conserved within several bacteria particularly in the domains responsible for binding and hydrolysis of GTP. However, this part contains also variable segments that are suitable for differentiation between groups of bacteria (Steffensen *et al.*, 1997; Hedegaard *et al.*, 1999). The length of the gene can vary between species and within the *Pasteurellaceae* it ranges between 2463 bp for *H. influenzae* KW20 Rd and 2526 bp for *A. pleuropneumoniae* L20.

Partial *infB* sequences have been used to investigate the phylogeny of the genus *Haemophilus* (Hedegaard *et al.*, 2001). Fragments of approximately 650 nucleotides were amplified from strains representing at the time the genus and 453-bp-long sequences covering the mentioned GTP-binding domain of IF2 were unambiguously determined and used for analysis of relationships within the group (Table 2.1). In an another study Christensen *et al.* (2004b) applied the same sequencing approach but used the deduced partial protein sequences for comparative phylogenetic analysis of 36 taxa of *Pasteurellaceae* representing nine genera. In order to obtain more information on the *infB* gene we developed a general PCR-based sequencing approach for the family covering approximately 1360 bp of the gene. This was used in the description of the genus *Nicoletella*, when it was phylogenetically compared with the type strains of nine established genera as well as type strains of species of the genus *Actinobacillus* (Kuhnert *et al.*, 2004). The same approach was later used to obtain a deeper understanding of genetic relationship within strains of the phenotypically closely related and therefore often misidentified [*Actinobacillus*] *rossii* and [*Pasteurella*] *aerogenes* (Mayor *et al.*, 2006).

A comprehensive tree derived from the 1360-bp-long sequences of 50 species representing all 13 currently named genera as well as candidates for potential new taxa within the family has been constructed for illustration of the phylogeny provided by *infB* (Fig. 2.2). The tree is in good agreement with the 16S rRNA gene-based tree and a good separation of genera and groups was obtained with the *infB* phylogeny. Similarities of partial *infB* sequences within the current family were more than 73% and in general more than 83% within the same genus. However, these values are an approximation and might also change while new taxa will be described.

Phylogeny based on *rpoB*

The *rpoB* gene fulfils the main expectations for being a good phylogene. It encodes the β-subunit of the DNA-dependent RNA polymerase, the highly conserved molecule that catalyses transcription of DNA into RNA. It is found among the *Bacteria* (Ovchinnikov *et al.*, 1981) as well as among the *Archeae* (Klenk and Zillig, 1994). So far, it has been found in only one copy in all investigated bacterial genomes and the size within members of the family *Pasteurellaceae* ranges from 4029 bp for *A. pleuropneumoniae* L20, [*H.*] *ducreyi* 35000HP, *Hist. somni* 129Pt, '*M. succiniciproducens*' MBEL55E and *P. multocida* Pm70 to 4032 bp for *H. influenzae* KW20 Rd. Similar to the 16S rRNA gene the *rpoB* sequence contains conserved and variable regions (Lisitsyn *et al.*, 1988; Palenik, 1992). This enables to select PCR and sequencing primers in conserved regions and interspacing variable regions provide very useful information about diversity of investigated strains at DNA level. The *rpoB* has been successfully applied as a molecular chronometer for investigation of phylogenetic relations within different groups of organisms (Klenk and Zillig, 1994; Nolte, 1995; Taillardat-Bisch *et al.*, 2003; Rowland *et al.*, 1993; Mollet *et al.*, 1997).

A comprehensive analysis of the use of *rpoB* for the phylogeny of *Pasteurellaceae* was published by Korczak *et al.* (2004). The universal approach for the family allowed sequence determination of about 560 bp from all 72 included strains by use of a single pair of primers (Table 2.1). The fragment covers the most variable polypeptide region 4 of the DNA-dependent β-subunit RNA polymerase in *Escherichia coli* and despite its limited size proved to be highly informative for phylogenetic analysis of *Pasteurellaceae*. Fig. 2.3 gives an updated *rpoB*-based tree. In general, *rpoB* provided higher resolution than 16S rRNA gene, however, it was also not sufficient enough for differentiation of very closely related species and subspecies. Currently estimated similarities of partial *rpoB* sequences within the family were more than 77% and generally more than 87%

Table 2.1 Primers for amplification and sequencing of target genes used for phylogenetic investigations within the family *Pasteurellaceae*

Target gene	PCR primers[a] Name	Sequence[b]	Product length[c]	Sequencing primers Name	Sequence[b]	References
16S rRNA gene	16SUNI-L	5′-AGAGTTTGATCATGGCTCAG-3′	1400	16SRNAI-S	5′-CTACGGGAGGCAGCAGTGGGG-3′	Kuhnert *et al.*, 2002
	16SUNI-R	5′-GTGTGACGGGCGGTGTGTAC-3′		16SRNA1-S	5′-CTACGGGAGGCAGCAGTGTAG-3′	
				16SRNAII-S	5′-GTGTAGCGGTGAAATGCGTAG-3′	
				16SRNA2-S	5′-GTGTAGGGGTAAAATCCGTAG-3′	
				16SRNAIV-S	5′-GGTTAAGTCCCGCAACGAGCGC-3′	
				16SRNA4-S	5′-GCTTAAGTGCCATAACGAGCGC-3′	
				16SRNAV-S	5′-CCCCACTGCTGCCTCCCGTAG-3′	
				16SRNAVI-S	5′-CTACGCATTTCACCGCTACAC-3′	
				16SRNA6-S	5′-CTACGGATTTTACCCCTACAC-3′	
				16SRNAVIII-S	5′-GCGCTCGTTGCGGGACTTAACC-3′	
				16SRNA8-S	5′-GCGCTCGTTATGGCACTTAAGC-3′	
infB	1186F	5′-ATYATGGGHCAYGTHGAYCAYGGHAARAC-3′	650	1748R	5′-GTAGCAACCGGACCACGACCTTTAT-3′	Hedegaard *et al.*, 1999
	1833R	5′-TATCCGACGCCGAACTCCGRTTNCGCATNGCNCGNAYNCGNCC-3′		or 1745R	5′-GCAACCGRICCICTICCTTTRTC-3′	
	infB-L	5′-ATGGGNCACGTTGACCACGGTAAAAC-3′	1400	infB-L	5′-ATGGGNCACGTTGACCACGGTAAAAC-3′	Kuhnert *et al.*, 2004
	infB-R	5′-CCGATACCACATTCCATACC-3′		infB-R	5′-CCGATACCACATTCCATACC-3′	
				infB-1	5′-CGTGAYGAGAARAAAGCACGTGAAG-3′	
				infB-2	5′-CTTCACGTGCTTTYTTCTCRTCACG-3′	
	infB-R1	5′-TCGTTGTAGTTTTTCACGCCGATACC-3′		infB-R1	5′-TCGTTGTAGTTTTTCACGCCGATACC-3′	Mayor *et al.*, 2006
rpoB	Pasrpob-L	5′-GCAGTGAAAGARTTCTTTGGTTC-3′	560	Pasrpob-L	5′-GCAGTGAAAGARTTCTTTGGTTC-3′	Korczak *et al.*, 2004
	Rpob-R	5′-GTTGCATGTTNGNACCCAT-3′		Rrpob-R	5′-GTTGCATGTTNGNACCCAT-3′	
sodA	d1	5′-CCITAYICITAYGAYGCIYTIGARCC-3′	510	d1	5′-CCITAYICITAYGAYGCIYTIGARCC-3′	Poyart *et al.*,1998,
	d2	5′-ARRTARTAIGCRTGYTCCCAIACRTC-3′		d2	5′-ARRTARTAIGCRTGYTCCCAIACRTC-3′	Gautier *et al.*, 2005, Cattoir *et al.*, 2006,
atpD	pmkatpf3w[d]	5′-GCTCGTCAAGCAAGTATYAC-3′	1430	pmkatpf3w[d]	5′-GCTCGTCAAGCAAGTATYAC-3′	Petersen *et al.*, 2001
	pmkatpr3w	5′-CTAAYACTTCRTCGATKGWAC-3′		pmkatpr3w	5′-CTAAYACTTCRTCGATKGWAC-3′	
				pmkatp4wf	5′-CGCAATGGGATCWTCWG-3′	
				pmkatpf5	5′-RAGYTATGAAGAACAAKC-3′	
				pmkatpf7	5′-TAYRAYGTTGCHCGTGGYG-3′	
				pmkatp4wr	5′-GWACCRTTAAATACTTCMGC-3′	
				pmkatpr5	5′-GTTCTTSACCMACRAC-3′	
rpoA	rpoA-L	5′-TCTGTRACAGAATTTTTAAARCC-3′	1000	rpoA-L	5′-TCTGTRACAGAATTTTTAAARCC-3′	Kuhnert and Korczak, 2006
	rpoA-R	5′-TTGCNGGNGGCCARTTTTCAAGG-3′		rpoA-R	5′-TTGCNGGNGGCCARTTTTCAAGG-3′	
recN	recN-L	5′-CAACTYACTATYAATMATTTTGC-3′	1400	recN-L	5′-CAACTYACTATYAATMATTTTGC-3′	
	recN-R	5′-CTAATGCCYACRTCYACTTCATC-3′		recN-R	5′-CTAATGCCYACRTCYACTTCATC-3′	

recN	recN-L2	5′-CATTTAACGGTTAATAATTTTGC-3′		recN-L2	5′-CATTTAACGGTTAATAATTTTGC-3′	Kuhnert and Korczak, 2006
	recN-R2	5′-CTAATYCCMACATCNACYTCATC-3′		recN-R2	5′-CTAATYCCMACATCNACYTCATC-3′	
	recN_Vp-R	5′-CTGATCCCTACATCAATTTCATC-3′		recN_Vp-R	5′-CTGATCCCTACATCAATTTCATC-3′	
				recN_Ga-1	5′-GCTATTTAGATGAATTGGTTGAGG-3′	
				recN_Ga-2	5′-CCTCAACCAATTCATCTAAATAGC-3′	
thdF	thdF_first-L2	5′-ATGAAAGANACNATYGTNGCWCARGC-3′	1400	thdF_first-L2	5′-ATGAAAGANACNATYGTNGCWCARGC-3′	
	thdF_first-R2	5′-TGTTTATTTNCCRATRCARAANGARC-3′		thdF_first-R2	5′-TGTTTATTTNCCRATRCARAANGARC-3′	
	thdF-L2	5′-AAAGANACNATYGTNGCWCARGC-3′		thdF-L2	5′-AAAGANACNATYGTNGCWCARGC-3′	
	thdF-R2	5′-TTATTTNCCRATRCARAANGARC-3′		thdF-R2	5′-TTATTTNCCRATRCARAANGARC-3′	
	thdF_MP-L	5′-AAAGAMACCATTGTTGCWCAAGC-3′		thdF_MP-L	5′-AAAGAMACCATTGTTGCWCAAGC-3′	
				thdF-1	5′-GGAATGAAAGTCGTGATTGCAGG-3′	
				thdF-2	5′-CCTGCAATCACGACTTTCATTCC-3′	
				thdF-3	5′-GGATGAAAGTGGTAATTGC-3′	
				thdF-4	5′-GCAATTACCACTTTCATCC-3′	
adk	adk-up	5′-GGTGCACCGGGTGCAGGTAA-3′	620	adk-up	5′-GGTGCACCGGGTGCAGGTAA-3′	Meats *et al*., 2003
	adk-dn	5′-CCTAAGATTTTATCTAACTC-3′		adk-dn	5′-CCTAAGATTTTATCTAACTC-3′	
atpG	atpG-up	5′-ATGGCAGGTGCAAAAGAGAT-3′	560	atpG-up	5′-ATGGCAGGTGCAAAAGAGAT-3′	
	atpG-dn	5′-TTGTACAACAGGCTTTTGCG-3′		atpG-dn	5′-TTGTACAACAGGCTTTTGCG-3′	
frdB	frdB-up	5′-CTTATCGTTGGTCTTGCCGT-3′	580	frdB-up	5′-CTTATCGTTGGTCTTGCCGT-3′	
	frdB-dn	5′-TTGGCACTTTCCACTTTTCC-3′		frdB-dn	5′-TTGGCACTTTCCACTTTTCC-3′	
fucK	fucK-up	5′-ACCACTTTCGGCGTGGATGG-3′	560	fucK-up	5′-ACCACTTTCGGCGTGGATGG-3′	
	fucK-dn	5′-AAGATTTCCCAGGTGCCAGA-3′		fucK-dn	5′-AAGATTTCCCAGGTGCCAGA-3′	
mdh	mdh-up	5′-TCATTGTATGATATTGCCCC-3′	550	mdh-up	5′-TCATTGTATGATATTGCCCC-3′	
	mdh-dn	5′-ACTTCTGTACCTGCATTTTG-3′		mdh-dn	5′-ACTTCTGTACCTGCATTTTG-3′	
pgi	pgi-up	5′-GGTGAAAAAATCAATCGTAC-3′	590	pgi-up	5′-GGTGAAAAAATCAATCGTAC-3′	
	pgi-dn	5′-ATTGAAAGACCAATAGCTGA-3′		pgi-dn	5′-ATTGAAAGACCAATAGCTGA-3′	
recA	recA-up	5′-ATGGCAACTCAAGAAGAAAA-3′	620	recA-up	5′-ATGGCAACTCAAGAAGAAAA-3′	
	recA-dn	5′-TTACCAAACATCACGCCTAT-3′		recA-dn	5′-TTACCAAACATCACGCCTAT-3′	
adk	adk.34f	5′-GGIAAAGGIACWCARGCICARTT-3′	580	adk.34f	5′-GGIAAAGGIACWCARGCICARTT-3′	Nørskov-Lauritsen *et al*., 2005
	adk.610r	5′-CTTCCACTTTTTKYGTMCCGTC-3′		adk.610r	5′-CTTCCACTTTTTKYGTMCCGTC-3′	
pgi	pgi.838f	5′-GATGGIAAAGAYGTIATGCC-3′	490	pgi.838f	5′-GATGGIAAAGAYGTIATGCC-3′	
	pgi.1331r	5′-GCTGACCAYAAIGARTAACG-3′		pgi.1331r	5′-GCTGACCAYAAIGARTAACG-3′	

Table 2.1 *continued*

Target gene	PCR primers[a] Name	Sequence[b]	Product length[c]	Sequencing primers Name	Sequence[b]	References
recA	recA.54f recA.617r	5′-GARAAACAATTTGGKAAAGGC-3′ 5′-TTACCRAACATMACRCCIAT-3′	560	recA.54f recA.617r	5′-GARAAACAATTTGGKAAAGGC-3′ 5′-TTACCRAACATMACRCCIAT-3′	Nørskov-Lauritsen *et al.*, 2005
6pgd	6pgdF 6pgdR	5′-TTATTACCGCACTTAGAAG-3′ 5′-CGTTGATCTTTGAATGAAGA-3′	630	6pgdF 6pgdR	5′-TTATTACCGCACTTAGAAG-3′ 5′-CGTTGATCTTTGAATGAAGA-3′	Olvera *et al.*, 2006
atpD	atpDF atpDR	5′-CAAGATGCAGTACCAAAAGTTTA-3′ 5′-ACGACCTTCATCACGGAAT-3′	610	atpDF atpDR	5′-CAAGATGCAGTACCAAAAGTTTA-3′ 5′-ACGACCTTCATCACGGAAT-3′	
frdB	frdBF frdBR	5′-CATATCGTTGGTCTTGCCGT-3′ 5′-TTGGCACTTTCGATCTTACCTT-3′	580	frdBF frdBR	5′-CATATCGTTGGTCTTGCCGT-3′ 5′-TTGGCACTTTCGATCTTACCTT-3′	
g3pd	3gpdF 3gpdR	5′-GGTCAAGACATCGTTTCTAAC-3′ 5′-TCTAATACTTTGTTTGAGTAACC-3′	570	3gpdF 3gpdR	5′-GGTCAAGACATCGTTTCTAAC-3′ 5′-TCTAATACTTTGTTTGAGTAACC-3′	
rpoB	rpoBF rpoBR	5′-TCACAACTTTCICAATTTATG-3′ 5′-ACAGAAACCACTTGTTGCG-3′	470	rpoBF rpoBR	5′-TCACAACTTTCICAATTTATG-3′ 5′-ACAGAAACCACTTGTTGCG-3′	
adk	adk forward adk reverse	5′-AAGGBACWCAAGCVCAAT-3′ 5′-CACTTTTTKYGTMCCGTC-3′	570	adk rorward adk reverse	5′-AAGGBACWCAAGCVCAAT-3′ 5′-CACTTTTTKYGTMCCGTC-3′	Davies *et al.*, 2004
aroA	aroA forward aroA reverse	5′-TTTACCGGGCTCCAAAAG-3′ 5′-CTTTTACGCGCCAGTTAT-3′	990	aroA forward aroA reverse	5′-TTTACCGGGCTCCAAAAG-3′ 5′-TGCATCATCTTAAGGGTG-3′	
deoD	deoD forward deoD reverse	5′-GTGCATTTGCCGATGTTG-3′ 5′-TGGTGTTGTTTGTTCGTG-3′	620	deoD forward deoD reverse	5′-GTGCATTTGCCGATGTTG-3′ 5′-TGGTGTTGTTTGTTCGTG-3′	
gdhA	gdhA forward gdhA reverse	5′-CTTAGTTGAACCTGAACG-3′ 5′-CTTGACCTTCAATCGTGC-3′	1100	gdhA forward gdhA reverse	5′-CGCGTTAACCACATTACC-3′ 5′-CCCTTCAGCCACTAATTG-3′	
g6pd	g6pd forward g6pd reverse	5′-CHGGYGAYYTMACTYATCG-3′ 5′-TTTBGCGATBARTTTRTCRGC-3′	1400	g6pd forward g6pd reverse	5′-GATGCTGCCGATTATGG-3′ 5′-CAAGACTTTTGCCACTTC-3′	
mdh	mdh forward mdh reverse	5′-AAGTTGCWGTWYTAGGTG-3′ 5′-CCTAATTCAATATCYGCACG-3′	910	mdh forward mdh reverse	5′-TGTCCAAAAGCTTGTGTG-3′ 5′-CCTAATTCAATATCYGCACG-3′	
pgi	pgi forward pgi feverse	5′-GCCGTGGTTGGTTGATGG-3′ 5′-TTGGCTTGGCGCGATGAA-3′	900	pgi forward pgi reverse	5′-TGATTTCTGGTGAATGGA-3′ 5′-GGAAATACGCTGCAAAAC-3′	

[a]Annealing temperature is not given since it might vary due to different conditions and equipment in different laboratories.

[b]I= deoxyinosine; W=A or T; R=G or A; K=G or T; Y= C or T; M= A or C.

[c]PCR product length might differ between species. Approximate size of amplicons is given.

[d]Primer aligns to *atpG*

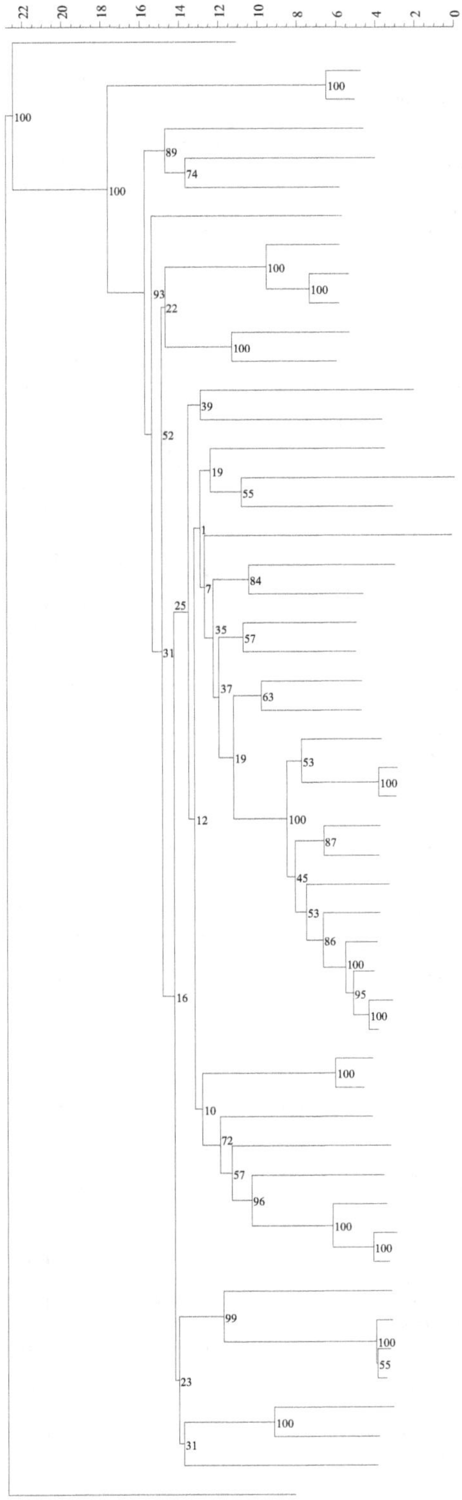

[*Pasteurella*] *testudinis*	CCUG19802^{T}	EU350942
Gallibacterium genomospecies 1	CCM5974	EU350939
Gallibacterium anatis	CCUG15563^{T}	AY508843
'*Mannheimia succiniciproducens*'	MBEL55E	NC_006300
[*Actinobacillus*] *succinogenes*	CCUG43843^{T}	EU350937
Aggregatibacter actinomycetemcomitans	HK1651	AA01922*
[*Actinobacillus*] *muris*	NCTC12432^{T}	EU350935
[*Pasteurella*] *aerogenes*	ATCC27883^{T}	DQ211791
[*Actinobacillus*] *seminis*	CCUG27187^{T}	DQ211793
[*Pasteurella*] *mairii*	CCUG27189^{T}	DQ211784
Avibacterium paragallinarum	NCTC11296^{T}	EU350938
Avibacterium gallinarum	NCTC11188^{T}	EF055560
Nicoletella semolina	CCUG43639^{T}	AY508835
Histophilus somni	HS8025^{T}	AY508848
Phocoenobacter uteri	NCTC12872^{T}	AY508849
[*Actinobacillus*] *scotiae*	CCUG46379^{T}	EU350936
[*Actinobacillus*] *delphinicola*	NCTC12870^{T}	EU350934
[*Haemophilus*] *ducreyi*	35000HP	NC_002940
[*Actinobacillus*] *rossii*	ATCC27072^{T}	DQ211763
[*Actinobacillus*] *porcinus*	NM319^{T}	EF059970
[*Actinobacillus*] *indolicus*	46KC2	EF059971
[*Haemophilus*] *parasuis*	CCUG3712^{T}	DQ410886
'*Actinobacillus porcitonsillarum*'	CCUG46996	EF055563
[*Actinobacillus*] *minor*	NM305	EF055562
Actinobacillus genomospecies 1	CCUG22229	AY508850
Actinobacillus pleuropneumoniae	4074^{T}	AY508851
Actinobacillus lignieresii	NCTC4189^{T}	AY508852
Actinobacillus arthritidis	CCUG24862^{T}	AY508854
Actinobacillus genomospecies 2	CCUG15571	AY508853
Actinobacillus capsulatus	CCUG12396^{T}	AY508855
Actinobacillus hominis	CCUG19800^{T}	AY508856
Actinobacillus ureae	CCUG2139^{T}	AY508860
Actinobacillus suis	ATCC33415^{T}	AY508857
Actinobacillus equuli subsp. *haemolyticus*	CCUG19799^{T}	AY508858
Actinobacillus equuli subsp. *equuli*	ATCC19392^{T}	AY508859
Haemophilus influenzae	Rd KW20	NC_000907
Haemophilus aegyptius	CCUG25716^{T}	EU350940
Bibersteinia trehalosi	NCTC10370^{T}	EF055561
Mannheimia granulomatis	ATCC49244^{T}	EF055564
Mannheimia varigena	CCUG38462^{T}	DQ410888
Mannheimia ruminalis	CCUG38470^{T}	DQ410889
Mannheimia haemolytica	NCTC9380^{T}	AY508847
Mannheimia glucosida	CCUG38457^{T}	DQ410887
Pasteurella canis	CCUG12400^{T}	EF055567
Pasteurella multocida subsp. *gallicida*	CCUG17978^{T}	EF055565
Pasteurella multocida subsp. *septica*	CCUG17977^{T}	EF055566
Pasteurella multocida subsp. *multocida*	CCUG17976^{T}	AY508844
Volucribacter psittacicida	CCUG47536^{T}	EF077221
Volucribacter amazonae	CCUG47537^{T}	EU350941
Lonepinella koalarum	ATCC700131^{T}	AY508846
Escherichia coli	MG1655	NC_00913

Figure 2.2 Phylogenetic tree based on the *infB* gene from members of *Pasteurellaceae*. *E. coli* was included as an outgroup to root the tree. The Neighbour-Joining tree was built from a Jukes–Cantor corrected distance matrix in Bionumerics v.4.6. Bootstrap values of 500 trees are indicated as per cent confidence values for the branching. The distance bar on top represents per cent differences. *Gene ID, Oralgen www.oralgen.lanl.gov

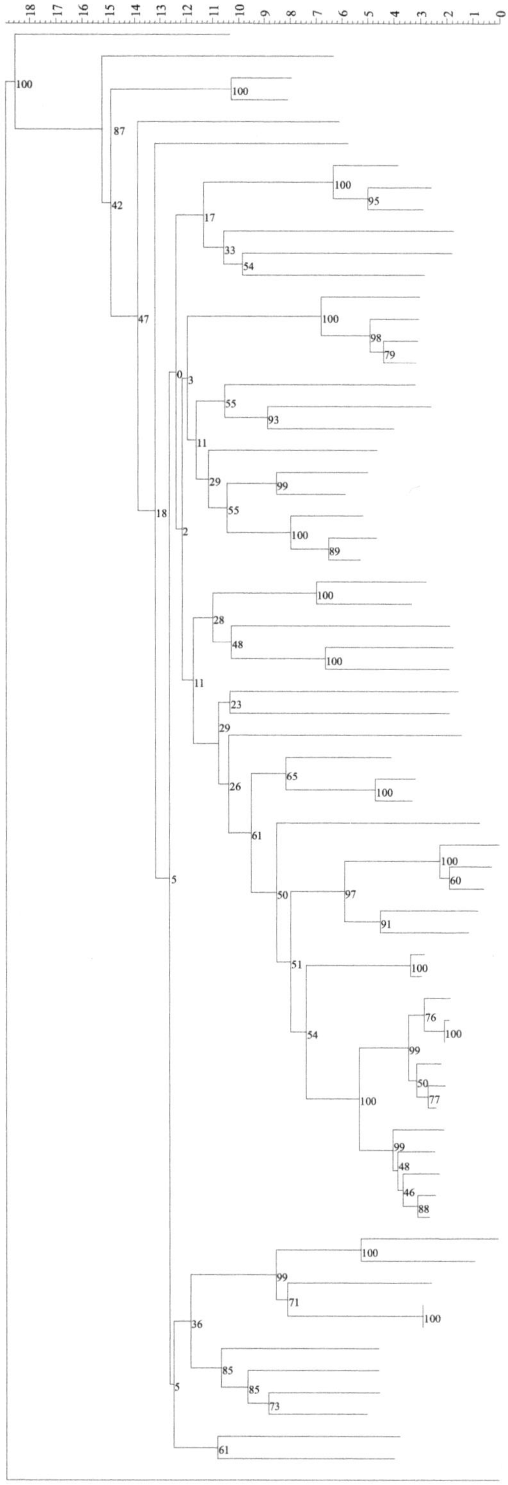

Bisgaard Taxon 14	CCUG16499	AY314043
[*Pasteurella*] *testudinis*	CCUG19802ᵀ	AY362972
Aggregatibacter aphrophilus	CCUG3715ᵀ	AY362953
Aggregatibacter actinomycetemcomitans	HK1651	AA00355*
Lonepinella koalarum	ATCC700131ᵀ	AY170218
[*Haemophilus*] *felis*	ATCC49733ᵀ	AY362954
[*Pasteurella*] *aerogenes*	ATCC27883ᵀ	AY314040
[*Pasteurella*] *mairii*	CCUG27189ᵀ	AY362968
[*Actinobacillus*] *seminis*	CCUG27187ᵀ	AY362945
[*Actinobacillus*] *succinogenes*	CCUG43843ᵀ	AY362946
'*Mannheimia succiniciproducens*'	MBEL55E	NC_006300
Bisgaard Taxon 5	CCUG16493	AY314039
Avibacterium paragallinarum	NCTC11296ᵀ	EU350943
Avibacterium volantium	CCUG3713ᵀ	AY362973
Avibacterium avium	CCUG12833ᵀ	AY362965
Avibacterium gallinarum	NCTC11188ᵀ	AY314041
[*Pasteurella*] *bettyae*	CCUG2042ᵀ	AY314042
Histophilus somni	HS8025ᵀ	AY170207
[*Haemophilus*] *haemoglobinophilus*	CCUG3714ᵀ	AY362955
Pasteurella canis	CCUG12400ᵀ	AY314038
Pasteurella dagmatis	CCUG12397ᵀ	AY362966
Pasteurella stomatis	CCUG17979ᵀ	AY362971
Pasteurella multocida subsp. *septica*	CCUG17977ᵀ	AY362970
Pasteurella multocida subsp. *multocida*	CCUG17976ᵀ	AY170216
Pasteurella multocida subsp. *gallicida*	CCUG17978ᵀ	AY362969
Gallibacterium genomospecies 1	CCM5974	AY314033
Gallibacterium anatis	CCUG15563ᵀ	AY314032
Bisgaard Taxon 7	CCUG24852	AY314030
Volucribacter psittacicida	CCUG47536ᵀ	EF077220
Volucribacter amazonae	CCUG47537ᵀ	EU350944
[*Actinobacillus*] *muris*	NCTC12432ᵀ	AY314029
Nicoletella semolina	CCUG43639ᵀ	AY508861
Phocoenobacter uteri	NCTC12872ᵀ	AY170219
Bibersteinia trehalosi	NCTC10370ᵀ	AY314028
[*Actinobacillus*] *indolicus*	46KC2	AY362938
[*Haemophilus*] *parasuis*	CCUG3712ᵀ	AY314027
[*Haemophilus*] *ducreyi*	35000HP	NC_002940
Mannheimia ruminalis	CCUG38470ᵀ	AY362961
Mannheimia haemolytica	ATCC33396ᵀ	AY170217
Mannheimia glucosida	CCUG38457ᵀ	AY362959
Mannheimia varigena	CCUG38462ᵀ	AY362962
Mannheimia granulomatis	ATCC49244ᵀ	AY362960
'*Actinobacillus porcitonsillarum*'	CCUG46996	AY362942
[*Actinobacillus*] *minor*	NM305	AY362939
Actinobacillus capsulatus	CCUG12396ᵀ	AY362932
Actinobacillus hominis	NCTC11529ᵀ	AY362937
Actinobacillus ureae	CCUG2139ᵀ	AY362950
Actinobacillus suis	ATCC33415ᵀ	AY362947
Actinobacillus equuli subsp. *equuli*	ATCC19392ᵀ	AY362936
Actinobacillus equuli subsp. *haemolyticus*	CCUG19799ᵀ	AY362935
Actinobacillus lignieresii	NCTC4189ᵀ	AY170215
Actinobacillus pleuropneumoniae	4074ᵀ	AY314026
Actinobacillus genomospecies 1	CCUG22229	AY362948
Actinobacillus genomospecies 2	CCUG15571	AY362951
Actinobacillus arthritidis	CCUG24862ᵀ	AY362974
Bisgaard Taxon 17	CCUG17206	AY314035
[*Pasteurella*] *pneumotropica*	NCTC8141ᵀ	AY314034
Haemophilus parainfluenzae	CCUG12836ᵀ	AY362956
Haemophilus aegyptius	CCUG25716ᵀ	AY362952
Haemophilus influenzae	Rd KW20	NC_000907
[*Actinobacillus*] *porcinus*	CCUG38924ᵀ	AY362941
[*Actinobacillus*] *rossii*	ATCC27072ᵀ	AY362940
[*Pasteurella*] *langaaensis*	CCUG15566ᵀ	AY362967
[*Pasteurella*] *caballi*	ATCC49197ᵀ	AY314036
[*Actinobacillus*] *delphinicola*	NCTC12870ᵀ	AY314037
[*Actinobacillus*] *scotiae*	CCUG46379ᵀ	AY362944
Escherichia coli	MG1655	NC_00913

Figure 2.3 Phylogenetic tree based on the *rpoB* gene from members of *Pasteurellaceae*. *E. coli* was included as an outgroup to root the tree. The Neighbour-Joining tree was built from a Jukes–Cantor corrected distance matrix in Bionumerics v.4.6. Bootstrap values of 500 trees are indicated as per cent confidence values for the branching. The distance bar on top represents per cent differences. *Gene ID, Oralgen www.oralgen.lanl.gov

within a genus. Again, these values are an approximation and may also change with more taxa of the family being described.

There was a good correlation between topologies in 16S rRNA and *rpoB*-based tree even though some discrepancies were observed. Phylogeny based on *rpoB* also correlated well with whole-genome DNA–DNA hybridization results and appeared to be valuable for dissection of genera within the family. Partial *rpoB* gene sequences were further used for clarification of the taxonomic position of strains identified as *Hist. somni* previously known as '*H. somnus*', '*H. agni*', or '*Hist. ovis*' (Angen *et al.*, 2003), including eight *Hist. somni* strains and the type strains of at this time existing genera. A sequence analysis of the *rpoB* fragments showed that strains identified as *Hist. somni* were clearly separated from the other genera within the family and should be referred to a new genus. It has also been shown that phylogeny obtained from parsimony analysis of the deduced *rpoB* partial protein sequence of representative taxa of the *Pasteurellaceae* allows clear discrimination of different genera in the family (Christensen *et al.*, 2004b).

Phylogeny based on *sodA*

The gene encoding manganese-dependent superoxide dismutase (*sodA*) is a further target that has been used for a better understanding of the phylogeny of *Pasteurellaceae*. The gene is rather small in size, ranging from 630 to 666 bp in investigated members of *Pasteurellaceae*. These differences in sequence length are the result of the presence of extra nucleotides corresponding to supplemental blocks of codons in the deduced amino acid sequences. Interestingly, additional amino acids define signatures which are specific for species of certain genera. The first study on *sodA* in particular comprised species belonging to the genus *Pasteurella* and related species (Gautier *et al.*, 2005). One year later this was complemented by a study giving a more detailed phylogenetic insight into the genus *Haemophilus* (Cattoir *et al.*, 2006). In these studies *sodA* fragments of 449–473 bp were generated from different species using conserved primers for PCR and sequencing (Poyart *et al.*, 1998); Table 2.1). In the analysis of the genera *Pasteurella*, *Gallibacterium* and *Mannheimia* the *sodA* provided higher resolution than 16S rRNA gene with a mean identity of 83.5% compared with 98.2% with the latter (Gautier *et al.*, 2005). In the study on the genus *Haemophilus* the mean identity value of *sodA* achieved 74.9% and 93.5% for 16S rRNA (Cattoir *et al.*, 2006).

Fig. 2.4 shows the phylogeny based on the currently available *sodA* sequences from *Pasteurellaceae*. Comparison of the similarity values of *sodA* between the 44 analysed *Pasteurellaceae* resulted in more than 64% sequence similarity within the family and more than 80% within genera. In the individual studies (Cattoir *et al.*, 2006; Gautier *et al.*, 2005), the *sodA*-based tree topology was in good agreement with the corresponding phylogeny provided by the16S rRNA gene; however, there were a few discrepancies observed in the full tree (Fig. 2.4). Clustering confirmed previous observations that certain taxa should be reclassified due to their phylogenetic position. However, [*Haemophilus*] *paracuniculus* interfered with the genus *Mannheimia*, members of the genus *Haemophilus sensu stricto* were split in two and [*H.*] *felis* interfered with the genus *Pasteurella sensu stricto*. The full tree contains many more entries than the individual studies what might have led to this changed topology. Moreover, due to the high discriminatory power of the *sodA* gene, resolution might be too high to delineate larger phylogenetic distances within the family, and complementary analysis of amino acid sequences might be indicated in this case. Nevertheless, the sequences were shown to be useful for diagnostic purposes and have a good resolution at the species level. Therefore it certainly has a great potential to help resolving classification within *Pasteurellaceae*, but further analysis extending sequencing of *sodA* to the other members of the family will be necessary to see if it will also be useful for the phylogeny at the family level.

Phylogeny based on *atpD*

The *atpD* gene encodes the β-subunit of the ATP synthase. Its role is to couple the electrochemical potential difference for H^+ across the inner membrane to synthesis of ATP from ADP and P_i. The enzyme is composed of the two subunits F_o and F_1. F_o traverses the membrane facilitating proton

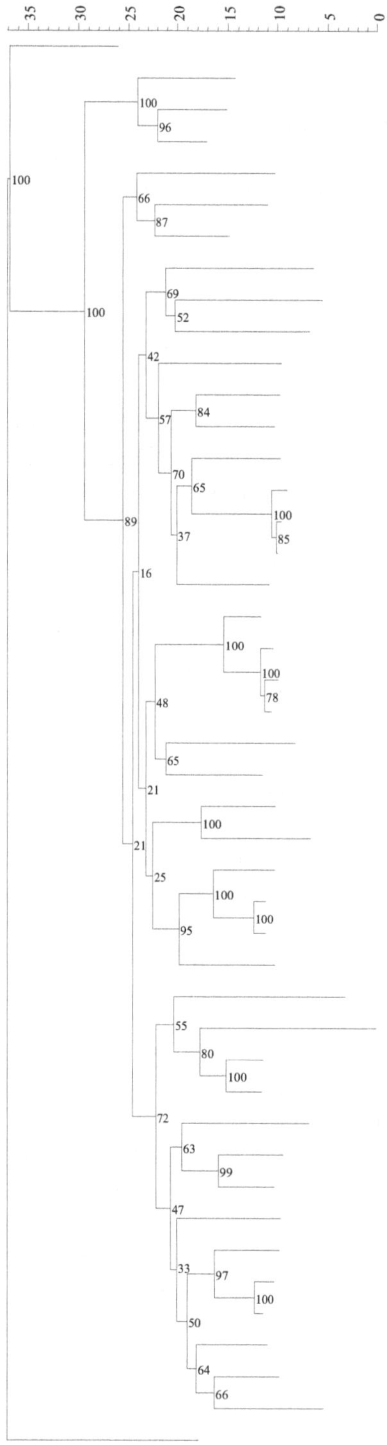

[Pasteurella] testudinis	CIP 82.117^T	AY702508
Aggregatibacter actinomycetemcomitans	CIP 52.106^T	DQ005612
Aggregatibacter aphrophilus	CIP 70.73^T	DQ005614
Aggregatibacter segnis	CIP 103292^T	DQ005625
Gallibacterium anatis	CIP 102679^T	AY702516
Haemophilus pittmaniae	CCUG48703^T	DQ267822
Haemophilus parainfluenzae	CIP 102513^T	AY702518
'Mannheimia succiniciproducens'	MBEL55E	NC_006300
[Pasteurella] caballi	CIP 103915^T	AY702495
[Pasteurella] bettyae	CIP 102636^T	AY702494
[Pasteurella] langaaensis	CIP 102678^T	AY702499
Pasteurella canis	CIP 103294^T	AY702496
Pasteurella stomatis	CIP 102680^T	AY702505
Pasteurella dagmatis	CIP103293^T	AY702497
Pasteurella multocida subsp. *septica*	CIP A125^T	AY702503
Pasteurella multocida subsp. *multocida*	CIP 103286^T	AY702502
Pasteurella multocida subsp. *gallicida*	CIP 103285^T	AY702501
[Haemophilus] felis	CIP 103890^T	DQ005616
Avibacterium avium	CIP 100919^T	AY702493
Avibacterium gallinarum	CIP 102676^T	AY702498
Avibacterium volantium	CIP 102677^T	AY702509
Avibacterium paragallinarum	CIP 103453^T	DQ005620
[Pasteurella] mairii	CIP 103902^T	AY702500
[Pasteurella] aerogenes	CIP 80.14^T	AY702492
Histophilus somni	129Pt	NC_008309
[Haemophilus] haemoglobinophilus	CIP 53.88^T	DQ005617
Haemophilus haemolyticus	CIP 103290^T	DQ005618
Haemophilus aegyptius	CIP 52.129^T	DQ005613
Haemophilus influenzae	Rd KW20	NC_000907
[Pasteurella] pneumotropica	CIP 66.16^T	AY702504
[Pasteurella] skyensis	CIP 107586^T	AY702506
[Haemophilus] parasuis	CIP 100918^T	DQ005624
[Haemophilus] parahaemolyticus	CIP 56.86^T	DQ005621
[Haemophilus] paraphrohaemolyticus	CIP 102512^T	DQ005622
[Haemophilus] ducreyi	35000HP	NC_002940
Actinobacillus pleuropneumoniae	CIP 100916^T	DQ005626
Actinobacillus ureae	CIP 62.8^T	AY702515
Bibersteinia trehalosi	CIP 103829^T	AY702507
Mannheimia ruminalis	CIP 106062^T	AY702513
Mannheimia haemolytica	CIP 103426^T	AY702512
Mannheimia glucosida	CIP 106063^T	AY702510
Mannheimia granulomatis	CIP 103433^T	AY702511
Mannheimia varigena	CIP 106064^T	AY702514
[Haemophilus] paracuniculus	CIP 107045^T	DQ005619
Escherichia coli	MG1655	NC_00913

Figure 2.4 Phylogenetic tree based on the *sodA* gene from members of *Pasteurellaceae*. *E. coli* was included as an outgroup to root the tree. The Neighbour-Joining tree was built from a Jukes–Cantor corrected distance matrix in Bionumerics v.4.6. Bootstrap values of 500 trees are indicated as per cent confidence values for the branching. The distance bar on top represents per cent differences.

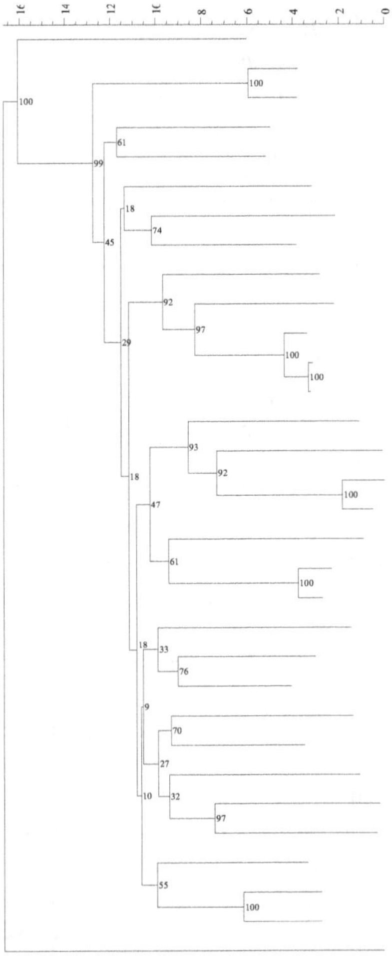

Bisgaard Taxon 14	CCUG16499	AY296242
Gallibacterium anatis	F149^{T}	AY296227
Gallibacterium genomospecies 1	CCUG23139	AY296226
Bisgaard Taxon 5	7.3	AY296238
Aggregatibacter actinomycetemcomitans	HK1651	AA01482*
[*Actinobacillus*] *delphinicola*	CCUG41362^{T}	AY296240
Volucribacter psittacicida	JEO101	AY296228
Avibacterium gallinarum	NCTC11188^{T}	AY296246
Bisgaard Taxon 7	7.4	AY296239
Pasteurella canis	NCTC11621^{T}	AF326331
Pasteurella multocida subsp. *septica*	HIM746-6^{T}	AF326329
Pasteurella multocida subsp. *multocida*	NCTC10322^{T}	AF326326
Pasteurella multocida subsp. *gallicida*	HIM830-7^{T}	AF326327
Bibersteinia trehalosi	NCTC10370^{T}	AY296231
[*Haemophilus*] *parasuis*	CCUG3712^{T}	AY296235
Mannheimia haemolytica	NCTC9380^{T}	AY296230
Mannheimia glucosida	P737	AY296229
[*Haemophilus*] *ducreyi*	35000HP	NC_002940
Actinobacillus lignieresii	NCTC4189^{T}	AY296232
Actinobacillus pleuropneumoniae	4074^{T}	AACK01000012
[*Pasteurella*] *bettyae*	CCUG2042^{T}	AY296236
[*Pasteurella*] *caballi*	ATCC49197^{T}	AY296244
[*Actinobacillus*] *rossii*	CCUG24851^{T}	AY296243
Histophilus somni	129Pt	NC_008309
[*Pasteurella*] *aerogenes*	ATCC27883^{T}	AY296233
Lonepinella koalarum	ACM3666^{T}	AY296241
'*Mannheimia succiniciproducens*'	MBEL55E	NC_006300
[*Actinobacillus*] *succinogenes*	CCUG43843^{T}	AY296234
Haemophilus influenzae	Rd KW20	NC_000907
[*Pasteurella*] *pneumotropica*	NCTC8141^{T}	AY296245
Bisgaard Taxon 17	CCUG17206	AY296237
Escherichia coli	MG1655	NC_00913

Figure 2.5 Phylogenetic tree based on the *atpD* gene from members of *Pasteurellaceae*. *E. coli* was included as an outgroup to root the tree. The Neighbour-Joining tree was built from a Jukes–Cantor corrected distance matrix in Bionumerics v.4.6. Bootstrap values of 500 trees are indicated as per cent confidence values for the branching. The distance bar on top represents per cent differences. *Gene ID, Oralgen www.oralgen.lanl.gov

transport and F_1 is located on the cytoplasmic side of the membrane responsible for ATPase activity. F_1 is composed of the five polypeptides α, β, γ and ε. The gene *atpD* is responsible for the β-subunit functioning in nucleotide binding and catalysis. Its ubiquitous distribution within *Bacteria* and high sequence conservation makes it suitable for phylogenetic analysis (Walker *et al.*, 1984). Within *Pasteurellaceae* from which data are available the gene is 1374 bp long. Sequences covering nearly whole length of *atpD* were generated and used for investigation of genetic diversity of *P. multocida* fowl cholera isolates (Petersen *et al.*, 2001). By sequence analysis of *atpD* it was possible to separate *P. multocida* subsp. *septica* from the other two subspecies, but similar to 16S rRNA gene-based analysis, *P. multocida* subsp. *multocida* and *P. multocida* subsp. *gallicida* could not be separated using *atpD* showing close relatedness of these two subspecies. More information about usability of the gene for clarification of phylogeny within *Pasteurellaceae* was provided by an analysis comprising a number of different representatives of the family (Christensen *et al.*, 2004b). The phylogeny based on analysis of deduced protein sequences from DNA indicated that *atpD* might be useful for recognition of genera within the family.

The *atpD* sequence of only a limited number of representatives of the family is available and primers used for determining them are listed in Table 2.1. Fig. 2.5 shows the phylogenetic tree based on *atpD* from 31 taxa including ten genera and a few representatives of potential new taxa. The tree is in good agreement with the 16S rRNA gene based phylogeny but better resolves some groups as e.g. *Bibersteinia* from *Mannheimia*. Therefore *atpD* is certainly also a promising phylogenetic marker for the family and more data including more representatives should be gained. Comparison of the sequence similarity of *atpD* between the limited number of analysed strains resulted in more than 78% within the family and more than 89% within genera.

Phylogeny based on *recN*, *thdF* and *rpoA* representing the whole-genome sequence

DNA relatedness of taxa as determined by DNA–DNA hybridization gives an overall estimate about phylogenetic relationship between species and is a taxonomic marker (Wayne *et al.*, 1987). However, the method is time-consuming requiring cross-comparisons of new taxa with established ones, technically challenging and sometimes not possible to perform with certain taxa. The reproducibility is questionable when comparing results generated in different laboratories under different conditions. Moreover, the method only has a resolution at the species level with 85% being the threshold for *Pasteurellaceae* species (compared to the generally used 70%) and therefore it is not useful to delineate relationship at the genus level (Mutters *et al.*, 1989; Stackebrandt and Goebel, 1994). Whole-genome sequencing provides the most precise information about genetic content of organisms and would allow cross-comparison with available genome sequences. However, this might be an option for the future since generating sequences from single strains is still expensive, comparison of the data using nowadays available bioinformatic tools is time-consuming and simple whole-genome analysis processes are not commonly available. An alternative to whole-genome sequence comparison was recently demonstrated by the work of Zeigler (2003) showing that comparison of only three genes *recN* (encoding DNA repair protein), *thdF* (encoding ATPase) and *rpoA* (encoding the α subunit of the RNA polymerase) between species has the potential to predict the whole-genome relatedness of bacteria thereby presenting a very promising solution for comprehensive phylogenetic studies of bacteria. Even more, only the *recN* gene might be sufficient for that purpose, however with less confidence. Whole-genome relatedness for two related bacterial strains expressed as a similarity value (SI) can be calculated from the similarities of the corresponding marker genes using the specific formulas

$SI_{genome} = -1.88 + 0.52(SI\mathit{recN}) + 1.78(SI\mathit{thdF}) + 0.52(SI\mathit{rpoA})$ and

$SI_{genome} = -1.30 + 2.25(SI\mathit{recN})$, respectively.

This approach was taken to establish a general sequence determination method for *Pasteurellaceae* based on 43 strains representing 37 species and subspecies including type strain of all the currently 13 genera from the family (Kuhnert and Korczak, 2006). Calculated genome similarity values based on *recN*, *thdF* and *rpoA* as well as of only *recN* correlated with available published DNA–DNA reassociation results. Moreover, it allowed deducing the per cent GC content of species from the three genes, a further marker used in taxonomy. Like that it was for the first time possible to carry out 'whole-genome' cross-comparisons of species and genera on a large scale and analysis of data allowed establishing threshold values for species (SI genome around 0.85) and genus (SI genome around 0.4) either based on all three genes or *recN* alone. The similarity values provide a broad view on genetic relatedness between species belonging to the family and showed clear allocation of species representing the same genus in particular groups. The approach was since used to clarify the taxonomic position of *A. capsulatus* as well as for description of the new species *Av. endocarditidis* (Bisgaard *et al.*, 2007; Kuhnert *et al.*, 2007).

DNA–DNA hybridization values as well as the similarity values calculated on the three genes also provide some phylogenetic information. However, the sequences of the three genes *recN*, *thdF* and *rpoA* can further be used to construct phylogenetic trees, thereby enhancing resolution

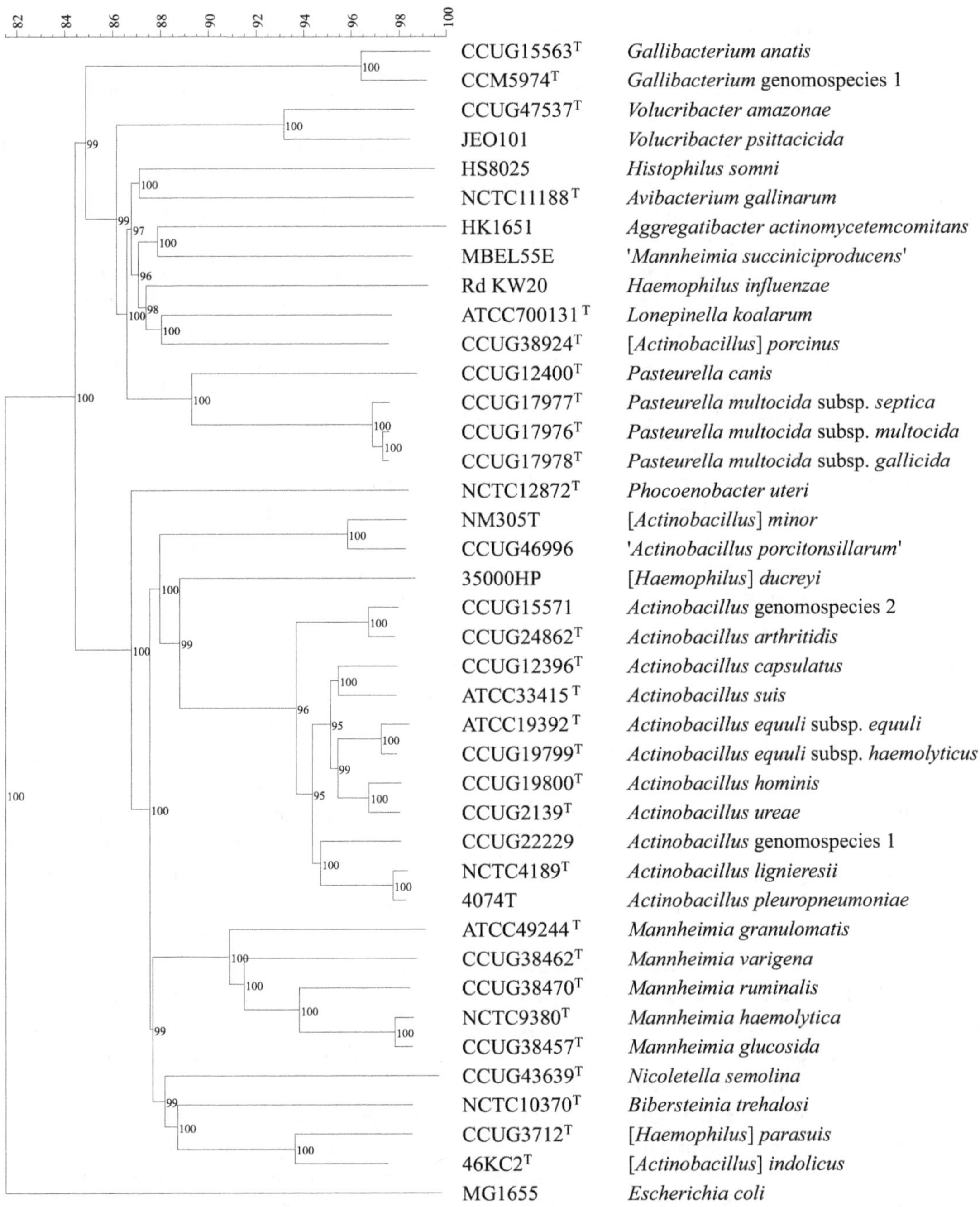

Figure 2.6 Consensus tree based on the *recN*, *thdF* and *rpoA* genes from members of *Pasteurellaceae*. *E. coli* was included as an outgroup to root the tree. Jukes–Cantor correction was applied for the distance matrix and Neighbour-Joining for tree construction in Bionumerics v.4.6. Cophenetic correlations are given, indicating the reliability of the branching compared with the actual genetic relatedness of the taxa. The distance bar on top represents per cent differences

gained from the sequence data compared to calculation of similarity values alone. Fig. 2.6 shows a combined tree based on the distance matrices of the three genes. The tree very nicely resolves all the genera and monophyletic groups. Moreover it also clearly distinguishes species within a genus. In the previous study we included 16S rRNA gene, *rpoB*, and *infB* in a multilocus sequence analysis (MLSA) and constructed a combined tree based on all six genes (Kuhnert and Korczak, 2006). The congruence between trees obtained from each of the six genes calculated by Pearson correlation showed that topologies of individual trees obtained from the *recN*, *thdF* and *rpoA* genes most properly represents the combined tree whereby the *recN* tree showed the highest similarity to the combined tree (Kuhnert and Korczak, 2006). Congruence values with the combined tree were 97% for *recN*, 95% for *thdF*, 93% for *rpoB*, 92% for *rpoA* and *infB* and only 83% with 16S rRNA gene. This is further evidence that these genes and in particular *recN* are highly representative for the entire genome of taxa.

Interestingly, the rough tree structure in Fig. 2.6 is congruent with the one described by Redfield *et al.* (2006). There a phylogenetic analysis of amino acid sequences of 12 well-conserved genes (*gapdH*, *lepA*, *ffh*, *serS*, *secD*, *dapA*, *ruvB*, *xerC*, *ispB*, *secA*, *crp*, and *dnaJ*) deduced from eight *Pasteurellaceae* genome sequences was carried out (see also Fig. 4.3). The two subclades represented by *H. influenzae* and *A. pleuropneumoniae* can be recognized in the *recN-thdF-rpoA* consensus tree. They are also specified by the DNA uptake signal sequences specific for the two large phylogenetic groups (Chapter 4). It will be interesting to see whether branches formed by the genus *Gallibacterium* and *Phocoenobacter* are also defined by specific DNA uptake signal sequences. One could speculate that also the Testudinis cluster, which forms a clear outgroup in all trees where this group was included (Figs. 2.1–2.5) has its specific uptake signal sequences.

Phylogeny based on multilocus sequence typing (MLST)

Multilocus sequence typing (MLST) is used to explore very close phylogenetic relationships and is mainly applied for defining genotypes within a species (Maiden *et al.*, 1998). The technique is based on amplification and sequencing of internal fragments of about 500 bp from 6–8 housekeeping genes. Every unique sequence of each gene is given a unique allele number, and the combination of allele numbers of all gene loci defines the allelic profile of a strain and is given as a sequence type (ST). In addition to the designation of ST, sequences can also be used for phylogenetic analysis. It furthermore allows assessing the recombination frequency within species by looking at the congruence of the individual trees.

MLST has been applied for only few members of the family *Pasteurellaceae* mainly for *Haemophilus*. Meats *et al.* (2003) used fragments of *adk* (adenylate kinase), *atpG* (ATP synthase F1 γ-subunit), *frdB* (fumarate reductase iron–sulphur protein), *fucK* (fuculokinase), *mdh* (malate dehydrogenase), *pgi* (glucose-6-phosphate isomerase) and *recA* (RecA protein) for characterization of encapsulated and non-capsulated *H. influenzae* strains. For phylogenetic purpose sequences of all seven gene fragments were concatenated in a sequence of totally 3057 bp and a distance matrix-based tree was constructed. The study allowed reconstructing phylogenetic relationships between different lineages of *H. influenzae*. Two major groups of encapsulated *H. influenzae* could be discerned from the strains analysed, whereas the non-capsulated isolates showed little structuring in the tree, probably as a result of higher recombination frequency in the latter. The MLST typing scheme for *H. influenzae* is available on the internet under www.mlst.net.

The three genes *adk*, *pgi*, *recA* from the work of Meats *et al.* (2003) and additionally the *infB* were used in a multilocus sequence analysis of the genus *Haemophilus* resulting in the description of a new species *H. pittmaniae* (Nørskov-Lauritsen *et al.*, 2005). This study also clearly showed the advantage of using housekeeping genes rather than the 16S rRNA gene for resolving the phylogenetic relatedness within the genus *Haemophilus*.

A third study was done on [*Haemophilus*] *parasuis* based on the genes *mdh*, *6pgd* (6-phosphogluconate dehydrogenase), *atpD*, *frdB*, *g3pd* (glyceraldehyde-3-phosphate dehydrogenase), *infB* and *rpoB* (Olvera *et al.*, 2006). This investigation revealed that this species undergoes recombination to a significant extent and that it is very heterogeneous with many sequence types and no clonal structure. Nevertheless, two major lineages could be discerned.

An MLST scheme based on the genes *adk*, *aroA* (3-phosphoshikimate 1-carboxyvinyl transferase), *deoD* (purine-nucleoside phosphorylase), *gdhA* (glutamate dehydrogenase), *g6pd* (glucose-6-phosphate 1-dehydrogenase), *mdh* and *pgi* was also developed for *P. multocida* and applied on bovine strains (Davies *et al.*, 2004). The study revealed that MLST gives only limited information on the phylogeny within the isolates from cattle, however it could still be a valuable tool for phylogenetic analysis of *P. multocida* from different sources.

MLST schemes have also been established for *Agg. actinomycetemcomitans* based on six of the genes used for *H. influenzae*. One study investigated the intragenomic recombination in the highly toxigenic clone JP2 (Eriksen *et al.*, 2005). Whereas MLST showed almost no discrimination and target genes showed only minor sequence difference between the five strains analysed, the authors surprisingly found high recombination of rRNA genes. In a second study, Haubek *et al.* (2007) looked at more geographically, ethnically and timely separated JP2 isolates and found a clonal population structure with evolutionary lineages corresponding to serotypes.

Phylogeny of the genera

In this section data gained from the various phylogenetic marker genes described in the previous part are summarized to get a closer look into the phylogeny between and within the genera. Generally there is good congruence between the trees and not much discussion about the structure of recently described genera as well as of course the monophyletic genera. Discrepancies between the topology of trees are most probably due to recombination events for certain genes in certain species. This is one argument more to include as many genes as possible in a phylogenetic comparison of species.

The genus *Actinobacillus sensu stricto* is composed of the species *A. lignieresii*, *A. pleuropneumoniae*, *A. suis*, *A. ureae*, *A. hominis*, *A. arthritidis*, *A. capsulatus*, the two subspecies of *A. equuli* and two genomospecies (Table 1.4). These species form a distinct cluster in the 16S rRNA gene, *infB*, *rpoB* and the *recN-thdF-rpoA* based trees (Figs. 2.1–2.3 and 2.6) with the exception of *A. capsulatus* in the 16S rRNA gene tree where it clusters clearly outside. This was the reason why the taxonomic position of *A. capsulatus* within the genus was questioned. However, as mentioned previously in a recent polyphasic study including 23 strains it was shown, that this species should definitely be left in the genus (Kuhnert *et al.*, 2007), confirming earlier DNA–DNA reassociation studies (Escande *et al.*, 1984; Mutters *et al.*, 1989). Furthermore, the type species *A. lignieresii* and *A. pleuropneumoniae* are very closely related in all trees where we have the corresponding sequence data available. Based on the genome similarity values of more than 0.9 of *A. lignieresii* to various *A. pleuropneumoniae* serotypes (Kuhnert and Korczak, 2006), these two species could phylogenetically be regarded as the same species.

Related to the genus *Actinobacillus* is the branch composed of the [*A.*] *minor* type strain and the '*A. porcitonsillarum*' reference strain. These are phenotypically very closely related and the main difference is in haemolysis (Moller *et al.*, 1996; Gottschalk *et al.*, 2003). The two form a distinct branch in the 16S rRNA, *infB*, *rpoB* and *recN-thdF-rpoA* trees, separated from but close to the cluster of *Actinobacillus*. The distinct branching, as well as the genome similarity value of only 0.79 between the two and less than 0.4 to other *Actinobacillus* species (Kuhnert and Korczak, 2006) favour the argument that [*A.*] *minor* and '*A. porcitonsillarum*' might represent separate species that form a new genus, adding that the species '*A. porcitonsillarum*' still awaits formal description. Strains belonging to this putative species pose a diagnostic problem since they can be misidentified as *A. pleuropneumoniae*, but can actually be regarded as normal flora of the pig (Gottschalk *et al.*, 2003). All other species currently assigned to the genus *Actinobacillus* are clearly misclassified based on their phylogenetic position in all trees.

The genus *Aggregatibacter* has only recently been described and contains the three species:

Agg. actinomycetemcomitans, *Agg. aphrophilus* and *Agg. segnis* (Nørskov-Lauritsen and Kilian, 2006). A monophyletic branching of the three species was shown in this paper; however, only six other taxa of *Pasteurellaceae* and *E. coli* as an outgroup were included in the analysis. Monophyly of the genus is also seen in all protein encoding gene-based trees where more taxa were included (Figs. 2.2–2.5) as well as in the *recN-thdF-rpoA* consensus tree (Fig. 2.6). The structure in the 16S rRNA gene tree is somewhat less clear since *Agg. actinomycetemcomitans* clusters with *Haemophilus*. However, when including one or more additional sequences from *Agg. actinomycetemcomitans* strains the genus forms also a monophyletic cluster (data not shown). This phenomenon can often be seen and shows once more that phylogenetic trees are dynamic structures with changing topology and relationships when including new and additional data (Ludwig and Schleifer, 1999).

The genus *Avibacterium* is also one of the recently established genera and includes the reclassified species *Av. avium*, *Av. gallinarum*, *Av. paragallinarum* and *Av. volantium* (Blackall *et al.*, 2005) as well as the new species *Av. endocarditidis* (Bisgaard *et al.*, 2007). The monophyletic nature of the genus is reflected in the 16S rRNA gene tree as well as in the other phylogenetic analyses.

The genus *Gallibacterium* currently comprises the species *G. anatis* and genomospecies 1 (Christensen *et al.*, 2003a) and will be expanded by new species formed by representatives of Bisgaard Taxon 2 and 3 (H. Christensen, personal communication). In all phylogenetic trees they form a monophyletic branch. Remarkable is that *Gallibacterium* together with *Avibacterium* and *Volucribacter* form the 'avian' cluster 18 of Olsen (Olsen *et al.*, 2005) in the 16S rRNA tree based on Neighbour-Joining (Fig. 2.1) as well as based on maximum-likelihood analysis (Christensen *et al.*, 2004b). However, this 'avian' cluster is not reflected in the other trees where *Volucribacter* as well as the other two genera *Avibacterium* and *Gallibacterium* form distinct and separated clusters.

The genus *Haemophilus* contains many misclassified species what is based on the long taxonomic history this genus has. Growth requirements, i.e. the need for haemin (X factor) and β-NAD (V factor) were the basis to assign species to this genus. Phylogeny impressively shows that the phenotypic criteria for classification are unrelated to genetic similarity of the species. In the herein presented analyses only a limited number of species from this genus were included, but the work of Hedegaard and collaborators showed that phylogeny of *infB* as well as DNA–DNA hybridizations suggest that based on genetic relationship the genus *sensu stricto* should be formed by *H. influenzae*, *H. aegyptius*, *H. haemolyticus*, *H. parainfluenzae* and *H. pittmaniae* (Hedegaard *et al.*, 2001; Nørskov-Lauritsen *et al.*, 2005). This is also in agreement with the 16S rRNA gene phylogeny, however it stays in contrast to the branching observed in the *sodA* tree, where *H. influenzae*, *H. aegyptius*, *H. haemolyticus* are on a quite distant branch to *H. parainfluenzae* and *H. pittmaniae* (Fig. 2.4). The presence of two lineages can also be seen in the 16S rRNA gene tree, however the *Haemophilus sensu stricto* are on the same branch in this case.

The genus *Mannheimia* was established on a polyphasic approach and the five species can be found on the same branch in all phylogenetic analyses. *M. haemolytica* and *M. glucosida* are always closely related, and this is most clearly seen in the *recN-thdF-rpoA* tree (Fig. 2.6). The genome similarity value of 0.88 (Kuhnert and Korczak, 2006) is also quite high and indicates the close genetic relationship between the two, which phylogenetically could be regarded the same species or at least represent subspecies of the same species. Also related to these two is *M. ruminalis*, whereas *M. granulomatis* and *M. varigena* seem to form a different lineage within the genus, as seen from all trees for which data from all five species are available. From all trees it becomes obvious that '*M. succiniciproducens*' is wrongly assigned to the genus and shows that classification on simple and limited 16S rRNA gene analysis, as in this case, can lead to wrong conclusions (Lee *et al.*, 2002). The *recN-thdF-rpoA* analysis impressively shows this misclassification, since '*M. succiniciproducens*' is on a different and most distant cluster to the genus *Mannheimia* (Fig. 2.6).

Like the other two old 'HAP' genera, the genus *Pasteurella* needs revision, since it still includes many wrongly classified species despite the description of former members as *Mannheimia, Gallibacterium, Avibacterium* and *Bibersteinia.* What is regarded as *sensu stricto* includes the *P. multocida* subspecies, *P. canis, P. stomatis* and *P. dagmatis.* Good congruence for the phylogenetic clustering of these species is seen in the trees. Based on 16S rRNA gene and *rpoB* analyses it was speculated that the *P. multocida* subsp. *septica* could be regarded a single species, since it is distinct from the other two subspecies (Kuhnert *et al.*, 2000; Davies, 2004) that were discussed in its original description by Mutters *et al.* (1985). However, based on the *recN-thdF-rpoA* tree topology and genome similarity values of about 0.87 it would make sense to keep the subspecies concept (Kuhnert and Korczak, 2006).

The genus *Volucribacter,* previously known as a group of organisms originating from psittacine birds called Bisgaard taxon 33 (Bisgaard *et al.*, 1999), currently comprises two species: *V. psittacicida* and *V. amazonae.* These two species appear to be genetically closely related to each other with 16S rRNA gene sequence similarity ranging from 99.5% to 99.8%, which makes their differentiation using this marker hardly possible. However, the DNA–DNA reassociation value between type strains of *V. psittacicida* and *V. amazonae* was only 47%, clearly separating the two species from each other. Investigation of the genetic relationships between members of the taxon and the other known *Pasteurellaceae* based on 16S rRNA gene sequences provided clear evidence that the genus *Volucribacter* forms a distinct monophyletic group within the family *Pasteurellaceae* (Christensen *et al.*, 2004a) and shares the highest 16S rRNA gene sequence similarity outside the genus with Bisgaard taxon 34 strain 69 (94.6%) and with the type strain of *Av. avium* (94.5%) (Blackall *et al.*, 2005; Christensen *et al.*, 2003a,b).

The other named genera are all defined by a single species and form monophyletic taxa in most of the trees. *Bibersteinia* is the newest genus within the *Pasteurellaceae,* which nowadays comprises only one species namely *B. trehalosi.* This organism was assigned to the family some time ago. In 1990, Sneath and Stevens (Sneath and Stevens, 1990) described a new species as [*Pasteurella*] *trehalosi* that had been classified as a part of the [*Pasteurella*] *haemolytica* complex (Angen *et al.*, 1999). The close phylogenetic relationship to the genus *Mannheimia* is reflected in all the trees (Figs. 2.1–2.6). However, in all trees *B. trehalosi* clearly separates from the *Mannheimia* species and previous 16S rRNA gene analysis indicated that the four recognized serovars within the species are closely related showing at least 98.7% similarity (Davies *et al.*, 1996).

The genus *Lonepinella,* represented by *L. koalarum,* was described based on seven strains representing three biovars (Osawa *et al.*, 1995). The 16S rRNA gene sequences of the seven strains showed some variability, up to more than 2%. The monotypic nature of the genus is well represented in the phylogenetic trees.

Nicoletella semolina was shown to be a phylogenetically very homogeneous species with very low intraspecies variability of 0.3%, 0.4% and 0.2% in the analysed *infB, rpoB* and 16S rRNA gene sequences, respectively (Kuhnert *et al.*, 2004). The genus is always located on a distinct branch in the trees with little association to other genera.

The description of *Phocoenobacter uteri* is only based on a single isolate and there is no report in the literature on the isolation of additional strains of this species (Foster *et al.*, 2000). Nevertheless, this genus also seems to form a monotypic branch in all the analysed trees. Interestingly, [*P.*] *skyensis* in the 16S rRNA gene tree collocates on the same branch, and this distinct lineage was also shown by Birkbeck *et al.* (2002). Since we do not have comparable sequence data for the two species for the other genes, we can only speculate that the two might be members of the same genus.

Strains reclassified as *Hist. somni* were also shown to be rather homogeneous with only 0.5% 16S rRNA gene divergence in 19 isolates from geographically different parts of the world (Angen *et al.*, 2003). Analysis of *rpoB* resulted in two subclusters within the monotypic main cluster (Angen *et al.*, 2003). The monotypic nature of *Histophilus* is also seen in the *recN-thdF-rpoA* based tree (Fig. 2.6).

Putative new genera

As outlined a few times in this chapter as well as in others, new genera and species will be described within *Pasteurellaceae*. Phylogenetic analyses indicate which of the named species are candidates to form a new genus and will help to classify unnamed (e.g. the Bisgaard taxa) or newly isolated taxa. In the following we list a few such groups based on the herein presented phylogenetic investigations. Some of these groups are rather clearly defined others less, and more thorough investigations for description of these putative new genera are certainly needed. However, it becomes obvious that the number of genera within the family can easily increase to over 30. In this context it should be remembered that good taxonomic practise includes more than a single strain for description of new taxa and we recommend to refer to the 'minimal standards' published for taxonomy of *Pasteurellaceae* (Christensen *et al.*, 2001, 2007).

A distinct branch is always formed by the [*P.*] *testudinis* and Bisgaard taxon 14. This Testudinis branch of the phylogenetic trees is certainly underrepresented by isolates analysed in this chapter as well as in the literature in general. What in diagnostics is identified as [*P.*] *testudinis* is probably a phylogenetically very heterogeneous group of isolates as 16S rRNA gene and *rpoB* sequence analyses indicate (P. Kuhnert and H. Christensen, unpublished data). It can therefore be assumed, that at least one new genus and several new species can be delineated from such strains in the future.

The Delphinicola cluster formed by the sea mammal isolates [*Actinobacillus*] *scotiae* and [*Actinobacillus*] *delphinicola* shows also deep branching in the 16S rRNA, *infB* and *rpoB* trees and could therefore form a distinct new genus.

The wrongly classified '*M. succiniciproducens*' was included in all herein constructed trees and forms a monophyletic branch in each. There is currently only one isolate reported which is deposited at the Culture Collection of the University of Gothenburg (CCUG) but due to patent rights not publicly available (Lee *et al.*, 2002). This makes further taxonomic investigations of this putative new genus difficult even though a lot of genomic and proteomic data are available (Hong *et al.*, 2004; Lee *et al.*, 2006). Another species isolated from bovine rumen, [*Actinobacillus*] *succinogenes*, clusters close to '*M. succiniciproducens*'; however, the branches are rather deep, making it unlikely that they belong to the same genus. The same holds true for Bisgaard taxon 10; however, only 16S rRNA gene data are available in this case.

A Rodent cluster is formed by [*Pasteurella*] *pneumotropica*, Bisgaard taxon 17 and [*A.*] *muris*. It is, however, difficult to assess if this corresponds to one genus or more and further data are needed to get a more precise picture.

The Seminis cluster formed of the species [*Actinobacillus*] *seminis*, [*P.*] *aerogenes* and [*Pasteurella*] *mairii* also forms a distinct branch in the 16S rRNA, the *infB* and the *rpoB* tree, suggesting a genus rank of this group as has been proposed earlier (Christensen *et al.*, 2005). However, more phenotypic and genotypic data are needed on the [*A.*] *seminis* to define this genus and separate the species within it. Interestingly, the RTX toxin operon *pax*, thought to be involved in abortion cases, can be found in this phylogenetic cluster (Mayor *et al.*, 2006).

The closely related species [*Haemophilus*] *parahaemolyticus* and [*Haemophilus*] *paraphrohaemolyticus* form the Parahaemolyticus group, a distinct branch as seen in the 16S rRNA gene and *sodA* trees.

The transfer of [*A.*] *minor* and '*A. porcitonsillarum*' to a separate genus has already been discussed. This is especially obvious when looking at the *recN-thdF-rpoA* tree, where they cluster distantly from the genus *Actinobacillus*.

The Parasuis group consists of the two species, [*H.*] *parasuis* and [*Actinobacillus*] *indolicus*, forming a monophyletic branch in the 16S rRNA, *infB*, *rpoB* and *recN-thdF-rpoA* trees. Interestingly the two type strains of the species showed a significant genome similarity value of 0.6, suggesting that they in fact belong to the same genus (Kuhnert and Korczak, 2006).

The three species [*H.*] *ducreyi*, [*Pasteurella*] *caballi* and [*Pasteurella*] *bettyae* are on the same branch in the 16S rRNA gene tree, but clearly separated by distance. The fact that they cluster separately in the *rpoB*, *sodA*, *atpD* trees could indicate that they represent separate genera.

[*H.*] *felis* and [*Haemophilus*] *haemoglobinophilus* each form a separate branch in the 16S rRNA, *rpoB* and *sodA* trees, suggesting that they represent two genera.

The phylogenetic position of Bisgaard taxon 7 in the 16S rRNA, *rpoB* and *atpD* trees also suggests that it forms a new genus. Of course, analysis of more strains representing this taxon is necessary to support this. The same holds true for Bisgaard taxon 5.

Difficult to interpret are the data for [*Pasteurella*] *langaaensis*, [*Actinobacillus*] *rossii* and [*Actinobacillus*] *porcinus*, which are related in the *rpoB* sequence but with the last being separated in the 16S rRNA gene tree. Data are not conclusive at the moment.

Limited data are also available for [*H.*] *paracuniculus*, which nevertheless forms a monotypic branch in the *sodA* tree and a distinct cluster with *A. capsulatus* in 16S rRNA gene tree.

Conclusions

Based on the different phylogenetic approaches presented, and the various genes they rely on, one wonders what would constitute the perfect phylogene. It is obvious that each of the genes has its advantages but this also results in difficult to interpret relationships in certain cases. The 16S rRNA gene phylogeny will certainly continue to be a *conditio sine qua non* for the investigation and description of new taxa. It might be useful to delineate the large phylogenetic relations among bacterial families very nicely, and in many cases it can also resolve species relationships. However, this gene has also many drawbacks, and in some cases it stands in contradiction to the phylogeny based on alternative targets and should therefore not be taken as the sole reference gene for phylogenetic analysis in *Pasteurellaceae*. It will be interesting to see if this holds true for other bacterial families as well. Housekeeping genes have proved very useful for investigating the phylogeny within *Pasteurellaceae*, and the ones described in this chapter (*infB*, *rpoB*, *sodA*, *atpD*) will probably be equally applicable. In general, the inclusion of several genes for phylogenetic analysis is in any case favourable and will give more precise ideas on the genetic relationship of taxa. In this respect the most promising candidates would appear to be genes representing whole-genome similarities such as *recN*, *thdF* and *rpoA*. If, based on current knowledge, any single gene should be recommended as 'the phylogene' for *Pasteurellaceae* it might be the *recN*.

References

Acinas, S.G., Marcelino, L.A., Klepac-Ceraj, V., and Polz, M.F. (2004). Divergence and redundancy of 16S rRNA sequences in genomes with multiple *rrn* operons. J. Bacteriol. *186*, 2629–2635.

Acinas, S.G., Sarma-Rupavtarm, R., Klepac-Ceraj, V., and Polz, M.F. (2005). PCR-induced sequence artifacts and bias: insights from comparison of two 16S rRNA clone libraries constructed from the same sample. Appl. Environ. Microbiol. *71*, 8966–8969.

Angen, O., Ahrens, P., Kuhnert, P., Christensen, H., and Mutters, R. (2003). Proposal of *Histophilus somni* gen. nov., sp. nov. for the three species incertae sedis '*Haemophilus somnus*', '*Haemophilus agni*' and '*Histophilus ovis*'. Int. J. Syst. Evol. Microbiol. 53, 1449–1456.

Angen, O., Mutters, R., Caugant, D.A., Olsen, J.E., and Bisgaard, M. (1999). Taxonomic relationships of the [*Pasteurella*] *haemolytica* complex as evaluated by DNA–DNA hybridizations and 16S rRNA sequencing with proposal of *Mannheimia haemolytica* gen. nov., comb. nov., *Mannheimia granulomatis* comb. nov., *Mannheimia glucosida* sp. nov., *Mannheimia ruminalis* sp. nov. and *Mannheimia varigena* sp. nov. Int. J. Syst. Bacteriol. *49*, 67–86.

Benson, D.A., Karsch-Mizrachi, I., Lipman, D.J., Ostell, J., and Wheeler, D.L. (2007). GenBank. Nucleic Acids Res. 35, D21–D25.

Birkbeck, T.H., Laidler, L.A., Grant, A.N., and Cox, D.I. (2002). *Pasteurella skyensis* sp. nov., isolated from Atlantic salmon (*Salmo salar* L.). Int. J. Syst. Evol. Microbiol. 52, 699–704.

Bisgaard, M., Christensen, J.P., Bojesen, A.M., and Christensen, H. (2007). *Avibacterium endocarditidis* sp. nov., isolated from valvular endocarditis in chickens. Int. J. Syst. Evol. Microbiol. *57*, 1729–1734.

Bisgaard, M., Hinz, K.-H., Petersen, K.D., and Christensen, J.P. (1999). Identification of members of the *Pasteurellaceae* isolated from birds and characterization of two new taxa isolated from psittacine birds. Avian Pathol. 28, 369–377.

Blackall, P.J., Bojesen, A.M., Christensen, H., and Bisgaard, M. (2007). Reclassification of [*Pasteurella*] *trehalosi* as *Bibersteinia trehalosi* gen. nov., comb. nov. Int. J. Syst. Evol. Microbiol. 57, 666–674.

Blackall, P.J., Christensen, H., Beckenham, T., Blackall, L.L., and Bisgaard, M. (2005). Reclassification of *Pasteurella gallinarum*, [*Haemophilus*] *paragallinarum*, *Pasteurella avium* and *Pasteurella volantium* as *Avibacterium gallinarum* gen. nov., comb. nov., *Avibacterium paragallinarum* comb. nov., *Avibacterium avium* comb. nov. and *Avibacterium volantium* comb. nov. Int. J. Syst. Evol. Microbiol. 55, 353–362.

Cattoir, V., Lemenand, O., Avril, J.L., and Gaillot, O. (2006). The *sodA* gene as a target for phylogenetic dissection of the genus *Haemophilus* and accurate identification of human clinical isolates. Int. J. Med. Microbiol. *296*, 531–540.

Christensen, H., Bisgaard, M., Aalbaek, B., and Olsen, J.E. (2004a). Reclassification of Bisgaard taxon 33 with proposal of *Volucribacter psittacicida* gen. nov., sp. nov. and *Volucribacter amazonae* sp. nov. as new members of *Pasteurellaceae*. Int. J. Syst. Evol. Microbiol. *54*, 813–818.

Christensen, H., Bisgaard, M., Bojesen, A.M., Mutters, R., and Olsen, J.E. (2003a). Genetic relationships among avian isolates classified as *Pasteurella haemolytica*, '*Actinobacillus salpingitidis*' or *Pasteurella anatis* with proposal of *Gallibacterium anatis* gen. nov., comb. nov. and description of additional genomospecies within *Gallibacterium* gen. nov. Int. J Syst. Evol. Microbiol. *53*, 275–287.

Christensen, H., Bisgaard, M., Frederiksen, W., Mutters, R., Kuhnert, P., and Olsen, J.E. (2001). Is characterization of a single isolate sufficient for valid publication of a new genus or species? Proposal to modify recommendation 30b of the Bacteriological Code (1990 Revision). Int. J Syst. Evol. Microbiol. *51*, 2221–2225.

Christensen, H., Foster, G., Christensen, J.P., Pennycott, T., Olsen, J.E., and Bisgaard, M. (2003b). Phylogenetic analysis by 16S rDNA gene sequence comparison of avian taxa of Bisgaard and characterization and description of two new taxa of *Pasteurellaceae*. J. Appl. Microbiol. *95*, 354–363.

Christensen, H., Kuhnert, P., Bisgaard, M., Mutters, R., Dziva, F., and Olsen, J.E. (2005). Emended description of porcine [*Pasteurella*] *aerogenes*, [*Pasteurella*] *mairii* and [*Actinobacillus*] *rossii*. Int. J. Syst. Evol. Microbiol. *55*, 209–223.

Christensen, H., Kuhnert, P., Busse, H.J., Frederiksen, W.C., and Bisgaard, M. (2007). Proposed minimal standards for the description of genera, species and subspecies of the *Pasteurellaceae*. Int. J. Syst. Evol. Microbiol. *57*, 166–178.

Christensen, H., Kuhnert, P., Olsen, J.E., and Bisgaard, M. (2004b). Comparative phylogenies of the housekeeping genes *atpD*, *infB* and *rpoB* and the 16S rRNA gene within the *Pasteurellaceae*. Int. J. Syst. Evol. Microbiol. *54*, 1601–1609.

Chuba, P.J., Bock, R., Graf, G., Adam, T., and Gobel, U. (1988). Comparison of 16S rRNA sequences from the family *Pasteurellaceae*: phylogenetic relatedness by cluster analysis. J. Gen. Microbiol. *134*, 1923–1930.

Davies, R.L. (2004). Genetic diversity among *Pasteurella multocida* strains of avian, bovine, ovine and porcine origin from England and Wales by comparative sequence analysis of the 16S rRNA gene. Microbiology *150*, 4199–4210.

Davies, R.L., MacCorquodale, R., and Reilly, S. (2004). Characterisation of bovine strains of *Pasteurella multocida* and comparison with isolates of avian, ovine and porcine origin. Vet. Microbiol. *99*, 145–158.

Davies, R.L., Paster, B.J., and Dewhirst, F.E. (1996). Phylogenetic relationship and diversity within the *Pasteurella haemolytica* complex based on 16S rRNA sequence comparison and outer membrane protein and lipopolysaccharide analysis. Int. J. Syst. Bacteriol. *46*, 736–744.

De Ley, J., Mannheim, W., Mutters, R., Piechulla, K., Tytgat, R., Segers, P., Bisgaard, M., Frederiksen, W., Hinz, K.H., and Vanhoucke, M. (1990). Inter- and intrafamilial similarities of rRNA cistrons of the *Pasteurellaceae*. Int. J. Syst. Bacteriol. *40*, 126–137.

Dewhirst, F.E., Paster, B.J., Olsen, I., and Fraser, G.J. (1992). Phylogeny of 54 representative strains of species in the family *Pasteurellaceae* as determined by comparison of 16S rRNA sequences. J. Bacteriol. *174*, 2002–2013.

Dewhirst, F.E., Paster, B.J., Olsen, I., and Fraser, G.J. (1993). Phylogeny of the *Pasteurellaceae* as determined by comparison of 16S ribosomal ribonucleic acid sequences. Zentralbl. Bakteriol. *279*, 35–44.

Eriksen, K.T., Haubek, D., and Poulsen, K. (2005). Intragenomic recombination in the highly leukotoxic JP2 clone of *Actinobacillus actinomycetemcomitans*. Microbiology *151*, 3371–3379.

Escande, F., Grimont, P.A.D., and Bercovier, H. (1984). Deoxyribonucleic acid relatedness among strains of *Actinobacillus* spp. and *Pasteurella ureae*. Int. J. Syst. Bacteriol. *34*, 309–315.

Foster, G., Ross, H.M., Malnick, H., Willems, A., Hutson, R.A., Reid, R.J., and Collins, M.D. (2000). *Phocoenobacter uteri* gen. nov., sp. nov., a new member of the family *Pasteurellaceae* Pohl (1979) 1981 isolated from a harbour porpoise (*Phocoena phocoena*). Int. J Syst. Evol. Microbiol *50*, 135–139.

Fox, G.E., Wisotzkey, J.D., and Jurtshuk, P. (1992). How close is close: 16S rRNA sequence identity may not be sufficient to guarantee species identity. Int. J. Syst. Bacteriol. *42*, 166–170.

Gautier, A.L., Dubois, D., Escande, F., Avril, J.L., Trieu-Cuot, P., and Gaillot, O. (2005). Rapid and accurate identification of human isolates of *Pasteurella* and related species by sequencing the *sodA* gene. J. Clin. Microbiol. *43*, 2307–2314.

Gottschalk, M., Broes, A., Mittal, K.R., Kobisch, M., Kuhnert, P., Lebrun, A., and Frey, J. (2003). Non-pathogenic *Actinobacillus* isolates antigenically and biochemically similar to *Actinobacillus pleuropneumoniae*: a novel species? Vet. Microbiol. 92, 87–101.

Gutell, R.R., Larsen, N., and Woese, C.R. (1994). Lessons from an evolving rRNA: 16S and 23S rRNA structures from a comparative perspective. Microbiol. Rev. *58*, 10–26.

Haubek, D., Poulsen, K., and Kilian, M. (2007). Microevolution and patterns of dissemination of the JP2 clone of *Aggregatibacter* (*Actinobacillus*) *actinomycetemcomitans*. Infect. Immun. *75*, 3080–3088.

Hedegaard, J., Okkels, H., Bruun, B., Kilian, M., Mortensen, K.K., and Nørskov-Lauritsen, N. (2001). Phylogeny of the genus *Haemophilus* as determined by comparison of partial *infB* sequences. Microbiology *147*, 2599–2609.

Hedegaard, J., Steffensen, S.A., Nørskov-Lauritsen, N., Mortensen, K.K., and Sperling-Petersen, H.U. (1999). Identification of *Enterobacteriaceae* by partial sequencing of the gene encoding translation initiation factor 2. Int. J. Syst. Bacteriol. *49*, 1531–1538.

Hong, S.H., Kim, J.S., Lee, S.Y., In, Y.H., Choi, S.S., Rih, J.K., Kim, C.H., Jeong, H., Hur, C.G., and Kim, J.J. (2004). The genome sequence of the capnophilic rumen bacterium *Mannheimia succiniciproducens*. Nat. Biotechnol. 22, 1275–1281.

Klenk, H.P. and Zillig, W. (1994). DNA-dependent RNA polymerase subunit B as a tool for phylogenetic reconstructions: branching topology of the archaeal domain. J. Mol. Evol. *38*, 420–432.

Korczak, B., Christensen, H., Emler, S., Frey, J., and Kuhnert, P. (2004). Phylogeny of the family *Pasteurellaceae* based on *rpoB* sequences. Int. J. Syst. Evol. Microbiol. *54*, 1393–1399.

Kuhnert, P., Boerlin, P., Emler, S., Krawinkler, M., and Frey, J. (2000). Phylogenetic analysis of *Pasteurella multocida* subspecies and molecular identification of feline *P. multocida* subsp. *septica* by 16S rRNA gene sequencing. Int. J. Med. Microbiol. *290*, 599–604.

Kuhnert, P., Frey, J., Lang, N.P., and Mayfield, L. (2002). A phylogenetic analysis of *Prevotella nigrescens, Prevotella intermedia* and *Porphyromonas gingivalis* field strains reveals a clear species clustering. Int. J. Syst. Evol. Microbiol. *52*, 1391–1395.

Kuhnert, P., Korczak, B., Falsen, E., Straub, R., Hoops, A., Boerlin, P., Frey, J., and Mutters, R. (2004). *Nicoletella semolina* gen.nov., sp.nov., a new member of *Pasteurellaceae* isolated from horses with airway disease. J. Clin. Microbiol. *42*, 5542–5548.

Kuhnert, P. and Korczak, B.M. (2006). Prediction of whole genome DNA–DNA similarity, determination of G+C content and phylogenetic analysis within the family *Pasteurellaceae* by multilocus sequence analysis (MLSA). Microbiology *152*, 2537–2548.

Kuhnert, P., Korczak, B.M., Christensen, H., and Bisgaard, M. (2007). Emended description of *Actinobacillus capsulatus* Arseculeratne 1962, 38AL. Int. J. Syst. Evol. Microbiol. *57*, 625–632.

Lawrence, J.G., Hartl, D.L., and Ochman, H. (1991). Molecular considerations in the evolution of bacterial genes. J. Mol. Evol. *33*, 241–250.

Lee, J.W., Lee, S.Y., Song, H., and Yoo, J.S. (2006). The proteome of *Mannheimia succiniciproducens*, a capnophilic rumen bacterium. Proteomics *6*, 3550–3566.

Lee, P.C., Lee, S.Y., Hong, S.H., and Chang, H.N. (2002). Isolation and characterization of a new succinic acid-producing bacterium, *Mannheimia succiniciproducens* MBEL55E, from bovine rumen. Appl. Microbiol. Biotechnol. *58*, 663–668.

Lisitsyn, N.A., Monastyrskaya, G.S., and Sverdlov, E.D. (1988). Genes coding for RNA polymerase β subunit in bacteria. Structure/function analysis. Eur. J. Biochem. *177*, 363–369.

Ludwig, W. and Schleifer, K.-H. (1999). Phylogeny of *Bacteria* beyond the 16S rRNA standard. ASM News *65*, 752–757.

Maiden, M.C., Bygraves, J.A., Feil, E., Morelli, G., Russell, J.E., Urwin, R., Zhang, Q., Zhou, J., Zurth, K., Caugant, D.A., Feavers, I.M., Achtman, M., and Spratt, B.G. (1998). Multilocus sequence typing: a portable approach to the identification of clones within populations of pathogenic microorganisms. Proc. Natl. Acad. Sci. USA *95*, 3140–3145.

Mayor, D., Korczak, B.M., Christensen, H., Bisgaard, M., Frey, J., and Kuhnert, P. (2006). Distribution of RTX toxin genes in strains of [*Actinobacillus*] *rossii* and [*Pasteurella*] *mairii*. Vet. Microbiol. *116*, 194–201.

Meats, E., Feil, E.J., Stringer, S., Cody, A.J., Goldstein, R., Kroll, J.S., Popovic, T., and Spratt, B.G. (2003). Characterization of encapsulated and noncapsulated *Haemophilus influenzae* and determination of phylogenetic relationships by multilocus sequence typing. J. Clin. Microbiol. *41*, 1623–1636.

Moller, K., Fussing, V., Grimont, P.A.D., Paster, B.J., Dewhirst, F.E., and Kilian, M. (1996). *Actinobacillus minor* sp. nov., *Actinobacillus porcinus* sp. nov., and *Actnobacillus indolicus* sp. nov., three new V factor-dependent species from the respiratory tract of pigs. Int. J. Syst. Bacteriol. *46*, 951–956.

Mollet, C., Drancourt, M., and Raoult, D. (1997). *rpoB* sequence analysis as a novel basis for bacterial identification. Mol. Microbiol. *26*, 1005–1011.

Mutters, R., Ihm, P., Pohl, S., Frederiksen, W., and Mannheim, W. (1985). Reclassification of the genus *Pasteurella* Trevisan 1887 on the basis of deoxyribonucleic acid homology, with proposals for the new species *Pasteurella dagmatis, Pasteurella canis, Pasteurella stomatis, Pasteurella anatis,* and *Pastuerella langaa*. Int. J. Syst. Bacteriol. *35*, 309–322.

Mutters, R., Mannheim, W., and Bisgaard, M. (1989). Taxonomy of the Group. In *Pasteurella* and Pasteurellosis, C.Adlam and J.M.Rutter, eds. (London: Academic Press), pp. 3–34.

Nolte, O. (1995). Nucleotide sequence and genetic variability of a part of the *rpoB* gene encoding the second largest subunit of DNA-directed RNA polymerase of *Neisseria meningitidis*. Med. Microbiol. Lett. *4*, 59–67.

Nørskov-Lauritsen, N., Bruun, B., and Kilian, M. (2005). Multilocus sequence phylogenetic study of the genus *Haemophilus* with description of *Haemophilus pittmaniae* sp. nov. Int. J. Syst. Evol. Microbiol. *55*, 449–456.

Nørskov-Lauritsen, N. and Kilian, M. (2006). Reclassification of *Actinobacillus actinomycetemcomitans, Haemophilus aphrophilus, Haemophilus paraphrophilus* and *Haemophilus segnis* as *Aggregatibacter actinomycetemcomitans* gen. nov., comb. nov., *Aggregatibacter aphrophilus* comb. nov. and *Aggregatibacter segnis* comb. nov., and emended description of *Aggregatibacter aphrophilus* to include V factor-dependent and V factor-independent isolates. Int. J. Syst. Evol. Microbiol. *56*, 2135–2146.

Olivier, A., Lee, H.Y., and Cote, J.C. (2005). Study of the heterogeneity of 16S rRNA genes in gamma-proteobacteria: implications for phylogenetic analysis. J. Gen. Appl. Microbiol *51*, 395–405.

Olsen, I., Dewhirst, F.E., Paster, B.J., and Busse, H.J. (2005). Family *Pasteurellaceae*. In Bergey's Manual of Systematic Bacteriology, R. Garrity, ed. (New York: Springer).

Olvera, A., Cerda-Cuellar, M., and Aragon, V. (2006). Study of the population structure of *Haemophilus parasuis* by multilocus sequence typing. Microbiology *152*, 3683–3690.

Osawa, R., Rainey, F.A., Fujisawa, T., Lang, E., Busse, H.J., Walsh, T., and Stackebrandt, E. (1995). *Lonepinella koalarum* gen. nov., sp. nov., a new tannin-protein complex degrading bacterium. Syst. Appl. Microbiol. *18*, 368–373.

Ovchinnikov, Y.A., Monastyrskaya, G.S., Gubanov, V.V., Guryev, S.O., Chertov, O.Y., Modyanov, N.N., Grinkevich, V.A., Makarova, I.A., Marchenko, T.V., Polovnikova, I.N., Lipkin, V.M., and Sverdlov, E.D. (1981). The primary structure of *Escherichia coli* RNA polymerase. Nucleotide sequence of the *rpoB* gene and amino-acid sequence of the β-subunit. Eur. J. Biochem. *116*, 621–629.

Palenik, B. (1992). Polymerase evolution and organism evolution. Curr. Opin. Genet. Dev. *2*, 931–936.

Petersen, K.D., Christensen, H., Bisgaard, M., and Olsen, J.E. (2001). Genetic diversity of *Pasteurella multocida* fowl cholera isolates as demonstrated by ribotyping and 16S rRNA and partial *atpD* sequence comparisons. Microbiology *147*, 2739–2748.

Pohl, S. (1981). DNA relatedness among members of *Haemophilus, Pasteurella* and *Actinobacillus*. In *Haemophilus, Pasteurella* and *Actinobacillus*, M. Kilian, W. Frederiksen, and E.L. Biberstein, eds. (London: Academic Press), p. 253.

Poyart, C., Quesne, G., Coulon, S., Berche, P., and Trieu-Cuot, P. (1998). Identification of streptococci to species level by sequencing the gene encoding the manganese-dependent superoxide dismutase. J. Clin. Microbiol. *36*, 41–47.

Redfield, R.J., Findlay, W.A., Bosse, J., Kroll, J.S., Cameron, A.D., and Nash, J.H. (2006). Evolution of competence and DNA uptake specificity in the *Pasteurellaceae*. BMC Evol. Biol. *6*, 82.

Rowland, G.C., Aboshkiwa, M., and Coleman, G. (1993). Comparative sequence analysis and predicted phylogeny of the DNA-dependent RNA polymerase β subunits of *Staphylococcus aureus* and other eubacteria. Biochem. Soc. Trans. *21*, 40S-41S.

Saitou, N. and Nei, M. (1987). The neighbor-joining method: a new method for reconstructing phylogenetic trees. Mol. Biol. Evol. *4*, 406–425.

Sneath, P.H. and Stevens, M. (1990). *Actinobacillus rossii* sp. nov., *Actinobacillus seminis* sp. nov., nom. rev., *Pasteurella bettii* sp. nov., *Pasteurella lymphangitidis* sp. nov., *Pasteurella mairi* sp. nov., and *Pasteurella trehalosi* sp. nov. Int. J. Syst. Bacteriol. *40*, 148–153.

Stackebrandt, E., Frederiksen, W., Garrity, G.M., Grimont, P.A., Kampfer, P., Maiden, M.C., Nesme, X., Rossello-Mora, R., Swings, J., Truper, H.G., Vauterin, L., Ward, A.C., and Whitman, W.B. (2002). Report of the ad hoc committee for the re-evaluation of the species definition in bacteriology. Int. J Syst. Evol. Microbiol. 52, 1043–1047.

Stackebrandt, E. and Goebel, B.M. (1994). Taxonomic note: a place for DNA–DNA reassociation and 16S rRNA sequence analysis in the present species definition in bacteriology. Int. J. Syst. Bacteriol. *44*, 846–849.

Steffensen, S.A., Poulsen, A.B., Mortensen, K.K., and Sperling-Petersen, H.U. (1997). *E. coli* translation initiation factor IF2 – an extremely conserved protein. Comparative sequence analysis of the *infB* gene in clinical isolates of *E. coli*. FEBS Lett. *419*, 281–284.

Taillardat-Bisch, A.V., Raoult, D., and Drancourt, M. (2003). RNA polymerase beta-subunit-based phylogeny of *Ehrlichia* spp., *Anaplasma* spp., *Neorickettsia* spp. and *Wolbachia pipientis*. Int. J Syst. Evol. Microbiol. *53*, 455–458.

Tamura, K., Dudley, J., Nei, M., and Kumar, S. (2007). MEGA4: Molecular Evolutionary Genetics Analysis (MEGA) software version 4.0. Mol. Biol. Evol. *24*, 1596–1599.

Thompson, J.D., Gibson, T.J., Plewniak, F., Jeanmougin, F., and Higgins, D.G. (1997). The CLUSTAL_X windows interface: flexible strategies for multiple sequence alignment aided by quality analysis tools. Nucleic Acids Res. *25*, 4876–4882.

Walker, J.E., Saraste, M., and Gay, N.J. (1984). The *unc* operon. Nucleotide sequence, regulation and structure of ATP-synthase. Biochim. Biophys. Acta *768*, 164–200.

Wayne, L.G., Brenner, D.J., Colwell, R.R., Grimont, P.A.D., Kandler, O., Krichevsky, M.I., Moore, L.H., Moore, W.E.C., Murray, R.G.E., Stackebrandt, E., Starr, M.P., and Trüper, H.G. (1987). Report of the ad hoc committee on reconciliation of approaches to bacterial systematics. Int. J. Syst. Bacteriol. *37*, 463–464.

Woese, C.R. (1987). Bacterial evolution. Microbiol. Rev. *51*, 221–271.

Yap, W.H., Zhang, Z., and Wang, Y. (1999). Distinct types of rRNA operons exist in the genome of the actinomycete *Thermomonospora chromogena* and evidence for horizontal transfer of an entire rRNA operon. J. Bacteriol. *181*, 5201–5209.

Zeigler, D.R. (2003). Gene sequences useful for predicting relatedness of whole genomes in bacteria. Int. J. Syst. Evol. Microbiol. 53, 1893–1900.

Comparative Genomics of *Pasteurellaceae*

3

Jean F. Challacombe and Thomas J. Inzana

Abstract

The family *Pasteurellaceae* contains diverse members, including opportunistic pathogens, human and animal pathogens as well as avirulent commensal strains. Although they are related, the diversity of the bacteria comprising this family makes them ideal candidates for comparative genome analysis. In this chapter we review and compare the available genomic information on the sequenced *Pasteurellaceae* members: *Aggregatibacter actinomycetemcomitans* strain HK1651, *Actinobacillus pleuropneumoniae* strains L20 and sv1 4074, [*Haemophilus*] *ducreyi* strain 35000HP, *Haemophilus influenzae* strains 86–028NP, R2846, R2866, and Rd, *Histophilus somni* strains 129Pt and 2336, *Mannheimia haemolytica* A1 strain ATCC BAA-410, '*Mannheimia succiniciproducens*' strain MBEL55E, and *Pasteurella multocida* strain Pm70. We also present some example comparisons to demonstrate the utility of the comparative genomics approach for studying this family of bacteria.

Introduction

Due to the ever-increasing number of complete sequenced genomes, whole-genome comparative analysis has become a preferred method for studying groups of related bacteria. Whole-genome comparisons provide a wealth of information about genetic similarities and differences among organisms on many levels. At the genomic level, such analyses can reveal general genome features, such as size, per cent GC content, number of coding sequences, pseudogenes, etc. Pairwise comparisons of orthologous sequences from genomes of interest can reveal or refine evolutionary relationships (May *et al.*, 2001). On a smaller gene-centric scale, whole-genome comparisons highlight similarities and differences in gene content and organization that can provide novel insights into the biology and lifestyle of the organisms. In order to more clearly understand the *Pasteurellaceae* family, which consists of diverse organisms that occupy a variety of niches in several different hosts, this chapter will review the results of comparative genomic studies applied to sequenced members of the *Pasteurellaceae*. The *Pasteurellaceae* genomes that have been sequenced thus far include *Aggregatibacter actinomycetemcomitans* strain HK1651 (http://www.genome.ou.edu/act.html; http://oralgen.lanl.gov), *Actinobacillus pleuropneumoniae* strains L20 (Wheeler *et al.*, 2008) and sv1 4074 (http://microgen.ouhsc.edu/a_pleuro/a_pleuro_home.htm), [*Haemophilus*] *ducreyi* strain 35000HP (Challacombe *et al.*, 2007; Ray *et al.*, 2001), *Haemophilus influenzae* strains 86–028NP (Harrison *et al.*, 2005; Munson *et al.*, 2004a), R2846, R2866, and Rd (Fleischmann *et al.*, 1995; Tatusov *et al.*, 1996), *Histophilus somni* strains 129Pt (Challacombe *et al.*, 2007) and 2336 (http://microgen.ouhsc.edu/h_somnus/h_somnus_home.htm), *Mannheimia haemolytica* A1 strain ATCC BAA-410 (Gioia *et al.*, 2006), '*Mannheimia succiniciproducens*' strain MBEL55E (Hong *et al.*, 2004), and *Pasteurella multocida* strain Pm70 (May *et al.*, 2001). In addition to compiling and examining the genomic data already published on these genomes, we performed our own analyses to identify common features among the genomes, and to more closely examine each of the genomes to identify groups of genes of interest in host colonization, persistence, and pathogenesis.

General genome features of the sequenced *Pasteurellaceae*

Table 3.1 summarizes the general features of the sequenced *Pasteurellaceae* genomes. These genomes range in size from 1.7 to 2.6 Mb, have from 37% to 44% GC content, and contain 3–6 complete ribosomal RNA (rRNA) operons. Some of the genomes include plasmids, some do not, and for some there was no information available as to whether or not they contain plasmid(s). The genomes that contain integrated plasmids include [*H.*] *ducreyi* 35000HP (Munson *et al.*, 2004b), *H. influenzae* R2866 (Erwin *et al.*, 2005), *Hist. somni* 129Pt and 2336 (Mohd-Zain *et al.*, 2004), while *Hist. somni* 129Pt (Siddaramappa *et al.*, 2006) and possibly *Agg. actinomycetemcomitans* (Galli *et al.*, 2001) contain separate plasmid replicons. The *H. influenzae* Rd (Fleischmann *et al.*, 1995), *H. influenzae* 86-028NP (Harrison *et al.*, 2005), '*M. succiniciproducens*' (Hong *et al.*, 2004), *P. multocida* Pm70 (May *et al.*, 2001), *Hist. somni* 2336 and *A. pleuropneumoniae* L20 genomes do not contain plasmid sequences. The genomes currently represented by draft sequences (*A. pleuropneumoniae* sv 1 4074, *H. influenzae* R2846, *M. haemolytica* A1) may contain plasmid sequences, but in the absence of literature references describing the presence of plasmids in these specific strains this cannot be confirmed until these genomes are finished.

Several of the *Pasteurellaceae* genomes contain prophage or candidate prophage regions; these genomes include *H. influenzae* strains 86-028NP (Harrison *et al.*, 2005; Munson *et al.*, 2004a), R2866 (Erwin *et al.*, 2005), 86-028NP (Harrison *et al.*, 2005), and Rd (Fleischmann *et al.*, 1995), *Hist. somni* 129Pt (Challacombe *et al.*, 2007), *Hist. somni* 2336 (Siddaramappa *et al.*, 2007) and *M. haemolytica* strain ATCC BAA-410 (Gioia *et al.*, 2006). A search of the annotated protein sets from each of the other genomes revealed numerous phage-related genes. A more detailed examination of these genes indicated that *A. pleuropneumoniae* strain L20 had at least one candidate prophage region (APL0488-APL0524), which may also be present in the draft sequence of *A. pleuropneumoniae* sv 1 4074, but the genes were split among several contigs; the draft genome sequence of *A. pleuropneumoniae* sv 1 4074 had at least one putative prophage region; *Agg. actinomycetemcomitans* HK1651 and *H. influenzae* R2846 each had four phage-related genes in different locations and no obvious candidate prophage regions; [*H.*] *ducreyi* 35000HP had at least three candidate prophage regions as well as some individual phage-related genes; the candidate regions included a Mu-like prophage (HD0095–HD0156), which was also present in *H. influenzae* Rd (from HI1478 to HI1523) and *Hist. somni* 2336; some of the genes in this region had counterparts in '*M. succiniciproducens*' MBEL55E (MS0114-MS0075), which appear to be part of a larger prophage region. Another putative prophage in [*H.*] *ducreyi* 35000HP (HD1517–HD1581) aligned well with the larger region in '*M. succiniciproducens*' (MS0075–MS0129) and part of the *H. influenzae* Rd Mu-like prophage (HI1478–HI1489). The third candidate prophage in [*H.*] *ducreyi* 35000HP (HD0486–HD0539) also aligned with the Mu-like prophage of *Hist. somni* 2336, which also had a P2-like prophage region, as well as another putative prophage region that aligned with part of a larger candidate prophage in *H. influenzae* R2866.

At the whole-genome level, comparative studies are commonly performed to compare genomic architectures and reveal duplications and inversions, identify syntenic regions among two or more related genomes, find regions (gene clusters) or individual genes conferring novel functions that are missing in certain genomes but present in others, and to provide further insight into evolutionary relationships. The identification of genomic features that contribute to an organism's biology has become commonplace, and there are many examples in the literature involving the *Pasteurellaceae*. For example, comparison of *H. influenzae* Rd with *Escherichia coli* (Tatusov *et al.*, 1996) demonstrated a lower number of paralogues in *H. influenzae* (about one third of genes had paralogues in *H. influenzae* compared to about half in *E. coli*), and also identified the genetic basis for differences in metabolism and physiology between the two organisms. Comparison of *P. multocida* Pm70 with *H.*

Table 3.1 General genome features of the *Pasteurellaceae*

Genome	Size (Mb)	% coding	%GC	CDSs	rRNA operons	tRNA	References
Aggregatibacter actinomycetemcomitans HK1651	2.1	?	44	2363	6 + 1 extra 5S-rRNA gene	55	Actis *et al*. (2003), http://www.genome.ou.edu/act.html, http://oralgen.lanl.gov
Actinobacillus pleuropneumoniae L20	2.3	86	41	2012	6	62	Wheeler *et al*. (2007)
Actinobacillus pleuropneumoniae sv 1 4074*	2.3	86	41	2132	7–23S 4–16S 6–5S **	ND	Markowitz *et al*. (2006), http://microgen.ouhc.edu/a_pleuro/a_pleuro_home.htm
[*Haemophilus*] *ducreyi* 35000HP	1.7	86.6	38	1781	6 + 1 extra 5S-rRNA gene	45	Challacombe *et al*. (2007), Wheeler *et al*. (2008)
Haemophilus influenzae 86–028NP	1.9	86	38	1791	6 + 1 extra 5S-rRNA gene	58	Harrison *et al*. (2005)
Haemophilus influenzae R2846*	1.8	87	37	1691	6–5S 7–23S 4–16S**	ND	Wheeler *et al*. (2008)
Haemophilus influenzae R2866*	1.9	87	38	1817	4–5S 4–23S 3–16S**	ND	Wheeler *et al*. (2008)
Haemophilus influenzae Rd KW20	1.8	89	38	1743	6 + 1 extra 5S-rRNA gene	54	Fleischmann *et al*. (1995)
Histophilus somni 129PT	2.0	89.5	37	1844	5	50	Challacombe *et al*. (2007)
Histophilus somni 2336*	2.3	87.6	37	2081	5–5S 6–23S 3–16S**	ND	Markowitz *et al*. (2006)
Mannheimia haemolytica A1*	2.6	92	41	2839	incomplete	42	Gioia *et al*. (2006)
'Mannheimia succiniciproducens' MBEL55E	2.3	89.9	42.5	2384	6 + 1 extra 5S-rRNA gene	60	Hong *et al*. (2004)
Pasteurella multocida subsp. *multocida* Pm70	2.3	89	41	2014	6 + 2 extra 5S-rRNA genes	57	May *et al*. (2001)

*Draft genome

**Putative rRNA genes identified by blastn analysis using as query sequences *A. pleuropneumoniae* L20 and/or *Hist. somni* 129Pt rRNA sequences.

ND, not determined.

influenzae and *E. coli* (May *et al.*, 2001) revealed that *P. multocida* diverged from *H. influenzae* about 270 million years ago, and also identified 104 potential virulence genes in *P. multocida*. A comparative study of the genomes of *H. influenzae* Rd and non-typeable strain 86–028NP (Harrison *et al.*, 2005; Munson *et al.*, 2004a) showed that the 86-028NP genome architecture contained large rearrangements compared to Rd and had a genomic island that was absent in Rd. Analysis of the '*M. succiniciproducens*' MBEL55E genome and comparison to the genome of *P. multocida* Pm70 and *H. influenzae* Rd (Hong *et al.*, 2004) revealed the extent of genome conservation among these three species; the '*M. succiniciproducens*' and *P. multocida* genomes have weak linear relationships in a dot plot, but no colinearity in genome sequences was found between '*M. succiniciproducens*' and *P. multocida*; these results suggest that '*M. succiniciproducens*' MBEL55E and *P. multocida* Pm70 are closely related and descended from a common ancestor, but were subjected to different evolutionary changes. Genomic analysis combined with *in silico* metabolic flux analysis of fermentation data provided further clues as to how '*M. succiniciproducens*' metabolism evolved and how its metabolism is regulated by rumen conditions (Hong *et al.*, 2004). In a very comprehensive study, *M. haemolytica* A1 was compared to all of the currently sequenced *Pasteurellaceae* (Gioia *et al.*, 2006). In addition to a wealth of specific information on the genes involved in virulence, natural competence and transcriptional regulation, these authors presented phylogenetic trees based upon 16S rRNA genes and the sequences of 50 highly conserved housekeeping genes. From these analyses, *M. haemolytica* and '*M. succiniciproducens*' appeared to be less related to each other than to other *Pasteurellaceae*. A comparative analysis of *Hist. somni* 129Pt, [*H.*] *ducreyi* 35000HP and *H. influenzae* Rd (Challacombe *et al.*, 2007) revealed some differences in gene content that may contribute to host specificity. Specifically, there were differences in the numbers and types of adhesins and outer membrane (OM) protein genes, including those encoding haemoglobin and transferrin receptors, which had previously been shown to play a role in the host specificity of *Hist. somni* strains (Ekins *et al.*, 2004; Ekins and Niven, 2001, 2003; Tremblay *et al.*, 2006; Yu *et al.*, 1992).

In addition to whole-genome comparisons, investigators also routinely perform targeted evaluations of smaller genomic regions or groups of genes to focus on one or more specific functions of interest. McGillivary *et al.* (2005) focused on a genomic island from a pathogenic strain of *H. influenzae*, which causes Brazilian purpuric fever (BPF). Although some of the island genes showed similarity to genes in *H. influenzae* Rd and [*H.*] *ducreyi* 35000HP, the entire island was only found in other BPF isolates, indicating that it may contribute to the highly invasive nature of these particular strains. The genetic and phenotypic diversity of invasive non-typeable *H. influenzae* strains was compared with non-invasive strains (Erwin *et al.*, 2005), and it was determined that *H. influenzae* R2866 contained genes encoding an autotransporter (*lav*), bacteriophage HP2, the *tna* gene cluster, and a 53-Kb integrative plasmid, that were not present in *H. influenzae* Rd and several non-typeable strains (Davis *et al.*, 2001; Williams *et al.*, 2001). These studies also found that several genetic traits, in particular the possession of the *hia* and *hmw* genes encoding adhesins, were correlated, but did not identify a gene or group of genes that were universally present in all invasive isolates (Erwin *et al.*, 2005).

Genome comparisons might reveal genetic bases for lifestyle, niche preferences and virulence properties

Comparisons of general genome features can provide insight into evolutionary relationships and identify similarities and differences in genome architecture and arrangement. However, to fully understand the common and unique features of two or more genomes of interest requires a detailed examination of the gene content of each of the genomes. By determining whether groups of organisms contain the same or analogous sets of genes required for certain functions or phenotypes, one can discover the genetic basis for those functions that an organism possesses or lacks. Of the group of sequenced *Pasteurellaceae* available at this time, two are commensals (*Hist. somni* 129Pt and '*M. succiniciproducens*'

MBEL55E), one is an avirulent laboratory strain (*H. influenzae* Rd), and the others are disease isolates. A strict classification and differentiation of pathogenic species/strains and non-pathogenic or commensals is in many cases difficult in the *Pasteurellaceae*. Most strains are commensals, some may act as opportunistic pathogens and only a few can be regarded as primary pathogens in some hosts, such as avian *P. multocida* infections. Nevertheless, taking into account this fact, comparative analyses of the genomes of these different *Pasteurellaceae* members can provide insight into their lifestyles, host and tissue preference, and mechanisms of virulence. For example, comparisons among the avirulent strains, which were isolated from different hosts and tissues, may reveal similarities and differences in gene content important for their survival in a given host and tissue niche. Identification of genes in common among pathogenic strains that reside in a similar tissue, but different hosts (i.e. bovine and pig lung), may lead to the identification of a common set of genes involved in infection and disease in a particular tissue across more than one host, as well as provide insight into the mechanism of host-specific pathogenesis. Comparisons between commensal and pathogenic strains that infect the same host may identify candidate virulence genes. The results of these comparisons could lead to therapies that target the specific organisms in their preferred host niche, as well as broader spectrum prophylactics and therapeutics that might be applied to a variety of hosts.

Comparison of commensal isolate genomes

From published studies comparing *'M. succiniciproducens'* MBEL55E with *H. influenzae* Rd (Hong *et al.*, 2004), and *Hist. somni* 129Pt with *H. influenzae* Rd (Challacombe *et al.*, 2007), we have learned about their metabolic characteristics. For example, all three of these organisms have incomplete tricarboxylic acid (TCA) cycles by virtue of missing genes encoding one or more enzymes: all three lack the gene encoding succinate dehydrogenase, as well as the entire glyoxalate bypass. However, *Hist. somni* 129Pt and *H. influenzae* are missing the gene encoding isocitrate dehydrogenase, and *H. influenzae* Rd also lacks the genes for citrate synthase and aconitase (Challacombe *et al.*, 2007; Tatusov *et al.*, 1996). While all three genomes possess the menaquinone biosynthesis pathway, *Hist. somni* 129Pt also has the complete pathway for ubiquinone biosynthesis. These investigations also revealed that the three genomes have the genes required to utilize glucose and xylose. In addition, *H. influenzae* Rd and *Hist. somni* 129Pt have genes involved in the utilization of ribose and fucose, *H. influenzae* and *'M. succiniciproducens'* can use fructose, *Hist. somni* and *'M. succiniciproducens'* can use mannitol, and *H. influenae* Rd can also utilize glycerol and galactose (Fleischmann *et al.*, 1995; Macfadyen and Redfield, 1996). *Hist. somni* 129Pt has genes involved in the utilization of galactitol, glucitol/sorbitol, and mannose (Challacombe *et al.*, 2007), while *'M. succiniciproducens'* can utilize glucose, fructose, xylose, sucrose, maltose, lactose and mannitol (Hong *et al.*, 2004). To identify further features that these three genomes have in common, we performed a tri-directional best hit analysis of the translated protein sequences using blastp with an E-value cut-off of 10–5. Fig. 3.1 shows the Venn diagram depicting the numbers of common and unique genes in each of these three genomes compared to the others. Results of these pairwise BLAST comparisons of protein sequences from *'M. succiniciproducens'* MBEL55E, *H. influenzae* Rd, and *Hist. somni* 129Pt revealed 709 coding sequences (CDSs) unique to *'M. succiniciproducens'*, 168 CDSs unique to *H. influenzae* Rd, and 366 CDSs unique to *Hist. somni* 129Pt (Fig. 3.1). Excluding paralogues, *'M. succiniciproducens'* MBEL55E and *H. influenzae* Rd had 1329 sequences in common, *Hist. somni* 129Pt and *H. influenzae* Rd had 1242 sequences in common (Challacombe *et al.*, 2007), and *'M. succiniciproducens'* MBEL55E and *Hist. somni* 129Pt had 1253 sequences in common. The three genomes had approximately 1124 to 1271 sequences in common. The exact number of sequences in common among the three genomes cannot be determined due to the presence of paralogous sequences.

The different numbers of unique genes in each genome may reflect differences in genome size, as *'M. succiniciproducens'* has a larger number of CDSs and a higher number of unique genes, while the smallest genome of

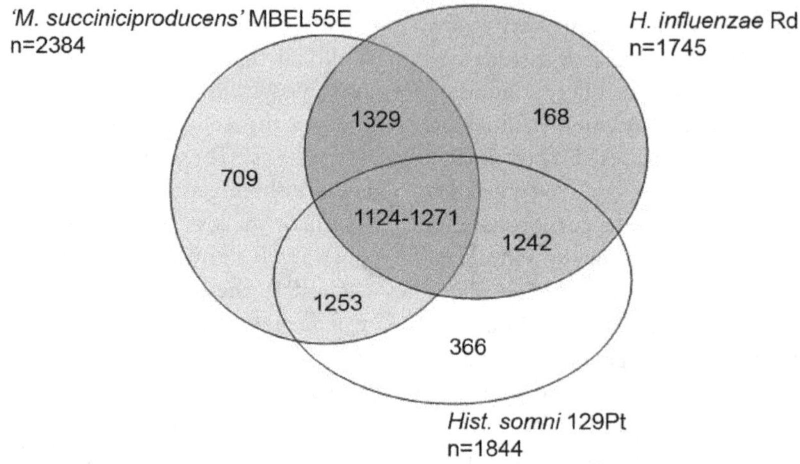

Figure 3.1 Common and unique genes in the genomes of the avirulent *Pasteurellaceae 'M. succiniciproducens'* MBEL55E, *H. influenzae* Rd and *Hist. somni* 129Pt.

H. influenzae Rd has the fewest unique genes (Fig. 3.1). However, it is also likely that many of the unique genes in *'M. succiniciproducens'* play roles in its adaptation to the bovine rumen. In fact, *'M. succiniciproducens'* has a genetic profile of respiratory system genes that enables it to survive in the O_2-deficient and CO_2-rich rumen environment (Hong *et al.*, 2004), in particular this genome lacks the *fumA* (aerobic fumarase) and *sdaA* (succinate dehydrogenase) genes of the TCA cycle, as well as *narI*, *narY* and *cyoA* genes for nitrate and oxygen-mediated respiration. The presence of NADH dehydrogenase genes was also noted. Nevertheless, we found that *Hist. somni* 129Pt and *H. influenzae* Rd lack the same genes as *'M. succiniciproducens'*, and also contain NADH dehydrogenase genes, so their respiratory systems appear to be very similar, and further analyses are necessary to discover which genetic features potentially contribute to their lifestyle and adaptation to a particular host niche.

Comparison of commensal vs. disease isolate genomes

A comparison of the genomes of *Hist. somni* 129Pt, *H. influenzae* Rd and [*H.*] *ducreyi* 35000HP (Challacombe *et al.*, 2007) showed differences in the numbers and presence of genes involved in host colonization and persistence, in particular, genes encoding adhesins, OM proteins, and haemoglobin- and transferrin-binding proteins. Since differences in the numbers and types of specific genes involved in these functions may play roles in host specificity and colonization of particular tissue environments (Challacombe *et al.*, 2007; Ekins and Niven, 2001, 2003; Kachlany *et al.*, 2001; Nika *et al.*, 2002), we also looked for these genes in *'M. succiniciproducens'*. Table 3.2 shows a comparison of the genomes of pathogenic representatives and the three commensal isolates concerning the presence/absence of putative virulence genes. All three of these genomes (*'M. succiniciproducens'*, *Hist. somni* 129Pt and *H. influenzae* Rd) had genes encoding adhesins and pilus components, as well as genes involved in iron utilization. However, *Hist. somni* 129Pt had many more adhesin genes than either *H. influenzae* Rd or *'M. succiniciproducens'* MBEL55E, which only had one (MS0748/HS_0209) of the 12 genes encoding large autotransporter adhesins that have been described previously (Challacombe *et al.*, 2007). As reported for *Hist. somni* 129Pt (Challacombe *et al.*, 2007), *'M. succiniciproducens'* did not have *hmw*, *hia* or *hsf* genes encoding adhesins present in pathogenic *H. influenzae* isolates, although *H. influenzae* Rd had one gene (HI1731a) annotated as an Hsf-like protein.

'M. succiniciproducens' had more genes encoding pilin/fimbriae components than either *H. influenzae* Rd or *Hist. somni* 129Pt (Table 3.2). Whereas the *pilABCD* genes in *H. influenzae* Rd and *'M. succiniciproducens'* MBEL55E were clustered together, only *pilCD* were co-localized in *Hist. somni* 129Pt. *'M. succiniciproducens'* had genes that were part of the *tad-flp-rcp* locus, which

Table 3.2 Comparison of representative pathogenic and commensal genomes regarding putative virulence genes

Gene	*Hist. somni* 2336*	*P. multocida* Pm70	*M. haemolytica* A1*[1]	*A. pleuropneumoniae* L20	In commensal strain(s)?
Adhesins					
Autotransporter adhesin	14 genes	No	Similarity to two *hsf*- and *hia*-like genes	No	Seven are present in *Hist. somni* 129Pt
Autotransporter adhesin *hsf*	No	PM0714	Possibly	APL_0104	HS_1085
Periplasmic zinc transporter/adhesin B precursor *znuA*	Yes	PM0926	Yes	APL_1440	Yes[2]
Autotransporter adhesin *hsf*	No	PM1570	Possibly	No	HI1731a
Autotransporter adhesin *hia*	No	No	Possibly	No	No. HI1731a is closest hit
IgA protease	No	No	Yes	No	No
HMW adhesin	No	No	No	No	No
Filamentous haemagglutinins					
pfhB1	Yes	PM0057	Yes	APL_0959	MS_1167
pfhB2	Yes	PM0059	No	No	No
Pili/fimbriae genes					
Prepilin peptidase *pilD*	Yes	PM0087	Yes	APL_0877	Yes
Pilin biogenesis protein *pilC*	Yes	PM0086	Yes	APL_0878	Yes
Pilin biogenesis protein *pilB*	Yes	PM0085	Yes	APL_0879	Yes
Prepilin peptidase *pilA*	Yes	PM0084	Yes	APL_0880	Yes
Type IV pilin assembly protein *pilF*	Yes	PM2008	Yes	APL_1273	Yes
pseudopilin *pulG/hofG*	Yes	PM0965	No	APL_1888	Yes
tadG	No	PM0843	No	APL_0546	MS1771
tadF	No	PM0844	No	APL_0547	MS1772
tadE	No	PM0845	No	APL_0548	MS1773
tadD	No	PM0846	No	APL_0549	MS1774
tadC	No	PM0847	No	APL_0550	MS1775
tadB	No	PM0848	No	APL_0551	MS1776

Table 3.2 *continued*

Gene	*Hist. somni* 2336*	*P. multocida* Pm70	*M. haemolytica* A1*[1]	*A. pleuropneumoniae* L20	In commensal strain(s)?
tadA	No	PM0849	No	APL_0552	MS1777
tadZ	No	PM0850	No	APL_0553	MS1778
rcpA	No	PM0852	No	APL_0555	MS1779
tadV/cpaA	No	PM0854	No	APL_0557	MS1781
flp1	No	PM0855	No	APL_0559	MS1782
flp2	No	No	No	APL_0558	No
flp3	No	No	No	No	No
Outer membrane proteins					
oapA	Yes	PM0098	Yes	APL_1405	Yes
oapB	No	No	Yes	APL_1404	HI0331 MS0505
P1/OMP47	Yes	PM1069	Yes	APL_0933	Yes
P2	Maybe –candidate gene is at the end of a contig	PM0389	Yes	APL_0649	Yes[3]
P4	No	PM1064	Yes	APL_0389	HI0693 MS1438
P5	Yes	PM0786	Yes	APL_1421 APL_1852	Yes
P6 *pal*	Yes	PM0966	Yes	APL_0304	HI0381 HS_0263
OmpW	No	PM0331	Yes	APL_1086	HS_1265 MS0740
TolC-like lipoprotein	Yes	PM0527	Yes	APL_0257	Yes
OmpV/OM24	No	PM0998	No	APL_1815	No
Phospholipase A	No	PM1426	No	No	No
Organic solvent tolerance protein *ostA*	Yes	PM1600	Yes	APL_0962	Yes
outer membrane protein	No	PM1809	Yes	APL_0829	Yes
antigen OMA87/ D15	Yes	PM1992	Yes	APL_0411	Yes
VacJ lipoprotein	Yes	PM1501	Yes	APL_1918	Yes
putative lipoprotein	No	PM1805	Yes	APL_0331	MS0803
SmpA lipoprotein	Yes	PM1886	Yes	APL_0428	HS_0227 MS1912

Table 3.2 *continued*

Gene	*Hist. somni* 2336*	*P. multocida* Pm70	*M. haemolytica* A1*[1]	*A. pleuropneumoniae* L20	In commensal strain(s)?
LolB lipoprotein	Yes	PM0246	Yes	APL_0777	Yes
MltA murein hydrolase A	Yes	PM0928	Yes	APL_0816	Yes
SrpC solvent efflux protein	No	PM1980	Possibly	APL_0840	MS1304
Rare lipoprotein B	Yes	PM1215	Yes	APL_0873	Yes
Lipoprotein 1 *plpA*	Yes	No?	Yes	APL_0910	Yes
Lipoprotein 2 *plpB*	No	PM1730	Yes. May have three copies	APL_0909	No
Lipoprotein comL	Yes	PM1720	Yes	APL_1121	Yes
LppC	Yes	PM0646	Yes	APL_1362	Yes
Capsule export protein cpxD/bexD	No	PM0778	Yes	APL_1582	MS1487
Neutral endopeptidase PepO	No	No	No	APL_1913	No
LppB	Yes	PM1613	Yes	APL_1928	Yes
Membrane protein HI0974	No	PM1515	Yes	APL_2002	HS_0446 HI0974
Outer membrane lipoprotein A OMLA	No	No	No	APL_1748	No
CutF copper homeostatis protein	No	No	maybe	No	No
Iron utilization					
Transferrin-binding proteins					
Transferrin-binding protein A *tbpA*	Yes	No	Yes	APL_1567	HS_0449 HI0994
Transferrin-binding protein A *tbpA*	Yes	No	No	APL_0246	HS_0582 HI0994
Transferrin-binding protein B *tbpB*	Yes	No	Yes	APL_0245	HS_0448 HI0995
Haemoglobin-binding proteins					
Haemoglobin-binding protein A *hgbA*	Yes	PM0300	Counterpart is HmbR1	APL_1047	HI0661
Haemoglobin-binding protein B *hgbB*	No	PM0337	Counterpart is HmbR2	No	HI0635 HI0712

Table 3.2 *continued*

Gene	*Hist. somni* 2336*	*P. multocida* Pm70	*M. haemolytica* A1*[1]	*A. pleuropneumoniae* L20	In commensal strain(s)?
Haem receptors					
TonB-dependent haem receptor A *tdhA/hemR*	No	PM0576	No	APL_1299?	Yes
Haem–haemopexin utilization protein C *hxuC*	No	No	Yes	APL_1299?	HI0262
Haem–haemopexin utilization protein B *hxuB*	No	No	Yes	No	HI0263 MS1169
Haem–haemopexin utilization protein A *hxuA*	No	No	Yes	No	HI0264
Haem-binding protein A *hbpA*	No	PM0592	Yes 3 copies	APL_2010 APL_0855	HI0853 MS2053
Haem utilization protein *hugZ*	Yes	PM0299	Yes	APL_1048	HS_0722
Haem utilization protein *hugX*	Yes	PM0298	Yes	No	HS_0723
Haem biosynthesis L*-glutamate to protohaem IX*					
gltX	Yes	PM1115	Yes	APL_1275	HS_0456 MS0490
hemA	Yes	PM0684	Yes	APL_0404	HS_0810 MS1324
hemL	Yes	PM0462	Yes	APL_1555	HS_1229 MS0910
hemB	Yes	PM1692	Yes	APL_1988	HS_0548 MS0503
hemC	Yes	PM1812	Yes	APL_1010	HS_0044 MS0276
hemD	Yes	PM1813	Yes	APL_1009	HS_0045 MS0275
hemE	Yes	PM1734	Yes	APL_0112	HS_1540 MS0183
hemN	Yes	PM1833 PM1669	Yes. Two copies	APL_0963 APL_1523 APL_0847	HS_0472 HS_0124 MS0228 MS1746
hemG	Yes	PM1499	Yes	APL_1591	HS_1569 MS0176

Table 3.2 *continued*

Gene	*Hist. somni* 2336*	*P. multocida* Pm70	*M. haemolytica* A1*[1]	*A. pleuropneumoniae* L20	In commensal strain(s)?
hemH	Yes	PM0789		APL_1937	HS_1047 MS1346
TonB system					
tonB	Yes	PM1188	Yes	APL_0076 APL_1571	Yes
exbD	Yes	PM1187	Yes	APL_0077 APL_1569	Yes
exbB	Yes	PM1186	Yes	APL_0078 APL_1570	Yes
Other iron-related					
Fur	Yes	PM0352	Yes	APL_1218	Yes
Iron-regulated outer membrane protein	Yes	PM1428 PM0745	Yes MHA_0860	APL_0919	HS_0069
Iron-regulated outer membrane protein	Yes	No	Yes MHA_2109	APL_1299?	HS_1306
Iron-regulated outer membrane protein	No	No	Yes MHA_1629	No	No
Iron-regulated outer membrane protein *frpB*	No	No	No	APL_0276	No
Iron transport					
Iron (Fe^{3+}) ABC transporter FbpC	Yes	PM0049	Yes MHA_0312	No	*hitC* HI0099 HS_0781
Iron (Fe^{3+}) ABC transporter FbpB	Yes	PM0050	Yes MHA_0313	No	*hitB* HI0098 HS_0782
Iron (Fe^{3+}) ABC transporter FbpA	Yes	PM0051	Yes MHA_0314	No	*hitA* HI0097 HS_0783
Possible haemin or iron (Fe^{3+}) ABC transporter, membrane protein	No	No	MHA_0463	APL_1792	No
Possible haemin or iron (Fe^{3+}) ABC transporter, ATPase	No	No	MHA_0464	APL_1793	No
Possible haemin or iron (Fe^{3+}) ABC transporter, substrate-binding protein	No	No	MHA_0465	APL_1795	No

Table 3.2 *continued*

Gene	*Hist. somni* 2336*	*P. multocida* Pm70	*M. haemolytica* A1*[1]	*A. pleuropneumoniae* L20	In commensal strain(s)?
iron (Fe^{3+}) transporter, substrate-binding protein	Three copies	PM0203	MHA_1442	APL_0563	HS_0641 HS_0643 HI0131
Possible iron (Fe^{3+}) ABC transporter, ATPase	No	No	MHA_1646	No	MS1012
Possible iron (Fe^{3+}) ABC transporter, membrane protein	No	No	MHA_1647	No	MS1013
AfuA	Yes	PM0953 PM0954 PM0955	MHA_2173	APL_1446	Yes
AfuB	Yes	PM0956	MHA_2174	APL_1447	Yes
AfuC	Yes	PM0957	MHA_2175	APL_1448	Yes
Possible haemin ABC superfamily ATP binding cassette transporter, ABC protein	No	No	MHA_2258	No	No
Possible haemin ABC superfamily ATP binding cassette transporter, membrane protein	No	No	MHA_2259	No	No
Possible haemin ABC superfamily ATP binding cassette transporter, binding protein	No	No	MHA_2260	No	No
YfeA	Yes	PM0400	MHA_2378	APL_0272	HS_0328 HI0362
YfeB	Yes	PM0399	MHA_2379	APL_0271	HS_0327 HI0361
YfeD	Yes	PM0397	MHA_2380	APL_0127	HS_0325 HI0359
YfeC	Yes	PM0398	MHA_2381	APL_0128	HS_0326 HI0360
CcmA haem exporter protein A	Yes	PM0005	Yes	APL_1372	Yes
CcmB haem exporter protein B	Yes	PM0006	Yes	APL_1371	Yes
CcmC haem exporter protein C	Yes	PM0007	Yes	APL_1370	Yes

Table 3.2 *continued*

Gene	*Hist. somni* 2336*	*P. multocida* Pm70	*M. haemolytica* A1*[1]	*A. pleuropneumoniae* L20	In commensal strain(s)?
CcmD haem exporter protein D	Yes	PM0008	Yes	APL_1369	Yes
CcmE hem exporter protein E	Yes	PM0009	Yes	APL_1368	Yes
ABC-type cobalamin/Fe^{3+}-siderophores transport systems, ATPase components	No	PM1147	Possibly	No	HI1470
ABC-type cobalamin/Fe^{3+}-siderophores transport systems, ATPase components	No	No	Maybe	APL_1793	No
ABC-type Fe^{3+}-siderophore transport system, permease component	No	PM1148	Yes	APL_1792	HI1471 MS1013
Putative periplasmic iron/ siderophore binding protein	No	PM1149	Yes	APL_1791	MS0795
Putative periplasmic iron/ siderophore binding protein	No	No	Yes	APL_1795	HI1472
Protein implicated in iron transport, frataxin homologue	Yes	PM1425	Yes	APL_1117	Yes
Outer membrane receptor proteins, mostly Fe transport	No	No	No	APL_1707	No
Siderophore uptake					
TonB-dependemt siderophore receptor	No	No	Yes MHA_2388	No	No
Ferric hydroxamate siderophore receptor	No	No	Yes MHA_1541	APL_2016??	No
Probable siderophore ABC superfamily ATP binding cassette transporter, membrane protein FhuB	No	No	Yes MHA_2389	APL_2015	MS1203

Table 3.2 *continued*

Gene	*Hist. somni* 2336*	*P. multocida* Pm70	*M. haemolytica* A1*[1]	*A. pleuropneumoniae* L20	In commensal strain(s)?
Probable siderophore ABC superfamily ATP binding cassette transporter, binding protein FhuD	No	No	Yes MHA_2390	APL_2014	MS1202
Possible siderophore ABC superfamily ATP binding cassette transporter, ABC protein FhuC	No	No	Yes MHA_2391	APL_2013	MS1201
ABC-type cobalamin/Fe^{3+}-siderophores transport systems, permease components	No	No	Yes MHA_1775	APL_0715	MS0792
ABC-type cobalamin/Fe^{3+}-siderophores transport systems, permease components	No	No	Yes MHA_1776	APL_0716	MS0793
ABC-type cobalamin/Fe^{3+}-siderophores transport systems, ATPase components	No	PM0128 FecE	Yes MHA_1777	APL_0717	MS0794
ABC-type Fe^{3+}-siderophores transport systems, periplasmic components	No	PM0131 FecB	MHA_1774	APL_0714	MS0795

*Draft genomes; some locus_tags are not listed.
[1]Majority of results from Gioia *et al*. (2006).
[2]Yes for commensals means gene is present in all three ('*M. succiniciproducens*', *H. influenzae* Rd and *Hist. somni* 129Pt) genomes, otherwise the locus_tags are listed for the genomes that have the gene.
[3]Gene in *Hist. somni* 129Pt was not previously identified (Challacombe *et al*., 2007).

is also found in [*H.*] *ducreyi* 35000HP (Nika *et al.*, 2002), *P. multocida* Pm70, *A. pleuropneumoniae* L20 and sv1 4074, and *Agg. actinomycetemcomitans* HK1651; this locus is important for colonization in a variety of environmental niches (Kachlany *et al.*, 2001). The genomes of *Hist. somni* 129Pt, *H. influenzae* Rd and '*M. succiniciproducens*' MBEL55E contained slightly different profiles of genes encoding OM proteins (Table 3.2): *Hist. somni* 129Pt was missing *oapB* and P4 genes, '*M. succiniciproducens*' was missing only the gene for P6, and *H. influenzae* Rd had all of the genes in this set.

Unlike *Hist. somni* 129Pt and *H. influenzae* Rd, '*M. succiniciproducens*' had no genes encoding haemoglobin- or transferrin-binding proteins, but all three genomes contained a gene encoding a TonB-dependent haem receptor A (*tdhA/*

hemR) as well as all of the TonB system genes, which function together with TonB-dependent receptors in the uptake of iron from host iron-containing proteins (see Chapter 8; reviewed also by Guerinot, 1994; Postle and Kadner, 2003); '*M. succiniciproducens*' and *H. influenzae* Rd also had *hbpA* (haem-binding protein A). In common with *Hist. somni* 129Pt, '*M. succiniciproducens*' had all of the genes required to make protohaem from L-glutamate (Table 3.2). '*M. succiniciproducens*' also contained several genes (MS0792–MS0795) that appear to encode a siderophore-iron uptake system, which were not present in the other two genomes, but were found in *A. pleuropneumoniae* strains L20 and sv1 4074. We and others previously provided evidence supporting the hypothesis that differences in the numbers and types of adhesins and OM protein genes, as well as haemoglobin and transferrin receptor genes, may contribute to host specificities and niche differences among *Pasteurellaceae* strains (Challacombe *et al.*, 2007; Ekins and Niven, 2001; Tremblay *et al.*, 2006). These findings in '*M. succiniciproducens*' add further support to that hypothesis. In addition, none of the commensal genomes contained genes encoding leukotoxin (Hong *et al.*, 2004) (present study) or any other known *Pasteurellaceae* toxins, which is consistent with their avirulent lifestyle.

Comparison of disease isolate genomes

Identification of common and unique virulence genes among pathogenic strains is a challenging problem, and has been addressed by many groups of investigators. The set of virulence genes identified thus far in the *Pasteurellaceae* genomes is large; therefore, elucidating the similarities and differences in virulence genes among the genomes of all disease-causing strains is not a trivial task. Potential virulence genes have been identified and analysed in *A. pleuropneumoniae* (Deslandes *et al.*, 2007; Sheehan *et al.*, 2003), *Agg. actinomycetemcomitans* (reviewed by Actis *et al.*, 2003; Henderson *et al.*, 2003), *P. multocida* (Harper *et al.*, 2006; May *et al.*, 2001), *M. haemolytica* A1 (Gioia *et al.*, 2006; Lo, 2001), many *H. influenzae* strains (Barenkamp and St Geme III, 1996; Ecevit *et al.*, 2004; Erwin and Smith, 2007; Gilsdorf *et al.*, 2004; Hood *et al.*, 1996; Rao *et al.*, 1999; St Geme III *et al.*, 1994), [*H.*] *ducreyi* 35000HP (AlTawfiq *et al.*, 2000; Bong *et al.*, 2001; Fortney *et al.*, 2000; Fulcher *et al.*, 2006; Janowicz *et al.*, 2004; Spinola *et al.*, 2003; Totten *et al.*, 1995) and *Hist. somni* (Corbeil, 1990; Czuprynski *et al.*, 2004; Siddaramppa and Inzana, 2004; Ward *et al.*, 1995). The genes identified in these studies are involved in capsule and lipopolysaccharide biosynthesis, pili/fimbriae biogenesis, adhesion, toxin production and export, iron utilization, iron regulation, phosphorylcholine and sialic acid metabolism, and OM protein synthesis. Because comparing all of the potential virulence genes in all sequenced disease-causing *Pasteurellaceae* family genomes is beyond the scope of this chapter, smaller groups of currently available genomes were compared. Furthermore, we focus on the pathogenic *Pasteurellaceae* strains that reside in similar tissues, and in different hosts.

The lung pathogens *Hist. somni* 2336, *M. haemolytica* A1, *A. pleuropneumoniae* L20, and *P. multocida* Pm70 infect and cause disease in bovine, porcine, and avian (chicken) hosts, respectively. Identifying a common set of virulence-associated genes may lead to novel targets for vaccines and therapies directed against these lung pathogens. To address these needs, gene sets involved in adhesion and host colonization, invasion, persistence as well as toxin genes were compared. Table 3.2 shows the results of these comparisons. As demonstrated for the avirulent isolate genomes, these four virulent isolates contained a gene (*znuA*) encoding a periplasmic zinc transporter/adhesin B precursor. However, only *P. multocida*, *M. haemolytica*, and *A. pleuropneumoniae* had genes encoding Hsf-like autotransporter adhesins. Only the *M. haemolytica* A1 genome contained genes that encode an *hia*-like adhesin and IgA protease (Gioia *et al.*, 2006). All four of the genomes contained at least one sequence encoding filamentous haemagglutinin. In addition to those genes listed in Table 3.2, *Hist. somni* 2336 contained 14 sequences encoding autotransporter adhesins that were not present in either *P. multocida* Pm70 or *A. pleuropneumoniae* L20. These *Hist. somni* 2336 sequences all showed the highest sequence similarity to the two *hsf*- and *hia*- like autotransporter adhesins of *M. haemolytica* A1.

None of these genomes had the *fimABCD* pilus genes of *H. influenzae*. However, all four (Doughty *et al.*, 2000; Ruffolo *et al.*, 1997; Stevenson *et al.*, 2003) contained a gene cluster homologous to the *pilABCD* genes present in the *Hist. somni* 129Pt, *H. influenzae* Rd and '*M. succinciproducens*'. These genes were also present in the other genomes described (*Agg. actinomycetemcomitans* HK1651, *A. pleuropneumoniae* sv1 4074, *H. influenzae* 86–028NP (Harrison *et al.*, 2005), *H. influenzae* R2866, *H. influenzae* R2846, and [*H.*] *ducreyi* 35000HP (Challacombe *et al.*, 2007). *P. multocida* and *A. pleuropneumoniae* L20 also had the *tad-rcp-flp* locus, as did '*M. succiniciproducens*'. This locus was also present in *A. pleuropneumoniae* sv1 4074, *Agg. actinomycetemcomitans* (Henderson *et al.*, 2003), and [*H.*] *ducreyi* (Brentjens *et al.*, 1996; Nika *et al.*, 2002).

OM proteins discussed in detail in Chapter 8 are present at the cell surface, and along with lipopolysaccharide and capsule, may protect pathogens against the host immune system (Boyce *et al.*, 2006; Chung *et al.*, 2007); OM proteins may also mediate infection by interacting with host cells (Chung *et al.*, 2007). The OM of *A. pleuropneumoniae* and *P. multocida* contain many proteins that have recently been identified experimentally (Boyce *et al.*, 2006; Chung *et al.*, 2007; Cooney and Lo, 1993; Gatto *et al.*, 2002; Kasten *et al.*, 1995; Lo *et al.*, 2004; Luo *et al.*, 1997; Ruffolo and Adler, 1996) and annotated by protein sequence similarity (Boyce *et al.*, 2006; Chung *et al.*, 2007). These OM proteins are included in Table 3.2. Many of these genes were also present in the three avirulent commensal genomes that we examined. However, a few were only present in the pathogenic strains: OM protein V (*P. multocida* Pm70 and *A. pleuropneumoniae* L20), phospholipase A (*P. multocida* Pm70), *plpB* (*A. pleuropneumoniae* L20 and *M. haemolytica* A1), *pepO* (*A. pleuropneumoniae* L20), OmlA (*A. pleuropneumoniae* L20) (Bunka *et al.*, 1995) and possibly *cutF* (*M. haemolytica* A1). Both OmlA and P6/PalA have been investigated as vaccine candidates in *A. pleuropneumoniae* (Frey *et al.*, 1996; van den Bosch and Frey, 2003), while OM proteins P1, P5, P6 and others show promise as vaccine targets in non-typeable *H. influenzae* (Berenson *et al.*, 2005; DeMaria *et al.*, 1996; Esmaily *et al.*, 2006; Giebink, 1997; Kurita *et al.*, 2006; Mes and van Putten, 2007; Novotny *et al.*, 2002; Wu and Gu, 1999).

All of the *Pasteurellaceae* genomes contained genes involved in iron utilization (Actis *et al.*, 2003; Challacombe *et al.*, 2007; Fleischmann *et al.*, 1995; Gioia *et al.*, 2006; Harrison *et al.*, 2005; Hong *et al.*, 2004; Munson *et al.*, 2004a; Ray *et al.*, 2001; Tatusov *et al.*, 1996), and many of these genes were present in both pathogenic and commensal strains. However, as for OM proteins, there were some genes for iron utilization that were present only in one or more of the pathogenic strains. For the respiratory pathogens, the proteins encoded by these genes included two iron-regulated OM proteins (one in *M. haemolytica* and another in *A. pleuropneumoniae*) and the components of two iron/haemin transporters (one was present in both *M. haemolytica* and *A. pleuropneumoniae* and another was only present in *M. haemolytica*). One CDS in *A. pleuropneumoniae* L20 (APL_1299) had significant similarity to *tdhA/hemR, hxuC*, and one of the genes that encode the iron-regulated OM protein in *M. haemolytica* was also present in *Hist. somni* 129Pt (HS_1306). APL_1299 was also highly similar to the gene encoding iron-regulated OM protein 64 from *Agg. actinomycetemcomitans.*

The *M. haemolytica* and *A. pleuropneumoniae* genomes had components of two siderophore uptake systems that were also present in '*M. succiniciproducens*'. All but one of the iron transporter genes that were present in *P. multocida* and *Hist. somni* 2336 were also present in one or more avirulent commensal strains. Other genes involved in iron utilization that were present in at least one of the genomes included *afuABC*, which encode an iron transport system (Chin *et al.*, 1996), *fbpABC* (Berish *et al.*, 1990), which is a locus similar to *hitABC* of *H. influenzae* (Adhikari *et al.*, 1996), *yfeABCD* for acquisition of chelated iron (Bearden *et al.*, 1998), *fecABCD*, encoding the ferric citrate transport system (Hussein *et al.*, 1981), haem and haemoglobin receptor genes, and genes encoding iron-regulated OM proteins. Many of these genes were also present in one or more of the commensal genomes. *Hist. somni* 2336, *A. pleuropneumoniae* L20 (Gerlach *et al.*, 1992) and *M. haemolytica* (Gioia *et al.*, 2006) had genes encoding transferrin-binding proteins, but *P. multocida* Pm70 did not. These genes were also present in *Hist. somni* 129Pt and *H. influenzae* Rd (Challacombe *et al.*, 2007).

All of the *Pasteurellaceae* genomes that we examined contained the TonB system genes, which are involved in the active transport of iron and haem from host proteins into the periplasmic space (reviewed by (Postle and Kadner, 2003)); *A. pleuropneumoniae* L20 has two TonB systems, one is transcriptionally associated with the *tbpAB* (transferrin-binding protein) genes (Beddek *et al.*, 2004).

The draft genome of *Hist. somni* 2336 contained 14 genes annotated as encoding for 'large exoproteins involved in haem utilization or adhesion'. One of these genes had a counterpart in 129Pt (HS_1739). Another was also present in *H. influenzae* R2866, and yet another was highly similar in sequence to APL_0959; three others were similar to sequences in '*M. succiniciproducens*' (MS1162–MS1164). No similar sequences were found in [*H.*] *ducreyi* or *P. multocida*. As most of these genes were unique to *Hist. somni* 2336 and the encoded proteins have a predicted cell surface location, they are good candidates for further characterization.

In addition to the adhesin, OM protein, and iron utilization genes that we have described, each of the respiratory pathogen genomes contains other genes that have been implicated in virulence. For *P. multocida* these genes include *pfhB1* and *pfhB2*, which encode filamentous haemagglutinins (May *et al.*, 2001). Counterparts of *pfhB1* that showed acceptable sequence similarity were present in '*M. succiniciproducens*' (MS1167 FhaB) and *Hist. somni* 2336. *Hist. somni* 2336 and [*H.*] *ducreyi* 35000HP (HD1156 Lsp2) also had genes homologous to *pfhB2*. *M. haemolytica* has a gene with high similarity to the haemagglutinin *fhaB* (Gioia *et al.*, 2006), which was also present in '*M. succiniciproducens*' (MS1167). In addition to the haemagglutinins, the *A. pleuropneumoniae* (Boekema *et al.*, 2004) and *M. haemolytica* (Gioia *et al.*, 2006; Highlander, 2001) genomes contain RTX toxin genes *apx* and *lkt*, respectively, which are necessary for virulence. Comparable RTX sequences were present in *Agg. actinomycetemcomitans*, but not in any of the other genomes. '*M. succiniciproducens*' MBEL55E had a homologue of the *lktC* gene (MS1168), which encodes a transacylase that converts pro-LktA to the biologically active leukotoxin (Thumbikat *et al.*, 2003), but none of the other leukotoxin operon genes were present. The genomes of [*H.*] *ducreyi* 35000HP and *Agg. actinomycetemcomitans* contain three genes encoding for a cytolethal distending toxin (Cope *et al.*, 1997; Mayer *et al.*, 1999; Sugai *et al.*, 1998), which were not present in the other *Pasteurellaceae* genomes examined.

Capsule and endotoxin (lipopolysaccharide (LPS) or lipooligosaccharide (LOS)) have been implicated in the virulence of *Pasteurellaceae* (Boyce and Adler, 2000; Chung *et al.*, 2001; Erwin *et al.*, 2005; Ewers *et al.*, 2006; Foxwell *et al.*, 1998; Harper *et al.*, 2006; Harrison *et al.*, 2005; Hood *et al.*, 1996; Lo *et al.*, 2001; Moxon, 1992; Page *et al.*, 1991; Potter and Lo, 1995, 1996; Rao *et al.*, 1999; Sheehan *et al.*, 2003; Siddaramppa and Inzana, 2004; Ward *et al.*, 1998; Zwahlen *et al.*, 1989). A capsule enables bacteria to avoid clearance by the host (Weller *et al.*, 1977), disseminate and cause systemic infections; endotoxin also plays a role in host colonization and persistence, such as by decreasing respiratory ciliary activity (Johnson and Inzana, 1986) and by facilitating host immune evasion and serum resistance (Rao *et al.*, 1999; Weiser *et al.*, 1990; Williams *et al.*, 2001). The genomes of *A. pleuropneumoniae* (Ward and Inzana, 1997), *M. haemolytica* (Highlander, 2001; Lo *et al.*, 2001), *P. multocida* (Boyce *et al.*, 2000; Chung *et al.*, 1998), and encapsulated strains of *H. influenzae* (Hoiseth *et al.*, 1985; Kroll *et al.*, 1988; Kroll and Moxon, 1988) contain capsule biosynthetic genes. These genes are typically organized into loci comprised of three regions: regions 1 and 3 are involved in export from the inner to outer membrane and to the cell surface, while region 2 contains serotype-specific capsule biosynthesis genes. The genes in regions 1 and 3 are homologous and in a similar configuration among *H. influenzae* type b, *M. haemolytica* and *P. multocida* (Bandara *et al.*, 2003; Boyce *et al.*, 2000; Kroll *et al.*, 1989; Lo *et al.*, 2001; Satola *et al.*, 2003; Sukupolvi-Petty *et al.*, 2006; Ward *et al.*, 1998). In *A. pleuropneumoniae*, region 3 is not adjacent to regions 1 and 2. *Hist. somni* is not encapsulated, but does make an exopolysaccharide (Inzana *et al.*, 2002). *Agg. actinomycetemcomitans* produces capsular components (Kamin *et al.*, 1986), and its genome contained at least two genes involved in capsule biosynthesis. Encapsulated strains can be assigned to serogroups based on capsule antigenicity (Boyce *et al.*, 2000; Kroll *et al.*, 1989), and evolutionary relationships among serotypes have been investigated by comparing capsulation loci (Kroll *et al.*, 1989; Musser *et al.*, 1988).

In some pathogenic *Pasteurellaceae*, capsule appears to be more important for virulence, while in others, especially the strains that do not produce capsule, LPS or LOS plays a key role (Melhus *et al.*, 1998). The genetics of LPS/LOS biosynthesis in the Pasteurellaceae is complicated. Some genera produce LPS with O-antigen (e.g. *Actinobacillus*), whereas other genera produce an LOS, which lacks O-antigen (e.g. *Haemophilus*), but may be very heterogeneous, subject to phase variation, and decorated with various substituents (such as sialic acid and phosphorylcholine) that have important interactions with the host. The reader is referred to Chapter 9, which deals with LPS/LOS in more detail.

Intraspecies variation of genomes

Comparative genome analyses can provide information about variation within a species, which is necessary to obtain a more complete understanding of the species as a whole. In addition, these analyses can pinpoint important differences among related genomes, providing insight into the individual organisms comprising a species and how they differ from each other. Several recent investigations have targeted *H. influenzae* (Erwin and Smith, 2007; Erwin *et al.*, 2005; Harrison *et al.*, 2005; Hogg *et al.*, 2007; Munson *et al.*, 2004a; Shen *et al.*, 2005), and we have compared *Hist. somni* 129Pt and 2336 in an attempt to gain insight into intraspecies variation.

Comparisons of non-typeable strains of *H. influenzae* with the laboratory strain Rd revealed that the clinically isolated non-typeable strains each contain about 10% novel genes compared to Rd, and that these novel genes are not uniformly distributed among the strains (Shen *et al.*, 2005). Taken together with clinical evidence that polyclonal populations of non-typeable *H. influenzae* are associated with chronic disease, and that horizontal gene transfer occurs among strains *in situ*, results of genome comparisons suggest that horizontal gene transfer may be responsible for the genetic diversity observed in disease-causing populations of *H. influenzae* (Hogg *et al.*, 2007). These studies led to a distributed genome hypothesis, which states that the genes available to a pathogenic bacterial species are not contained in all members, but are present in a population of naturally transformable strains (Hogg *et al.*, 2007). Results of whole-genome alignments between *H. influenzae* Rd and clinical strains revealed that each clinical strain had a significant number of genomic insertions (127) and deletions (on average 147) greater than 90 basepairs that were not present in the Rd genome (Hogg *et al.*, 2007).

We performed tridirectional best blast hits analyses of the protein sequences from four *H. influenzae* genomes (strain Rd and three non-typeable isolates), and found that the number of genes in common among the strains is high and the number of genes unique to each strain is low (Fig. 3.2), compared to our comparison of the three non-pathogenic isolates, which are not members of the same species (Fig. 3.1). The number of unique genes present in each *H. influenzae* genome ranged from 3% to 11%, compared with the number of unique genes in the avirulent isolate genomes, which ranged from 10% to 30%. As described by other investigators, the genes unique to the disease-causing isolates of *H. influenzae* include virulence-associated genes encoding autotransporters, adhesins, hemagglutinating pili, and other sequences expressed during infection and disease (Harrison *et al.*, 2005; Shen *et al.*, 2005).

Compared with *H. influenzae*, there are only two completely sequenced *Hist. somni* genomes. However, comparison of these genomes is very informative, since one is an avirulent commensal isolate (strain 129Pt) and the other a pneumonia isolate (strain 2336). Results of whole-genome comparisons revealed that the *Hist. somni* 129Pt and 2336 chromosomes have undergone numerous insertions, deletions and duplications (Siddaramappa *et al.*, 2007). These genomes also contain multiple bacteriophage insertions, which partially contribute to the genetic diversity

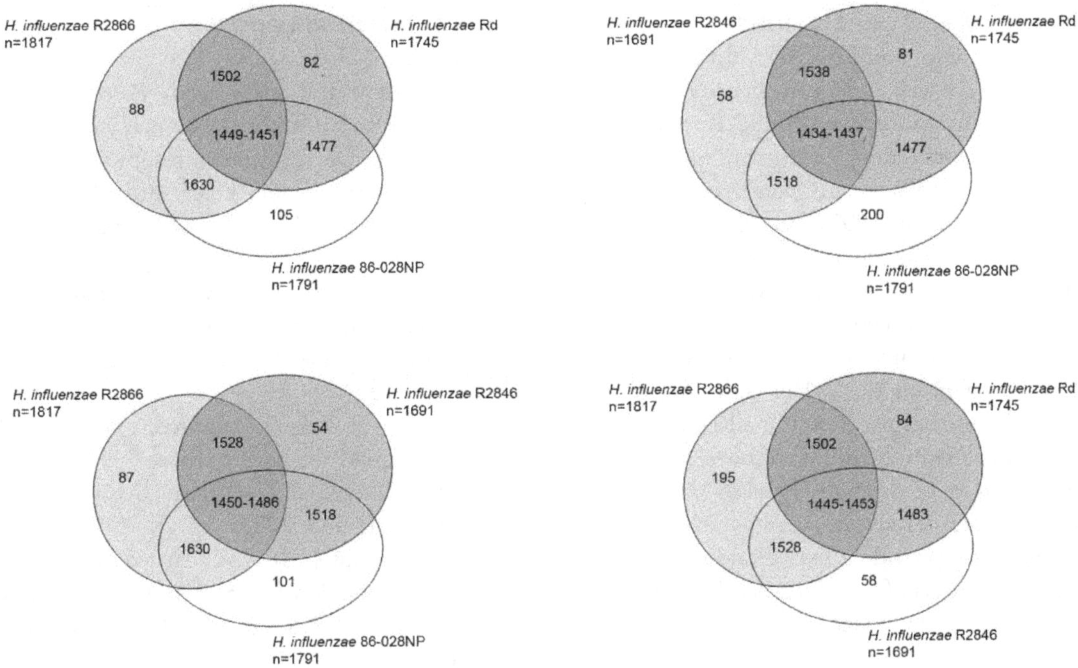

Figure 3.2 Common and unique genes in four *H. influenzae* genomes compared three at a time.

of the two isolates. Fig. 3.3 shows the common and unique genes identified by bidirectional best blast hit analysis of the *Hist. somni* 129Pt and 2336 protein sequences. This analysis revealed a higher number of unique genes in strain 2336 (16%) than in 129Pt (8%). A majority of the genes unique to strain 129Pt are not functionally characterized (annotated as hypothetical and conserved hypothetical), while others encode phage-related functions. Among the genes unique to strain 2336 are some LOS biosynthesis genes, autotransporter adhesins, 11 homologues of filamentous haemagglutinins, and prophage-like sequences (Siddaramappa *et al.*, 2007).

Results of these studies show that although the genomes of the individual species have a high proportion of genes in common, the pathogenic strains of *H. influenzae* and *Hist. somni* contain many unique genes associated with virulence, compared with their avirulent relatives.

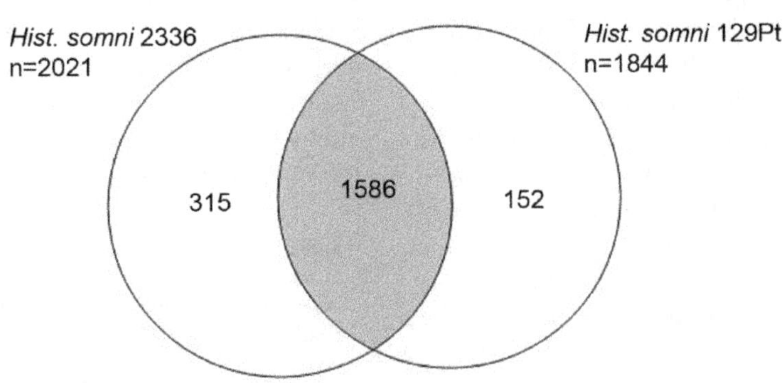

Figure 3.3 Common and unique genes in the genomes of commensal and pathogenic strains of *Hist. somni.*

Summary and conclusions

Whole-genome comparative studies, such as the ones reviewed here, can identify promising novel vaccine candidates in the *Pasteurellaceae*. For example, potential vaccine candidates might include the novel adhesin genes that were identified in *Hist. somni* 129Pt and 2336, but appear to be absent in the other *Pasteurellaceae* genomes. Autotransporter proteins have been used as components of vaccines (Wells *et al.*, 2007), and are often implicated in virulence by virtue of their adhesion functions. This class of sequences should be further examined in all *Pasteurellaceae* for their utility in the design of novel vaccine targets. As more of the non-typeable *H. influenzae* genomes are sequenced, comparative genome analyses may provide a better understanding of the pathogenesis of these strains (Munson *et al.*, 2004a), which may lead to the development of novel vaccine strategies.

Results of whole-genome comparisons of several *Pasteurellaceae* species (Challacombe *et al.*, 2007; Gioia *et al.*, 2006; Harrison *et al.*, 2005; Hong *et al.*, 2004; May *et al.*, 2001; Munson *et al.*, 2004a; Tatusov *et al.*, 1996), as well as targeted studies focusing on specific groups of genes in one or more *Pasteurellaceae* species, have increased our understanding of the organisms that comprise the *Pasteurellaceae* in terms of global genomic features. These studies also revealed underlying similarities and differences in gene content and organization that are responsible for features that these organisms have in common, as well as those that contribute to their uniqueness. The genomes of non-pathogenic and pathogenic strains have many genes in common that are required for host colonization and persistence, while the genomes of pathogenic strains have additional genes encoding virulence factors, which are not found in the avirulent commensals. Future comparisons between commensal–pathogen pairs should be able to further identify genes that contribute to organism preferences for different tissue environments in the same host, as well as genes that contribute to virulence in the pathogenic strain.

Future directions and web resources

In comparing the sequenced *Pasteurellaceae* genomes across hosts and niches, many investigations have uncovered promising similarities and differences in the genomes that provide insight into mechanisms of virulence, host-specificity, and preferred niche. Results of these and subsequent genomic analyses will potentially lead to the development of new prophylactic and therapeutic strategies for diseases caused by the *Pasteurellaceae*.

Many web-based resources were used for comparative analyses. These included the Integrated Microbial Genomes (IMG) system (Markowitz *et al.*, 2006) at the DOE Joint Genome Institute, BLAST, the database resources at the National Center for Biotechnology Information (Wheeler *et al.*, 2008), MetaCyc (Caspi *et al.*, 2006), UniProt (http://www.expasy.uniprot.org/) and the Oralgen Pathogen Sequence Databases at Los Alamos National Laboratory (http://oralgen.lanl.gov), the *A. pleuropneumoniae* sequencing project (David Dyer and Allison Gillaspy: http://microgen.ouhsc.edu/a_pleuro/a_pleuro_home.htm), and the *Agg. actinomycetemcomitans* Genome Sequencing Project (Bruce A. Roe, Fares Z. Najar, Allison Gillaspy, Sandra Clifton, Tom Ducey, Lisa Lewis and David W. Dyer: http://www.genome.ou.edu/act.html).

Acknowledgements

We thank Sarah Highlander for providing us with the annotated protein translations of *M. haemolytica* A1 strain BAA-410, and Shivakumara Siddaramappa for review of the manuscript. This work was supported by the U.S. Department of Energy under contract no. W-7405-ENG-36 and by USDA/NSF microbial sequencing grant 2001-52100-11314.

References

Actis, L.A., Rhodes, E.R., and Tomaras, A.P. (2003). Genetic and molecular characterization of a dental pathogen using genome-wide approaches. Adv. Dent Res. *17*, 95–99.

Adhikari, P., Berish, S.A., Nowalk, A.J., Veraldi, K.L., Morse, S.A., and Mietzner, T.A. (1996). The fbpABC locus of *Neisseria gonorrhoeae* functions in the periplasm-to-cytosol transport of iron. J. Bacteriol. *178*, 2145–2149.

Al-Tawfiq, J.A., Fortney, K.R., Katz, B.P., Hood, A.F., Elkins, C., and Spinola, S.M. (2000). An isogenic hemoglobin receptor-deficient mutant of *Haemophilus ducreyi* is attenuated in the human model of experimental infection. J. Infect. Dis. *181*, 1049–1054.

Bandara, A.B., Lawrence, M.L., Veit, H.P., and Inzana, T.J. (2003). Association of *Actinobacillus pleuropneumoniae* capsular polysaccharide with virulence in pigs. Infect. Immun. *71*, 3320–3328.

Barenkamp, S.J., and St Geme III, J.W. (1996). Identification of a second family of high-molecular-weight adhesion proteins expressed by non-typeable *Haemophilus influenzae*. Mol. Microbiol. *19*, 1215–1223.

Bearden, S.W., Staggs, T.M., and Perry, R.D. (1998). An ABC transporter system of *Yersinia pestis* allows utilization of chelated iron by *Escherichia coli* SAB11. J. Bacteriol. *180*, 1135–1147.

Beddek, A.J., Sheehan, B.J., Bosse, J.T., Rycroft, A.N., Kroll, J.S., and Langford, P.R. (2004). Two TonB systems in *Actinobacillus pleuropneumoniae*: their roles in iron acquisition and virulence. Infect. Immun. *72*, 701–708.

Berenson, C.S., Murphy, T.F., Wrona, C.T., and Sethi, S. (2005). Outer membrane protein P6 of nontypeable *Haemophilus influenzae* is a potent and selective inducer of human macrophage proinflammatory cytokines. Infect. Immun. *73*, 2728–2735.

Berish, S.A., Mietzner, T.A., Mayer, L.W., Genco, C.A., Holloway, B.P., and Morse, S.A. (1990). Molecular cloning and characterization of the structural gene for the major iron-regulated protein expressed by *Neisseria gonorrhoeae*. J. Exp. Med. *171*, 1535–1546.

Boekema, B.K., Kamp, E.M., Smits, M.A., Smith, H.E., and Stockhofe-Zurwieden, N. (2004). Both ApxI and ApxII of *Actinobacillus pleuropneumoniae* serotype 1 are necessary for full virulence. Vet. Microbiol. *100*, 17–23.

Bong, C.T., Throm, R.E., Fortney, K.R., Katz, B.P., Hood, A.F., Elkins, C., and Spinola, S.M. (2001). DsrA-deficient mutant of *Haemophilus ducreyi* is impaired in its ability to Infect. human volunteers. Infect. Immun. *69*, 1488–1491.

Boyce, J.D., and Adler, B. (2000). The capsule is a virulence determinant in the pathogensis of *Pasteurella multocida* M1404 (B:2). Infect. Immun. *68*, 3463–3468.

Boyce, J.D., Chung, J.Y., and Adler, B. (2000). *Pasteurella multocida* capsule: composition, function and genetics. J. Biotechnol. *83*, 153–160.

Boyce, J.D., Cullen, P.A., Nguyen, V., Wilkie, I., and Adler, B. (2006). Analysis of the *Pasteurella multocida* outer membrane sub-proteome and its response to the *in vivo* environment of the natural host. Proteomics *6*, 870–880.

Brentjens, R.J., Ketterer, M., Apicella, M.A., and Spinola, S.M. (1996). Fine tangled pili expressed by *Haemophilus ducreyi* are a novel class of pili. J. Bacteriol. *178*, 808–816.

Bunka, S., Christensen, C., Potter, A.A., Willson, P.J., and Gerlach, G.F. (1995). Cloning and characterization of a protective outer membrane lipoprotein of *Actinobacillus pleuropneumoniae* serotype 5. Infect. Immun. *63*, 2797–2800.

Caspi, R., Foerster, H., Fulcher, C.A., Hopkinson, R., Ingraham, J., Kaipa, P., Krummenacker, M., Paley, S., Pick, J., Rhee, S.Y., *et al.* (2006). MetaCyc: A multiorganism database of metabolic pathways and enzymes. Nucleic Acids Res. *34*, D511-D516.

Challacombe, J.F., Duncan, A.J., Brettin, T.S., Bruce, D., Chertkov, O., Detter, J.C., Han, C.S., Misra, M., Richardson, P., Tapia, R., *et al.* (2007). Complete genome sequence of *Haemophilus somnus (Histophilus somni)* strain 129Pt and comparison to *Haemophilus ducreyi* 35000HP and *Haemophilus influenzae* Rd. J. Bacteriol. *189*, 1890–1898.

Chin, N., Frey, J., Chang, C.F., and Chang, Y.F. (1996). Identification of a locus involved in the utilization of iron by *Actinobacillus pleuropneumoniae*. FEMS Microbiol. Lett. *143*, 1–6.

Chung, J.W., Ng-Thow-Hing, C., Budman, L.I., Gibbs, B.F., Nash, J.H., Jacques, M., and Coulton, J.W. (2007). Outer membrane proteome of *Actinobacillus pleuropneumoniae*: LC-MS/MS analyses validate in silico predictions. Proteomics *7*, 1854–1865.

Chung, J.Y., Wilkie, I., Boyce, J.D., Townsend, K.M., Frost, A.J., Ghoddusi, M., and Adler, B. (2001). Role of capsule in the pathogenesis of fowl cholera caused by *Pasteurella multocida* serogroup A. Infect. Immun. *69*, 2487–2492.

Chung, J.Y., Zhang, Y., and Adler, B. (1998). The capsule biosynthetic locus of *Pasteurella multocida* A:1. FEMS Microbiol. Lett. *166*, 289–296.

Cooney, B.J., and Lo, R.Y. (1993). Three contiguous lipoprotein genes in *Pasteurella haemolytica* A1 which are homologous to a lipoprotein gene in *Haemophilus influenzae* type b. Infect. Immun. *61*, 4682–4688.

Cope, L.D., Lumbley, S., Latimer, J.L., Klesney-Tait, J., Stevens, M.K., Johnson, L.S., Purven, M., Munson, R.S. Jr., Lagergard, T., Radolf, J.D., *et al.* (1997). A diffusible cytotoxin of *Haemophilus ducreyi*. Proc. Natl. Acad. Sci. USA *94*, 4056–4061.

Corbeil, L.B. (1990). Molecular aspects of some virulence factors of *Haemophilus somnus*. Can J. Vet. Res. *54 Suppl*, S57-S62.

Czuprynski, C.J., Leite, F., Sylte, M., Kuckleburg, C., Schultz, R., Inzana, T., Behling-Kelly, E., and Corbeil, L. (2004). Complexities of the pathogenesis of *Mannheimia haemolytica* and *Haemophilus somnus* infections: challenges and potential opportunities for prevention? Anim Health Res. Rev. *5*, 277–282.

Davis, J., Smith, A.L., Hughes, W.R., and Golomb, M. (2001). Evolution of an autotransporter: domain shuffling and lateral transfer from pathogenic *Haemophilus* to *Neisseria*. J. Bacteriol. *183*, 4626–4635.

DeMaria, T.F., Murwin, D.M., and Leake, E.R. (1996). Immunization with outer membrane protein P6 from nontypeable *Haemophilus influenzae* induces bactericidal antibody and affords protection in the chinchilla model of otitis media. Infect. Immun. *64*, 5187–5192.

Deslandes, V., Nash, J.H., Harel, J., Coulton, J.W., and Jacques, M. (2007). Transcriptional profiling of *Actinobacillus pleuropneumoniae* under iron-restricted conditions. BMC Genomics *8*, 72–91.

Doughty, S.W., Ruffolo, C.G., and Adler, B. (2000). The type 4 fimbrial subunit gene of *Pasteurella multocida*. Vet. Microbiol. *72*, 79–90.

Ecevit, I.Z., McCrea, K.W., Pettigrew, M.M., Sen, A., Marrs, C.F., and Gilsdorf, J.R. (2004). Prevalence of the *hifBC, hmw1A, hmw2A, hmwC*, and *hia* genes in *Haemophilus influenzae* Isolates. J. Clin. Microbiol. *42*, 3065–3072.

Ekins, A., Bahrami, F., Sijercic, A., Maret, D., and Niven, D.F. (2004). *Haemophilus somnus* possesses two systems for acquisition of transferrin-bound iron. J. Bacteriol. *186*, 4407–4411.

Ekins, A., and Niven, D.F. (2001). Production of transferrin receptors by *Histophilus ovis*: three of five strains require two signals. Can J. Microbiol. *47*, 417–423.

Ekins, A., and Niven, D.F. (2003). Transferrin-dependent expression of TbpA by *Histophilus ovis* involves a poly G tract within *tbpA*. FEMS Microbiol. Lett. *220*, 95–98.

Erwin, A.L., and Smith, A.L. (2007). Nontypeable *Haemophilus influenzae*: understanding virulence and commensal behavior. Trends Microbiol. *15*, 355–362.

Erwin, A.L., Nelson, K.L., Mhlanga-Mutangadura, T., Bonthuis, P.J., Geelhood, J.L., Morlin, G., Unrath, W.C., Campos, J., Crook, D.W., Farley, M.M., *et al.* (2005). Characterization of genetic and phenotypic diversity of invasive nontypeable *Haemophilus influenzae*. Infect. Immun. *73*, 5853–5863.

Esmaily, F., Vaezzadeh, F., and Sharifi-Yazdi, M.K. (2006). Efficacy of immunization with outer membrane proteins for induction of pulmonary clearance of nontypeable *Haemophilus influenzae* in a rat respiratory model. Iran J. Allergy Asthma Immunol. *5*, 57–61.

Ewers, C., Lubke-Becker, A., Bethe, A., Kiebling, S., Filter, M., and Wieler, L. (2006). Virulence genotype of *Pasteurella multocida* strains isolated from different hosts with various disease status. Vet. Microbiol. *114*, 304–317.

Fleischmann, R.D., Adams, M.D., White, O., Clayton, R.A., Kirkness, E.F., Kerlavage, A.R., Bult, C.J., Tomb, J.F., Dougherty, B.A., Merrick, J.M., *et al.* (1995). Whole-genome random sequencing and assembly of *Haemophilus influenzae* Rd. Science *269*, 496–498, 507–512.

Fortney, K.R., Young, R.S., Bauer, M.E., Katz, B.P., Hood, A.F., Munson R.S., Jr., and Spinola, S.M. (2000). Expression of peptidoglycan-associated lipoprotein is required for virulence in the human model of *Haemophilus ducreyi* infection. Infect. Immun. *68*, 6441–6448.

Foxwell, A.R., Kyd, J.M., and Cripps, A.W. (1998). Nontypeable *Haemophilus influenzae*: Pathogenesis and prevention. Microbiol. Mol. Biol. Rev. *62*, 294–308.

Frey, J., Kuhnert, P., Villiger, L., and Nicolet, J. (1996). Cloning and characterization of an *Actinobacillus pleuropneumoniae* outer membrane protein belonging to the family of PAL lipoproteins. Res. Microbiol. *147*, 351–361.

Fulcher, R.A., Cole, L.E., Janowicz, D.M., Toffer, K.L., Fortney, K.R., Katz, B.P., Orndorff, P.E., Spinola, S.M., and Kawula, T.H. (2006). Expression of *Haemophilus ducreyi* collagen binding outer membrane protein NcaA is required for virulence in swine and human challenge models of chancroid. Infect. Immun. *74*, 2651–2658.

Galli, D.M., Chen, J., Novak, K.F., and Leblanc, D.J. (2001). Nucleotide sequence and analysis of conjugative plasmid pVT745 J. Bacteriol. *183*, 1585–1594.

Gatto, N.T., Dabo, S.M., Hancock, R.E., and Confer, A.W. (2002). Characterization of, and immune responses of mice to, the purified OmpA-equivalent outer membrane protein of *Pasteurella multocida* serotype A:3 (Omp28). Vet. Microbiol. *87*, 221–235.

Gerlach, G.F., Klashinsky, S., Anderson, C., Potter, A.A., and Willson, P.J. (1992). Characterization of two genes encoding distinct transferrin-binding proteins in different *Actinobacillus pleuropneumoniae* isolates. Infect. Immun. *60*, 3253–3261.

Giebink, G.S. (1997). Vaccination against middle-ear bacterial and viral pathogens. Ann. NY Acad. Sci. *830*, 330–352.

Gilsdorf, J.R., Marrs, C.F., and Foxman, B. (2004). *Haemophilus influenzae*: Genetic variability and natural selection to identify virulence factors. Infect. Immun. *72*, 2457–2461.

Gioia, J., Qin, X., Jiang, H., Clinkenbeard, K., Lo, R., Liu, Y., Fox, G.E., Yerrapragada, S., McLeod, M.P., McNeill, T.Z., *et al.* (2006). The genome sequence of *Mannheimia haemolytica* A1: insights into virulence, natural competence, and *Pasteurellaceae* phylogeny. J. Bacteriol. *188*, 7257–7266.

Guerinot, M.L. (1994). Microbial iron transport. Annu. Rev. Microbiol. *48*, 743–772.

Harper, M., Boyce, J.D., and Adler, B. (2006). *Pasteurella multocida* pathogenesis: 125 years after Pasteur. FEMS Microbiol. Lett. *265*, 1–10.

Harrison, A., Dyer, D.W., Gillaspy, A., Ray, W.C., Mungur, R., Carson, M.B., Zhong, H., Gipson, J., Gipson, M., Johnson, L.S., *et al.* (2005). Genomic sequence of an otitis media isolate of nontypeable *Haemophilus influenzae*: Comparative Study with *H. influenzae* Serotype d, Strain KW20. J. Bacteriol. *187*, 4627–4636.

Henderson, B., Nair, S.P., Ward, J.M., and Wilson, M. (2003). Molecular pathogenicity of the oral opportunistic pathogen *Actinobacillus actinomycetemcomitans*. Annu. Rev. Microbiol. *57*, 29–55.

Highlander, S.K. (2001). Molecular genetic analysis of virulence in *Mannheimia (pasteurella) haemolytica*. Front. Biosci *6*, D1128–1150.

Hogg, J.S., Hu, F.Z., Janto, B., Biossy, R., Hayes, J., Keefe, R., Post, J.C., and Ehrlich, G.D. (2007). Characterization and modeling of the *Haemophilus influenzae* core and supragenomes based on the complete genomic sequences of Rd and 12 clinical nontypeable strains. Genome Biol. *8*, R103.

Hoiseth, S.K., Connelly, C.J., and Moxon, E.R. (1985). Genetics of spontaneous, high-frequency loss of b capsule expression in *Haemophilus influenzae*. Infect. Immun. *49*, 389–395.

Hong, S.H., Kim, J.S., Lee, S.Y., In, Y.H., Choi, S.S., Rih, J.K., Kim, C.H., Jeong, H., Hur, C.G., and Kim, J.J. (2004). The genome sequence of the capnophilic rumen bacterium *Mannheimia succiniciproducens*. Nature Biotechnol. 22, 1275–1281.

Hood, D.W., Deadman, M.E., Jennings, M.P., Bisercic, M., Fleischmann, R.D., Venter, J.C., and Moxon, E.R. (1996). DNA repeats identify novel virulence genes in *Haemophilus influenzae*. Proc. Natl. Acad. Sci. USA 93, 11121–11125.

Hussein, S., Hantke, K., and Braun, V. (1981). Citrate-dependent iron transport system in *Escherichia coli* K-12. Eur. J. Biochem. *117*, 431–437.

Inzana, T.J., Cox, A.E., and Glindemann, G. (2002). Exopolysaccharide production by *Haemophilus somnus*. Paper presented at: 102nd general meeting of the American Society for Microbiology, Salt Lake City, UT.

Janowicz, D.M., Fortney, K.R., Katz, B.P., Latimer, J.L., Deng, K., Hansen, E.J., and Spinola, S.M. (2004). Expression of the LspA1 and LspA2 proteins by *Haemophilus ducreyi* is required for virulence in human volunteers. Infect. Immun. *72*, 4528–4533.

Johnson, A.P., and Inzana, T.J. (1986). Loss of ciliary activity in organ cultures of rat trachea treated with lipo-oligosaccharide from *Haemophilus influenzae*. J. Med. Microbiol. 22, 265–268.

Kachlany, S.C., Planet, P.J., DeSalle, R., Fine, D.H., and Figurski, D.H. (2001). Genes for tight adherence of *Actinobacillus actinomycetemcomitans*: from plaque to plague to pond scum. Trends Microbiol. 9, 429–437.

Kamin, S., Harvey, W., Wilson, M., and Scutt, A. (1986). Inhibition of fibroblast proliferation and collagen synthesis by capsular material from *Actinobacillus actinomycetemcomitans*. J. Med. Microbiol. 22, 245–249.

Kasten, R.W., Hansen, L.M., Hinojoza, J., Bieber, D., Ruehl, W.W., and Hirsh, D.C. (1995). *Pasteurella multocida* produces a protein with homology to the P6 outer membrane protein of *Haemophilus influenzae*. Infect. Immun. *63*, 989–993.

Kroll, J.S., Hopkins, I., and Moxon, E.R. (1988). Capsule loss in *H. influenzae* type b occurs by recombination-mediated disruption of a gene essential for polysaccharide export. Cell *53*, 347–356.

Kroll, J.S., and Moxon, E.R. (1988). Capsulation and gene copy number at the cap locus of *Haemophilus influenzae* type b. J. Bacteriol. *170*, 859–864.

Kroll, J.S., Zamze, S., Loynds, B., and Moxon, E.R. (1989). Common organization of chromosomal loci for production of different capsular polysaccharides in *Haemophilus influenzae*. J. Bacteriol. *171*, 3343–3347.

Kurita, S., Koyama, J., Onizuka, S., Motomura, K., Watanabe, H., Watanabe, K., Senba, M., Apicella, M.A., Murphy, T.F., Yoneyama, H., *et al.* (2006). Dynamics of dendritic cell migration and the subsequent induction of protective immunity in the lung after repeated airway challenges by nontypeable *Haemophilus influenzae* outer membrane protein. Vaccine *24*, 5896–5903.

Lo, M., Boyce, J.D., Wilkie, I.W., and Adler, B. (2004). Characterization of two lipoproteins in *Pasteurella multocida*. Microbes Infect. *6*, 58–67.

Lo, R.Y., McKerral, L.J., Hills, T.L., and Kostrzynska, M. (2001). Analysis of the capsule biosynthetic locus of *Mannheimia (Pasteurella) haemolytica* A1 and proposal of a nomenclature system. Infect. Immun. *69*, 4458–4464.

Lo, R.Y. (2001). Genetic analysis of virulence factors of *Mannheimia (Pasteurella) haemolytica* A1. Vet. Microbiol. *83*, 23–35.

Luo, Y., Glisson, J.R., Jackwood, M.W., Hancock, R.E., Bains, M., Cheng, I.H., and Wang, C. (1997). Cloning and characterization of the major outer membrane protein gene (ompH) of *Pasteurella multocida* X-73. J. Bacteriol. *179*, 7856–7864.

Macfadyen, L.P., and Redfield, R.J. (1996). Life in mucus: sugar metabolism in *Haemophilus influenzae*. Res. Microbiol. *147*, 541–551.

McGillivary, G., Tomaras, A.P., Rhodes, E.R., and Actis, L.A. (2005). Cloning and sequencing of a genomic island found in the Brazilian purpuric fever clone of *Haemophilus influenzae* biogroup aegyptius. Infect. Immun. *73*, 1927–1938.

Markowitz, V.M., Korzeniewski, F., Palaniappan, K., Szeto, E., Werner, G., Padki, A., Zhao, X., Dubchak, I., Hugenholtz, P., Anderson, I., *et al.* (2006). The integrated microbial genomes (IMG) system. Nucleic Acids Res. *34*, D344–348.

May, B.J., Zhang, Q., Li, L.L., Paustian, M.L., Whittam, T.S., and Kapur, V. (2001). Complete genomic sequence of *Pasteurella multocida*, Pm70. Proc. Natl. Acad. Sci. USA *98*, 3460–3465.

Mayer, M.P., Bueno, L.C., Hansen, E.J., and DiRienzo, J.M. (1999). Identification of a cytolethal distending toxin gene locus and features of a virulence-associated region in *Actinobacillus actinomycetemcomitans*. Infect. Immun. *67*, 1227–1237.

Melhus, A., Hermansson, A., Forsgren, A., and Prellner, K. (1998). Intra- and interstrain differences of virulence among nontypeable *Haemophilus influenzae* strains. APMIS *106*, 858–868.

Mes, T.H., and van Putten, J.P. (2007). Positively selected codons in immune-exposed loops of the vaccine candidate OMP-P1 of *Haemophilus influenzae*. J. Mol. Evol. *64*, 411–422.

Mohd-Zain, Z., Turner, S.L., Cerdeño-Tárraga, A.M., Lilley, A.K., Inzana, T.J., Duncan, A.J., Harding, R.M., Hood, D.W., Peto, T.E., and Crook, D.W. (2004). Transferable antibiotic resistance elements in *Haemophilus influenzae* share a common evolutionary origin with a diverse family of syntenic genomic islands. J. Bacteriol. *186*, 8114–8122.

Moxon, E.R. (1992). Molecular basis of invasive *Haemophilus influenzae* type b disease. J. Infect. Dis. *165 Suppl 1*, S77-S81.

Munson, R.S., Jr., Harrison, A., Gillaspy, A., Ray, W.C., Carson, M.B., Armbruster, D., Gipson, J., Gipson, M., Johnson, L.S., Lewis, L., *et al.* (2004a). Partial analysis of the genomes of two nontypeable *Haemophilus influenzae* otitis media isolates. Infect. Immun. *72*, 3002–3010.

Munson, R.S., Jr., Zhong, H., Mungur, R., Ray, W.C., Shea, R.J., Mahairas, G.G., and Mulks, M.H. (2004b). *Haemophilus ducreyi* strain ATCC 27722 contains a genetic element with homology to the vibrio RS1 element that can replicate as a plasmid and confer NAD independence on *Haemophilus influenzae*. Infect. Immun. 72, 1143–1146.

Musser, J.M., Kroll, J.S., Moxon, E.R., and Selander, R.K. (1988). Evolutionary genetics of the encapsulated strains of *Haemophilus influenzae*. Proc. Natl. Acad. Sci. USA *85*, 7758–7762.

Nika, J.R., Latimer, J.L., Ward, C.K., Blick, R.J., Wagner, N.J., Cope, L.D., Mahairas, G.G., Munson, R.S., Jr., and Hansen, E.J. (2002). *Haemophilus ducreyi* requires the *flp* gene cluster for microcolony formation in vitro. Infect. Immun. *70*, 2965–2975.

Novotny, L.A., Pichichero, M.E., Denoël, P.A., Neyt, C., Vanderschrick, S., Dequesne, G., and Bakaletz, L.O. (2002). Detection and characterization of pediatric serum antibody to the OMP P5-homologous adhesin of nontypeable *Haemophilus influenzae* during acute otitis media. Vaccine *20*, 3590–3597.

Page, R.C., Sims, T.J., Engel, L.D., Moncla, B.J., Bainbridge, B., Stray, J., and Darveau, R.P. (1991). The immunodominant outer membrane antigen of *Actinobacillus actinomycetemcomitans* is located in the serotype-specific high-molecular-mass carbohydrate moiety of lipopolysaccharide. Infect. Immun. *59*, 3451–3462.

Postle, K., and Kadner, R.J. (2003). Touch and go: tying TonB to transport. Mol. Microbiol. *49*, 869–882.

Potter, M.D., and Lo, R.Y. (1995). Cloning and characterization of a gene from *Pasteurella haemolytica* A1 involved in lipopolysaccharide biosynthesis. FEMS Microbiol. Lett. *129*, 75–81.

Potter, M.D., and Lo, R.Y. (1996). Cloning and characterization of the *galE* locus of *Pasteurella haemolytica* A1. Infect. Immun. *64*, 855–860.

Rao, V.K., Krasan, G.P., Hendrixson, D.R., Dawid, S., and St Geme 3rd, J.W. (1999). Molecular determinants of the pathogenesis of disease due to nontypeable *Haemophilus influenzae*. FEMS Microbiol. Rev. *23*, 99–129.

Ray, W.C., Forst, C., Mahairas, G.G., Sabo, P., Johnson, L., Nguyen, D., Hood, L., and Munson, R.S., Jr. (2001). The *Haemophilus ducreyi* genome project. Paper presented at: 101st General Meeting of the American Society for Microbiology (Orlando, FL, USA).

Ruffolo, C.G., and Adler, B. (1996). Cloning, sequencing, expression, and protective capacity of the oma87 gene encoding the *Pasteurella multocida* 87-kilodalton outer membrane antigen. Infect. Immun. *64*, 3161–3167.

Ruffolo, C.G., Tennent, J.M., Michalski, W.P., and Adler, B. (1997). Identification, purification, and characterization of the type 4 fimbriae of *Pasteurella multocida*. Infect. Immun. *65*, 339–343.

Satola, S.W., Schirmer, P.L., and Farley, M.M. (2003). Complete sequence of the *cap* locus of *Haemophilus influenzae* serotype b and nonencapsulated b capsule-negative variants. Infect. Immun. *71*, 3639–3644.

Sheehan, B.J., Bosse, J.T., Beddek, A.J., Rycroft, A.N., Kroll, J.S., and Langford, P.R. (2003). Identification of *Actinobacillus pleuropneumoniae* genes important for survival during infection in its natural host. Infect, Immun. *71*, 3960–3970.

Shen, K., Antalis, P., Gladitz, J., Sayeed, S., Ahmed, A., Yu, S., Hayes, J., Johnson, S., Dice, B., Dopico, R., *et al.* (2005). Identification, distribution, and expression of novel genes in 10 clinical isolates of nontypeable *Haemophilus influenzae*. Infect. Immun. *73*, 3479–3491.

Siddaramappa, S., Duncan, A.J., Brettin, T., and Inzana, T.J. (2006). Comparative analyses of two cryptic plasmids from Haemophilus somnus (*Histophilus somni*). Plasmid 55, 227–234.

Siddaramappa, S., Duncan, A.J., Challacombe, J.F., Dyer, D., Gillaspy, A., Gipson, J., Carson, M., Zaitshik, J., Rainey, D., and Inzana, T.J. (2007). Comparative Genomics of *Histophilus somni* strains 2336 and 129Pt. Paper presented at: Microbial Genomes 2007 (Wellcome Trust Conference Centre, Hinxton, Cambridgeshire, UK).

Siddaramppa, S., and Inzana, T.J. (2004). *Haemophilus somnus* virulence factors and resistance to host immunity. Anim Health Res. Rev. 5, 79–93.

Spinola, S.M., Fortney, K.R., Katz, B.P., Latimer, J.L., Mock, J.R., Vakevainen, M., and Hansen, E.J. (2003). *Haemophilus ducreyi* requires an intact *flp* gene cluster for virulence in humans. Infect. Immun. *71*, 7178–7182.

St Geme III, J.W., de la Morena, M.L., and Falkow, S. (1994). A *Haemophilus influenzae* IgA protease-like protein promotes intimate interaction with human epithelial cells. Mol. Microbiol. *14*, 217–233.

Stevenson, A., Macdonald, J., and Roberts, M. (2003). Cloning and characterization of type 4 fimbrial genes from *Actinobacillus pleuropneumoniae*. Vet. Microbiol. *92*, 121–134.

Sugai, M., Kawamoto, T., Peres, S.Y., Ueno, Y., Komatsuzawa, H., Fujiwara, T., Kurihara, H., Suginaka, H., and Oswald, E. (1998). The cell cycle-specific growth-inhibitory factor produced by *Actinobacillus actinomycetemcomitans* is a cytolethal distending toxin. Infect. Immun. *66*, 5008–5019.

Sukupolvi-Petty, S., Grass, S., and St Geme 3rd, J.W. (2006). The *Haemophilus influenzae* Type b *hcsA* and *hcsB* gene products facilitate transport of capsular polysaccharide across the outer membrane and are essential for virulence. J. Bacteriol. *188*, 3870–3877.

Tatusov, R.L., Mushegian, A.R., Bork, P., Brown, N.P., Hayes, W.S., Borodovsky, M., Rudd, K.E., and Koonin, E.V. (1996). Metabolism and evolution of *Haemophilus influenzae* deduced from a whole-genome comparison with *Escherichia coli*. Curr. Biol. *6*, 279–291.

Thumbikat, P., Briggs, R.E., Kannan, M.S., and Maheswaran, S.K. (2003). Biological effects of two genetically defined leukotoxin mutants of *Mannheimia haemolytica*. Microb. Pathog. *34*, 217–226.

Totten, P.A., Norn, D.V., and Stamm, W.E. (1995). Characterization of the hemolytic activity of *Haemophilus ducreyi*. Infect. Immun. *63*, 4409–4416.

Tremblay, Y.D., Bahrami, F., and Niven, D.F. (2006). Acquisition of haemoglobin-bound iron by *Histophilus somni*. Vet. Microbiol. *114*, 104–114.

van den Bosch, H., and Frey, J. (2003). Interference of outer membrane protein PalA with protective immunity against *Actinobacillus pleuropneumoniae* infections in vaccinated pigs. Vaccine *21*, 3601–3607.

Ward, A.C., Jaworski, M.D., Eddow, J.M., and Corbeil, L.B. (1995). A comparative study of bovine and ovine *Haemophilus somnus* isolates. Can J. Vet. Res. *59*, 173–178.

Ward, C.K., and Inzana, T.J. (1997). Identification and characterization of a DNA region involved in the export of capsular polysaccharide by *Actinobacillus pleuropneumoniae* serotype 5a. Infect. Immun. *65*, 2491–2496.

Ward, C.K., Lawrence, M.L., Veit, H.P., and Inzana, T.J. (1998). Cloning and mutagenesis of a serotype-specific DNA region involved in encapsulation and virulence of *Actinobacillus pleuropneumoniae* serotype 5a: concomitant expression of serotype 5a and 1 capsular polysaccharides in recombinant *A. pleuropneumoniae* serotype 1. Infect. Immun. *66*, 3326–3336.

Weiser, J.N., Williams, A., and Moxon, E.R. (1990). Phase-variable lipopolysaccharide structures enhance the invasive capacity of *Haemophilus influenzae*. Infect. Immun. *58*, 3455–3457.

Weller, P.F., Smith, A.L., Anderson, P., and Smith, D.H. (1977). The role of encapsulation and host age in the clearance of *Haemophilus influenzae* bacteremia. J. Infect. Dis. *135*, 34–41.

Wells, T.J., Tree, J.J., Ulett, G.C., and Schembri, M.A. (2007). Autotransporter proteins: novel targets at the bacterial cell surface. FEMS Microbiol. Lett. *274*, 163–172.

Wheeler, D.L., Barrett, T., Benson, D.A., Bryant, S.H., Canese, K., Chetvemin, V., Church, D.M., DiCuccio, M., Edgar, R., Federhen, S., *et al.* (2008). Database resources of the National Center for Biotechnology Information. Nucleic Acids Res. *36*, D13–21.

Williams, B.J., Morlin, G., Valentine, N., and Smith, A.L. (2001). Serum resistance in an invasive, nontypeable *Haemophilus influenzae* strain. Infect. Immun. *69*, 695–705.

Wu, T.H., and Gu, X.X. (1999). Outer membrane proteins as a carrier for detoxified lipooligosaccharide conjugate vaccines for nontypeable *Haemophilus influenzae*. Infect. Immun. *67*, 5508–5513.

Yu, R.H., Gray-Owen, S.D., Ogunnariwo, J., and Schryvers, A.B. (1992). Interaction of ruminant transferrins with transferrin receptors in bovine isolates of *Pasteurella haemolytica* and *Haemophilus somnus*. Infect. Immun. *60*, 2992–2994.

Zwahlen, A., Kroll, J.S., Rubin, L.G., and Moxon, E.R. (1989). The molecular basis of pathogenicity in *Haemophilus influenzae*: comparative virulence of genetically-related capsular transformants and correlation with changes at the capsulation locus *cap*. Microb. Pathog. *7*, 225–235.

Competence, DNA Uptake and Transformation in *Pasteurellaceae*

4

Heather Maughan, Sunita Sinha, Lindsay Wilson and Rosemary Redfield*

Abstract

The ability to take up DNA from the environment and recombine it into the chromosome appears to be ancestral to the *Pasteurellaceae*, although only some isolates do this efficiently under laboratory conditions. Studies of readily transformable isolates have shown that competence for DNA uptake is regulated by the cyclic AMP-dependent regulatory protein CRP and by Sxy, a competence-specific transcriptional activator. Once cells are competent, DNA uptake is promoted by recognition of an uptake signal sequence motif that is highly over-represented in the genomes of all *Pasteurellaceae*, including those that cannot be transformed. Transport of the DNA across the cell envelope uses components of the type 4 pilus machinery, homologous to those used by other naturally competent bacteria. Once in the cytoplasm this DNA may be degraded or, if sequence similarity permits, it may be recombined into the chromosome. Although such recombination can have important evolutionary consequences, DNA uptake is likely to serve primarily as a source of nucleotides for the cell.

Introduction

The family *Pasteurellaceae* has historically drawn much attention because of its many pathogenic members. However, another trait described in several of its members has also attracted the interest of biologists from fields including pathogenesis and molecular and evolutionary biology. This is natural competence, the ability of intact, living cells to actively take up DNA from their extracellular environment. If the sequences are sufficiently similar, this DNA may recombine with the host genome. When such a recombination event changes the cell's genotype, the cell is said to be transformed.

It is important to begin by clarifying the difference between natural competence and two other phenomena, artificially induced competence and transformation. Natural competence differs from artificially induced competence, where cells are chemically treated or exposed to high electrical currents to allow entry of DNA into the cell. Second, transformation is not a necessary consequence of DNA uptake, both because much of the DNA taken up by competent cells is degraded and because cells that take up unrelated or genetically identical DNAs will not be transformed.

In this chapter, we first give a brief overview of competence development, DNA uptake and transformation. We then provide detailed reviews of each of these topics and consider the evolutionary implications of competence. We then conclude with suggestions for improving transformability of problematic strains. Historically, *Haemophilus influenzae* is the best-studied *Pasteurellaceae* species with respect to competence; unless otherwise stated, gene names and numbers in this chapter refer to the sequenced *H. influenzae* Rd strain.

Like most other bacteria, *Pasteurellaceae* species develop competence only under certain conditions (i.e. competence is not constitutive).

*all authors contributed equally.

In *H. influenzae*, competence is induced under conditions of depleted carbon and energy sources (Macfadyen *et al.*, 1996). Competence induction causes an initial increase in cAMP (cofactor of the cAMP receptor protein CRP, also known as CAP). Active CRP first causes increased transcription of the gene encoding the competence activator protein Sxy. CRP and Sxy together then induce the expression of the genes in the competence regulon.

The products of many of these genes play essential roles in DNA uptake, with proteins homologous to components of the type IV pilus machinery (Tfp) thought to be responsible for binding DNA and pulling it across the outer membrane into the periplasm. DNA normally crosses the outer membrane in its double-stranded form, but one strand is then degraded during translocation across the inner membrane into the cytoplasm; the nucleotides released are rapidly reused for new DNA synthesis (Goodgal, 1982). If sequence similarity permits, the other strand may recombine with a homologous sequence in the chromosome; otherwise it too will be degraded.

Although much of this machinery appears to be shared by other bacteria, one aspect of DNA uptake in *Pasteurellaceae* is shared only with the *Neisseriaceae*: they do not indiscriminately take up external DNA, but instead show a strong bias for DNA fragments containing specific uptake signal sequence (USS) motifs. The motifs of the two families are unrelated. Because USSs are present at more than thousand sites in *Pasteurellaceae* genomes, the bias leads to preferential uptake of DNA derived from relatives.

Competence and transformation have medical relevance at several levels, because antibiotic resistance genes, virulence determinants and capsular serotype genes are spread by transformation (Kroll and Moxon, 1990; Kroll *et al.*, 1998; Maiden, 1998). Tfp genes are specifically induced upon host cell contact by *Actinobacillus pleuropneumoniae* (Boekema *et al.*, 2004) and during biofilm formation by *H. influenzae* (Bakaletz *et al.*, 2005; Jurcisek and Bakaletz, 2007). The DNA abundantly present in respiratory mucus may therefore be an important nutrient for these bacteria (Lethem *et al.*, 1990). It follows that, in order to understand how *Pasteurellaceae* pathogens exploit their specific environments, we must understand the regulation and mechanism of competence.

Distribution of competence in *Pasteurellaceae*

In the laboratory, the most sensitive way to detect natural competence is to measure transformation by genetically marked donor DNA; in the *Pasteurellaceae* this is most conveniently done with DNA carrying an antibiotic resistance gene. These assays express competence as a 'transformation frequency', the ratio of bacteria that have recombined the marked DNA to the total bacteria present. Competence can also be directly assessed by measuring the uptake of radiolabelled donor DNA, but this is less sensitive by several orders of magnitude.

Transformation has been observed in laboratory cultures of at least some isolates of three of the eight sequenced *Pasteurellaceae* species: *H. influenzae* (Alexander and Leidy, 1951), *Aggregatibacter actinomycetemcomitans* (Wang *et al.*, 2002) and *A. pleuropneumoniae* (Bosse *et al.*, 2004). *[Haemophilus] parasuis* (Bigas *et al.*, 2005) and *Haemophilus parainfluenzae* (Nickel and Goodgal, 1964) are also transformable. *Pasteurella multocida* has not been demonstrated to be competent, but its genome sequence (May *et al.*, 2001) contains all of the genes known to be necessary for competence development in *H. influenzae*. Homologues of *H. influenzae* competence genes in the sequenced *Pasteurellaceae* genomes are listed in Table 4.1, with footnotes indicating genes with defects likely to preclude function. All of the genes known to be needed for DNA uptake by *H. influenzae* are present in all of the other genomes, although one or more are obviously defective in *[Haemophilus] ducreyi*, *Histophilus somni*, *Mannheimia haemolytica* and '*Mannheimia succiniciproducens*' (Challacombe *et al.*, 2007; Gioia *et al.*, 2006; Hong *et al.*, 2004). These strains (though perhaps not all strains of these species) are thus unlikely to be able to develop full competence under any conditions.

Different strains of *Pasteurellaceae* species are known to exhibit very different transformation frequencies, as do different strains in other bacte-

Table 4.1 Orthologues of *H. influenzae* Rd KW20 genes in the CRP-S regulon

Name	*Haemophilus influenzae* Rd KW20	*'Mannheimia succinici-producens'* MBEL55E	*[Actinobacillus] succinogenes* 130z	*Pasteurella multocida* Pm70	*Histophilus somni* 129PT	*[Haemophilus] ducreyi* 35000HP	*Actinobacillus pleuro-pneumoniae* L20	*Escherichia coli* K-12 MG1665
sxy/tfoX (217 aa)	HI_0601 NP_438758	MS2301 (*tfoX*) YP_089493	Asuc_0283 YP_001343596	PM1558 (*tfoX*) NP_246497	HS_1715 (*tfoX*) YP_719920	HD1985 (*tfoX*) NP_874313	APL_1758 YP_001054447	B0959 (*tfoX/sxy*) NP_415479
comA (265 aa)	HI_0439 NP_438600	MS1974 YP_089166	Asuc_0692 (*pilM*) YP_001344000	PM1229 (*comA*) NP_246166	HS_1109 (*comA*) YP_719321	HD0427 (*comA*) NP_873003 *1	APL_0196 YP_001052909	B3395 (*hofM*) NP_417854
comB (168 aa)	HI_0438 NP_438599	MS1973 YP_089165	Asuc_0691 YP_001343999	PM1228 (*comB*) NP_246165	HS_1110 (*comB*) YP_719322	HD0429 (NP_873005) to HD0431 (NP_873007) 339758–340231 *2	APL_0197 YP_001052910 *3	B3394 (*hofN*) NP_417853
comC (173 aa)	HI_0437 NP_438598	MS1972 YP_089164	Asuc_0690 (*comC*) YP_001343998	PM1227 (*comC*) NP_246164	HS_1111 (*comC*) YP_719323	HD0432 NP_873008	APL_0198 YP_001052911	B3393 (*hofO*) NP_417852
comD (137 aa)	HI_0436 NP_438597	MS1971 YP_089163	Asuc_0689 YP_001343997	PM1226 (*comD*) NP_246163	Defective *4	HD0433 NP_873009	APL_0199 YP_00105912	B3392 (*hofP*) NP_417851
comE (445 aa)	HI_0435 NP_438596	MS1970 (*gspD*) YP_089162	Asuc_0688 (*pilQ*) YP_001343996	PM1225 (*comE*) NP_246162	HS_1112 (*comE*) YP_719324 *5	HD0434 (*comE*) NP_873010	APL_0200 (*hofQ*) YP_001052913	B3391 (*hofQ*) NP_417850
comF/ com101 (228 aa)	HI_0434 NP_438595	MS2234 (*comFC*) YP_089426	Asuc_0293 (*comF*) YP_001343606	PM1556 (*comF*) NP_246495	HS_1490 (*comF*) YP_719700	HD0209 (*comF*) NP_872810	APL_2004 (*comF*) YP_001054683	B3413 (*gntX*) NP_417872
comE1 (112 aa)	HI_1008 NP_439169	MS1748 (*comEA*) YP_088940	Asuc_1873 (*comEA*) YP_001345159	PM1665 NP_246604	HS_0126 (*comEA*) YP_718330	HD0650 NP_873186	APL_1406 (*comEA*) YP_001054095	B0442 (*comEA*) NP_414976
pilA (148 aa)	HI_0299 NP_438466	MS0364 (*hofG*) YP_087556	Asuc_0613 YP_001343922	PM0084 NP_245021	HS_0250 (*pilA*) YP_718458	HD1123 NP_873597	APL_0880 (*apfA*) YP_001053581	B0108 (*ppdD*) NP_414650
pilB (464 aa)	HI_0298 NP_438465	MS0363 (*pilB/gspE*) YP_087555 *6	Asuc_0612 YP_001343921	PM0085 (*hofB*) NP_245022	HS_1430 (*pilB*) YP_719642	HD1124 (*hofB*) NP_873598	APL_0879 (*apfB*) YP_001053580	B0107 (*hofB*) NP_414649

Table 4.1 *continued*

Name	*Haemophilus influenzae* Rd KW20	*'Mannheimia succinici-producens'* MBEL55E	*[Actinobacillus] succinogenes* 130z	*Pasteurella multocida* Pm70	*Histophilus somni* 129PT	*[Haemophilus] ducreyi* 35000HP	*Actinobacillus pleuro-pneumoniae* L20	*Escherichia coli* K-12 MG1665
pilC (406 aa)	HI_0297 NP_438464	MS0361 YP_087553 (*hofF*) and MS0362 YP_087554 *7	Asuc_0611 YP_001343920	PM0086 (*hofC*) NP_245023	HS_0457 (*pilC*) YP_718667	HD1125 (*hofC*) NP_873599	APL_0878 (*apfC*) YP_001053579	B0106 (*hofC/ pulF*) NP_414648
pilD (230 aa)	HI_0296 NP_438463	MS0360 (*pppA*) YP_087552	Asuc_0610 YP_001343919	PM0087 (*hopD*) NP_245024	HS_0458 (*pilD*) YP_718668	HD1126 NP_873600	APL_0877 (*apfD*) YP_001053578	B0105 *pulO*) NP_417446
pulG /comN (170 aa)	HI_0938 NP_439098	MS0724 (*hofG*) YP_087916	Asuc_1789 YP_001345076	PM0965 NP_245902	HS_0264 (*pulG*) YP_718472	HD0182 (*pulG*) NP_872786	APL_1888 (*pulG*) YP_001054573	B3328 (*hofG/ gspG*) NP_417787
pulJ /comO (238 aa)	HI_0939 NP_439099	MS0725 (*pulJ*) YP_087917	Asuc_1788 YP_001345075	PM0964 NP_245901	HS_0265 YP_718473	HD0183 NP_872787	APL_1887 (*pulJ*) YP_001054572 *8	B3329 *hofH/gspH*) NP_417788
HI0940/ comP (227 aa)	HI_0940 NP_439100	MS0726 YP_087918	Asuc_1787 YP_001345074	PM0963 NP_245900	HS_0266 YP_718474	HD0184 NP_872788	APL_1886 YP_001054571	B3330 (*gspI*) NP_417789
HI0941 /*comQ* (101 aa)	HI_0941 NP_439101	MS0727 YP_087919	Asuc_1786 YP_001345073	PM0962 NP_245899	HS_0267 YP_718475	HD0185 NP_872789	APL_1885 YP_001054570	B3331 (*gspJ*) NP_417790
rec2 (788 aa)	HI_0061 NP_438234	MS0931 (*comEC*) YP_088123	Asuc_1729 (*comEC/rec2*) YP_001345016	PM0862 (*rec2*) NP_245799	HS_1022 (*rec2*) YP_719234	HD1256 (*rec2*) NP_873699	APL_0766 (*rec2*) YP_001053469	B0913 (*ycaI/ comEC*) NP_415433
HI0365 (390 aa)	HI_0365 NP_438526	MS1916 YP_089108	Asuc_2030 YP_001345311	PM2007 NP_246946	HS_1198 YP_719410	HD0319 NP_872907	APL_1274 YP_001053965	B2517 (*yfgB*) NP_417012
pilF2 (179 aa)	HI_0366 NP_438527	MS1917 (*pilF*) YP_089109	Asuc_2029 YP_001345310	PM2008 NP_246947	HS_1197 (*pilF*) YP_719409	HD0318 NP_872906	APL_1273 YP_001053964	Not present
comM (509 aa)	HI_1117 NP_439274	Defective *9	Asuc_0505 YP_001343815	PM1510 (*comM*) NP_246449	HS_0331 (*comM*) YP_718539	HD1870 NC_002940 *10	APL_1747 (*comM*) YP_001054436	B3765 (*yifB*) NP_418214

dprA (373 aa)	HI_0985 NP_439148	MS0041 (*smf*) YP_087233	Asuc_0566 (*smf*) YP_001343876	PM1599 (*dprA*) NP_246538	HS_0412 (*dprA*) YP_718622	HD1888 (*smf*) NP_874228	APL_1712 (*smf*) YP_001054401	B4473 (*smf*) YP_026211
radC (234 aa)	HI_0952 NP_439113	MS1939 YP_089132 and MS1940 YP_089133 *11	Asuc_0013 YP_001343330	PM1152 (*radC*) NP_246089	HS_0144 (*radC*) YP_718349	HI0732 (*radC*) NP_873255	APL_1970 (*radC*) YP_001054649	B3638 (*radC*) NP_418095
ssb (168 aa)	HI_0250 NP_438419	MS0585 (*ssb*) YP_087777	Asuc_1880 (*ssb*) YP_001345166	PM1950 (*ssb*) NP_246889	HS_1729 (*ssb*) YP_719934 & HS_1129 (ssb) YP_719341	HD1285 (*ssb*) NP_873718 and HD0962 (ssb) NP_873454	APL_0783 (*ssb*) YP_001053486	B4059 (*ssb*) NP_418483
ligA (185 aa)	HI_1182 AAC22836	MS0939 YP_088131	Asuc_1188 YP_001344487	Not present	HS_1324 (*ligA*) YP_719536	Not present	Not present	Not present
HI0660 (119 aa)	HI_0660 NP_438820	Not Present	Not present	Not present	Not present	Not present	APL_1358 YP_001054047	Not present
HI0659 (98 aa)	HI_0659 NP_438819	Not Present	Not present	Not present	Not present	Not present	APL_1357 YP_001054046	Not present
HI1631 (190 aa)	HI_1631 NP_438791	Not Present	Not present	Not present	Not present	Not present	Not present	Not present

Only orthologues from fully sequenced *Pasteurellaceae* genomes available as a single contig (as of August 2007) and *E. coli* K-12 are presented. Row headings are the names used for the *H. influenzae* Rd KW20 genes, with their length (aa) shown in brackets. Cells whose ORFs are unlikely to be functional are shaded. Where genes are annotated with a common name, this is shown in parentheses.

*1 Loss of ~67 aa from C-terminus.

*2 Two frameshifts generate three ORFs.

*3 Protein likely to be 30 aa longer at N terminus than annotated (start 217,981 bp).

*4 Terminal 50 aa fused to adjacent *comE*

*5 End 115 aa lost.

*6 Start may be at aa 11 of annotated protein i.e. from MQSIK (based on alignments).

*7 May be an additional 62 aa at N-terminal encompassing annotated YP_087553 (based on alignments).

*8 Protein is likely to be 50 aa longer than annotated due to extension of N-terminus (based on alignments).

*9 ORF disrupted by 17,976 bp insert: H. influenzae 1–287 aa match 1,970,356–1,971,216 bp and Hin 287508 aa match 1,989,192–19,89,857 bp.

*10 Annotated as *comM* pseudogene due to multiple frameshifts.

*11 Annotated as two ORFs because of a disruption by a frameshift.

rial families. Some strains are completely non-transformable in the laboratory. Transformability varied over three orders of magnitude in different isolates of *Agg. actinomycetemcomitans*, with transformation frequencies ranging from 5×10^{-3} to 4×10^{-6} (Fujise *et al.*, 2004). In addition, transformants were never observed in 49 of the 67 strains tested. In *H. parainfluenzae*, transformation frequencies of 19 strains ranged from 1×10^{-2} to 1×10^{-6} with transformants never observed in an additional 54 strains examined (Gromkova *et al.*, 1998). Strains of *H. influenzae* also transform with variable efficiency. Rowji *et al.* (1989) attempted to transform 31 clinical isolates of *H. influenzae* serotype b and found that only 18 of them were transformable, with transformation frequencies ranging from 8×10^{-4} to 1×10^{-6}. The culture conditions used in these studies differed: *Agg. actinomycetemcomitans* was assayed on Trypticase Soy agar while the two *Haemophilus* species were assayed in static supplemented Brain Heart Infusion broth (sBHI) cultures. Nevertheless, substantial variation in transformation clearly exists within the species of *Pasteurellaceae* that have been examined. Thus the inability to transform a specific strain should not be taken as evidence that the species is not naturally competent.

The conservation of competence genes across the *Pasteurellaceae* strongly suggests that the ancestor of all *Pasteurellaceae* was competent and transformable. Furthermore, the similarity of competence genes and their regulation between *Pasteurellaceae* and other gamma-proteobacteria in the *Vibrionaceae* and *Enterobacteriaceae* (Cameron and Redfield, 2006) may be evidence of an even deeper ancestry. If competence is indeed ancestral, strains that are not able to take up DNA must have lost competence. However, finding that a strain fails to transform in laboratory culture does not necessarily mean that it has lost competence. The standard transformation assays will be a poor indicator of natural competence if internalized DNA is rapidly degraded. Additionally, competence will not be detected if test conditions do not include the environmental cues that trigger its development.

The regulation of competence

Most naturally competent bacteria express DNA uptake genes only in response to certain environmental conditions (Solomon and Grossman, 1996). This section describes the competence-inducing conditions identified in the *Pasteurellaceae* and the response to these signals that leads to competence development.

Competence development in the laboratory

Our understanding of competence regulation in the *Pasteurellaceae* comes predominantly from studies of *H. influenzae* growing in the laboratory; the relationship between growth phase and competence levels is depicted in Fig. 4.1. Cells growing exponentially in a nutrient-rich medium such as sBHI do not take up DNA: competence genes are not expressed, DNA uptake cannot be detected, and transformation frequencies with genetically marked chromosomal DNA are below the usual detection threshold of 10^{-8} (Fig. 4.1a) (Redfield, 1991; Redfield *et al.*, 2005). As such cultures approach stationary phase and nutrients become depleted, transformation frequency increases to approximately 10^{-4} (Fig. 4.1b) (Redfield, 1991).

This increase is enhanced if cells in exponential growth are transferred abruptly from rich medium to the starvation medium MIV ('M-four'), commonly used to induce maximal competence. The composition of medium MIV was established empirically; it contains amino acids but lacks purines, pyrimidines, sugars and required cofactors and thus allows protein synthesis but not DNA replication or cell division (Herriott *et al.*, 1970; Poje and Redfield, 2003). Most cells in MIV-induced cultures become competent within 60–100 minutes, giving transformation frequencies of 10^{-3} to 10^{-2} (Fig. 4.1c). Transfer to MIV has been used to induce competence in other *Pasteurellaceae*, including *Agg. actinomycetemcomitans* (Wang *et al.*, 2002) and *H. parainfluenzae* (Gromkova *et al.*, 1998). However, conditions that limit growth (either entering stationary phase or being shifted to a minimal growth medium) do not reproducibly induce natural competence in *A. pleuropneumoniae* (Bosse *et al.*, 2004).

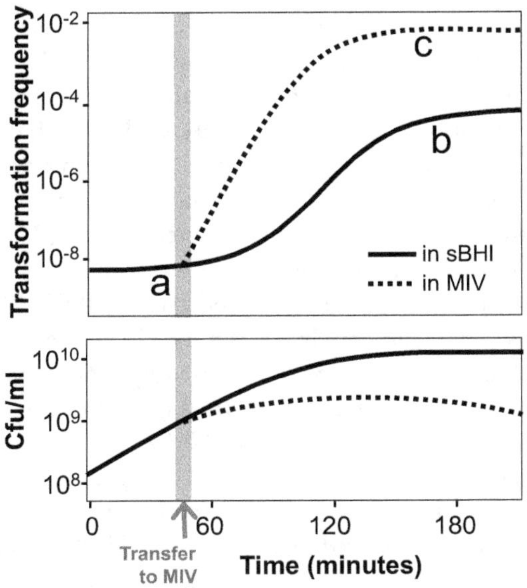

Figure 4.1 Relation between growth phase and transformation frequency in laboratory cultures of *H. influenzae*. Transformation frequencies are those typically measured with chromosomal DNA carrying a point mutation conferring antibiotic resistance. Grey bars: time when cells growing exponentially in sBHI are transferred to MIV. Solid line: cells in sBHI; dotted line: cells in MIV. (a) exponential growth in sBHI; (b) stationary phase in sBHI; (c) cells in MIV.

Mechanisms of competence induction

Sugar depletion and cAMP

All of the steps in competence induction are summarized in Fig. 4.2. The first step is a rise in intracellular cAMP levels; this rise is controlled by the bacterial phosphoenol:sugar phosphotransferase system (PTS) and signals a shortage of carbon and energy (Barabote and Saier, 2005; Macfadyen *et al.*, 1996). The PTS is an enzyme complex that transports and phosphorylates preferred sugars (PTS sugars) and also detects changes in their availability. When PTS sugars become unavailable, PTS proteins stimulate

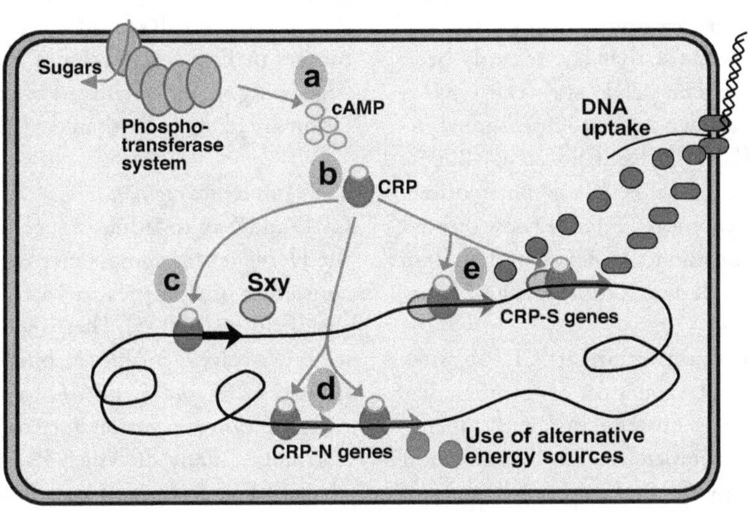

Figure 4.2 Model of the CRP-dependent induction of competence in *H. influenzae*. (a) Depletion of phosphoenol:sugar phosphotransferase system (PTS) sugars stimulates cAMP production. Rising cAMP levels then activate CRP (b), enabling it to activate transcription first at the *sxy* promoter (c) and then at CRP-N sites (d) and, with the assistance of Sxy, at CRP-S sites in the competence regulon (e).

cAMP production by adenylate cyclase (Dorocicz *et al.*, 1993; Levy *et al.*, 1990) (Fig. 4.2a). Rising cAMP levels then activate the cAMP receptor protein CRP, a transcriptional activator (Fig. 4.2b).

Amongst the *Pasteurellaceae*, a cAMP-dependent competence system has also been demonstrated in *Agg. actinomycetemcomitans* (Wang *et al.*, 2002) and [*H.*] *parasuis* (Bigas *et al.*, 2005). The components for regulation by cAMP are also present in *M. haemolytica* even though natural competence has not so far been reported in this organism (Gioia *et al.*, 2006). On the other hand, *A. pleuropneumoniae* is naturally transformable at low levels but cAMP does not induce the high levels of competence observed in *H. influenzae* (Bosse *et al.*, 2004).

The competence activators CRP and Sxy

Upon activation by cAMP, CRP binds with high affinity to a 22-bp sequence (CRP site) in promoter regions, and activates transcription (Fig. 4.2c–e) by making direct contact with RNA polymerase (Savery *et al.*, 1996). CRP is best known for its role as a global regulator that activates transcription of sugar utilization genes (Fig. 4.2d). In *H. influenzae*, CRP also activates transcription of competence genes (the competence regulon, see below) (Fig. 4.2e). The products of these genes allow the cell to bind, take up and process extracellular DNA. The promoters of these genes are united by possession of a conserved DNA motif that has recently been recognized as a variant CRP site (the CRP-S motif, previously known as the competence regulatory element (CRE)) (Redfield *et al.*, 2005). The core half-site of the CRP-S motif differs from that of the canonical CRP site (now termed CRP-N) at one position (TGCGA rather than TGTGA) (Cameron and Redfield, 2006; Macfadyen, 2000).

Transcriptional activation at CRP-S sites requires the second regulator of competence, Sxy. The *sxy* gene is up-regulated early during the induction of competence, its transcription directly activated by CRP (Fig. 4.2c) (Cameron *et al.*, 2008; Redfield *et al.*, 2005; Zulty and Barcak, 1995). Additional regulation is provided at the translational level, by the formation of a stem–loop structure in *sxy* mRNA (Cameron *et al.*, 2008). *H. influenzae* mutants lacking *sxy* fail to induce transcription of competence genes and therefore are not transformable (Williams *et al.*, 1994; Zulty and Barcak, 1995). Consistent with Sxy's role as an activator, overexpression of *sxy* from a multicopy plasmid results in constitutive competence in *H. influenzae* and *Agg. actinomycetemcomitans* (Williams *et al.*, 1994; Zulty and Barcak, 1995; Bhattacharjee *et al.*, 2007). A similar though less extreme hypercompetence is caused by mutations that disrupt basepairing in the regulatory stem of *sxy* mRNA (Cameron *et al.*, 2007).

Sxy's precise role in the activation of CRP-S promoters is still undefined. Sxy is unlikely to act as a DNA-binding protein at CRP-S sites because it lacks any DNA-binding motifs and because CRP-S promoters show no sequence consensus other than the CRP-S site. CRP has been shown to specifically bind CRP-S sites in the absence of Sxy (Redfield *et al.*, 2005). Sxy is instead thought to recruit CRP to CRP-S promoters (Fig. 4.2e), modifying its ability to activate transcription of competence genes at promoters which are otherwise unfavourable sites for CRP activation (Cameron *et al.*, 2008; Redfield *et al.*, 2005).

The *crp* and *sxy* genes are present in all species of the *Pasteurellaceae*, *Enterobacteriaceae* and *Vibrionaceae*, as are the CRP-S sites in the promoters of competence-regulon homologues (Cameron and Redfield, 2006). CRP-S promoters in *E. coli* also require Sxy for activation, suggesting that mechanisms of competence regulation are likely to be shared by these groups.

The competence regulon

CRP and Sxy together induce the expression of the *H. influenzae* competence regulon (Fig. 4.2d), consisting of 25 genes in 13 transcription units (Redfield *et al.*, 2005). These include genes known to be involved in DNA binding and uptake (*comA, C, E and F, pilA-D, comE1*) and DNA translocation and processing (*rec-2, dprA, comM, ssb, radC*), many of which had been previously identified in screens of transformation-deficient mutants (Gwinn *et al.*, 1998; Karudapuram and Barcak, 1997; Karudapuram *et al.*, 1995; Larson and Goodgal, 1992; McCarthy, 1989; Tomb *et al.*, 1991). More detail about these genes and the products they encode is provided in the following sections.

A small number of genes previously reported to play a role in DNA uptake or recombination are not competence-induced (Redfield *et al.*, 2005). These include *rec-1* (the *H. influenzae recA* homologue), which is required for recombination and thus for transformation (Notani *et al.*, 1972), and genes likely to have general effects on transcription, translation or protein processing [*topA* (Chandler and Smith, 1996), *atpA* (Gwinn *et al.*, 1997) and *por* (Tomb, 1992)].

Other signals influencing competence induction

In addition to lack of PTS sugars, full competence induction in *H. influenzae* requires depletion of nucleotides pools. Nucleotides and precursors are lacking in MIV, and conversely the addition of purine nucleotides to the growth medium or to MIV prevents induction of competence genes (MacFadyen *et al.*, 2001). This repression is most likely to be mediated by effects of nucleotide pools on Sxy translation or by the PurR repressor, which prevents transcription of purine biosynthesis genes when purines are abundant in the cytoplasm.

H. influenzae strains with certain amino acid substitution mutations in the peptidoglycan biosynthesis gene *murE* also display greatly elevated competence at all stages of growth (Ma and Redfield, 2000). This phenotype is not due to altered permeability of the cell envelope but to induction of the competence regulon, as expression of CRP-S genes is constitutively induced. The mechanism of this induction is not understood, which suggests that an additional component of competence regulation remains to be characterized.

Mechanisms of DNA uptake by competent *Pasteurellaceae*

Those studying the mechanisms of DNA uptake in the *Pasteurellaceae* can draw on experimental work from a wide range of competent bacterial species thought to employ similar mechanisms and genes. In particular, much of our understanding of uptake machinery in the *Pasteurellaceae* comes from studies of other Gram-negative bacteria, especially *N. gonorrhoeae* and *N. meningitidis*. However, some caution is needed in cross-species comparisons for two reasons. First, the nomenclature across these species is not consistent and can sometimes be misleading, and second, there are important differences in presence or absence of genes.

Because all *Pasteurellaceae* contain homologues of the *H. influenzae* competence genes, they are predicted to use a common mechanism to take up DNA. Below we first present a description of the DNA sequence motifs that promote DNA uptake (Fig. 4.3). We then discuss the details of the two stages of the uptake process (Fig. 4.4): (1) DNA uptake across the outer membrane into the periplasm and (2) translocation of periplasmic DNA across the inner membrane into the cytoplasm. This distinction is appropriate because the two steps are known to be able to operate independently of each other.

Uptake signal sequences

As mentioned in the introduction, the DNA uptake machinery of competent *Pasteurellaceae* does not bind DNA indiscriminately. Instead, uptake depends on sequence-specific recognition between external DNA and the uptake proteins at the cell surface (Danner *et al.*, 1980). The motif recognized by the uptake machinery is termed the uptake signal sequence (USS). *Pasteurellaceae* USS motifs consist of a 7–9 bp 'core' sequence and two adjacent AT-rich regions situated 1–2 bp downstream of the core, separated from one another by 4–6 bp. Typical motifs are shown above (*H. influenzae*) and below (*A. pleuropneumoniae*) the phylogenetic tree in Fig. 4.3 (Redfield *et al.*, 2006).

Sequences matching the USS motifs are abundant in the genomes of all *Pasteurellaceae* species that have been examined, including those that are not transformable. USSs are not mobile elements, each sequence arising independently by mutation. One section of this chapter considers an evolutionary explanation for their distribution. The USS motifs of different species are very similar, but fall into two sub-classes that correspond to the two *Pasteureurellaceae* subclades identified by phylogenetic analysis (Redfield *et al.*, 2006). These two sub-classes differ at several core positions and in the length of the AT-rich flanking sequences. The 29-bp motif, abundant

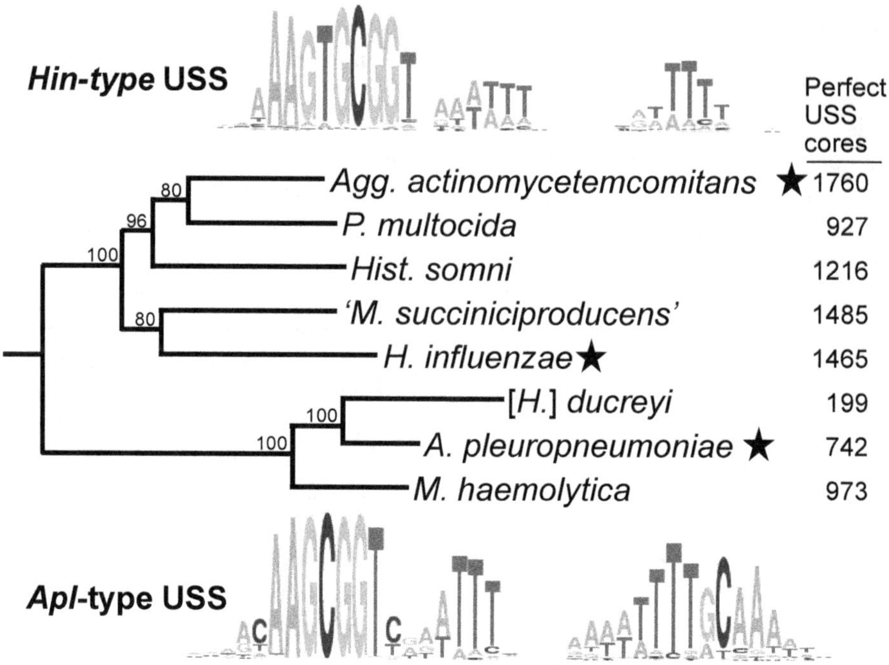

Figure 4.3 *Hin*-type and *Apl*-type USSs in *Pasteurellaceae* genomes. USS motifs (*H. influenzae* above and *A. pleuropneumoniae* below) are shown with the *Pasteurellaceae* phylogeny from Redfield *et al*. (2006). Stars indicate species in which laboratory competence has been demonstrated. Numbers on the right give the number of genomic sequences perfectly matching the 9-bp USS cores for that subclade.

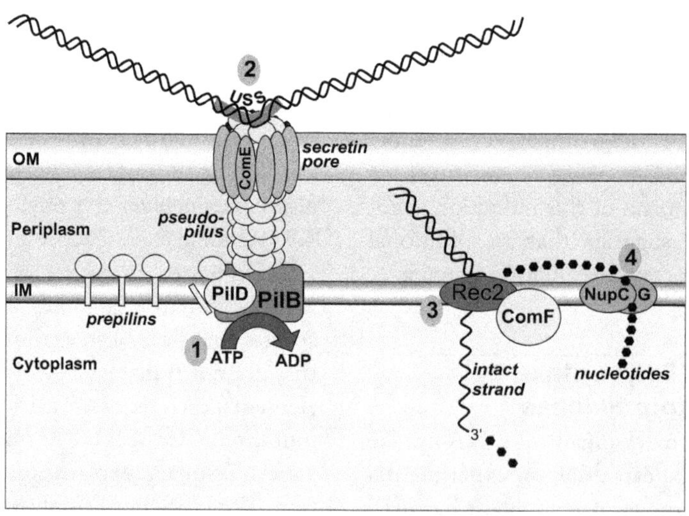

Figure 4.4 Model of DNA uptake by a *H. influenzae* competent cell. Step 1. Cells assemble the uptake machinery; PilD processes PilA prepilin subunits for ATP-dependent assembly into the pseudopilus by PilB. The pseudopilus then traverses the outer membrane through the secretin pore, composed of ComE subunits. Step 2. USS-containing DNA binds the uptake machinery at the cell surface. The pseudopilus is then retracted, bringing DNA into the periplasm. Step 3. A single strand of DNA is translocated across the inner membrane by Rec-2 and ComF. Step 4. The other strand is simultaneously degraded to its component nucleotides, which may enter the cytoplasm through the NupC and NupG pathways.

in *H. influenzae* and its relatives, is termed the Hin-type USS, while the 35-bp motif, abundant in the clade containing *A. pleuropneumoniae*, is known as the Apl-type USS (Fig. 4.3). The number of USS genomic copies with 9-bp core sequences that exactly match their USS consensus varies from 199 in [*H.*] *ducreyi* to 1760 in *Agg. actinomycetemcomitans* (see Fig. 4.3 for numbers). Experiments in *H. influenzae* and *A. pleuropneumoniae* have shown that these species preferentially take up DNA containing their own USS subtype (Redfield *et al.*, 2006).

Transport of DNA into the periplasm

The USS is predicted to bind to sites on the cell surface where specific DNA uptake complexes initiate the uptake process (step 2 in Fig. 4.4). Competent *H. influenzae* cells may develop only 4–8 such DNA uptake complexes, each able to act only once or a small number of times (Barouki and Smith, 1986; Deich and Smith, 1980; Stuy and Stern, 1964). However 50–100 fragments can be taken up in DNA uptake assays using very short USS-containing fragments (Wilson and Redfield, unpublished).

Several older papers on the subject of DNA uptake in *H. influenzae* describe specialized membrane extensions or surface vesicles/blebs known as 'transformasomes'. These were thought to be points of uptake of DNA, within which DNA is protected both from external DNase and from cytoplasmic restriction enzymes (Kahn *et al.*, 1979; Kahn *et al.*, 1983; Kahn *et al.*, 1982; Kahn and Smith, 1984). However, such membrane extensions have not been reported in other studies of competent cells. Because similar vesicles are both common in non-competent cells and a frequent fixation artefact in electron microscopy, the postulated transformasomal space is likely to equate to the Gram-negative periplasm.

Under normal conditions *H. influenzae* cells take up only double-stranded DNAs (dsDNAs); uptake of single-stranded DNA occurs only under low-pH conditions unlikely to be found in nature (Postel and Goodgal, 1966). Transporting dsDNA molecules across the outer membrane presents physical challenges to the cell. Not only are dsDNA molecules both hydrophilic and quite stiff (persistence length ~50 nm), but *H. influenzae* takes up circular molecules as efficiently as linear ones (Kahn *et al.*, 1982), indicating that a free DNA end is not required for uptake. DNA is thought to cross the outer membrane through a channel formed by the secretin protein encoded by *comE* (see below). Passage across the outer membrane may require the DNA to be bent back on itself, perhaps at the intrinsic kink of 45 degrees predicted to form across the AT-rich flanking sequences of the USS.

A role for the type IV pilus machinery Tfp and its associated proteins in DNA uptake has been established in most naturally competent bacteria (Dubnau, 1999). Pili are protein filaments that extend from the cell surface – Tfp are distinguished by their ability to exert strong pulling forces (Burrows, 2005). In different bacterial species, attachment and retraction of Tfp has been shown to enable the cell to adhere to surfaces, to exhibit some forms of motility, and/or to take up DNA. However, not all bacteria with Tfp exhibit all of these properties. Furthermore, many competent bacteria do not produce pili detectable by electron microscopy, and evidence for Tfp involvement in their competence comes primarily from mutational analysis.

Our knowledge of Tfp assembly and function comes primarily from studies in *Neisseria* and *Pseudomonas* (Burrows, 2005). The principal pilus subunit is pilin, which is synthesized as prepilin, transported to the periplasmic face of the inner membrane and processed by the prepilin peptidase (Nunn and Lory, 1993; Tonjum and Koomey, 1997). PilB then incorporates the subunits at the base of the growing pilus using energy from ATP (step 1 in Fig. 4.4). The pilus forms 6-nm-diameter helical fibres that cross the outer membrane through a gated pore composed of 12 secretin subunits (Collins *et al.*, 2003; Collins *et al.*, 2005; Collins *et al.*, 2004). The diameter of the secretin pore is predicted to be just wide enough (6.5 nm) to accommodate the elongating pilus (Collins *et al.*, 2005). Pilus retraction is driven by the PilT ATPase, which catalyses the disassembly of pilin subunits from the base at a rate of 1500 pilin residues per second (Kaiser, 2000).

The force generated by ATP-driven retraction of Tfp is thought to be responsible for movement of DNA across the outer membrane, but little is known about how this might actually happen. Isolated *Pseudomonas aeruginosa* pili and *N. meningitidis* pili bind DNA only weakly, although the N-terminal of secretin binds DNA with high affinity (Assalkhou *et al.*, 2007; van Schaik *et al.*, 2005). However, because the 6 nm pilus is thought to fit so snugly into the secretin pore, it is unclear how a pilus that has bound DNA can be retracted through the pore. In principle this could be circumvented by having DNA bound only to the tip of the pilus. This would allow pilus retraction to pull DNA to the cell surface, but the length of DNA this would bring into the periplasm would be very short.

Homologues of the genes required by other bacteria for production and activity of Tfp are widespread in the *Pasteurellaceae* (Table 4.1). In *H. influenzae* the principal pilin subunit is encoded by the prepilin *pilA* gene, the peptidase by *pilD*, and secretin by *comE*. Several additional genes within the *Pasteurellaceae* (*comNOPQ, HI0660, HI0659*) (Table 4.1) encode other pilin-like proteins identified by the presence of prepilin peptidase-dependent signal sequences, which may be incorporated as major or minor ('accessory') constituents of the Tfp to modify its function or length (Molnar, 2001; VanWagoner *et al.*, 2004). The *Hist. somni comE* homologue encodes a protein that lacks the terminal 115 amino acids seen in the *H. influenzae* homologue, and the [*H.*] *ducreyi comA* and *comB* genes have deletions and frameshift mutations predicted to render them non-functional. Surprisingly, although very rapid internalization of DNA is observed in *H. influenzae* (approximately 500–1000 nucleotides/s) (Kahn and Smith, 1984), no *pilT* homologues have been identified in any sequenced *Pasteurellaceae*. Mechanisms of pilus retraction in this organism remain unclear. It is possible that the *pilB* ATPase is responsible for both assembly and disassembly of Tfp; alternatively, an unidentified protein may drive disassembly.

Pili have been observed in *A. pleuropneumoniae* and in *P. multocida* serogroups A, B & D (Boekema *et al.*, 2004; Doughty *et al.*, 2000; Zhang *et al.*, 2000). Because competent *H. influenzae* Rd cells do not have visible pili under electron microscopy, Tfp involved in DNA uptake are predicted to be very short fibres termed 'pseudopili' which may not extend beyond the outer membrane. The involvement of Tfp in competence has been confirmed genetically in *Agg. actinomycetemcomitans* and *H. influenzae*, where Tfp mutants show a 100, 000 fold drop in transformation frequency (Dougherty and Smith, 1999; Wang *et al.*, 2002).

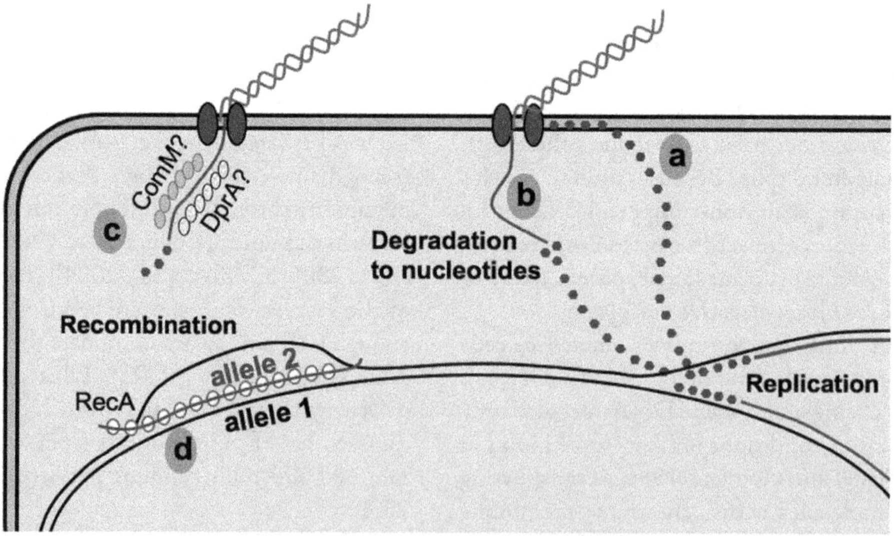

Figure 4.5 Fate of DNA inside the cytoplasm. (a) The incoming DNA strand whose 5′ end is leading is degraded to its component nucleotides, which can then be used in DNA replication. (b and c) The strand whose 3′ end is leading can be degraded into its component nucleotides (b) or can be protected from degradation by ComM and/or DprA (c) and undergo RecA-mediated recombination (d).

Translocation of DNA from the periplasm into the cytoplasm

Once inside the *H. influenzae* periplasm, DNA is translocated across the inner membrane into the cytoplasm (step 3 in Fig. 4.4). For this step, studies of DNA uptake by Gram-positive species such as *Bacillus subtilis* are informative, because DNA uptake components at their membrane appear to have homologues at the inner membrane of *Pasteurellaceae* species.

Translocation of DNA across the inner membrane requires a free end. Only the strand whose 3′ end is leading enters the cytoplasm intact; the other is degraded to its component nucleotides (Barany *et al.*, 1983) (Fig. 4.5a). *H. influenzae* cells with knockout mutations in *rec-2* or *comF* take up normal amounts of DNA, but this DNA remains trapped in the periplasm (Barouki and Smith, 1985; Larson and Goodgal, 1992; Zulty and Barcak, 1995). The *rec-2* homologue in *B. subtilis*, *comEC*, is predicted to form a water-filled channel through which the single intact strand of DNA enters the cytoplasm (Draskovic and Dubnau, 2005; Dubnau, 1999). If the same role is maintained by Rec-2 in the *Pasteurellaceae*, then the nucleotides of the degraded strand may accumulate in the periplasm prior to dephosphorylation and transport into the cytoplasm via the *nupC* and *nupG* pathways, as shown in step 4 of Fig. 4.4.

Because *rec-2* and *comF* mutants take up normal amounts of DNA, the force required to translocate DNA across the inner membrane must be independent of that used to bring DNA across the outer membrane. In *N. gonorrhoeae*, a transmembrane proton motive force is needed for translocation (Maier *et al.*, 2004). This may also be the case in *H. influenzae*, as transformation frequency has been reported to increase with proton motive force (Bremer *et al.*, 1984).

Fate of internalized DNA

The long term consequences of DNA uptake critically depend on whether the DNA that enters the cytoplasm is degraded (Fig. 4.5b) or integrated into the chromosome (Fig. 4.5d). Degradation will be the fate of all DNA of insufficiently high similarity for homologous recombination. Even when incoming DNA is perfectly homologous to the *H. influenzae* Rd chromosome, less than half of it escapes degradation and recombines (Barany *et al.*, 1983). Nevertheless, recombination is a very common consequence of DNA uptake in laboratory cultures, and natural populations show evidence of substantial recombination in their recent history. Below we describe the events leading to these two outcomes.

Degradation

Strains lacking DprA or ComM in *H. influenzae* take up DNA normally but do not transform. These proteins may prevent degradation and/or facilitate integration of donor DNA (Fig. 4.5c). Knockout mutations in *comM* and *dprA* reduce *H. influenzae* transformation frequencies by 2 and 4 orders of magnitude, respectively (Gwinn *et al.*, 1998; Karudapuram *et al.*, 1995). Mutations in *dprA* have also been shown to reduce transformation frequencies in *Campylobacter jejuni*, *Helicobacter pylori*, and *Streptococcus pneumoniae* (Ando *et al.*, 1999; Berge *et al.*, 2003; Takata *et al.*, 2005); the *S. pneumoniae* mutants rapidly degrade incoming DNA (Berge *et al.*, 2003). Excessive DNA degradation may thus explain the non-transformability of the sequenced strains of '*M. succiniciproducens*' and [*H.*] *ducreyi*, as their *comM* genes are disrupted by a 17-kb insertion and a frameshift, respectively.

Homologues of ComM and DprA are found in almost all bacteria, suggesting that these proteins have a function other than transformational recombination. One such function could be recombination repair. Recent work in *B. subtilis* shows that DprA and RecA bind to single stranded DNA and facilitate homologous recombination (Mortier-Barriere *et al.*, 2007). However, *E. coli* cells that are mutant for *dprA* have normal conjugation, recombination and DNA repair (Smeets *et al.*, 2006). On the other hand, the *H. influenzae comM* mutant showed an 8-fold recombination defect in a phage-based recombination assay, although this provided only a partial explanation of its 300-fold transformation defect (Gwinn *et al.*, 1998).

Recombination

Like mutants in *comM* or *dprA*, *H. influenzae* mutants lacking the *recA* homologue *rec-1* take up DNA but do not recombine it (Notani *et al.*, 1972). Unlike DprA and ComM, RecA has known primary functions outside of transformation; it both regulates the cell's response to DNA damage and mediates recombinational repair (reviewed in Cox, 1991). RecA-mediated homologous recombination is reduced if the sequences are more than 5–10% divergent (Shen and Huang, 1986). In *H. influenzae* donor fragments smaller than 3.5 kb do not recombine with high efficiency, probably because most of them are degraded before recombination can occur (Pifer and Smith, 1985). For DNA fragments between 3.5 and 11 kb, transformation frequency increases linearly with increasing size of donor fragment (Pifer and Smith, 1985).

In *H. influenzae* RecA is the only protein known to be specifically needed for homologous recombination of incoming DNA. It is not a member of the competence regulon, so the basal levels of RecA maintained for DNA repair must be sufficient to account for the observed recombination of incoming DNA (Redfield *et al.*, 2005). Although the RecBCD protein complex is required in other recombination systems (e.g. conjugation), mutations inactivating it in *H. influenzae* do not significantly reduce transformation frequencies, and the corresponding genes are not competence-induced (Redfield *et al.*, 2005; Wilcox and Smith, 1976a; Wilcox and Smith, 1976b).

Recombination in nature

As described above, recombination of incoming DNA with the cell's chromosome is required for transformation. The consequence at the population level is the redistribution of alleles between individuals. This in turn can affect the course of evolution by creating new combinations of alleles and by changing the amount of genetic variation in the population. Similar effects arise from the other common processes that mediate genetic exchange, conjugation and transduction and studies of the distribution of genetic variation in bacterial populations (e.g. Multi-Locus Sequencing Typing (MLST)) can give information about the combined contribution of all three of these processes.

In MLST studies, several loci (typically 6–7) are sequenced in many strains of a given bacterial species (Feil and Enright, 2004; Maiden *et al.*, 1998). The sequence data from these projects are used to examine the nucleotide variation present at these loci. Finding that alleles occur in different combinations in different strains is evidence for recombination, although sophisticated statistical tests are performed to rule out alternative causes such as convergent mutation. MLST studies in *H. influenzae* have sequenced seven housekeeping loci from more than 700 strains (Feil *et al.*, 2001; Meats *et al.*, 2003). Patterns of nucleotide variation at these loci show that recombination between *H. influenzae* strains does indeed occur in nature (Feil *et al.*, 2001; Perez-Losada *et al.*, 2006). However, the analysis is sufficiently complex that we do not yet have a good estimate of its frequency. MLST studies of other *Pasteurellaceae* are mentioned in Chapter 2.

Our understanding of the contribution of transformation to between-strain recombination can be refined by analyzing the locations of homologous USSs in *Pasteurellaceae* genomes. Because DNA uptake occasionally leads to recombination, the USSs preferentially recognized and taken up by the uptake machinery will have greater opportunity to be recombined into the genome than other sequences. Over evolutionary time, this should result in the accumulation in the genome of sequences corresponding to the motif. Bakkali *et al.* (2004) found a very high frequency of invariant USS motif sites across several *Pasteurellaceae* genomes, indicating that transformational recombination occurs frequently enough in nature to maintain USS not only within but between species.

Evolutionary consequences of competence

This chapter has discussed our knowledge of the mechanisms of DNA uptake and transformational recombination. The immediate consequences of this DNA uptake are not in question. First, the cell obtains the nucleotides in the DNA it has taken up, independent of whether or not this DNA undergoes recombination. Second, incoming DNA does recombine with the chromosome. However the long-term evolutionary significance of these consequences is less clear.

Although the recombination made possible by DNA uptake can produce the beneficial new combinations of alleles and/or genes that we see when bacterial genomes are sequenced, these are not necessarily the reason that competence evolved. Rather, three hypotheses (not mutually exclusive) have been presented to explain the benefit responsible for the origin and maintenance of competence and transformation. First, DNA may be taken up to serve as a source of genetic novelty. Second, it may be used as a template for DNA repair. Third, DNA may provide the cell with nutrients. The evidence for and against each of these hypotheses will be discussed in turn.

DNA as a source of genetic novelty

It has usually been assumed that competence and transformation evolved for genetic exchange (see for example Mortier-Barrière *et al.* (2007)). Although horizontal transfer of genetic material is often assumed to be generally beneficial for the recipient cell, whether by transformation, conjugation or transduction, most transfer events that are not functionally neutral are likely to be deleterious. Because the examples of horizontal transfer seen in genome sequences have passed through the filter of natural selection, the observation that beneficial changes predominate tells us nothing about the relative frequency of beneficial and harmful changes. The rare integration of non-homologous DNA is also much more likely to disrupt existing functions than to add a beneficial new function. In the same way, homologous recombination will more often break up beneficial combinations of alleles than create beneficial new ones.

The abundance of USS in *Pasteurellaceae* genomes is often taken as evidence that DNA uptake has evolved for genetic exchange, because it is assumed that USS could only arise by selection for uptake of conspecific DNA. The distribution of USS types in the *Pasteurellaceae* argues against USS having this role. Because species within each of the two subclades shown in Fig. 4.3 have indistinguishable USS motifs (Redfield *et al.*, 2006), USS are unable to function as species-specific identity tags. Furthermore, the combination of bias in the uptake machinery and occasional homologous recombination explains the abundance of USS in *Pasteurellaceae* genomes. The uptake bias results in USS-containing DNA being overrepresented in the intracellular pool of incoming DNA, which will inevitably lead to USS accumulation because more USS-containing fragments will recombine with the chromosome, replacing regions that do not contain USS. Over many millions of years of biased uptake and recombination, USS-containing sequences are thus expected to accumulate in genomes.

DNA as a template for repair

Because DNA damage is both more common and more immediately threatening to survival than lack of genetic variation, a role for DNA uptake in DNA repair is more plausible than a role in the generation of variation. Double stranded breaks and other DNA lesions lacking an intact template strand are lethal unless repaired with an intact homologous strand of DNA. It is thus reasonable to hypothesize that because DNA taken up by cells could be useful as a template for recombinational repair, competence originated for this purpose. However, initial reports of competent bacteria being able to survive DNA damage better than cells that were not competent (Hoelzer and Michod, 1991; Michod *et al.*, 1988; Wojciechowski *et al.*, 1989) have been experimentally refuted (Mongold, 1992; Redfield, 1993). Furthermore, DNA damage plays no role in the regulation of competence in either *H. influenzae* or *B. subtilis* (Redfield, 1993), suggesting that the cell's need for DNA repair has not influenced the evolution of competence regulation.

DNA as a nutrient

As DNA is abundant in natural environments and bases, nucleosides and nucleotides are expensive to synthesize, competence may have evolved to provide cells with nucleotides, thereby sparing energy resources for other needs (Redfield, 2001). The regulation of DNA uptake in *Pasteurellaceae*, described previously, is consistent with this, as expression of competence genes is induced during starvation stress and requires CRP, a global regulator that optimizes the carbon-energy balance of the cell.

Competence may be part of a broader response to nucleotide depletion. The induction of a competence regulon during this time can serve two purposes: first DNA uptake machinery is expressed so that the cell can obtain nucleotides and second, other genes in the competence regulon can mitigate the harmful consequences of stalled replication forks (Cox *et al.*, 2000). For example, at stalled replication forks, Ssb can help reinitiate replication while RadC assists in recombinational recovery of replication (Cadman and McGlynn, 2004; Saveson and Lovett, 1999). Similarly, DprA and ComM may protect all single stranded DNA in the cell (i.e. incoming DNA as well as DNA at stalled replication forks). As mentioned before, the genes encoding all of these are induced in competent cells.

Practical implications: improving transformation in the laboratory

Natural transformation is a very useful procedure for genetic manipulation, and many *Pasteurellaceae* researchers have been frustrated by their inability to transform strains of interest. However there are good reasons to be optimistic.

Transformability is not a recently acquired property of a few select strains but is an ancestral property of the family. All of the genes known to be needed for DNA uptake and recombination in *H. influenzae* are present in the genomes of all sequenced *Pasteurellaceae*. Some of these genomes have one or more genes that are obviously defective, but other non-transformable strains appear to have complete intact sets of competence genes (see Table 4.1). Thus, simple differences in gene regulation or in the activity of a single gene are likely to be responsible for the non-transformability of at least some strains.

Because all species with transformable isolates also have non-transformable isolates, if one strain of a species cannot be transformed, other isolates should be tested. The most useful DNA for such tests is likely to be a large PCR fragment of chromosomal DNA carrying a point mutation conferring antibiotic resistance (streptomycin resistant mutants are easy to isolate). If this is not suitable a cloned construct containing an antibiotic-resistance cassette flanked on each side by at least 1 kb of chromosomal DNA can be used.

However, if a particular strain is of interest, there are several approaches that may increase the probability of transformation. One problem may be that competence genes are not induced under laboratory conditions. The regulation of *H. influenzae*'s competence genes is conserved in the *Enterobacteraceae* and *Vibrionaceae*, so treatments that stimulate *H. influenzae* competence are likely to act similarly in other *Pasteurellaceae*. One easy intervention is to add cAMP (1 mM) to growing cells; in *H. influenzae*, this causes competence to peak within about 45 min. If plasmids can be introduced by electroporation, a plasmid carrying a *sxy* gene from the same or another species, under its own promoter, is likely to cause near-constitutive induction of the competence regulon.

If a genome sequence is available, homologues of the known competence genes should be identified. If there are any obvious defects in competence genes, such as frame-shifts or large insertions or deletions, it may be possible to introduce a cloned functional copy of the gene from a closely related species, using electroporation.

If none of the competence genes has an obvious defect, or if no genome sequence is available, the ability of the cells to take up radioactively labelled DNA might also be tested. Such tests should use cells treated in a way that maximizes competence in other species, and, because of the existence of at least two variants of the USS, chromosomal DNA or a PCR fragment from the same species is the best choice for these tests. If cells do not take up DNA, functional copies of genes required for DNA uptake in *H. influenzae* can be introduced on a plasmid by electroporation. If cells are taking up DNA but transformation is not occurring, plasmids containing *dprA* or *comM* should be introduced using electroporation.

Web resources

Biocyc: http://www.biocyc.org. Biocyc has metabolic information on genes from the completed genomes of [*H.*] *ducreyi*, *Hist. somni* and two strains of *H. influenzae*.

National Center for Biotechnology Information: http://www.ncbi.nlm.nih.gov/. NCBI has complete genome sequences and contigs from genome sequencing in progress from 10 and 23 *Pasteurellaceae* species/strains, respectively. NCBI also has web tools to make comparisons between genes and genomes.

Comprehensive microbial resource: http://cmr.tigr.org/tigr-scripts/CMR/CmrHomePage.

cgi. CMR contains complete genome data from five *Pasteurellaceae* species. Tools for gene and genome comparison are also available.

References

Alexander, H.E., and Leidy, G. (1951). Determination of inherited traits of *H. influenzae* by desoxyribonucleic acid fractions isolated from type-specific cells. J. Exp. Med. *93*, 345–359.

Ando, T., Israel, D.A., Kusugami, K., and Blaser, M.J. (1999). HP0333, a member of the *dprA* family, is involved in natural transformation in *Helicobacter pylori*. J. Bacteriol. *181*, 5572–5580.

Assalkhou, R., Balasingham, S., Collins, R.F., Frye, S.A., Davidsen, T., Benam, A.V., Bjoras, M., Derrick, J.P., and Tonjum, T. (2007). The outer membrane secretin PilQ from *Neisseria meningitidis* binds DNA. Microbiology *153*, 1593–1603.

Bakaletz, L.O., Baker, B.D., Jurcisek, J.A., Harrison, A., Novotny, L.A., Bookwalter, J.E., Mungur, R., and Munson, R.S., Jr. (2005). Demonstration of Type IV pilus expression and a twitching phenotype by *Haemophilus influenzae*. Infect. Immun. *73*, 1635–1643.

Bakkali, M., Chen, T.Y., Lee, H.C., and Redfield, R.J. (2004). Evolutionary stability of DNA uptake signal sequences in the Pasteurellaceae. Proc. Natl. Acad. Sci. USA *101*, 4513–4518.

Barabote, R.D., and Saier, M.H., Jr. (2005). Comparative genomic analyses of the bacterial phosphotransferase system. Microbiol. Mol. Biol. Rev. *69*, 608–634.

Barany, F., Kahn, M.E., and Smith, H.O. (1983). Directional transport and integration of donor DNA in *Haemophilus influenzae* transformation. Proc. Natl. Acad. Sci. USA *80*, 7274–7278.

Barouki, R., and Smith, H.O. (1985). Reexamination of phenotypic defects in *rec-1* and *rec-2* mutants of *Haemophilus influenzae* Rd. J. Bacteriol. *163*, 629–634.

Barouki, R., and Smith, H.O. (1986). Initial steps in *Haemophilus influenzae* transformation. Donor DNA binding in the *com10* mutant. J. Biol. Chem. *261*, 8617–8623.

Berge, M., Mortier-Barriere, I., Martin, B., and Claverys, J.P. (2003). Transformation of *Streptococcus pneumoniae* relies on DprA- and RecA-dependent protection of incoming DNA single strands. Mol. Microbiol. *50*, 527–536.

Bhattacharjee, M.K., Fine, D.H., and Figurski, D.H. (2007). *tfoX (sxy)*-dependent transformation of *Aggregatibacter (Actinobacillus) actinomycetemcomitans*. Gene 399, 53–64.

Bigas, A., Garrido, M.E., de Rozas, A.M., Badiola, I., Barbe, J., and Llagostera, M. (2005). Development of a genetic manipulation system for *Haemophilus parasuis*. Vet. Microbiol. *105*, 223–228.

Boekema, B.K., Van Putten, J.P., Stockhofe-Zurwieden, N., and Smith, H.E. (2004). Host cell contact-induced transcription of the type IV fimbria gene cluster of *Actinobacillus pleuropneumoniae*. Infect. Immun. 72, 691–700.

Bosse, J.T., Nash, J.H., Kroll, J.S., and Langford, P.R. (2004). Harnessing natural transformation in *Actinobacillus pleuropneumoniae*: a simple method for allelic replacements. FEMS Microbiol. Lett. *233*, 277–281.

Bremer, W., Kooistra, J., Hellingwerf, K.J., and Konings, W.N. (1984). Role of the electrochemical proton gradient in genetic transformation of *Haemophilus influenzae*. J. Bacteriol. *157*, 868–873.

Burrows, L.L. (2005). Weapons of mass retraction. Mol. Microbiol. *57*, 878–888.

Cadman, C.J., and McGlynn, P. (2004). PriA helicase and SSB interact physically and functionally. Nucleic Acids Res. *32*, 6378–6387.

Cameron, A., Volar, M., Bannister, L., and Redfield, R. (2008). RNA secondary structure regulates *sxy* expression and competence development in *Haemophilus influenzae*. Nucleic Acids Res. *36*, 10–20.

Cameron, A.D., and Redfield, R.J. (2006). Non-canonical CRP sites control competence regulons in *Escherichia coli* and many other gamma-proteobacteria. Nucleic Acids Res. *34*, 6001–6014.

Challacombe, J.F., Duncan, A.J., Brettin, T.S., Bruce, D., Chertkov, O., Detter, J.C., Han, C.S., Misra, M., Richardson, P., Tapia, R., *et al.* (2007). Complete genome sequence of *Haemophilus somnus (Histophilus somni)* strain 129Pt and comparison to *Haemophilus ducreyi* 35000HP and *Haemophilus influenzae* Rd. J. Bacteriol. *189*, 1890–1898.

Chandler, M.S., and Smith, R.A. (1996). Characterization of the *Haemophilus influenzae topA* locus: DNA topoisomerase I is required for genetic competence. Gene *169*, 25–31.

Collins, R.F., Ford, R.C., Kitmitto, A., Olsen, R.O., Tonjum, T., and Derrick, J.P. (2003). Three-dimensional structure of the *Neisseria meningitidis* secretin PilQ determined from negative-stain transmission electron microscopy. J. Bacteriol. *185*, 2611–2617.

Collins, R.F., Frye, S.A., Balasingham, S., Ford, R.C., Tonjum, T., and Derrick, J.P. (2005). Interaction with type IV pili induces structural changes in the bacterial outer membrane secretin PilQ. J. Biol. Chem. *280*, 18923–18930.

Collins, R.F., Frye, S.A., Kitmitto, A., Ford, R.C., Tonjum, T., and Derrick, J.P. (2004). Structure of the *Neisseria meningitidis* outer membrane PilQ secretin complex at 12 A resolution. J. Biol. Chem. *279*, 39750–39756.

Cox, M.M. (1991). The RecA protein as a recombinational repair system. Mol. Microbiol. *5*, 1295–1299.

Cox, M.M., Goodman, M.F., Kreuzer, K.N., Sherratt, D.J., Sandler, S.J., and Marians, K.J. (2000). The importance of repairing stalled replication forks. Nature *404*, 37–41.

Danner, D.B., Deich, R.A., Sisco, K.L., and Smith, H.O. (1980). An eleven-base-pair sequence determines the specificity of DNA uptake in *Haemophilus* transformation. Gene *11*, 311–318.

Deich, R.A., and Smith, H.O. (1980). Mechanism of homospecific DNA uptake in *Haemophilus influenzae* transformation. Mol. Gen. Genet. *177*, 369–374.

Dorocicz, I.R., Williams, P.M., and Redfield, R.J. (1993). The *Haemophilus influenzae* adenylate cyclase gene: cloning, sequence, and essential role in competence. J. Bacteriol. *175*, 7142–7149.

Dougherty, B.A., and Smith, H.O. (1999). Identification of *Haemophilus influenzae* Rd transformation genes using cassette mutagenesis. Microbiology *145 (Pt 2)*, 401–409.

Doughty, S.W., Ruffolo, C.G., and Adler, B. (2000). The type 4 fimbrial subunit gene of *Pasteurella multocida*. Vet. Microbiol. *72*, 79–90.

Draskovic, I., and Dubnau, D. (2005). Biogenesis of a putative channel protein, ComEC, required for DNA uptake: membrane topology, oligomerization and formation of disulphide bonds. Mol. Microbiol. *55*, 881–896.

Dubnau, D. (1999). DNA uptake in bacteria. Annu. Rev. Microbiol. *53*, 217–244.

Feil, E.J., and Enright, M. =C. (2004). Analyses of clonality and the evolution of bacterial pathogens. Curr. Opin. Microbiol. *7*, 308–313.

Feil, E.J., Holmes, E. =C., Bessen, D.E., Chan, M.S., Day, N.P., Enright, M.C., Goldstein, R., Hood, D.W., Kalia, A., Moore, C.E., *et al.* (2001). Recombination within natural populations of pathogenic bacteria: short-term empirical estimates and long-term phylogenetic consequences. Proc. Natl. Acad. Sci. USA *98*, 182–187.

Fujise, O., Lakio, L., Wang, Y., Asikainen, S., and Chen, C. (2004). Clonal distribution of natural competence in *Actinobacillus actinomycetemcomitans*. Oral Microbiol. Immunol. *19*, 340–342.

Gioia, J., Qin, X., Jiang, H., Clinkenbeard, K., Lo, R., Liu, Y., Fox, G.E., Yerrapragada, S., McLeod, M.P., McNeill, T.Z., *et al.* (2006). The genome sequence of *Mannheimia haemolytica* A1: insights into virulence, natural competence, and Pasteurellaceae phylogeny. J. Bacteriol. *188*, 7257–7266.

Goodgal, S.H. (1982). DNA uptake in *Haemophilus* transformation. Annu. Rev. Genet. *16*, 169–192.

Gromkova, R.C., Mottalini, T.C., and Dove, M.G. (1998). Genetic transformation in *Haemophilus parainfluenzae* clinical isolates. Curr. Microbiol. *37*, 123–126.

Gwinn, M.L., Ramanathan, R., Smith, H.O., and Tomb, J.F. (1998). A new transformation-deficient mutant of *Haemophilus influenzae* Rd with normal DNA uptake. J. Bacteriol. *180*, 746–748.

Gwinn, M.L., Stellwagen, A.E., Craig, N.L., Tomb, J.F., and Smith, H.O. (1997). In vitro Tn7 mutagenesis of *Haemophilus influenzae* Rd and characterization of the role of *atpA* in transformation. J. Bacteriol. *179*, 7315–7320.

Herriott, R.M., Meyer, E.M., and Vogt, M. (1970). Defined nongrowth media for stage II development of competence in *Haemophilus influenzae*. J. Bacteriol. *101*, 517–524.

Hoelzer, M.A., and Michod, R.E. (1991). DNA repair and the evolution of transformation in *Bacillus subtilis*. III. Sex with damaged DNA. Genetics *128*, 215–223.

Hong, S.H., Kim, J.S., Lee, S.Y., In, Y.H., Choi, S.S., Rih, J.K., Kim, C.H., Jeong, H., Hur, C.G., and Kim, J.J. (2004). The genome sequence of the capnophilic rumen bacterium *Mannheimia succiniciproducens*. Nat. Biotechnol. *22*, 1275–1281.

Jurcisek, J.A., and Bakaletz, L.O. (2007). Biofilms formed by nontypeable *Haemophilus influenzae* in vivo contain both double-stranded DNA and type IV pilin protein. J. Bacteriol. *189*, 3868–3875.

Kahn, M., Concino, M., Gromkova, R., and Goodgal, S. (1979). DNA binding activity of vesicles produced by competence deficient mutants of *Haemophilus*. Biochem. Biophys Res. Commun. *87*, 764–772.

Kahn, M.E., Barany, F., and Smith, H.O. (1983). Transformasomes: specialized membranous structures that protect DNA during *Haemophilus* transformation. Proc. Natl. Acad. Sci. USA *80*, 6927–6931.

Kahn, M.E., Maul, G., and Goodgal, S.H. (1982). Possible mechanism for donor DNA binding and transport in *Haemophilus*. Proc. Natl. Acad. Sci. USA *79*, 6370–6374.

Kahn, M.E., and Smith, H.O. (1984). Transformation in *Haemophilus*: a problem in membrane biology. J. Membr Biol. *81*, 89–103.

Kaiser, D. (2000). Bacterial motility: how do pili pull? Curr. Biol. *10*, R777–780.

Karudapuram, S., and Barcak, G.J. (1997). The *Haemophilus influenzae dprABC* genes constitute a competence-inducible operon that requires the product of the *tfoX* (*sxy*) gene for transcriptional activation. J. Bacteriol. *179*, 4815–4820.

Karudapuram, S., Zhao, X., and Barcak, G.J. (1995). DNA sequence and characterization of *Haemophilus influenzae dprA+*, a gene required for chromosomal but not plasmid DNA transformation. J. Bacteriol. *177*, 3235–3240.

Kroll, J.S., and Moxon, E.R. (1990). Capsulation in distantly related strains of *Haemophilus influenzae* type b: genetic drift and gene transfer at the capsulation locus. J. Bacteriol. *172*, 1374–1379.

Kroll, J.S., Wilks, K.E., Farrant, J.L., and Langford, P.R. (1998). Natural genetic exchange between *Haemophilus* and *Neisseria*: intergeneric transfer of chromosomal genes between major human pathogens. Proc. Natl. Acad. Sci. USA 95, 12381–12385.

Larson, T.G., and Goodgal, S.H. (1992). Donor DNA processing is blocked by a mutation in the *com101A* locus of *Haemophilus influenzae*. J. Bacteriol. *174*, 3392–3394.

Lethem, M.I., James, S.L., Marriott, C., and Burke, J.F. (1990). The origin of DNA associated with mucus glycoproteins in cystic fibrosis sputum. Eur. Respir J. *3*, 19–23.

Levy, S., Zeng, G.Q., and Danchin, A. (1990). Cyclic AMP synthesis in *Escherichia coli* strains bearing known deletions in the pts phosphotransferase operon. Gene *86*, 27–33.

Ma, C., and Redfield, R.J. (2000). Point mutations in a peptidoglycan biosynthesis gene cause competence induction in *Haemophilus influenzae*. J. Bacteriol. *182*, 3323–3330.

McCarthy, D. (1989). Cloning of the *rec-2* locus of *Haemophilus influenzae*. Gene *75*, 135–143.

Macfadyen, L.P. (2000). Regulation of competence development in *Haemophilus influenzae*. J. Theor Biol. *207*, 349–359.

MacFadyen, L.P., Chen, D., Vo, H.C., Liao, D., Sinotte, R., and Redfield, R.J. (2001). Competence development by *Haemophilus influenzae* is regulated by the availability of nucleic acid precursors. Mol. Microbiol. *40*, 700–707.

Macfadyen, L.P., Dorocicz, I.R., Reizer, J., Saier, M.H., Jr., and Redfield, R.J. (1996). Regulation of competence development and sugar utilization in *Haemophilus influenzae* Rd by a phosphoenolpyruvate:fructose phosphotransferase system. Mol. Microbiol. *21*, 941–952.

Maiden, M.C. (1998). Horizontal genetic exchange, evolution, and spread of antibiotic resistance in bacteria. Clin. Infect. Dis. *27* Suppl. 1, S12–20.

Maiden, M.C., Bygraves, J.A., Feil, E., Morelli, G., Russell, J.E., Urwin, R., Zhang, Q., Zhou, J., Zurth, K., Caugant, D.A., *et al.* (1998). Multilocus sequence typing: a portable approach to the identification of clones within populations of pathogenic microorganisms. Proc. Natl. Acad. Sci. USA *95*, 3140–3145.

Maier, B., Koomey, M., and Sheetz, M.P. (2004). A force-dependent switch reverses type IV pilus retraction. Proc. Natl. Acad. Sci. USA *101*, 10961–10966.

May, B.J., Zhang, Q., Li, L.L., Paustian, M.L., Whittam, T.S., and Kapur, V. (2001). Complete genomic sequence of *Pasteurella multocida*, Pm70. Proc. Natl. Acad. Sci. USA *98*, 3460–3465.

Meats, E., Feil, E.J., Stringer, S., Cody, A.J., Goldstein, R., Kroll, J.S., Popovic, T., and Spratt, B.G. (2003). Characterization of encapsulated and noncapsulated *Haemophilus influenzae* and determination of phylogenetic relationships by multilocus sequence typing. J. Clin. Microbiol. *41*, 1623–1636.

Michod, R.E., Wojciechowski, M.F., and Hoelzer, M.A. (1988). DNA repair and the evolution of transformation in the bacterium *Bacillus subtilis*. Genetics *118*, 31–39.

Molnar, S. (2001) Identification and characterisation of DNA receptor candidates in *H. influenzae*, University of British Columbia, Vancouver.

Mongold, J.A. (1992). DNA repair and the evolution of transformation in *Haemophilus influenzae*. Genetics *132*, 893–898.

Mortier-Barriere, I., Velten, M., Dupaigne, P., Mirouze, N., Pietrement, O., McGovern, S., Fichant, G., Martin, B., Noirot, P., Le Cam, E., *et al.* (2007). A key presynaptic role in transformation for a widespread bacterial protein: DprA conveys incoming ssDNA to RecA. Cell *130*, 824–836.

Nickel, L., and Goodgal, S.H. (1964). Effect of interspecific transformation on linkage relationships of markers in *Haemophilus influenzae* and *Haemophilus parainfluenzae*. J. Bacteriol. *88*, 1538–1544.

Notani, N.K., Setlow, J.K., Joshi, V.R., and Allison, D.P. (1972). Molecular basis for the transformation defects in mutants of *Haemophilus influenzae*. J. Bacteriol. *110*, 1171–1180.

Nunn, D.N., and Lory, S. (1993). Cleavage, methylation, and localization of the *Pseudomonas aeruginosa* export proteins XcpT, -U, -V, and -W. J. Bacteriol. *175*, 4375–4382.

Perez-Losada, M., Christensen, R.G., McClellan, D.A., Adams, B.J., Viscidi, R.P., Demma, J.C., and Crandall, K.A. (2006). Comparing phylogenetic codivergence between polyomaviruses and their hosts. J. Virol. *80*, 5663–5669.

Pifer, M.L., and Smith, H.O. (1985). Processing of donor DNA during *Haemophilus influenzae* transformation: analysis using a model plasmid system. Proc. Natl. Acad. Sci. USA *82*, 3731–3735.

Poje, G., and Redfield, R.J. (2003). Transformation of *Haemophilus influenzae*. Methods Mol. Med. *71*, 57–70.

Postel, E.H., and Goodgal, S.H. (1966). Uptake of 'single-stranded' DNA in *Hemophilus influenzae* and its ability to transform. J. Mol. Biol. *16*, 317–327.

Redfield, R.J. (1991). *sxy-1*, a *Haemophilus influenzae* mutation causing greatly enhanced spontaneous competence. J. Bacteriol. *173*, 5612–5618.

Redfield, R.J. (1993). Evolution of natural transformation: testing the DNA repair hypothesis in *Bacillus subtilis* and *Haemophilus influenzae*. Genetics *133*, 755–761.

Redfield, R J. (2001). Do bacteria have sex? Nat. Rev. Genet. *2*, 634–639.

Redfield, R.J., Cameron, A.D., Qian, Q., Hinds, J., Ali, T.R., Kroll, J.S., and Langford, P.R. (2005). A novel CRP-dependent regulon controls expression of competence genes in *Haemophilus influenzae*. J. Mol. Biol. *347*, 735–747.

Redfield, R.J., Findlay, W.A., Bosse, J., Kroll, J.S., Cameron, A.D., and Nash, J.H. (2006). Evolution of competence and DNA uptake specificity in the *Pasteurellaceae*. BMC Evol. Biol. *6*, 82.

Rowji, P., Gromkova, R., and Koornhof, H. (1989). Genetic transformation in encapsulated clinical isolates of *Haemophilus influenzae* type b. J. Gen. Microbiol. *135*, 2775–2782.

Savery, N., Rhodius, V., and Busby, S. (1996). Protein–protein interactions during transcription activation: the case of the *Escherichia coli* cyclic AMP receptor protein. Phil. Trans. R. Soc. Lond B Biol. Sci. *351*, 543–550.

Saveson, C.J., and Lovett, S.T. (1999). Tandem repeat recombination induced by replication fork defects in *Escherichia coli* requires a novel factor, RadC. Genetics *152*, 5–13.

Shen, P., and Huang, H.V. (1986). Homologous recombination in *Escherichia coli*: dependence on substrate length and homology. Genetics *112*, 441–457.

Smeets, L.C., Becker, S.C., Barcak, G.J., Vandenbroucke-Grauls, C.M., Bitter, W., and Goosen, N. (2006). Functional characterization of the competence protein DprA/Smf in *Escherichia coli*;. FEMS Microbiol. Lett. *263*, 223–228.

Solomon, J.M., and Grossman, A.D. (1996). Who's competent and when: regulation of natural genetic competence in bacteria. Trends Genet. *12*, 150–155.

Stuy, J.H., and Stern, D. (1964). The kinetics of DNA uptake by *Haemophilus Influenzae*. J. Gen. Microbiol. *35*, 391–400.

Takata, T., Ando, T., Israel, D.A., Wassenaar, T.M., and Blaser, M.J. (2005). Role of *dprA* in transformation of *Campylobacter jejuni*. FEMS Microbiol. Lett. *252*, 161–168.

Tomb, J.F. (1992). A periplasmic protein disulfide oxidoreductase is required for transformation of *Haemophilus influenzae* Rd. Proc. Natl. Acad. Sci. USA *89*, 10252–10256.

Tomb, J.F., el-Hajj, H., and Smith, H.O. (1991). Nucleotide sequence of a cluster of genes involved in the transformation of *Haemophilus influenzae* Rd. Gene *104*, 1–10.

Tonjum, T., and Koomey, M. (1997). The pilus colonization factor of pathogenic neisserial species: organelle biogenesis and structure/function relationships – a review. Gene *192*, 155–163.

van Schaik, E.J., Giltner, C.L., Audette, G.F., Keizer, D.W., Bautista, D.L., Slupsky, C.M., Sykes, B.D., and Irvin, R.T. (2005). DNA binding: a novel function of *Pseudomonas aeruginosa* type IV pili. J. Bacteriol. *187*, 1455–1464.

VanWagoner, T.M., Whitby, P.W., Morton, D.J., Seale, T.W., and Stull, T.L. (2004). Characterization of three new competence-regulated operons in *Haemophilus influenzae*. J. Bacteriol. *186*, 6409–6421.

Wang, Y., Goodman, S.D., Redfield, R.J., and Chen, C. (2002). Natural transformation and DNA uptake signal sequences in *Actinobacillus actinomycetemcomitans*. J. Bacteriol. *184*, 3442–3449.

Wilcox, K.W., and Smith, H.O. (1976a). Binding of the ATP-dependent DNase from *Hemophilus influenzae* to duplex DNA molecules. J. Biol. Chem. *251*, 6122–6126.

Wilcox, K.W., and Smith, H.O. (1976b). Mechanism of DNA degradation by the ATP-dependent DNase from *Hemophilus influenzae* Rd. J. Biol. Chem. *251*, 6127–6134.

Williams, P.M., Bannister, L.A., and Redfield, R.J. (1994). The *Haemophilus influenzae sxy-1* mutation is in a newly identified gene essential for competence. J. Bacteriol. *176*, 6789–6794.

Wojciechowski, M.F., Hoelzer, M.A., and Michod, R.E. (1989). DNA repair and the evolution of transformation in *Bacillus subtilis*. II. Role of inducible repair. Genetics *121*, 411–422.

Zhang, Y., Tennent, J.M., Ingham, A., Beddome, G., Prideaux, C., and Michalski, W.P. (2000). Identification of type 4 fimbriae in *Actinobacillus pleuropneumoniae*. FEMS Microbiol. Lett. *189*, 15–18.

Zulty, J.J., and Barcak, G.J. (1995). Identification of a DNA transformation gene required for *com101A+* expression and supertransformer phenotype in *Haemophilus influenzae*. Proc. Natl. Acad. Sci. USA *92*, 3616–3620.

Proteomic Analyses in *Pasteurellaceae* 5

John D. Boyce and Ben Adler

Abstract

Proteomics is the simultaneous identification and/or analysis of large groups of proteins expressed by an organism. Here we review the range of proteomic analyses which have been reported for members of the family *Pasteurellaceae*. These include the non-quantitative identification of proteins from whole organisms or from sub-cellular fractions, as well as the quantitative analysis of how the expression of certain proteins changes in response to particular growth conditions. As many members of the *Pasteurellaceae* are important pathogens, studies have focussed primarily on protein expression under conditions likely to be encountered during natural infections. The large variation in experimental techniques used and the conditions analysed has made it difficult to make meaningful comparisons about protein expression across the species within the *Pasteurellaceae*.

Introduction

What is proteomics?

Proteomics is the study of all the proteins expressed by an organism or literally the analysis of the proteome. However, the current usage of the term has extended to include any of the following: (1) the identification and quantification of large numbers of proteins from an organism or part of an organism, (2) the determination of protein sets with a particular modification (e.g. phosphorylated proteins), (3) determination of protein sets that interact, (4) large-scale analysis of protein activities and/or functions. While classical biochemistry has lead to the intricate and precise characterization of a huge number of cellular processes, and the roles that individual proteins play in these processes, these advances have mostly derived from detailed analyses of single or small numbers of proteins. Whilst not supplanting these detailed biochemical analyses the field of proteomics complements these specific studies by studying more general properties of large to very large groupings of proteins.

Over the last ten years a range of technological advances has led to the maturation of the field we now call proteomics. In its simplest sense proteomics is just the separation of proteins in a complex mixture and the identification of those proteins. Thus, the evolution of the field has been driven by the simultaneous improvements in protein separation and detection technologies.

Protein separation techniques

An overview of proteomics methods is shown in Fig. 5.1. Historically separation of complex protein mixtures has been done using either two-dimensional gel electrophoresis (2DGE) or by solution chromatographic techniques. By far the majority of early proteomic experiments used 2DGE to separate the proteins and give a visual reference map of the proteins present in the mixture. Such experiments allowed for visual comparison of the total set of proteins produced under different conditions even before it was possible to identify each of the proteins in the mixture.

2DGE usually uses isoelectric focusing (IEF) gradients to separate proteins by p*I* in the

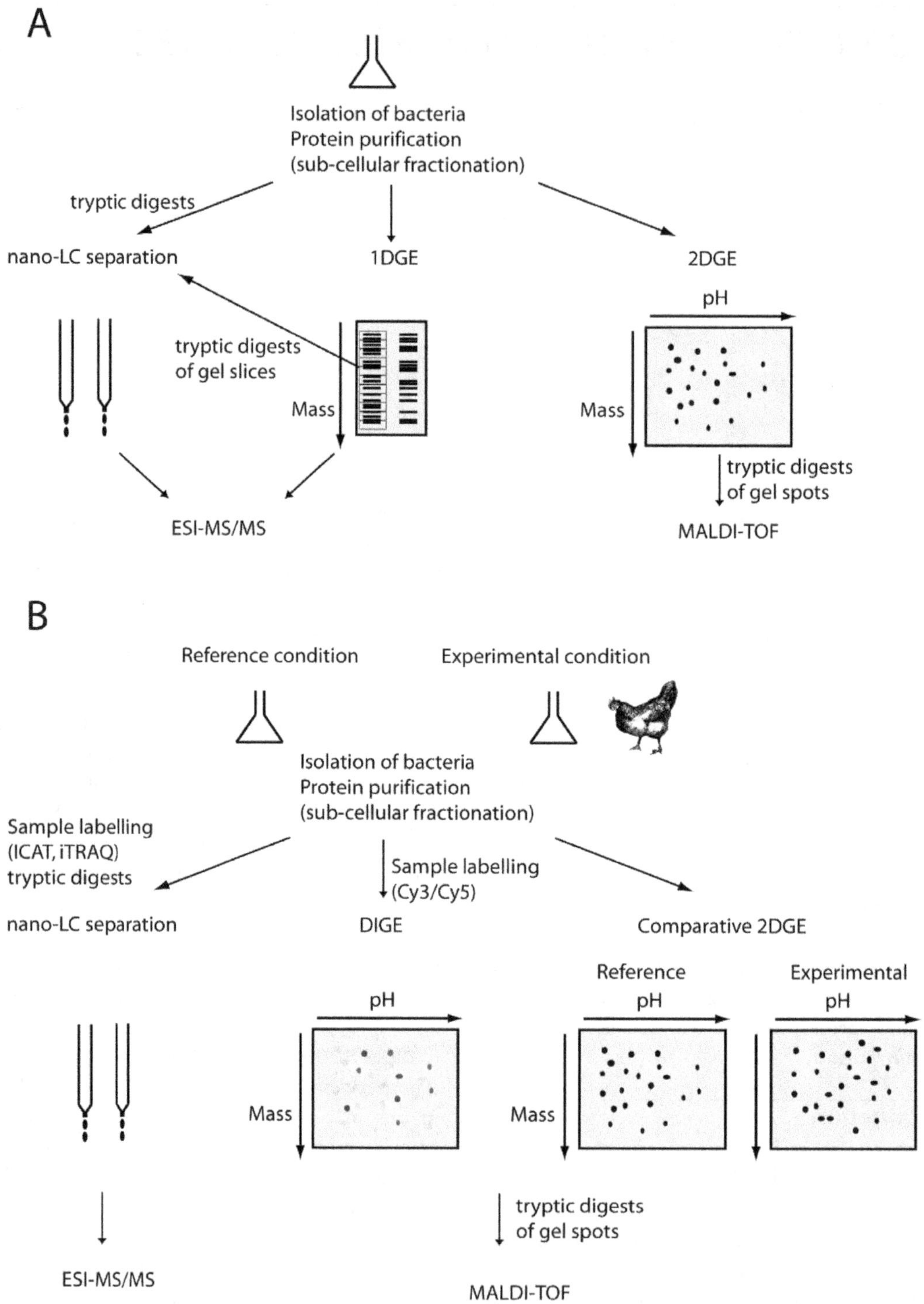

Figure 5.1 Generalized workflow for the three main qualitative (A) and quantitative (B) proteomics methods. The qualitative methods are: liquid chromatographic separation followed by tandem MS peptide identification (MUDPIT), 1D SDS-PAGE followed by tandem MS peptide identification and 2DGE separation followed by MALDI-TOF protein identification. The quantitative methods are liquid chromatographic separation of labelled peptides followed by tandem MS peptide identification, differential gel electrophoresis (DIGE) of fluorescently labelled samples and comparative 2DGE followed by MALDI-TOF protein identification.

first dimension and then uses sodium dodecyl sulfate polyacrylamide gel electrophoresis (SDS PAGE) to separate proteins by size in the second dimension. After separation, proteins can be visualized by any of a number of methods including staining with Coomassie brilliant blue, silver or the use of fluorescent stains such as Sypro Ruby. Protein spots, normally corresponding to single proteins can then be cut out of the gel and digested to tryptic peptides. These peptides can then be identified by accurate mass analysis by mass spectrometry (MS). As all of the peptides from a single 2DGE spot have come from the same protein the set of peptide masses identified by MS forms a peptide mass fingerprint (PMF). This PMF can then be compared to the *in silico* prediction of the PMF for every protein produced by the organism under study.

While 2DGE has proved to be a robust and useful method for analysis of large numbers of proteins in a sample it also has some limitations. Firstly, many biological samples have a very complex protein profile and this means that often certain proteins will not be adequately separated, thus making later identification difficult. Secondly, the concentrations of proteins in the sample may differ by many orders of magnitude so that only the more abundant proteins can be visualized and identified. Many technological improvements have been made to 2DGE to improve the coverage of proteins separated. Importantly the IEF step has been significantly improved by the introduction of immobilized pH gradient (IPG) strips. However, while these IPG strips give excellent separating power and allow separation of up to 10,000 spots on a single 2DGE run some samples may still be too complex for complete separation. Furthermore, many proteins, especially membrane proteins, are insoluble or of limited solubility in the loading buffers used for IEF and do not enter the gels. Also, as the protein concentration in the sample is increased to allow for identification of low abundance proteins precipitation during the IEF run can become a significant problem. While these problems can be partially overcome by reducing the total sample protein concentration, by using different loading buffers or by reducing the sample complexity, either by sample pre-fractionation or by using narrow range IPG strips, other separation methods have been developed to overcome the limitations inherent in the IEF step. However, at this stage 2DGE is still the best method for large-scale protein separation so these alternative methods should be seen as complementary rather than as alternatives to 2DGE.

Liquid chromatography techniques can also be used to separate proteins prior to identification by MS (for review see Frohlich and Arnold, 2006). Given the small amounts of biological samples usually available for analysis, these methods generally require micro-scale separations such as microcapillary reverse phase liquid chromatography (μLC or nano-LC). In these methods the protein sample is first digested to tryptic peptides prior to electrophoresis and then the eluted peptides are introduced directly onto the mass spectrometer. Again the complexity of biological samples means that single chromatographic separation methods rarely give adequate separation to allow individual peptides to enter the mass spectrometer at a given time. This complexity problem is compounded as the biological samples are digested with trypsin prior to separation which increases the sample complexity by many orders of magnitude. To address this problem multiple liquid chromatographic techniques can be used in series and this has been termed multidimensional peptide separation. While any variation of liquid chromatography techniques can be used, the original published method used initial separation through a strong cation exchange resin coupled with reverse phase nano-LC (Link *et al.*, 1999). As these liquid chromatography techniques result in peptides from many different proteins entering the MS detector simultaneously, accurate protein identification generally requires two rounds of MS (tandem MS or MS/MS). MS/MS generally relies on separation of ionized peptides initially on mass/charge (m/z) ratio, then the separated individual ions are allowed to decompose under certain conditions to produce a range of fragmentation ions (for review see Canas *et al.*, 2006). These fragmentation ions are then separated in a second round of MS to generate a peptide fragmentation fingerprint (PFF). These PFF spectra can be compared against the theoretical *in silico* derived PFF for all proteins in a genome or can be used to predict actual peptide sequences. Liquid chromatography separation coupled with automated tandem MS/MS is called multidimensional protein identification technology (MudPit) and can allow identification of some proteins not identified efficiently by 2DGE, especially membrane-associated and low-abundance proteins (Link *et al.*, 1999). However, even MudPit analyses fall short of the separation power required for complex eukaryotic biological samples.

As an alternative to multiple liquid chromatography steps some workers have employed hybrid gel and liquid chromatography separation procedures. Initial sample separation is generally done by SDS PAGE and the resulting gel is cut into thin horizontal strips. The proteins within each of the gel strips are then digested to give tryptic peptides and the peptides separated further by nano-LC prior to identification by MS or MS/MS. This method has a lower resolving power than MudPit and is more labour intensive but is very suitable for the identification of membrane proteins and those proteins of low solubility, as these are generally solubilized in the SDS sample buffer used for the initial SDS PAGE separation step.

Protein identification techniques

The ability to identify individual proteins forms the basis of all proteomic techniques. Identification relies almost exclusively on MS methods (see review by Canas *et al.*, 2006) but is also highly dependent on availability of genome sequences against which accurate peptide masses determined from MS can be searched and matched. The use of MS as a means of identifying peptides and other large biological molecules has only been routinely possible since the development of two novel ionization techniques; electrospray ionization (ESI) and matrix-assisted laser desorption ionization (MALDI). For ESI a liquid sample under high voltage is passed through a very thin needle at very low flow rates allowing the generation of a fine spray with charged droplets. As the droplets pass through the MS the solvent evaporates leaving charged peptide ions. Alternatively for MALDI the peptide samples are spotted onto a solid matrix which is then pulsed by laser. The laser radiation is primarily absorbed by the matrix which leads to ionization and release of charged peptides from the sample. Once peptide ions have been generated by either ESI or MALDI the charged molecules, now present in the gas phase, can be separated through electric or magnetic fields based on their m/z ratio. This results in very accurate mass estimates for individual peptides.

There is now a range of mass spectrometry instruments including time of flight (TOF), quadrupole, ion traps and tandem instruments (for review see Canas *et al.*, 2006). TOF instruments pass ions with equal energies through a long flight tube with high vacuum. The velocity of the ions is inversely proportional to mass so lighter ions impact the detector first. TOF instruments are often coupled with MALDI for ion formation and MALDI-TOF is often used to identify PMF from 2DGE samples. Quadrupole instruments separate ions using electrical fields generated by four parallel rods. Opposite rods have either a direct current or alternating current applied and the voltages can be manipulated to allow ions with different m/z ratios to impact the detector. Ion traps allow for the trapping of individual ions while certain procedures are carried out including ion fragmentation and product ion selection. Therefore, individual ions can be interrogated in multiple ways. This is also possible in tandem instruments.

Quantitative methods

An essential part of almost all proteomic analyses is the quantification of proteins and the determination of how expression levels of certain proteins change in response to different environmental cues. In its simplest form proteomic analysis of protein expression changes can be determined by densitometric analysis of duplicate 2D gels. In this way protein preparations derived from samples that have been prepared from cells grown under the different conditions to be compared, are separated identically by 2DGE and the separated proteins visualized, spot density determined and the concentration and position of each protein compared. Given the complex nature of the samples and the variability inherent in 2DGE it is often necessary to use special image analysis software to identify and compare the spots between the replicate gels. These types of analyses, while informative, suffer from the problems of reproducibility between different 2D runs. These reproducibility problems can be overcome by using 2D difference gel electrophoresis (DIGE) (Unlu *et al.*, 1997) where the two samples to be compared are labelled with a different fluorophore and then the samples are separated on the same gel. Furthermore, a third fluorophore can be used to label an internal control to allow for sample to sample normalization (for a review see Marouga *et al.*, 2005). The labelling of proteins for DIGE requires the presence of either lysine or cysteine residues which means that proteins with low content of these amino acids may be poorly visualized.

The major initial limitation with liquid chromatographic separation and identification techniques such as nano-LC MS/MS was the inability for protein/peptide quantification. However, recently a number of modifications to this technique have allowed for accurate comparison of protein concentration in different samples (for reviews see Gevaert *et al.*, 2007; Roe and Griffin, 2006). These methods rely on differential labelling of proteins in each sample with specific tags which can be discriminated on MW or fragmentation pattern during MS identification. The first quantitative liquid chromatographic method developed (Gygi *et al.*, 1999) used paired tags that are cysteine reactive and differ only in their isotopic composition; with one tag containing a heavy nucleotide and one tag a light nucleotide (e.g. $^{1}H/^{2}H$). These isotope-coded affinity tags (ICAT) also contained a biotin moiety allowing for selective enrichment of the tagged proteins. Using this system, the two protein samples to be compared are labelled with either the light or heavy ICAT reagent and the two labelled samples are combined. The samples are then digested (e.g. with trypsin) to generate peptide fragments and the fragments purified on an avidin affinity column. The peptide fragments are then identified by nano-LC MS/MS. For the original ICAT reagent (Gygi *et al.*, 1999) the difference in mass between the heavy and light tag was 8 Da. Therefore, peptide peaks differing by an m/z ratio of 8 Da (charge state of 1) or 4 Da (charge state of 2) indicate the same protein from each of the two samples. The relative quantities of any particular protein can be determined by comparing the intensities of the paired peaks. Thus sequence identity and quantity are determined simultaneously during MS.

A limitation of the ICAT reagent is that the tags are coupled to the proteins via cysteine residues so peptides without cysteine can never be labelled. To overcome this problem other protein tags and labelling methods have been developed. One widely applicable technique is stable isotope labelling by amino acids in cell culture (SILAC) (Ong *et al.*, 2002). This technique uses an isotopically marked essential amino acid to differentiate proteins from different samples. Thus, bacteria or cell lines are grown in a minimal medium supplemented with either a natural essential amino acid (e.g. leucine) or an isotopically labelled version of the essential amino acid (leucine-d3). These amino acids will be incorporated into all newly synthesized proteins and after a few rounds of replication all proteins will contain only the incorporated (i.e. either natural or labelled) amino acid. Protein samples from cells grown under the two conditions can then be differentiated again during MS analysis because of the mass difference between the isotopically light (natural) and heavy (isotopically labelled) amino acids. As this method relies on the ability to grow the bacteria in a minimal medium with or without the addition of a particular labelled amino acid, its application to some model systems (such as *in vivo* whole animal models) is limited.

Another recently developed set of tags is the iTRAQ reagents (Wiese *et al.*, 2007). These are a set of four reagents with identical masses (isobaric) of 145 Da but different fragmentation patterns ($m/z = 114–117$). These tags are coupled to the proteins by linkage to amine groups so they are not specific for particular amino acids. Furthermore, the fact that there are four tags means that there is the possibility for higher level multiplexing including internal controls. Comparative analyses suggest that ICAT, iTRAQ and DIGE are all accurate quantitative methods while each method has different coverage and sensitivity issues (Wu *et al.*, 2006).

Specialized protein visualization techniques

Standard protein visualization following 2DGE uses dyes such as Sypro Ruby, Colloidal Coomassie Blue, or silver which are highly sensitive protein stains that bind relatively equally to all the proteins in a mixture. However, sometimes specialized visualization reagents or techniques may be used to specifically stain only certain types of proteins. For example glycoproteins can be identified by staining methods based on the periodic-acid Schiff base reaction and the fluorescent dye Pro-Q Emerald 488 (Hart *et al.*, 2003). Immunostaining can also be used to identify specific antigenic proteins on 2D gels. Generally immunoblots are run in parallel with conventional 2DGE so that they can be overlaid and the reactive spots on the immunoblot identified on the conventionally stained gel and the spots excised for identification by MS.

Other techniques

While not generally seen as a proteomic method, *in vivo* induced antigen technology (IVIAT) is a method that allows for the large scale identification of pathogen antigens specifically expressed *in vivo* during natural infections (Handfield *et al.*, 2000). We have included discussions of this technique here as we believe it fulfils the criteria of a proteomic method as it allows for the large-scale identification of proteins specifically expressed under conditions important for pathogenesis. In this method sera taken from infected animals, or patients suffering disease, are adsorbed with whole cells of *in vitro* grown pathogen cultures. This allows for the removal of antibodies specific for antigens expressed by the pathogen during *in vitro* growth and thus not specifically expressed during infections. The absorbed sera are then used to probe an arrayed library of bacteria expressing each of the bacterial proteins from a recombinant expression plasmid. This expression library is usually made in a heterologous host such as *Escherichia coli* so that antibody binding is specific for the expressed protein. Clones which react with the absorbed sera are likely to express pathogen proteins which are specifically expressed only during infection.

Proteomic analyses of bacteria from the *Pasteurellaceae*

While proteomic analyses have been carried out on a range of members of the *Pasteurellaceae*, the types of analyses and methods used vary significantly from organism to organism. Therefore we will describe the experiments for each genus separately and then attempt to synthesize the findings and compare and contrast the results.

Proteomic analyses of species within the genus *Pasteurella*

The genus *Pasteurella* contains several species but proteomic analyses have only been carried out on the type species *Pasteurella multocida*.

The outer membrane (OM) subproteome of *P. multocida* strain x-73 (a highly virulent fowl cholera-causing isolate) was characterized using a combined 2DGE and nano-LC MS/MS approach (Boyce *et al.*, 2006). *P. multocida* outer membranes were prepared from cells grown either *in vitro* in brain heart infusion (BHI) broth, in BHI supplemented with 200 μM 2,2′-dipyridyl (to chelate free iron) or from cells purified from the blood of infected chickens near the terminal stages of infection (~10^{10} CFU/ml). Bacteria were disrupted by sonication after three freeze–thaw cycles and OM material purified as the sarkosyl insoluble fraction of concentrated membranes (Boyce *et al.*, 2006).

To maximize the number of OM proteins identified in these preparations both 2DGE and nano-LC MS/MS was used to identify the proteins present. In the BHI grown cells a total of 19 unique proteins was identified by 2DGE, of which 18 were predicted by bioinformatics to be true components of the *P. multocida* OM. A total of 29 unique proteins was identified by nano-LC MS/MS following 1D SDS PAGE of which 22 were predicted to be true components of the *P. multocida* OM.

The expression of the *P. multocida* OM proteins in rich (BHI) or iron-restricted medium (BHI supplemented with 200 μM 2,2′-dipyridyl) was quantified by comparative 2DGE. Only two proteins were differentially expressed under low-iron conditions, with the OM porin Pm0803 significantly up-regulated and OmpW significantly down-regulated. Furthermore, the expression of the *P. multocida* OM proteins during growth in the blood of chickens with fowl cholera was also quantified by 2DGE. Three proteins (Pm0803, TufA and TufB) were significantly up-regulated *in vivo* while one protein OmpW was significantly down-regulated. Thus Pm0803 and OmpW expression responded similarly to growth *in vivo* and growth under *in vitro* iron-limiting conditions. This is likely indicative of the *in vivo* environment of the chicken bloodstream displaying low free iron. The gene encoding Pm0803 has been shown by DNA microarray analysis to display increased transcription, compared to growth in BHI, both *in vivo* and during growth under iron-limiting conditions, while the gene encoding OmpW displayed reduced transcription under both of these conditions (Boyce *et al.*, 2002; Boyce *et al.*, 2004; Paustian *et al.*, 2001). Thus, the differential expression of these two genes is likely controlled mostly at the level of transcription in response to low free iron. TufA and TufB were up-regulated only under *in vivo* growth conditions. However, no difference in transcription was observed for the genes encoding these proteins during *in vivo* growth (Boyce *et al.*, 2002; Boyce *et al.*, 2004). While TufA and TufB are generally seen as cytoplasmic proteins (being components of elongation factor-Tu) in *E. coli* TufA and TufB have been shown to be preferentially targeted to the periplasm under stress conditions (Berrier *et al.*, 2000). Thus, it is likely that in *P. multocida* these proteins are present at increased levels in the OM during growth *in vivo* due to differential targeting and not due to altered transcriptional regulation (Boyce *et al.*, 2006).

The OM proteins of *P. multocida* strain P-1059 were analysed by 1D SDS PAGE followed by MALDI-TOF (Tabatabai and Zehr, 2004). Only five proteins were identified: PlpB, HemY, PhyA, Pm1050 and PM1819. PlpB was predicted to be a protein capable of generating protective antibodies against both homologous and heterologous serotypes of *P. multocida* (Rimler, 2001; Tabatabai and Zehr, 2004). However, vaccination with recombinant PlpB from *P. multocida* strain x-73 does not elicit protective immunity against infection (Wu *et al.*, 2007).

The response of *P. multocida* to growth in the presence of subinhibitory concentrations of the antibiotics amoxicillin, chlortetracycline and enrofloxacin has been analysed using ICAT 2D LC MS/MS (Nanduri *et al.*, 2006). A total of 245 proteins was identified from cells grown in the presence of each of the antibiotics. Thus, this method identified approximately 12% of the total *P. multocida* proteins. The expression of 11 proteins was significantly altered in response to growth in the presence of one-quarter the minimum inhibitory concentration (MIC) of amoxicillin. Amoxicillin acts by inhibiting peptidoglycan synthesis and three of the differentially regulated genes (*nag*C, PM1003 and *pckA*) were predicted to be involved in polysaccharide biosynthesis. Other differences observed included reduced expression of two proteins involved in nucleotide metabolism, ThyA (thymidylate synthase) and DeoD (purine-nucleoside phosphorylase) and two proteins involved in translation and RNA degradation, PnP (polynucleotide phosphorylase) and RpS4 (ribosomal protein S4) (Nanduri *et al.*, 2006). Furthermore, the protein DeaD, which is a homologue of the *E. coli* cold shock protein CsdA, showed a 33-fold increase in expression and is predicted to be involved in ribosomal assembly and mRNA processing.

Chlortetracycline acts by inhibiting protein synthesis at the 30S ribosomal subunit and the expression of 20 proteins was significantly altered in response to growth in the presence of ¼ MIC of chlortetracycline. These proteins included four (RpS12, PrfA, GlnS and GidA) predicted to be involved in ribosomal structure or translation. A number of genes involved in central metabolic functions were also differentially expressed including Pgi (glucose-6-phosphate isomerase) and PykA (pyruvate kinase) which were down-regulated and PflB (pyruvate formate lyase), AceE (pyruvate dehydrogenase complex E1 component) and Eno (enolase) which were all up-regulated (Nanduri *et al.*, 2006).

Enrofloxacin acts by inhibiting DNA replication through DNA gyrase and topoisomerase IV. The expression of 19 proteins was significantly altered in response to growth in the presence of ¼ MIC of enrofloxacin. It is predicted that the action of enrofloxacin will require the expression of genes involved in DNA repair; indeed, RecA was expressed at 13-fold higher levels in response to the antibiotic. Again a range of genes involved in transcription, translation and mRNA degradation were differentially expressed including RpoA (RNA polymerase α-subunit), DeoD, PM0900 (oligoribonuclease), RpS13 (ribosomal protein 13 small subunit) and PrfA (protein chain release factor) (Nanduri *et al.*, 2006).

Proteomic analyses of species within the genus *Actinobacillus* and *Aggregatibacter*

The genus *Actinobacillus* includes a large number of species but proteomic analyses have only been carried out on *Actinobacillus pleuropneumoniae* which is an obligate pathogen of pigs. Proteomic analyses have also been carried out on the human oral pathogen *Aggregatibacter actinomycetemcomitans* which we have included here together with *A. pleuropneumoniae* as it has only recently been reclassified from the genus *Actinobacillus* (Nørskov-Lauritsen and Kilian, 2006).

Actinobacillus pleuropneumoniae

The OM proteins of *A. pleuropneumoniae* have been identified by nano-LC MS/MS (Chung *et al.*, 2007). *A. pleuropneumoniae* was cultured aerobically in BHI and harvested at late exponential or early stationary phase. The harvested bacteria were lysed by disruption in a French pressure cell. The non-OM material was then removed either directly by 1% sarkosyl solubilization of the material or by washing with either 1% sarkosyl, 0.1 M sodium carbonate, 2.5 M sodium bromide or 2.5 M sodium bromide followed by 0.1 M sodium carbonate (NBSC treatment) after sucrose gradient centrifugation. The purified proteins from these preparations were used directly for solution-based LC MS/MS or were separated by 1D SDS PAGE, then the gel cut into thin segments and tryptic digests of each of the gels slices used for LC MS/MS. More than 100 proteins were identified by these combined analyses but a number of these were likely to be inner membrane or cytoplasmic contaminants. To overcome these problems a comprehensive bioinformatics analysis of the *A. pleuropneumoniae* genome was undertaken using BLAST (Altschul *et al.*, 1990), PSORTB (Gardy *et al.*, 2005), Proteome Analyst (Lu *et al.*, 2004), LipoP (Juncker *et al.*, 2003) and BOMP (Berven *et al.*, 2004). These combined analyses predicted 93 proteins as part of the *A. pleuropneumoniae* OM proteome. These predictions were compared with the combined LC MS/MS results and 47 probable OM proteins shown to be expressed under *in vitro* growth conditions in rich medium. These proteins included five previously characterized *A. pleuropneumoniae* proteins FhuA, HgbA, TbpA, TbpB and PalA (Gerlach *et al.*, 1992; Gonzalez *et al.*, 1995; Mikael *et al.*, 2002; Srikumar *et al.*, 2004; Wilke *et al.*, 1997). Furthermore, even though these proteomic analyses were carried out on bacteria grown under iron-replete growth conditions approximately one-fifth of the identified OM proteins (9/47) were predicted to be involved in iron uptake. This clearly shows the importance of iron uptake in this organism (see Chapter 8).

The differential expression of *A. pleuropneumoniae* secreted and surface-associated proteins in response to the addition of porcine bronchoalveolar lavage fluid (BALF) has been analysed by 2DGE (Jacobsen *et al.*, 2005b). This analysis identified 12 proteins up-regulated in response to BALF but only eight of these proteins could be identified. The identified proteins included the heat shock protein GroES, an OM lipoprotein OmlA, a polynucleotide phosphorylase, a parvulin-like peptidyl-prolyl isomerase, two components of ABC transport systems, a [Cu,Zn]-superoxide dismutase and a conserved hypothetical protein. Interestingly, four of the proteins up-regulated in the presence of BALF were also up-regulated in a *fur* mutant strain but this up-regulation did not appear to be a response to low iron levels as protein expression was still up-regulated in response to BALF that had been supplemented with iron. The two ABC transport systems were predicted to be involved in dipeptide transport and metal ion transport. These proteins are likely important for nutrient acquisition in BALF. However, a mutant in metal ion transport was not attenuated for survival in the porcine host so clearly expression of this protein is not critical for survival in this medium.

In a related study (Jacobsen *et al.*, 2005a) the cytoplasmic protein AspA (aspartate ammonia lyase) was shown to be expressed more highly when cells were grown in the presence of BALF compared to growth in rich medium. This was shown, using a *ΔaspA::luxAB* fusion, to be the result of increased transcription of the *aspA* gene in the presence of BALF. AspA is predicted to be involved in the synthesis of fumarate which functions as an alternate electron acceptor when cells are grown under anaerobic conditions. Indeed, an *aspA* mutant was impaired in its ability to grow under anaerobic conditions and gave lower lesion scores in a porcine aerosol challenge model (Jacobsen *et al.*, 2005a).

Aggregatibacter actinomycetemcomitans
Proteomic reference maps for *Agg. actinomycetemcomitans* strain Y4 grown *in vitro* at 37°C in 5% CO_2 atmosphere have been reported (Nakano *et al.*, 2005). A total of 121 unique proteins was identified (see http://myamagu.dent.kyushu-u.ac.jp/~bioinformatics/AA/index.html). The majority of these were predicted to be cytoplasmic proteins and expression was analysed only under a single condition.

The secreted subproteome of *Agg. actinomycetemcomitans* strain NCTC9710 was assessed under a number of environmental conditions, namely (1) biofilm cells grown anaerobically on semi-defined solid medium; (2) biofilm cells grown in a CO_2 enriched environment on semi-defined solid medium; (3) planktonic cells grown anaerobically in semi-defined liquid medium; (4) planktonic cells grown anaerobically in semi-defined liquid medium supplemented with 10% horse serum; (5) biofilm cells grown anaerobically on semi-defined solid medium containing 100 μM 2,2′-dipyridyl and (6) biofilm cells grown anaerobically on BHI solid medium supplemented with 6% horse serum (Fletcher *et al.*, 2001). In total more than 100 proteins were identified as secreted under the range of conditions assessed. The protein expression profile observed in each case was compared with the standard condition (condition 1). A number of proteins was shown to alter their expression profile between the various conditions. Indeed, compared to condition 1 condition 2 showed increased expression of five proteins and decreased expression of one protein. Condition 3 showed an additional four proteins, three proteins with increased expression and two proteins with decreased expression. Condition 4 showed decreased expression of three proteins and the absence of one protein. Condition 5 showed the presence of six additional proteins and increased expression of three proteins. Condition 6 showed the presence of five proteins, increased levels of four proteins and decreased levels of one protein. This analysis was carried out before routine protein identification by mass spectrometry was available. However a small number of proteins was identified by N-terminal sequencing. These included triose phosphate isomerase, thiol peroxidase, [Cu, Zn]-superoxide dismutase and the hypothetical protein YfiD. The secretion of the [Cu, Zn]-superoxide dismutase was reduced during *Agg. actinomycetemcomitans* planktonic growth as compared to biofilm growth; this protein was also up-regulated during *Agg. actinomycetemcomitans* growth in the presence of BALF (Jacobsen *et al.*, 2005b). The secretion of the thiol peroxidase was increased during anaerobic growth (condition 3), while the secretion of the hypothetical protein YfiD was increased during anaerobic biofilm growth in the presence of 6% horse serum (Fletcher *et al.*, 2001).

The antigenic proteins of *Agg. actinomycetemcomitans* expressed during *in vivo* growth within the human oral cavity have been analysed using IVIAT (Cao *et al.*, 2004; Song *et al.*, 2002). This study identified 116 antigenic proteins across a range of functional categories. Further analysis of nine of these proteins indicated that human antibodies against six of the nine were significantly elevated in periodontal patients as compared to healthy controls. Mutants with defects in two of these genes (*orf1402* and *orf859*) showed reduced intracellular survival in a HeLa competitive growth assay confirming that these proteins are likely to be critical for normal progression of *Agg. actinomycetemcomitans* periodontal disease (Cao *et al.*, 2004).

Proteomic analysis within the genus *Haemophilus*

The genus *Haemophilus* contains at least 12 species, many of which cause a range of diseases in humans and animals. However, substantial proteomic analyses have mainly focused on the major human pathogen *Haemophilus influenzae*, which was the first bacterial species whose entire genome was sequenced and annotated (Fleischmann *et al.*, 1995). At least a partial, but accurate, genome sequence is of course a prerequisite for any MS-based identification of proteins, whether by peptide mass fingerprinting or by determination of peptide or N-terminal sequences.

Not surprisingly, 2D gel based protein reference maps started to appear soon after the release of the genome sequence (Cash *et al.*, 1995).

However, in the absence of protein identification data, such studies were limited to reporting numbers of proteins and to comparisons of protein profiles between different strains, in this case between Hib and NTHi strains. The degree of variability was reported as similar, regardless of whether comparisons were made between Hib and NTHi strains or amongst NTHi strains alone. Claims that these analyses could be useful for epidemiology or molecular typing have not been realized, probably due to the technically more difficult methods involved when compared with PCR or DNA sequence-based methods such as multilocus sequence typing (MLST).

The first map with protein identification data was published in 1997 (Langen *et al.*, 1997), with the map restricted to the soluble cytoplasmic subproteome, a fraction more amenable to analysis by 2DGE and MS. The *H. influenzae* genome contains approximately 1700 predicted open reading frames, from which 119 corresponding protein spots were identified. Not surprisingly, metabolic enzymes and hypothetical proteins predominated. A comparison of the observed and theoretical molecular masses of all the proteins showed excellent correlation. However, the predicted p*I* values differed to a greater extent from the experimentally determined values, with differences as high as 2 p*I* units. Both post translational modification of proteins, such as phosphorylation, or deficiencies in the p*I* prediction algorithms may account for such discrepancies.

The use of affinity chromatography to enrich the soluble sub-proteome for heparin-binding proteins allowed the identification of an additional 110 proteins (Fountoulakis *et al.*, 1997). Pre-enrichment for a component of the bacterial cell proteome by affinity chromatography or by fractionation facilitates the identification of low abundance proteins as well as allowing the definition of the components of a particular cell fraction.

Further refinement by the same group resulted in the identification of novel basic and low molecular mass proteins (Fountoulakis *et al.*, 1998a; Fountoulakis *et al.*, 1998b). Restriction of the first isoelectric focusing dimension to the p*I* range of 6–11 identified 58 new proteins; not surprisingly, a major proportion of these proteins represented nucleic acid binding or ribosomal proteins (Fountoulakis *et al.*, 1998b). Similarly, the use of tricine gels and the presence of urea in the second dimension facilitated the improved separation and subsequent identification of proteins in the molecular mass range of 5–20 kDa (Fountoulakis *et al.*, 1998a). These techniques allowed the identification of 40 additional proteins, including several ribosomal proteins, but the majority were of unknown function.

More recent approaches have utilized gel-free techniques to overcome some of the limitations of gel-based systems discussed earlier in this chapter. Kolker *et al* (2003) used liquid chromatography coupled with tandem mass spectrometry (LC MS/MS) to identify definitively 25% of all predicted proteins, including some expressed only under aerobic (146 proteins) or anaerobic growth (65 proteins) conditions. Significantly, a number of membrane proteins, which present particular challenges for proteomics, were identified. Proteomics technology may also be of value in the identification of so-called hypothetical genes and proteins. Using both transcriptional and proteomics analysis, 54 genes previously termed hypothetical were shown to be expressed by *H. influenzae in vitro*, with subsequent analysis attributing a function to 43 of them (Kolker *et al.*, 2004). In combination with enzymatics and bioinformatics modelling, proteomics analysis has also been used to define intermediary metabolism of *H. influenzae* (Raghunathan *et al.*, 2004). Such integrated analyses were used to construct a model of the genome-wide metabolic proteome; the reader is referred to that paper for details.

As discussed earlier, membrane proteins present some problems for proteomics studies; their hydrophobic nature makes solubilization in a form compatible with isoelectric focusing more difficult. However, Thoren *et al.* (2002) applied proteomics analyses following a strategy of sequential protein extraction from bacterial cell pellet (Molloy *et al.*, 1998). Although only 34 proteins were identified, the set included at least seven predicted membrane proteins. Interestingly, the protein elongation factor, TufB, co-fractionated with membrane proteins, similar to the finding in *P. multocida* where proteomics studies showed the presence of TufAB in the

OM under *in vivo* conditions (Boyce *et al.*, 2006).

Bacterial biofilms are ubiquitous in nature, and they exist in a wide variety of environments, as well as on the skin and epithelial surfaces of humans and animals. As a mucosal pathogen, it is therefore not unexpected that *H. influenzae* readily forms biofilms and that this capacity is associated with the ability to colonize both inert and mucosal surfaces (Erwin and Smith, 2007). A role for biofilm formation in *H. influenzae* has been implicated in chronic infections such as otitis media, bronchitis and sinusitis. Proteomics offers the potential to identify all proteins whose expression is differentially regulated in biofilms compared with planktonic cells. In a preliminary study, a total of 265 proteins present in the extracellular matrix of an NTHI biofilm was identified by LC MS/MS following standard SDS-PAGE (Gallaher *et al.*, 2006). However, the proteins spanned all COGs and many of them had either been identified previously in planktonic cells or had functions (such as essential housekeeping roles) which would preclude their presence uniquely in a biofilm matrix. Accordingly, it is not possible to determine whether their presence in the matrix is a result of cellular contamination of the preparation or of a legitimate role in matrix formation or maintenance. The use of comparative techniques such as DIGE would allow a direct quantitative comparison of biofilm and planktonic proteins.

The human pathogen [*Haemophilus*] *ducreyi* causes the sexually transmitted disease chancroid. The whole cell proteomes of several isolates of [*H.*] *ducreyi* showed only limited strain variability. However, only 116 proteins could be identified out of 800 spots observed on 2D gels from an estimated total of over 1700 (Scheffler *et al.*, 2003).

Proteomic analysis within the genus *Mannheimia*

M. haemolytica is the primary cause of bovine respiratory disease. Using an unfinished (at the time) genome sequence database Nanduri *et al.* (2005) used LC MS/MS to quantify 495 *M. haemolytica* proteins. Notably, the protein reference map was used to examine the changes in protein profile in response to sub-MICs of chlortetracycline and sulfamethazine and identified approximately 30 proteins with altered expression. As might be expected, many of these were proteins related to the site of interaction of the antibiotic with the bacterial cell; ribosomal proteins and translation factors were highly represented. Interestingly, an effect on the expression of known virulence factors was observed. The secreted leukotoxin LktA and its activator protein LktC were significantly down-regulated. In addition, putative virulence factors like SodA also showed reduced expression.

The capnophilic (CO_2-loving) species '*Mannheimia succiniciproducens*' is used for the industrial production of succinic acid. Determination of the complete genome sequence (Hong *et al.*, 2004) identified 2384 ORFs and allowed subsequent establishment of a reference map (Lee *et al.*, 2006) encompassing the whole cell lysate, total membranes and secreted proteomes, but with only 200 proteins identified. A comparison of the proteomes at stationary and exponential growth phases identified changes in key metabolic enzymes, offering the future potential of metabolic engineering for improved industrial performance and highlighted the need for proteomics studies to complement bioinformatics analysis.

Conclusions

Ideally, it would be desirable to be able to compare and contrast proteome reference maps and expression profiles from different studies and across different genera, species and strains. However, the wide variability in techniques used makes any realistic comparison of this type almost impossible. Differences in the bacterial strains used, growth conditions, fractionation methods, separation techniques, analysis software and MS hardware and techniques, all contribute to uncertainty in interpretation and therefore to potential errors in any attempts at comparative proteomics across experiments not conducted in the same laboratory. A further major issue is the fact that it is impossible to cover the entire cell proteome. Even with recent developments in staining techniques, the dynamic range available for detection of proteins cannot hope to cover the range of concentrations of the total compliment of cellular proteins. The general consensus is that at a genome wide level each ORF will correspond to approximately 1.6 protein spots, yet no studies on any members of the *Pasteurellaceae* have approached that level of genome/proteome coverage.

Another issue which commonly arises is the comparison between microarray-based transcription profiles and proteomics-generated protein profiles. Which gives a more accurate picture of cellular activity? Why do they not concur? In a limited number of studies on bacterial species (Boyce and Adler, 2006; Boyce *et al.*, 2006; Chong *et al.*, 2007; Nie *et al.*, 2006a,b) there is significant variability in the concordance between transcriptional and translational profiles. Reasons for poor concordance likely include differences in mRNA and protein half-life, post-transcriptional regulatory mechanisms and post-translational modifications such as phosphorylation and glycosylation. Accordingly, it is essential that both transcriptional and translational aspects are considered in any bacterial global expression studies. Both are appropriate and complementary.

Acknowledgement

Original work in the authors' laboratory was supported by the Australian Research Council, Canberra, Australia.

References

Altschul, S.F., Gish, W., Miller, W., Myers, E.W., and Lipman, D.J. (1990). Basic local alignment search tool. J. Mol. Biol. *215*, 403–410.

Berrier, C., Garrigues, A., Richarme, G., and Ghazi, A. (2000). Elongation factor Tu and DnaK are transferred from the cytoplasm to the periplasm of *Escherichia coli* during osmotic downshock presumably via the mechanosensitive channel MscL. J. Bacteriol. *182*, 248–251.

Berven, F.S., Flikka, K., Jensen, H.B., and Eidhammer, I. (2004). BOMP: a program to predict integral beta-barrel outer membrane proteins encoded within genomes of Gram-negative bacteria. Nucleic Acids Res. *32*, W394–399.

Boyce, J.D., and Adler, B. (2006). How does *Pasteurella multocida* respond to the host environment? Curr, Opin, Microbiol, *9*, 117–122.

Boyce, J.D., Cullen, P., Nguyen, V., Wilkie, I., and Adler, B. (2006). Analysis of the *Pasteurella multocida* outer membrane sub-proteome and its response to the in vivo environment of the natural host. Proteomics *6*, 870–880.

Boyce, J.D., Wilkie, I., Harper, M., Paustian, M.L., Kapur, V., and Adler, B. (2002). Genomic scale analysis of *Pasteurella multocida* gene expression during growth within the natural chicken host. Infect. Immun. *70*, 6871–6879.

Boyce, J.D., Wilkie, I., Harper, M., Paustian, M.L., Kapur, V., and Adler, B. (2004). Genomic-scale analysis of *Pasteurella multocida* gene expression during growth within liver tissue of chickens with fowl cholera. Microbes Infect. *6*, 290–298.

Canas, B., Lopez-Ferrer, D., Ramos-Fernandez, A., Camafeita, E., and Calvo, E. (2006). Mass spectrometry technologies for proteomics. Brief Funct Genomic Proteomic *4*, 295–320.

Cao, S.L., Progulske-Fox, A., Hillman, J.D., and Handfield, M. (2004). In vivo induced antigenic determinants of *Actinobacillus actinomycetemcomitans*. FEMS Microbiol. Lett. *237*, 97–103.

Cash, P., Argo, E., and Bruce, K. (1995). Characterisation of *Haemophilus influenzae* proteins by two-dimensional gel electrophoresis. Electrophoresis *16*, 135–148.

Chong, P., Burja, A.M., Radianingtyas, H., Fazeli, A., and Wright, P.C. (2007). Translational and transcriptional analysis of *Sulfolobus solfataricus* P2 to provide insights into alcohol and ketone utilisation. Proteomics *7*, 424–435.

Chung, J.W., Ng-Thow-Hing, C., Budman, L.I., Gibbs, B.F., Nash, J.H., Jacques, M., and Coulton, J.W. (2007). Outer membrane proteome of *Actinobacillus pleuropneumoniae*: LC-MS/MS analyses validate in silico predictions. Proteomics *7*, 1854–1865.

Erwin, A.L., and Smith, A.L. (2007). Nontypeable *Haemophilus influenzae*: understanding virulence and commensal behavior. Trends Microbiol. *15*, 355–362.

Fleischmann, R.D., Adams, M.D., White, O., Clayton, R.A., Kirkness, E.F., Kerlavage, A.R., Bult, C.J., Tomb, J.F., Dougherty, B.A., Merrick, J.M., *et al.* (1995). Whole-genome random sequencing and assembly of *Haemophilus influenzae* Rd. Science *269*, 496–512.

Fletcher, J.M., Nair, S.P., Ward, J.M., Henderson, B., and Wilson, M. (2001). Analysis of the effect of changing environmental conditions on the expression patterns of exported surface-associated proteins of the oral pathogen *Actinobacillus actinomycetemcomitans*. Microb Pathog *30*, 359–368.

Fountoulakis, M., Juranville, J.F., Roder, D., Evers, S., Berndt, P., and Langen, H. (1998a). Reference map of the low molecular mass proteins of *Haemophilus influenzae*. Electrophoresis *19*, 1819–1827.

Fountoulakis, M., Langen, H., Evers, S., and Takács, C.G.B. (1997). Two-dimensional map of *Haemophilus influenzae* following protein enrichment by heparin chromatography. Electrophoresis *18*, 1193–1202.

Fountoulakis, M., Takacs, B., and Langen, H. (1998b). Two-dimensional map of basic proteins of *Haemophilus influenzae*. Electrophoresis *19*, 761–766.

Frohlich, T., and Arnold, G.J. (2006). Proteome research based on modern liquid chromatography – tandem mass spectrometry: separation, identification and quantification. J. Neural Transm. *113*, 973–994.

Gallaher, T.K., Wu, S., Webster, P., and Aguilera, R. (2006). Identification of biofilm proteins in non-typeable *Haemophilus Influenzae*. BMC Microbiol. *6*, 65.

Gardy, J.L., Laird, M.R., Chen, F., Rey, S., Walsh, C.J., Ester, M., and Brinkman, F.S. (2005). PSORTb v.2.0: expanded prediction of bacterial protein subcellular localization and insights gained from comparative proteome analysis. Bioinformatics *21*, 617–623.

Gerlach, G F., Klashinsky, S., Anderson, C., Potter, A.A., and Willson, P.J. (1992). Characterization of two genes encoding distinct transferrin-binding proteins in different *Actinobacillus pleuropneumoniae* isolates. Infect. Immun. *60*, 3253–3261.

Gevaert, K., Van Damme, P., Ghesquiere, B., Impens, F., Martens, L., Helsens, K., and Vandekerckhove, J. (2007). A la carte proteomics with an emphasis on gel-free techniques. Proteomics *7*, 2698–2718.

Gonzalez, G.C., Yu, R.H., Rosteck, P.R., Jr., and Schryvers, A.B. (1995). Sequence, genetic analysis, and expression of *Actinobacillus pleuropneumoniae* transferrin receptor genes. Microbiology *141*, 2405–2416.

Gygi, S.P., Rist, B., Gerber, S.A., Turecek, F., Gelb, M.H., and Aebersold, R. (1999). Quantitative analysis of complex protein mixtures using isotope-coded affinity tags. Nat. Biotechnol. *17*, 994–999.

Handfield, M., Brady, L.J., Progulske-Fox, A., and Hillman, J.D. (2000). IVIAT: a novel method to identify microbial genes expressed specifically during human infections. Trends Microbiol. *8*, 336–339.

Hart, C., Schulenberg, B., Steinberg, T.H., Leung, W.Y., and Patton, W.F. (2003). Detection of glycoproteins in polyacrylamide gels and on electroblots using Pro-Q Emerald 488 dye, a fluorescent periodate Schiff-base stain. Electrophoresis *24*, 588–598.

Hong, S.H., Kim, J.S., Lee, S.Y., In, Y.H., Choi, S., Rih, J.K., Kim, C.H., Jeong, H., Hur, C.G., and Kim, J.J. (2004). The genome sequence of the capnophilic rumen bacterium *Mannheimia succiniciproducens*. Nat. Biotechnol. *22*, 1275–1281.

Jacobsen, I., Hennig-Pauka, I., Baltes, N., Trost, M., and Gerlach, G.F. (2005a). Enzymes involved in anaerobic respiration appear to play a role in *Actinobacillus pleuropneumoniae* virulence. Infect. Immun. *73*, 226–234.

Jacobsen, I.D., Meens, J., Baltes, N., and Gerlach, G.F. (2005b). Differential expression of non-cytoplasmic *Actinobacillus pleuropneumoniae* proteins induced by addition of bronchoalveolar lavage fluid. Vet. Microbiol. *109*, 245–256.

Juncker, A.S., Willenbrock, H., Von Heijne, G., Brunak, S., Nielsen, H., and Krogh, A. (2003). Prediction of lipoprotein signal peptides in Gram-negative bacteria. Protein Sci. *12*, 1652–1662.

Kolker, E., Makarova, K.S., Shabalina, S., Picone, A.F., Purvine, S., Holzman, T., Cherny, T., Armbruster, D., Munson, R.S., Jr., Kolesov, G., *et al.* (2004). Identification and functional analysis of 'hypothetical' genes expressed in *Haemophilus influenzae*. Nucleic Acids Res. *32*, 2353–2361.

Kolker, E., Purvine, S., Galperin, M.Y., Stolyar, S., Goodlett, D.R., Nesvizhskii, A.I., Keller, A., Xie, T., Eng, J.K., Yi, E., *et al.* (2003). Initial proteome analysis of model microorganism *Haemophilus influenzae* Strain Rd KW20. J. Bacteriol. *185*, 4593–4602.

Langen, H., Gray, C., Roder, D., Juranville, J.F., Takacs, B., and Fountoulakis, M. (1997). From genome to proteome: protein map of *Haemophilus influenzae*. Electrophoresis *18*, 1184–1192.

Lee, J.W., Lee, S.Y., Song, H., and Yoo, J.S. (2006). The proteome of *Mannheimia succiniciproducens*, a capnophilic rumen bacterium. Proteomics *6*, 3550–3566.

Link, A.J., Eng, J., Schieltz, D.M., Carmack, E., Mize, G.J., Morris, D.R., Garvik, B.M., and Yates, J.R., 3rd (1999). Direct analysis of protein complexes using mass spectrometry. Nat. Biotechnol. *17*, 676–682.

Lu, Z., Szafron, D., Greiner, R., Lu, P., Wishart, D.S., Poulin, B., Anvik, J., Macdonell, C., and Eisner, R. (2004). Predicting subcellular localization of proteins using machine-learned classifiers. Bioinformatics *20*, 547–556.

Marouga, R., David, S., and Hawkins, E. (2005). The development of the DIGE system: 2D fluorescence difference gel analysis technology. Anal. Bioanal Chem. *382*, 669–678.

Mikael, L.G., Pawelek, P.D., Labrie, J., Sirois, M., Coulton, J.W., and Jacques, M. (2002). Molecular cloning and characterization of the ferric hydroxamate uptake (*fhu*) operon in *Actinobacillus pleuropneumoniae*. Microbiology *148*, 2869–2882.

Molloy, M.P., Herbert, B.R., Walsh, B.J., Tyler, M.I., Traini, M., Sanchez, J.-C., Hochstrasser, D.F., Williams, K.L., and Gooley, A.A. (1998). Extraction of membrane proteins by differential solubilization for separation using two-dimensional gel electrophoresis. Electrophoresis *19*, 837–844.

Nakano, Y., Shibata, Y., Kawada, M., Kojima, M., Fukamachi, H., Shibata, Y., Okano, S., Matsushima, K., Abiko, Y., and Yamashita, Y. (2005). A searchable database for proteomes of oral microorganisms. Oral Microbiol. Immunol. *20*, 344–348.

Nanduri, B., Lawrence, M.L., Vanguri, S., and Burgess, S.C. (2005). Proteomic analysis using an unfinished bacterial genome: the effects of subminimum inhibitory concentrations of antibiotics on *Mannheimia haemolytica* virulence factor expression. Proteomics *5*, 4852–4863.

Nanduri, B., Lawrence, M.L., Boyle, C.R., Ramkumar, M., and Burgess, S.C. (2006). Effects of subminimum inhibitory concentrations of antibiotics on the *Pasteurella multocida* proteome. J. Proteom Res. *5*, 572–580.

Nie, L., Wu, G., and Zhang, W. (2006a). Correlation between mRNA and protein abundance in *Desulfovibrio vulgaris*: A multiple regression to identify sources of variations. Biochemical and Biophysical Research Communications *339*, 603–610.

Nie, L., Wu, G., and Zhang, W. (2006b). Correlation of mRNA expression and protein abundance affected by multiple sequence features related to translational efficiency in *Desulfovibrio vulgaris*: A quantitative analysis. Genetics *174*, 2229–2243.

Nørskov-Lauritsen, N., and Kilian, M. (2006). Reclassification of *Actinobacillus actinomycetemcomitans, Haemophilus aphrophilus, Haemophilus paraph-*

rophilus and Haemophilus segnis as *Aggregatibacter actinomycetemcomitans* gen. nov., comb. nov., *Aggregatibacter* aphrophilus comb. nov. and *Aggregatibacter segnis* comb. nov., and emended description of *Aggregatibacter aphrophilus* to include V factor-dependent and V factor-independent isolates. Int. J. Syst Evol. Microbiol. *56*, 2135–2146.

Ong, S.E., Blagoev, B., Kratchmarova, I., Kristensen, D B., Steen, H., Pandey, A., and Mann, M. (2002). Stable isotope labeling by amino acids in cell culture, SILAC, as a simple and accurate approach to expression proteomics. Mol. Cell Proteomics *1*, 376–386.

Paustian, M.L., May, B.J., and Kapur, V. (2001). *Pasteurella multocida* gene expression in response to iron limitation. Infect. Immun. *69*, 4109–4115.

Raghunathan, A., Price, N.D., Galperin, M.Y., Makarova, K.S., Purvine, S., Picone, A.F., Cherny, T., Xie, T., Reilly, T.J., Munson, R., *et al.* (2004). In silico metabolic model and protein expression of *Haemophilus influenzae* Strain Rd KW20 in rich medium. OMICS *8*, 25–41.

Rimler, R.B. (2001). Purification of a cross-protective antigen from *Pasteurella multocida* grown in vitro and in vivo. Avian Dis. *45*, 572–580.

Roe, M.R., and Griffin, T.J. (2006). Gel-free mass spectrometry-based high throughput proteomics: tools for studying biological response of proteins and proteomes. Proteomics *6*, 4678–4687.

Scheffler, N.K., Falick, A.M., Hall, S.C., Ray, W.C., Post, D.M., Munson, R.S., and Gibson, B.W. (2003). Proteome of *Haemophilus ducreyi* by 2-D SDS-PAGE and mass spectrometry: strain variation, virulence, and carbohydrate expression. J. Proteome Res. *2*, 523–533.

Song, Y.H., Kozarov, E.V., Walters, S.M., Cao, S.L., Handfield, M., Hillman, J.D., and Progulske-Fox, A. (2002). Genes of periodontopathogens expressed during human disease. Ann. Periodontol. *7*, 38–42.

Srikumar, R., Mikael, L.G., Pawelek, P.D., Khamessan, A., Gibbs, B.F., Jacques, M., and Coulton, J.W. (2004). Molecular cloning of haemoglobin-binding protein HgbA in the outer membrane of *Actinobacillus pleuropneumoniae*. Microbiology *150*, 1723–1734.

Tabatabai, L.B., and Zehr, E.S. (2004). Identification of five outer membrane-associated proteins among cross-protective factor proteins of *Pasteurella multocida*. Infect. Immun. *72*, 1195–1198.

Thoren, K., Gustafsson, E., Clevnert, A., Larsson, T., Bergstrom, J., and Nilsson, C.L. (2002). Proteomic study of non-typeable *Haemophilus influenzae*. J. Chromatogr B Analyt Technol Biomed Life Sci. *782*, 219–226.

Unlu, M., Morgan, M.E., and Minden, J.S. (1997). Difference gel electrophoresis: a single gel method for detecting changes in protein extracts. Electrophoresis *18*, 2071–2077.

Wiese, S., Reidegeld, K.A., Meyer, H.E., and Warscheid, B. (2007). Protein labeling by iTRAQ: a new tool for quantitative mass spectrometry in proteome research. Proteomics *7*, 340–350.

Wilke, M., Franz, B., and Gerlach, G.F. (1997). Characterization of a large transferrin-binding protein from *Actinobacillus pleuropneumoniae* serotype 7. Zentralbl. Veterinarmed. B *44*, 73–86.

Wu, J.R., Shien, J.H., Shieh, H.K., Chen, C.F., and Chang, P.C. (2007). Protective immunity conferred by recombinant *Pasteurella multocida* lipoprotein E (PlpE). Vaccine *25*, 4140–4148.

Wu, W.W., Wang, G., Baek, S.J., and Shen, R.F. (2006). Comparative study of three proteomic quantitative methods, DIGE, cICAT, and iTRAQ, using 2D gel- or LC-MALDI TOF/TOF. J. Proteome Res. *5*, 651–658.

Protein Secretion in *Pasteurellaceae*

6

Scott C. Kachlany and Nataliya V. Balashova

Abstract

Protein secretion is an essential process for all cells. Gram-negative bacteria use different secretion systems to transport a variety of proteins across their double membrane and into the extracellular environment. In general, bacterial protein secretion systems are complex machineries made up of many components that act together to transport molecules out of the cell. There are essentially three classes of proteins that the members of the *Pasteurellaceae* secrete: toxins, adhesins and proteases. Toxins produced by the members of the *Pasteurellaceae* include the highly host-specific RTX toxins and cytolethal distending toxins. Adhesins include the self-secreting autotransporters, the widespread Flp pili, and type IV pili. The proteases produced by these members include the antibody-cleaving immunoglobulin and bifunctional HAP proteases.

Toxins

Protein toxins are important virulence factors for many bacteria. While some toxins require direct contact with host cells or are injected directly into cells, others are secreted as soluble proteins where they act from a distance. The two major families of secreted toxins within the *Pasteurellaceae* family are the RTX and CDT toxins.

RTX toxins

RTX (repeats-in-toxin) toxins are large, secreted proteins that contain C-terminal glycine-rich repeats that bind calcium. The toxins act on the host cell surface and damage cell membranes by pore formation (Lally *et al.*, 1999; Welch, 2001) or displacement of lipid molecules in the membrane (Soloaga *et al.*, 1999). The prototypical members of this toxin family include *Escherichia coli* α-haemolysin (HlyA) and *Bordetella pertussis* adenylate cyclase (CyaA). RTX toxins that are produced by members of the *Pasteurellaceae* include *Aggregatibacter actinomycetemcomitans* leukotoxin (LtxA), *M. haemolytica* leukotoxin (LktA), and *Actinobacillus pleuropneumoniae* Apx toxins (ApxI, ApxII, ApxIII, ApxIV). While there exist many similarities among all RTX toxins and the way they are secreted, RTX toxins of the *Pasteurellaceae* are unique in several ways. For example, the RTX toxins discussed here display a high degree of host cell specificity, in contrast to HlyA and CyaA, which act on cells from many different species.

RTX toxins are exported via a type I secretion machinery (Fig. 6.1) (Delepelaire, 2004). Secretion of RTX toxins does not require a specific signal peptide or consensus sequence and is sec-independent (see below). Rather, the C-terminal region of RTX toxins appears to assume a three-dimensional conformation that is recognized by the ATP-binding cassette (ABC) protein of the secretion apparatus. Interestingly, RTX toxins are secreted in an unfolded state and periplasmic intermediates are not detected.

In general, secretion of RTX toxins requires three proteins. The 'RtxB' and 'RtxD' components of the system are encoded by the *rtx* operon, which also codes for the actual toxin ('RtxA') and a lipid-activating protein ('RtxC') (Fig. 6.1). Another required component, a TolC-like protein, is unlinked to the *rtx* genes and plays a

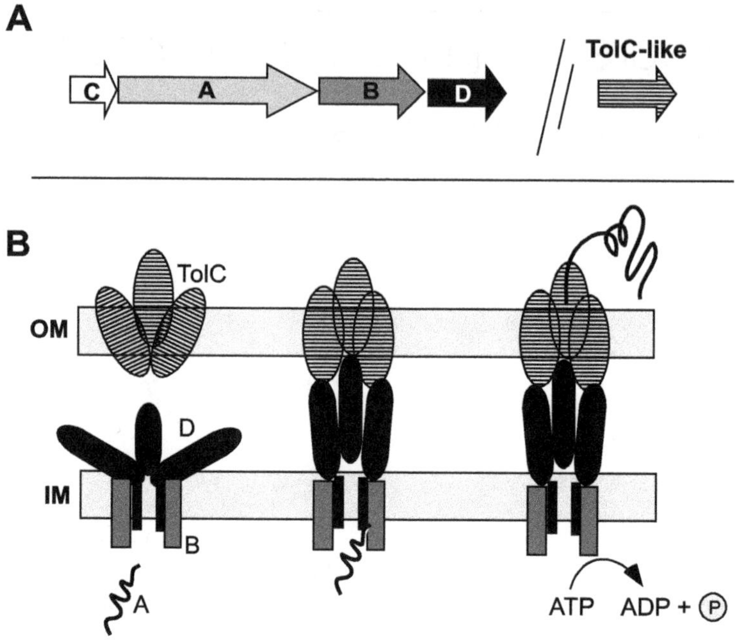

Figure 6.1 The RTX operon and type I secretion system. (A) Generic arrangement of the genes encoding RTX proteins. (B) A model for secretion of RTX toxins via the type I secretion system. Modified from (Andersen *et al.*, 2000). Upon contact of the toxin substrate with the IM secretion complex, RtxD makes contact with TolC and forms a continuous channel to the outside. The process requires ATP hydrolysis, which is carried out by RtxB.

role in the secretion of many types of molecules from the bacterial cell.

RtxB is a ~80-kDa ABC protein that is associated with the inner membrane (IM) as dimers. RtxB proteins contain two nucleotide binding domains (NBD) and possess ATPase activity that provides energy to power toxin secretion. Mutations in the NBD Walker boxes result in reduced levels of secretion (Saier, 2006). The proteins also possess transmembrane domains that anchor them in the IM. Evidence suggests that the C-terminal domain of the RTX toxin directly contacts one of the NBDs of RtxB. Indeed, specificity for the toxin to be secreted appears to reside in the RtxB component.

Located in the IM and in the periplasm and associated with RtxB is the membrane fusion protein, RtxD (Fig. 6.1). RtxD is a ~54-kDa protein that assumes a trimeric state. The N-terminus of the protein is cytoplasmic and interacts directly with the toxin. This interaction between the toxin and the RtxB/D apparatus results in a drastic conformational change of RtxD. Prior to toxin secretion, RtxD is folded toward the IM and does not traverse the entire periplasm. However, when the toxin is bound, RtxD folds inwards and forms a protein bridge to the outer membrane (OM). In the OM, RtxD contacts the OM secretion channel, a TolC protein (Andersen *et al.*, 2000)

TolC is a ~54-kDa multifaceted protein that has been extensively studied and plays a role in several cellular processes, including protein secretion and export of antimicrobial agents (Koronakis *et al.*, 2004). TolC was initially identified in a screen for bacterial mutants that were resistant to killing by colicin E1 (Hill and Holland, 1967; Nagel de Zwaig and Luria, 1967; Nomura and Witten, 1967). Mutations in these colicin-resistant bacteria mapped to *tolC*. Similarity at the amino acid level among *tolC* gene products is limited; however, secondary structure predictions show high similarity among all TolC proteins. In addition, a common feature to these proteins is internal sequence similarity between the N-terminal and C-terminal halves of the protein. The crystal structure of *E. coli* TolC has been solved and reveals a trimeric structure that forms a pore in the OM (Koronakis *et al.*, 2004). The trimeric assembly forms a 140 Å structure that is composed primarily of α-helices. Approximately 100 Å extends into the periplasm. The opening on the extracellular side of the trimeric structure is relatively wide while the periplasmic opening is narrow or closed. Interaction between toxin and the RtxB/D complex results in the opening of the periplasmic side using an iris-like mechanism. Once opened, the toxin can move from the intracellular space to the extracellular environment through a continuous channel. Thus, the model proposes that binding of the toxin to RtxB/D results in the formation of a continuous channel from the cytosol to the extracellular environment (Andersen *et al.*, 2000).

Agg. actinomycetemcomitans *leukotoxin (Ltx)*

Agg. actinomycetemcomitans produces a leukotoxin (LtxA) that is specific for blood cells from humans and Old World primates (Taichman *et al.*, 1987; Tsai *et al.*, 1984). The toxin targets LFA-1 on leukocytes and causes apoptosis at low concentrations and necrosis at high concentrations. It was recently discovered that LtxA can also lyse erythrocytes from a variety of species, although a higher concentration of toxin is required than for leukocytes (Balashova *et al.*, 2006a). Since erythrocytes lack LFA-1, interaction between LtxA and host erythrocytes proceeds by a mechanism distinct from leukocytes and the nature of an erythrocyte receptor remains unknown (Balashova *et al.*, 2006a).

LtxA is considered one of the major virulence factors of *Agg. actinomycetemcomitans*. However, because the toxin does not affect rat or mouse cells, it is difficult to assess its contribution to virulence in an animal model for periodontitis. Nonetheless evidence that strongly supports its important role in disease includes antibody to the toxin can be detected in patients with severe periodontitis, strains of *Agg. actinomycetemcomitans* that over-express the toxin may be more frequently associated with severe disease, and its similarity to other significant RTX toxins.

Prior to 2000, it was believed that LtxA was the only RTX toxin that was not secreted into culture supernatants. It was concluded that the toxin was found only in the cell membrane and in preparations of OM vesicles. However, it was discovered that in early stages of growth (early-log to mid-log), *Agg. actinomycetemcomitans* secretes abundant toxin into culture supernatants and this was essentially the only protein present (Diaz *et al.*, 2006; Kachlany *et al.*, 2000a). At later stages of growth, the toxin appears to be degraded almost completely, which may explain why leukotoxin secretion was not recognized earlier. Subsequent work has resulted in purification schemes that allow for the isolation of large quantities of active leukotoxin from both laboratory strains and clinical isolates.

Because the toxin was originally not known to be secreted, early work questioned the role of the 'putative' secretion genes, *ltxB* and *ltxD*. In 1991, Lally *et al.* (1991) expressed the *ltx* operon in *E. coli* and reported that the toxin could be found on the surface of cells only when *ltxBD* were also present. However, at the time, they did not examine culture supernatants for secretion. Thus, they concluded that *ltxBD* were required for localization of the toxin to the cell surface. In 1995, Kolodrubetz *et al.* (1995) generated mutations in *ltxB* and *ltxD* in *Agg. actinomycetemcomitans* and observed a decrease in LtxA protein, but not *ltxA* mRNA. But because they did not realize that LtxA was secreted, culture supernatants were never examined for the presence of the toxin. Recently, we isolated mutants of *ltxB* and *ltxD* using both directed and random transposon mutagenesis approaches (unpublished, M. P. Isaza and S. C. Kachlany). As one would predict, these mutants no longer secrete LtxA into culture supernatants, confirming the roles of LtxB and LtxD in LtxA secretion in *Agg. actinomycetemcomitans*.

Because TolC proteins show limited similarity to each other at the amino acid level, a standard search of the *Agg. actinomycetemcomitans* genome with *E. coli* TolC yields no significant matches. Using a set of criteria that included TolC–like domain similarity, intramolecular similarity, and protein size, a TolC-like candidate was recently identified in *Agg. actinomycetemcomitans* (Crosby and Kachlany, 2007). Genetic characterization of this gene revealed that it was required for both LtxA secretion and resistance to toxic compounds and antibiotics, similar to TolC from other bacteria. This gene product, TdeA (for toxin and drug export), is the third and likely final component of the *Agg. actinomycetemcomitans* LtxA type I secretion system and is also part of a drug efflux pump that has yet to be thoroughly studied.

Secretion of leukotoxin is regulated by iron, which further supports the notion that LtxA-mediated haemolysis may be a mechanism for the bacterium to obtain iron (Balashova *et al.*, 2006b). At high concentrations of free iron, LtxA is produced, but remains intracellular. In contrast, under low iron conditions, LtxA is exclusively secreted into culture supernatants. The mechanism of regulation is unknown, but iron could be affecting the size of the TdeA pore.

Mannheimia haemolytica *leukotoxin (Lkt)*

M. haemolytica produces a 102-kDa leukotoxin, LktA (Highlander *et al.*, 1990; Strathdee and Lo, 1989a; Zecchinon *et al.*, 2005). Like *Agg. actinomycetemcomitans* LtxA, LktA has high specificity for certain cell types. LktA specifically targets white blood cells from ruminants which parallels the disease causing ability of the bacterium. In addition, macrophages are more resistant to the toxin than neutrophils. LktA is considered the main virulence factor of *M. haemolytica* and inactivation of LktA results in significantly fewer pulmonary lesions in the upper respiratory tract of calves in a disease model (Zecchinon *et al.*, 2005). Because the cow is a natural host for *M. haemolytica*, this ideal 'animal model' provides strong evidence that this RTX toxin is indeed an important virulence factor.

Physiologically, production of LktA appears to be most abundant during log phase of growth and doesn't occur at 30°C (Strathdee and Lo, 1989b). Nutritional composition of growth media also affects toxin production (van Rensburg and du Preez, 2007). Production of LktA is significantly enhanced in amino acid limited medium. Addition of cysteine and glutamine to the amino acid-limited medium resulted in an even greater yield of LktA. In addition, manganese and iron also stimulated production of LktA. While these studies examine the overall production of LktA, and not necessarily secretion, it can be assumed that secretion must also be enhanced under stimulatory conditions where a greater yield of toxin is found in culture supernatants.

Bacteria heterologous host expression studies of LktA have lead to two different observations. Chang *et al.* (1989) found that LktA could be expressed in *E. coli* and secreted from cells when *lktB* and *lktD* were present. In contrast, Highlander *et al.* (1990) reported that LktA was not secreted from *E. coli* cells when *lktB* and *lktD* were present but that expression of *hlyB* and *hlyD* allowed secretion of LktA. This latter group found that the lack of LktB and LktD function in *E. coli* was likely the result of little or no transcription of these genes in the heterologous host. The discrepancy in outcomes between these two studies could be due to differences in cloning strategies and vectors used for *lktB* and *lktD*. To date, a TolC protein has not yet been experimentally identified in *M. haemolytica*; however, a predicted ORF in the completed genome shows significant similarity (56% amino acid identity; 73% similarity) to *Agg. actinomycetemcomitans* TdeA. Thus, it seems likely that LktB, LktD, and a TolC-like protein function as secretion proteins for LktA (Zecchinon *et al.*, 2005).

A. pleuropneumoniae *Apx toxins*

The porcine pathogen, *A. pleuropneumoniae*, produces four RTX toxins, ApxI, ApxII, ApxIII, and ApxIV, with molecular masses of 105 kDa, 105 kDa, 120 kDa, and 200 kDa, respectively (Bosse *et al.*, 2002). The toxins differ in their activity such that ApxI is strongly haemolytic while ApxII, ApxIII, and ApxIV are weakly or not haemolytic. In addition, ApxI and ApxIII are strongly cytotoxic and ApxII displays moderate cytotoxicity. ApxIV was discovered relatively recently and appears to be expressed only *in vivo* (Schaller *et al.*, 1999). The distribution of the toxins within *A. pleuropneumoniae* serotypes also varies with ApxII and ApxIV being the most widely distributed.

Analogous to *M. haemolytica*, the natural disease for *A. pleuropneumoniae* also serves as an effective animal model in which to study the contribution of putative virulence factors (Baarsch *et al.*, 1995; Chung *et al.*, 1994). Interestingly, preparations of the toxins themselves can cause lesions similar to those observed during natural infection. Because there are four different *apx* operons, the genetics of the system turns out to be very interesting. Secretion genes are found only in the *apxI* and *apxIII* operons, yet all the Apx toxins are secreted. In cases where *apxII* is present, the *apxIB* and *apxID* genes must also be present in order for ApxII to be secreted from cells. Apparently, though, *apxIIIB* and *apxIIID* are unable to complement the *apxII* system and ApxII is not secreted in serotypes where the *apxI* secretion genes are not encoded. This is indeed the scenario for serotype 3, which incidentally displays little haemolysis. Hence, secretion is required for haemolytic activity, similar to other RTX toxins. Further, just production of the toxin is not sufficient for pathogenesis, as secretion of the toxin must also occur for virulence in animals. To date little is known about the *apxIV* system, but ApxIV appears to be expressed by all serotypes of *A. pleuropneumoniae*, suggesting that it can employ the secretion components from either the *apxI* or *apxIII* operon. A BLAST search with *Agg. actinomycetemcomitans* TdeA revealed a predicted ORF in both sequenced strains of *A. pleuropneumoniae* (L20 and 4074) that showed high similarity (~55% identity, 73% similarity). Therefore, a TolC-like protein is probably also involved in the secretion of Apx toxins.

In addition to being secreted into culture supernatants, ApxI and ApxII can be found in membrane vesicles that bleb from the cell surface (Negrete-Abascal *et al.*, 2000), similar to LtxA of *Agg. actinomycetemcomitans*. The production of such vesicles may serve as a mechanism for delivery of the toxins to host cells. While an attractive hypothesis in pathogenesis, experimental evidence supporting the model in *A. pleuropneumoniae* is awaiting.

Cytolethal distending toxin

Cytolethal distending toxin (CDT) is a bacterial genotoxin that initiates a eukaryotic cell cycle block at the G2 stage prior to mitosis. The toxin causes cell distension, nucleus swelling and chromatin fragmentation. CDT is produced by a number of bacterial pathogens including *Campylobacter* species, *Escherichia coli*, *Salmonella enterica* serovar Typhi, *Shigella dysenteriae*, and enterohepatic *Helicobacter* species (Thelestam and Frisan, 2004). Two *Pasteurellaceae* family members, *Agg. actinomycetemcomitans* and [*Haemophilus*] *ducreyi*, also produce CDT and secrete it into culture supernatants (Stevens *et al.*, 1999; Sugai *et al.*, 1998).

Except for *S. enterica* serovar Typhi, the functional toxin is composed of three proteins: CdtA, CdtB, and CdtC (Fig. 6.2) (Pickett *et al.*, 1994; Scott and Kaper, 1994). All three of these protein subunits are required to constitute the fully active holotoxin by forming a tripartite complex (Lara-Tejero and Galan, 2001). CdtB is the functional component leading to cell cycle block while CdtA and CdtC bind CdtB and deliver it to the mammalian cell interior (Deng and Hansen, 2003; Lara-Tejero and Galan, 2000; Lee *et al.*, 2003). It has been proposed that the CDT holotoxin is internalized into target cells by means of receptor-mediated endocytosis (Cortes-Bratti *et al.*, 2001). Once inside the cell, CdtB enters the nucleus and exhibits a DNase I-like activity that results in DNA double-strand breaks (Elwell and Dreyfus, 2000; Lara-Tejero and Galan, 2000). CdtB subunit is presumably transported from the endoplasmic reticulum to the cytosol by retrotranslocation, and finally to the nucleus where it causes DNA lesions (Alby *et al.*, 2001; Cortes-Bratti *et al.*, 2001). The eukaryotic cell responds to the DNA damage by initiating a regulatory cascade that results in cell cycle arrest, cellular distension, and cell death (Elwell and Dreyfus, 2000; Ohara *et al.*, 2004). Therefore, CDT is a potential virulence factor and may play an important role in disease progression and immunosuppression (Purdy *et al.*, 2000; Shenker *et al.*, 1999; Young *et al.*, 2004).

A model for the secretion of CDT is shown in Fig. 6.2 and is based on the work of several groups. Unmodified CdtA, immature CdtB, and immature CdtC are secreted into the periplasmic space through a sec-dependent secretion pathway (see below). After translocation to the IM, immature CdtB and CdtC undergo processing by signal peptidase I. Unmodified CdtA then undergoes processing by lipoprotein-specific signal peptidase II followed by lipid modification. A fatty acid moiety is added to an N-terminal cysteine residue via amide linkage. The lipid-modified CdtA is then transported to the OM where CdtA, CdtB, and CdtC form a complex and processing of CdtA results in the removal of ~40 amino

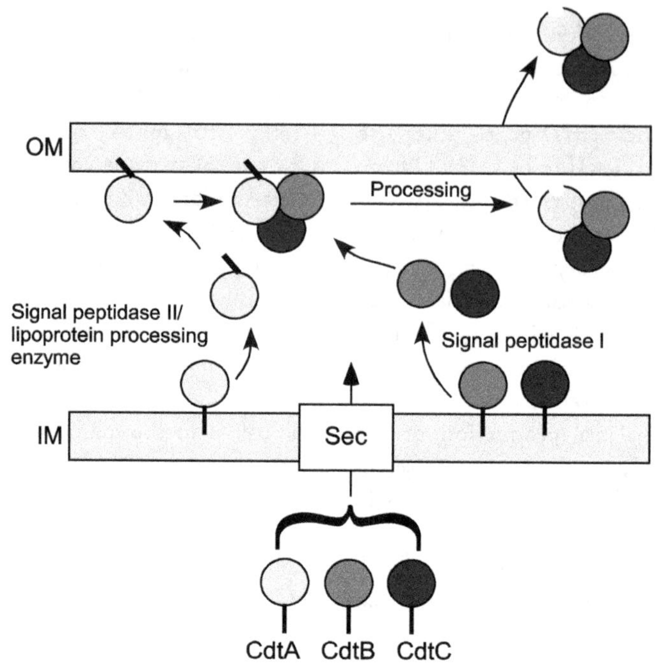

Figure 6.2 A model for the export of cytolethal distending toxin. The CDT toxins are secreted through the IM via the Sec machinery and then processed in the periplasm and OM before being transported across the OM.

acids from the N-terminus. Processing is carried out by an unidentified protease and evidently takes place in the periplasm. Membrane anchoring of lipid-modified CdtA may be important for the efficient formation of the CDT complex or for efficient secretion. The resulting mature holotoxin is finally secreted across the OM by an unknown mechanism (Hu *et al.*, 2006).

The Sec translocation system

To be secreted out of the cell, CDT must first cross the IM. A general mechanism for IM transport has evolved in Gram negative bacteria called the Sec system. Because this is a general system that is used by several secreted proteins in the *Pasteurellaceae*, we will discuss it here and refer back to this in subsequent discussions.

The Sec translocation system (also Sec secretion or Sec machinery) is a major pathway of protein translocation across the IM in bacteria (Saier, 2006). The system has been characterized most extensively in *E. coli* (Manting and Driessen, 2000) and is made up of the proteins SecA, SecB, SecD, SecE, SecF, SecG, SecY, YajC, and YidC (Fig. 6.3). To prevent newly synthesized proteins from misfolding and aggregating, the homotetrameric molecular chaperone SecB interacts with the newly synthesized protein in the cytoplasm and targets the protein to the SecA-SecYEG translocase at the membrane surface (Fekkes and Driessen, 1999; Manting *et al.*, 2000). The SecYEG complex is ubiquitous, being present in every bacterium, archaeon and eukaryote whose genome has been fully sequenced. The homodimer SecA is an ATPase and provides energy for the translocation of the unfolded protein across the membrane (Schmidt and Kiser, 1999). The pore through which the secreted protein passes is formed by the integral membrane proteins SecY, SecE and SecG (Manting and Driessen, 2000). The role of three additional membrane-bound proteins, SecD, SecF and YajC, is believed to improve translocation efficiency and to regulate the opening of the pore between translocation events (Nouwen *et al.*, 2001).

After the targeting and translocation steps have occurred, the secreted protein is released from the SecYEG pore and is bound to the membrane via its signal peptide. Type I signal peptidase, a membrane-bound endopeptidase, cleaves the signal off thereby releasing the mature protein into the periplasm. Sec-dependent signal sequences in the precursor proteins are characterized by a hydrophobic core of ten or more residues

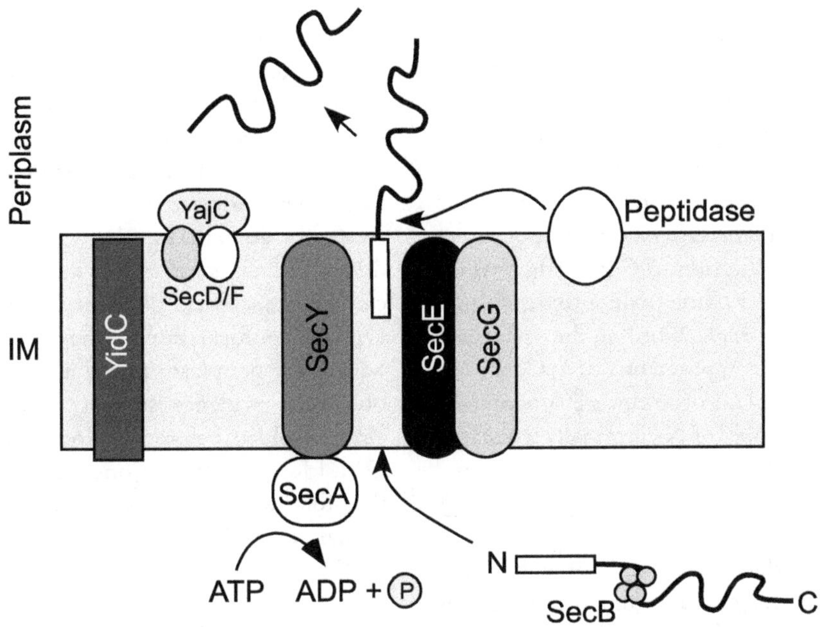

Figure 6.3 The Sec-translocation system. The IM components of the Sec machinery recognize an N-terminal secretion signal on the substrate. ATP hydrolysis by SecA drives the transport of the substrate through the Sec channel and into the periplasm where its N-terminal sequence is cleaved by a peptidase.

flanked by a positively charged amino-terminal region and a hydrophilic carboxy-terminal region containing a consensus sequence for signal peptidase cleavage. Some proteins are targeted to the IM after translocation via the signal recognition particle (SRP) protein-RNA complex. In *E. coli*, SRP is composed of a protein component, Ffh (Duong and Wickner, 1998; Kihara and Ito, 1998). SRP may also play a role in the export of some secreted proteins, such as autotransporters (Sijbrandi *et al.*, 2003). YidC, another membrane component of Sec machinery, has been shown to be involved in protein integration into membrane (Urbanus *et al.*, 2001; van der Laan *et al.*, 2001).

[H.] ducreyi *CDT*

[*H.*] *ducreyi* CDT was found in approximately 85% of [*H.*] *ducreyi* clinical isolates from different geographical origins (Ahmed *et al.*, 2001; Kulkarni *et al.*, 2003; Purven *et al.*, 1995). [*H.*] *ducreyi* CDT is chromosomally encoded by the *cdtABC* gene cluster whose protein products are CdtA (25 kDa), CdtB (32 kDa), and CdtC (21 kDa) (Cope *et al.*, 1997).

The crystal structure of the holotoxin from [*H.*] *ducreyi* has been determined and shows that CdtA, CdtB and CdtC form a ternary complex in a 1:1:1 stoichiometry. CdtA and CdtC form ricin-like lectin domains that may play a role in recognizing the cellular receptor that delivers the CdtB subunit into the target cells. CdtB shares conserved residues with the active sites of phosphodiesterase, DNase I and sphingomyelinase, and is structurally similar to the DNase I fold (Nesic *et al.*, 2004). Studies suggest that the active subunit, CdtB undergoes significant conformational changes when disassembled from the tripartite complex (Hu *et al.*, 2006).

Agg. actinomycetemcomitans *CDT*

The genes for CDT were originally identified in *Agg. actinomycetemcomitans* by Sugai *et al.* (1998) who reported a set of genes that resembled the [*H.*] *ducreyi cdt* locus. Two other groups reported on the toxin soon after this initial find (Mayer *et al.*, 1999; Shenker *et al.*, 1999). Most clinical strains harbour the locus, suggesting that it is important for survival or pathogenesis of the organism in the host (Ahmed *et al.*, 2001; Fabris *et al.*, 2002; Sugai *et al.*, 1998). The protein is highly similar to CDT from [*H.*] *ducreyi*, both at the amino acid and structural levels (Hu *et al.*, 2006; Yamada *et al.*, 2006). However, crystal structures have revealed that the molecular packing of *Agg. actinomycetemcomitans* CDT and [*H.*] *ducreyi* CDT are quite different due to variation of the amino acid residues in a non-conserved loop of CdtB. The loop of *Agg. actinomycetemcomitans* CdtB assumes a conformation to form a dimer, while the analogous loop *of [H.] ducreyi* CdtB prevents CDT from forming a dimer.

CdtA from *Agg. actinomycetemcomitans* was shown to be a membrane-associated lipoprotein and the lipid modification of CdtA is important for the export of CDT holotoxin into the culture supernatant (Ueno *et al.*, 2006). In the periplasm, CDT is a complex composed of CdtA, CdtB, and CdtC, whereas CDT in the culture supernatant is a complex composed of N-terminally truncated CdtA, CdtB and CdtC.

Adhesins

Adherence to surfaces is essential for all bacteria. In the host, pathogens use complex structures to adhere to, or colonize, cells and tissues. Proteinaceous appendages, generically termed adhesins, are produced and secreted by bacteria and mediate attachment to surfaces. Three major categories of adhesins are found in members of the *Pasteurellaceae*: the autotransporters, Flp pili and type IV pili.

Autotransporters

An extraordinary feature of the autotransporter (type V) secretion pathway is a self-directed OM transport step allowing the protein to be secreted autonomously, even from foreign Gram-negative host cells (Henderson *et al.*, 2004). Pohlner *et al.* (1987) were the first who proposed the model for type V secretion based on IgA1 protease secretion (Fig. 6.4). The autotransporter protein is initially translated as a precursor protein with an N-terminal signal peptide, the passenger domain and the carboxy-terminal helper domain. The signal peptide is necessary to target the protein into the periplasmic space via the Sec-pathway (see above). The C-terminal domain consists of an α-helical region and β-core region. The β-core region is involved in secretion of the protein by forming a pore in the OM. In the so-called trimeric autotransporters (also called the oligomeric coiled coil adhesins – OCA), three proteins oligomerize to form a trimeric pore (Cotter *et al.*, 2006).

The mechanism of secretion involves translocation through two membranes. In some cases, the passenger domain is then cleaved from the helper domain by an autoproteolytic event or by a membrane-bound protease and released into the extracellular milieu (Fig. 6.4, scenario A). The autoproteolytic digestion is carried out by a serine protease motif that is present in the passenger domain. In other cases, the passenger domain remains associated with the OM either covalently (scenario B) or peripherally (scenario C). The conformation of the passenger domain within the periplasm is still a subject of debate, but some evidence suggests that it might be already folded before translocation through the OM. However, at least one other study suggests that folding does not occur until the passenger domain is on the surface of the cell (Oliver *et al.*, 2003).

Haemophilus influenzae *Hap*

Hap is an autotransporter protein that mediates attachment to and entry into epithelial cells (St Geme *et al.*, 1994). All strains of *H. influenzae* contain the *hap* gene and the protein is 60% identical at the amino acid level to its IgA1 protease (see below). After translocation, the passenger domain may be cleaved via autoproteolysis and released. This extracellular form of Hap apparently acts as an IgA1 protease (see below). However, the cleavage reaction is not complete and the passenger domain may also remain attached to the helper domain, existing as a cell-associated protein. This cell-associated form of the protein is responsible for mediating cell attachment and entry (Hendrixson and St Geme, 1998).

H. influenzae *Hia*

A second autotransporter adhesin that is present in *H. influenzae* is Hia (St Geme and Cutter, 2000). In contrast to Hap, Hia is present in only 20–25% of strains, but acts as a major adhesin when it is present. Interestingly, no protease motif exists in the protein and so there is no cleavage event between the passenger and helper domains. Thus, the adhesin remains associated with the cell surface through covalent attachment to the passenger domain. Hia is part of the trimeric autotransporter subfamily (Cotter *et al.*, 2006). The recently determined crystal structure of Hia revealed that each monomer forms four β-strands in the OM, instead of 12 that is found

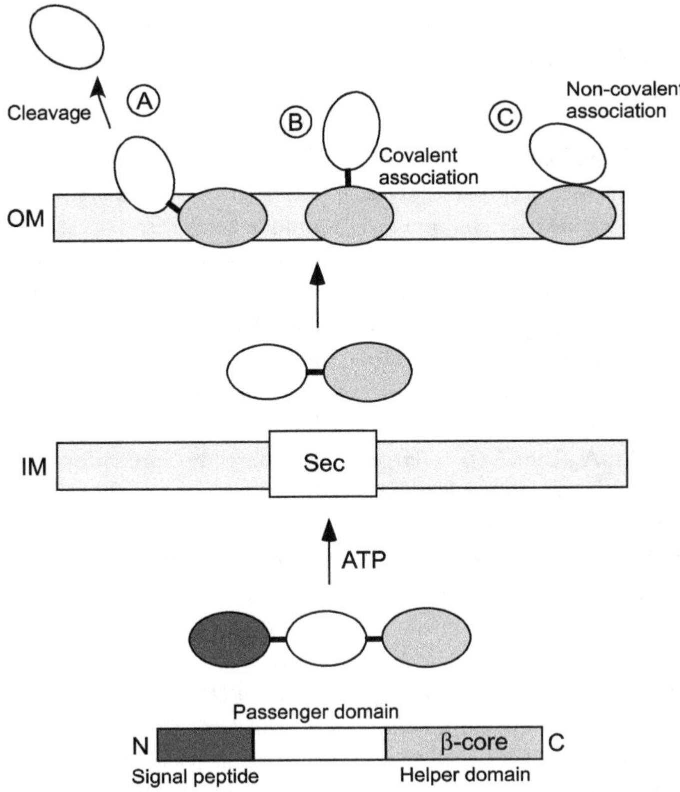

Figure 6.4 Secretion of autotransporter proteins via the type V secretion system. The single polypeptide is transported to the periplasm via the Sec machinery. The helper domain anchors the passenger domain to the OM. The three different fates of the passenger domain are indicated by A, B and C.

in conventional autotransporters (Meng *et al.*, 2006). Therefore, three monomers assembling together would reconstitute the conventional 12 β-strand configuration. Based on the structure of the protein extracted directly from the OM, Hia forms a channel that is 1.8 nm in diameter.

Agg. actinomycetemcomitans *Aae*

Aae was first identified based on its sequence similarity to other autotransporter proteins (Rose *et al.*, 2003). Aae mediates attachment to epithelial cells and while initial studies were carried out with KB cells, subsequent studies using buccal epithelial cells likely represent the more physiological scenario (Fine *et al.*, 2005). An *aae* mutant bound to buccal epithelial cells at levels that were reduced 2–3 logs compared to the wild type strain. Interestingly, specificity of Aae appears to mimic that of *Agg. actinomycetemcomitans* leukotoxin for humans and Old World primates (Fine *et al.*, 2005). In addition, Aae does not mediate binding to bronchial, palatal, or tongue epithelial cells.

Based on sequence analysis and structural predictions, Aae lacks the typical autoproteolytic site found in other autotransporters. However, the passenger domain may be cleaved by human lactoferrin (Rose *et al.*, 2003), which is interesting in light of the fact that human lactoferrin inhibits binding of *Agg. actinomycetemcomitans* cells to buccal epithelial cells (Fine and Furgang, 2002). A putative host cell receptor for Aae is still unknown but would represent a novel target to inhibit binding of *Agg. actinomycetemcomitans* to host tissue in the oral cavity.

Agg. actinomycetemcomitans *ApiA*

ApiA (also Omp100) is a 100-kDa OM protein that mediates attachment to epithelial cells, invasion of epithelial cells, and confers serum resistance (Asakawa *et al.*, 2003; Li *et al.*, 2004). Expression of ApiA in *E. coli* results in autoaggregation suggesting that the protein may also play a role in cell-cell adhesion (Li *et al.*, 2004). Little is known about ApiA secretion, but the protein is a predicted trimeric autotransporter and most closely resembles NadA of *Neisseria meningitidis* and the UspA adhesins of *Moraxella catarrhalis* (Cotter *et al.*, 2006).

Agg. actinomycetemcomitans *EmaA*

Binding to the eukaryotic extracellular matrix requires a protein in *Agg. actinomycetemcomitans* termed EmaA (extracellular matrix protein adhesin A). *emaA* was originally identified as a gene that is important for *Agg. actinomycetemcomitans* to bind to collagen (Mintz, 2004). The protein is a member of the OCA-trimeric autotransporters, which includes *H. influenza* Hia and *Agg. actinomycetemcomitans* ApiA. EmaA is a large (200 kDa) protein that is found on the OM of cells. The presence of an N-terminal Sec-dependent secretion signal suggests that secretion of EmaA to the OM is Sec dependent; however, no direct evidence exists to support this. An interesting feature of the EmaA 56-amino-acid putative signal sequence is that it is about twice as long as other members in the OCA family, suggesting that EmaA may represent a new OCA subtype. At the C-terminus of the protein is a domain that may be responsible for anchoring the protein to the OM. Ultrastructural examination of EmaA reveals approximately 155-nm structures that resemble antenna-like stalks capped with spherical appendages (Ruiz *et al.*, 2006).

[H.] ducreyi *DsrA*

[H.] ducreyi produces a protein that is a member of the trimeric autotransporter subfamily (Elkins *et al.*, 2000). DsrA confers resistance to serum and mediates attachment to keratinocytes (Cole *et al.*, 2002; Elkins *et al.*, 2000). In the swine model of chancroid pathogenesis, the *dsrA* mutant displayed less virulence than the wild type strain, demonstrating the importance of this autotransporter protein.

The Tad secretion system

The tight adherence (Tad) secretion system was first discovered in *Agg. actinomycetemcomitans* (Kachlany *et al.*, 2000b) and *Caulobacter crescentus* (Skerker and Shapiro, 2000). In *Agg. actinomycetemcomitans*, mutations that rendered the bacterium non-adherent were mapped to a 14-gene locus (Fig. 6.5). Genes within this locus have similarity to and predicted properties of genes from other macromolecular transport systems. The components of the Tad secretion machinery distantly resemble both type II secretion and type IV secretion systems. In addition, many of the proteins do not belong to any known system and appear to be novel and unique to the Tad system (RcpC, RcpB, TadZ, TadD, TadE, TadF, and TadG). A study of the evolutionary history of this locus has revealed a complex series of events that has led to its genesis including horizontal transfer, gene duplications and losses (Planet *et al.*, 2003). Indeed, the *tad* locus is present on a genomic island termed the widespread colonization island (WCI). Based on functional and phylogenetic evidence, Tomich *et al.* (2007) have characterized the Tad system as a major new type of prokaryotic secretion system (Fig. 6.5).

A hallmark feature of the WCI is its widespread distribution among *Bacteria* and *Archaea* (Planet *et al.*, 2003). More than half of the *Bacteria* that have thus far been sequenced contain at least one copy of the WCI, in addition to all of the sequenced *Archaea*. The WCI is present in both pathogens and non-pathogens and likely plays a role in colonization to surfaces, whether in an animal or in the environment.

The function of the Tad apparatus appears to be the secretion of pili. The pilus subunit is encoded by the *flp* (fimbrial low-molecular-weight protein) gene (Inoue *et al.*, 1998; Kachlany *et al.*, 2001). Flp is an ~75 amino acid protein and contains a ~20 amino acid cleavable signal peptide. Individual pili have a diameter of approximately 4–7 nm. Several bacteria contain more than one *flp* gene (*flp-1*, *flp-2*, etc.) and the exact reason for this is unknown but may allow differential antigenic expression, or some copies could represent pseudogenes (Perez *et al.*, 2006). Once produced, Flp undergoes proteolytic processing by TadV, which results in the cleavage of the ~20 amino acid N-terminal signal peptide (de Bentzmann *et al.*, 2006; Tomich *et al.*, 2006). The processed protein interacts with TadA, an ABC protein at the IM that possesses ATPase activity (Bhattacharjee *et al.*, 2001). Like other ABC proteins, TadA probably provides the energy for secretion and forms hexameric ring structures. TadG can also be found in preparations of pili and may serve to anchor the pilus in the mem-

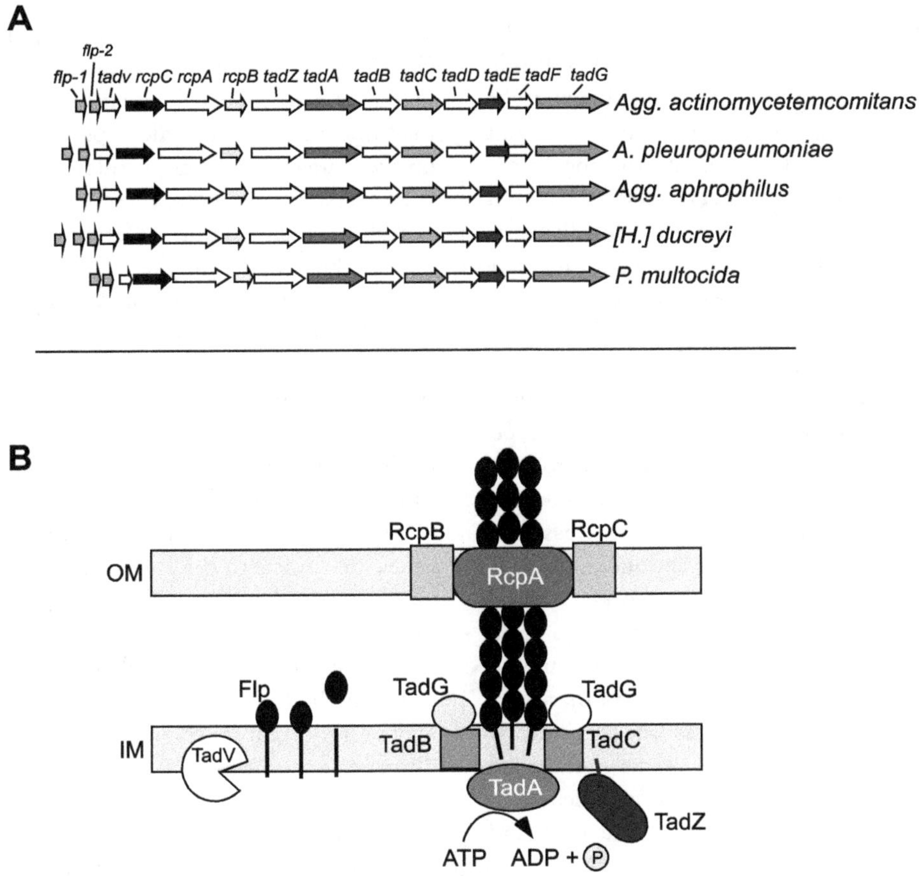

Figure 6.5 The Tad secretion system. (A) Arrangement of the *tad* genes that are found in the *Pasteurellaceae* family. (B) A model for assembly of the Tad proteins and secretion of pili. Redrawn from (Tomich *et al*., 2007). Flp proteins are processed by TadV and then secreted through the IM via a complex of Tad A, B, C, and G. The complex may be localized by TadZ. Mature Flp pili are extruded through a pore composed of RcpA.

brane. TadB and TadC are both predicted to be localized to the IM and may channel the energy from TadA-mediated ATP hydrolysis into pilus assembly or serve as a structural component of the secretion apparatus.

Flp pili probably begin assembling within the periplasm. The N-terminus of mature Flp contains a hydrophobic motif that may be responsible for interaction with other Flp monomers to hold the mature pilus together. TadE and TadF are similar to each other but their function is unknown. Interestingly, these proteins share several properties with Flp and may represent pseudopilins. In other pilus systems, the pseudopilins act as pistons to help assemble the growing pilus.

As the pilus grows, it is extruded through an OM pore. RcpA is an OM protein that is a member of the secretin family of proteins (Haase *et al*., 1999). Thus, RcpA encodes an OM pore through which the Flp proteins are exported. RcpA may interact with another protein in the OM, RcpB. While the function is not yet known, RcpB may play a role in the assembly or stability of the OM complex. In addition to RcpA and RcpB, TadD is predicted to be localized to the OM and may help assemble the OM complex.

Not all organisms that encode a *tad* locus produce visible pili. In these cases, the pili may be short or Flp may simply coat the surface of the bacterial cells. Alternatively, the genes may not be expressed under laboratory conditions, and may require the physiological environment for optimal expression.

Agg. actinomycetemcomitans

In *A. actinomycetemcomitans*, the *tad* locus is responsible for the production of bundled pili, tenacious adhesion to solid surfaces, and biofilm formation (Fine *et al.*, 1999; Inouye *et al.*, 1990; Kachlany *et al.*, 2000b; Planet *et al.*, 2003; Rosan *et al.*, 1988). These pili have been shown to be glycosylated and glycosylation may be carried out by RcpC based on its similarity to other sugar modifying enzymes (Inoue *et al.*, 2000; Tomich *et al.*, 2006). The role of glycosylation in adherence is not yet understood, but sugars could mediate specific interaction with host cell receptors.

Schreiner *et al.* (2003) reported that the Tad secretion system was required for virulence of *Agg. actinomycetemcomitans* in an animal model for aggressive periodontitis. Wild-type, *flp-1*, or *tadA* mutant bacteria were fed to rats, which represents the natural route of infection. Bacterial colonization, maxillary bone loss, and *Agg. actinomycetemcomitans*-specific antibody were then determined. These workers found that wild type bacteria could be cultured from the oral cavities of rats after 12 weeks while no bacteria could be obtained from rats that were fed mutant bacteria. Bone loss was significantly greater in wild-type-fed rats versus mutant-fed rats. In addition, the wild type strain elicited an immune response in contrast to the *flp-1* and *tadA* mutants.

Pasteurella multocida

Two separate studies have confirmed that the *tad* genes are important for pathogenesis of *P. multocida*. Fuller *et al.* (2000) used signature-tagged mutagenesis to identify mutants that are attenuated for disease in a septicaemic mouse model. They identified a *tadD* mutant that was reduced in virulence by 10^5-fold. In addition, Harper *et al.* (2003) isolated a transposon insertion mutation that may have disrupted expression of the whole *tad* locus due to its insertion just upstream of the *flp-1* gene. The mutation resulted in a strain that was attenuated for persistence in both mice and chickens.

[H.] ducreyi

In contrast to the other members of this family, the [*H.*] *ducreyi tad* locus contains three *flp* genes. Whether they are all functional is still unknown. Mutations within this locus prevented the bacterium from forming microcolonies in the presence of human foreskin fibroblasts, a property that is important for pathogenesis of the organism (Nika *et al.*, 2002). Spinola *et al.* (2003) tested wild type and *tadA* mutant strains in the human model for chancroid and found that, in contrast to the wild-type strain, the mutant was unable to form pustules on the skin of volunteers. Furthermore, these strains exhibited a slight, yet significant, difference in virulence in the rabbit model for chancroid. These results indicate that the *tad* genes play an important role in [*H.*] *ducreyi* pathogenesis.

A. pleuropneumoniae *and* Aggregatibacter aphrophilus

Both *A. pleuropneumoniae* and *Agg. aphrophilus* contain genes of the *tad* locus (Tomich *et al.*, 2007). The number and arrangement of the genes are similar to the *Agg. actinomycetemcomitans* locus. In *Agg. aphrophilus*, the genes are also required for tight adherence, biofilm formation, and the production of bundled pili (P.J. Planet, C.T. Sheth, D.H. Figurski, unpublished) and may be important for colonization of heart valves during endocarditis.

Type IV pili

Bacterial type IV pili (fimbriae) are involved in several important processes including twitching motility, attachment, and DNA uptake (Hansen and Forest, 2006; Mattick, 2002). Proteins of the type IV pilus system share components with the type II secretion system, which has been thoroughly studied in *Klebsiella oxytoca* (Peabody *et al.*, 2003). Type IV pili are ~5–7 nm in diameter and can extend several microns in length. The mature pilus can be glycosylated or phosphorylated, which is involved in host cell receptor recognition. The pilus is composed primarily of the PilA protein, which is a 145–160 amino acid protein containing an N-terminal signal peptide.

Depending on the organism, six or more proteins are required for the production and assembly of type IV pili (Fig. 6.6) (Mattick, 2002). The *pilABCD* operon encodes the prepilin subunit and essential IM components. PilA is considered the major pilin subunit, although minor pilins are also required in some systems. PilB has NTP binding activity and may be required for assembly and retraction of the pilus and twitching motility. PilC is an IM protein through which processed pilin subunits are extruded. PilD is a

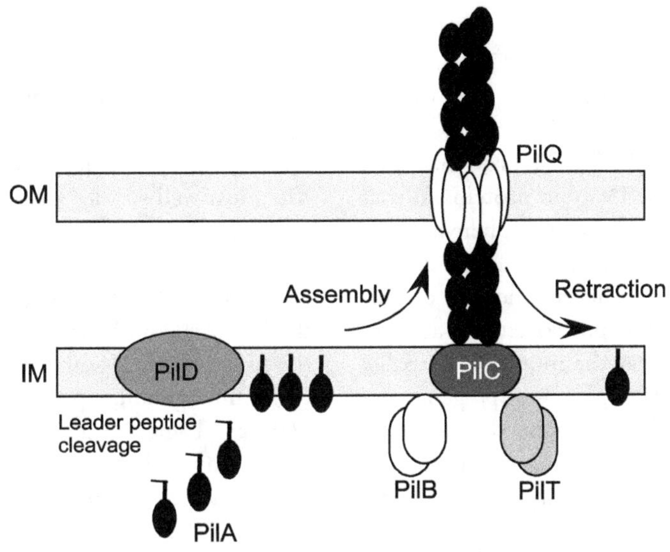

Figure 6.6 Assembly and secretion of type IV pili. Redrawn from (Mattick, 2002). The leader sequence of PilA is cleaved by PilD and mature pili are assembled and extruded through the PilQ OM channel. Twitching motility is due to the assembly and retraction of type IV pili.

bifunctional protein that is responsible for both the cleavage of the PilA N-terminal signal peptide and methylation of the resulting N-terminal residue. In addition to PilABCD, two other proteins are often required for assembly of type IV pili. PilT is involved in the retraction of the pilus and recycling of pilin subunits. Like PilB, PilT is a nucleotide binding protein and thus uses energy during retraction. PilQ is a well-studied secretin protein that forms an OM channel for extrusion of the pilus. PilQ assembles into a dodecamer doughnut-shaped structure that has a pore diameter of approximately 53–65 Å. Type IV pili are assembled from the base, in contrast to flagella, which are assembled from the tip. In several pathogens, type IV pili are important for colonization and pathogenesis.

H. influenzae

Bakaletz *et al.* (2005) described type IV pili in *H. influenzae* and the role in twitching motility. Prior to this report, the expression of type IV pili or twitching motility had not been reported for any *H. influenzae* isolate. *H. influenzae* contains two loci encoding type IV pilus genes, the *pilABCD* and *comABCDEF* loci. The ComE protein has significant similarity to PilQ and probably represents the OM pore-forming secretin.

Under growth conditions that *H. influenzae* is normally grown under, cells do not express type IV pili. However, on nutrient-defined conditions at alkaline pH (pH 8.5–9.0), the type IV pili genes are expressed and pili structures can be visualized (Bakaletz *et al.*, 2005). The cells also undergo twitching motility and both pili production and twitching require *pilA*. In *H. influenzae*, *pilA*, and thus type IV pili, is required for biofilm formation and colonization of the upper respiratory tract as demonstrated in the chinchilla model (Jurcisek *et al.*, 2007).

Agg. actinomycetemcomitans

In *Agg. actinomycetemcomitans*, genes encoding type IV pili (*pilABCD*) are required for competence (Wang *et al.*, 2003). Some strains of *Agg. actinomycetemcomitans* are naturally competent for DNA uptake and a common laboratory isolate, JP2, is unable to be transformed with DNA apparently due to very low expression levels of the *pil* genes. Type IV pili have not yet been observed on the surface of *Agg. actinomycetemcomitans* cells, suggesting that these genes do not encode a traditional pilus structure, or that physiological conditions are required for proper assembly. PilABCD from *Agg. actinomycetemcomitans* are highly similar to the *H. influenzae* Pil proteins, and whether *Agg. actinomycetemcomitans* also expresses other type IV pilus genes has not yet been determined.

A. pleuropneumoniae

A. pleuropneumoniae encodes *apfABCD*, which are homologous to *pilABCD* (Stevenson *et al.*, 2003). *apfA* encodes a 16-kDa protein that is processed to yield a 14.5-kDa protein. ApfA differs from other type IV pilin proteins in that the amino acid that precedes the phenylalanine at the predicted cleavage site is an alanine instead of the typical glycine. The significance of this change is unknown, but pili structures are still produced indicating that the mutation probably offers a selected advantage. The *apf* genes are tightly regulated and are expressed when cells are in contact with porcine lung epithelial cells (Boekema *et al.*, 2004). In addition, a pig model for *A. pleuropneumoniae* pathogenesis revealed that the type IV pili genes are turned on during natural infection. Thus, as occurs for many bacterial virulence genes, physiological conditions are required for significant gene expression.

P. multocida

Ruffolo *et al.* (1997) reported the purification of pili on the surface of *P. multocida* cells. Increased expression of pili occurred under microaerophilic conditions and amino acid sequencing of the N-terminus of the major pilin protein revealed significant similarity to other type IV pili. The gene for the major pilin subunit, *ptfA*, encodes a 18-kDa protein that has the general properties of type IV pilin proteins, including a signal peptide, a hydrophobic amino terminus of mature protein, and a pair of cysteine residues at the C-terminus (Doughty *et al.*, 2000). Interestingly, the signal sequence of PtfA is absolutely conserved among *P. multocida* strains, in contrast to the mature protein, which can differ by nearly 25% at the amino acid level (Siju *et al.*, 2007). Since bundled pili can be visualized on cells, it can be assumed that the other genes required for pilus production and assembly (e.g. *pilBCD*) must also be encoded in the *P. multocida* genome.

Proteases

Immunoglobulin (Ig) proteases

Microbial proteases produced by pathogens can cause direct pathological effects by destroying host tissues and inactivation of immunoglobulins, complement factors, and complement derived chemotactic factors (Maeda and Yamamoto, 1996; Miyoshi and Shinoda, 2000). Several members of the *Pasteurellaceae* family produce immunoglobulin-degrading proteases and secrete these enzymes into their environment. The most well studied protein in this class is *H. influenzae* IgA1 protease (Grundy *et al.*, 1987a; Poulsen *et al.*, 1989). IgA1 proteases are extracellular bacterial enzymes that specifically cleave the heavy chain of the human IgA1 molecule in the hinge region. IgA1 is the predominant class of immunoglobulin present at human mucosal surfaces (Tomasi and Yurchak, 1972). Therefore production of IgA1 proteases allows bacteria to infect and invade mucosal membranes (Plaut, 1983).

IgA1 proteases belong to the family of serine protease autotransporters and released by the type V secretion pathway that was discussed earlier (Fig. 6.4) (Henderson *et al.*, 2004).

H. influenzae *IgA1 protease*

The IgA1 protease of H. influenzae is genetically related to those of *Neisseria gonorrhoeae* and *Neisseria* meningitidis (Lomholt *et al.*, 1995). IgA1 protease is a serine protease and cleavage of IgA1 molecules by the protein prevents the antibody from blocking adhesion of bacteria to surfaces and complement activation. In *H. influenzae*, IgA1 protease is released into the extracellular environment. When the IgA1 protease genes (*iga*) from *H. influenzae* serotypes c and d were cloned and expressed in *E. coli*, they were secreted efficiently into the growth medium (Grundy *et al.*, 1987a; Grundy *et al.*, 1987b).

More than 90% of *H. influenzae* strains and all *Haemophilus aegyptius* strains have detectable IgA1 protease activity. The majority of strains produced a single IgA1 protease that cleaved a prolyl-seryl or a prolyl-threonyl peptide bond in the alpha 1 hinge region, whereas some *H. influenzae* strains possessed two separate IgA1 proteases (Kilian and Thomsen, 1983). Interestingly, non-pathogenic *Haemophilus* species lack IgA1 protease activity.

H. influenzae *HAP protease*

Another example of a protease secreted by *H. influenzae* is the adhesion protein, Hap, discussed above (St Geme *et al.*, 1994). Hap harbours serine protease activity that acts as an extracellular protease when the passenger domain is cleaved off (Fink *et al.*, 2001; Hendrixson *et al.*, 1997).

Other bacterial proteases

Several other pathogenic members of *Pasteurellaceae* family are reported to secrete proteases but these are less well characterized. Clinical isolates of *P. multocida* strains secrete proteases that cleaved human IgA1, IgA2 and IgG (Negrete-Abascal *et al.*, 1999). Some of these proteases cleaved IgA and IgG heavy chains outside the hinge region (Pouedras *et al.*, 1992), which differs from the other bacterial Ig proteases. *M. haemolytica*, secretes an O-sialoglycoprotein endopeptidase (Watt *et al.*, 1997) and *A. pleuropneumoniae* releases metalloproteases that hydrolyse porcine IgA and IgG (Negrete-Abascal *et al.*, 1994, 1998). *Actinobacillus suis* secretes IgG-degrading metalloproteases that target pig and bovine IgG proteins (Negrete-Abascal *et al.*, 2004). *Avibacterium paragallinarum*, the causative agent of infectious coryza, secretes metalloproteases that can degrade chicken IgG (Rivero-Garcia *et al.*, 2005).

Conclusions

Members of the *Pasteurellaceae* harbour several secretion systems for the export of proteins important for survival and virulence. These proteins include toxins, adhesins, and proteases. Many of these secretion systems have significant similarities to homologous systems of other bacteria. Specifically, *E. coli* has served as a model organism for many of the systems discussed here. However, important differences exist among these systems and across different bacteria and it will be crucial to study protein secretion in the context of the pathogens that are part of the *Pasteurellaceae* family.

References

Ahmed, H.J., Svensson, L.A., Cope, L.D., Latimer, J.L., Hansen, E.J., Ahlman, K., Bayat-Turk, J., Klamer, D., and Lagergard, T. (2001). Prevalence of cdtABC genes encoding cytolethal distending toxin among *Haemophilus ducreyi* and *Actinobacillus actinomycetemcomitans* strains. J. Med. Microbiol. *50*, 860–864.

Alby, F., Mazars, R., de Rycke, J., Guillou, E., Baldin, V., Darbon, J.M., and Ducommun, B. (2001). Study of the cytolethal distending toxin (CDT)-activated cell cycle checkpoint. Involvement of the CHK2 kinase. FEBS Lett. *491*, 261–265.

Andersen, C., Hughes, C., and Koronakis, V. (2000). Chunnel vision. Export and efflux through bacterial channel-tunnels. EMBO Rep. *1*, 313–318.

Asakawa, R., Komatsuzawa, H., Kawai, T., Yamada, S., Goncalves, R.B., Izumi, S., Fujiwara, T., Nakano, Y., Suzuki, N., Uchida, Y., *et al.* (2003). Outer membrane protein 100, a versatile virulence factor of *Actinobacillus actinomycetemcomitans*. Mol. Microbiol. *50*, 1125–1139.

Baarsch, M.J., Scamurra, R.W., Burger, K., Foss, D.L., Maheswaran, S.K., and Murtaugh, M.P. (1995). Inflammatory cytokine expression in swine experimentally infected with *Actinobacillus pleuropneumoniae*. Infect. Immun. *63*, 3587–3594.

Bakaletz, L.O., Baker, B.D., Jurcisek, J.A., Harrison, A., Novotny, L.A., Bookwalter, J.E., Mungur, R., and Munson, R.S., Jr. (2005). Demonstration of Type IV pilus expression and a twitching phenotype by *Haemophilus influenzae*. Infect. Immun. *73*, 1635–1643.

Balashova, N.V., Crosby, J.A., Al Ghofaily, L., and Kachlany, S.C. (2006a). Leukotoxin confers beta-hemolytic activity to *Actinobacillus actinomycetemcomitans*. Infect. Immun. *74*, 2015–2021.

Balashova, N.V., Diaz, R., Balashov, S.V., Crosby, J.A., and Kachlany, S.C. (2006b). Regulation of *Aggregatibacter (Actinobacillus) actinomycetemcomitans* leukotoxin secretion by iron. J. Bacteriol. *188*, 8658–8661.

Bhattacharjee, M.K., Kachlany, S.C., Fine, D.H., and Figurski, D.H. (2001). Nonspecific adherence and fibril biogenesis by *Actinobacillus actinomycetemcomitans*: TadA protein is an ATPase. J. Bacteriol. *183*, 5927–5936.

Boekema, B.K., Van Putten, J.P., Stockhofe-Zurwieden, N., and Smith, H.E. (2004). Host cell contact-induced transcription of the type IV fimbria gene cluster of *Actinobacillus pleuropneumoniae*. Infect. Immun. 72, 691–700.

Bosse, J.T., Janson, H., Sheehan, B.J., Beddek, A.J., Rycroft, A.N., Kroll, J.S., and Langford, P.R. (2002). Actinobacillus pleuropneumoniae: pathobiology and pathogenesis of infection. Microbes and infection/ Institut Pasteur 4, 225–235.

Chang, Y.F., Young, R., Moulds, T.L., and Struck, D.K. (1989). Secretion of the *Pasteurella* leukotoxin by *Escherichia coli*;. FEMS Microbiol. Lett. *51*, 169–173.

Chung, W.B., Backstrom, L.R., and Collins, M.T. (1994). Experimental model of swine pneumonic pasteurellosis using crude *Actinobacillus pleuropneumoniae* cytotoxin and *Pasteurella multocida* ;given endobronchially. Can. J. Vet. Res. *58*, 25–30.

Cole, L.E., Kawula, T.H., Toffer, K.L., and Elkins, C. (2002). The *Haemophilus ducreyi* serum resistance antigen DsrA confers attachment to human keratinocytes. Infect. Immun. *70*, 6158–6165.

Cope, L.D., Lumbley, S., Latimer, J.L., Klesney-Tait, J., Stevens, M.K., Johnson, L.S., Purven, M., Munson, R.S., Jr., Lagergard, T., Radolf, J.D., *et al.* (1997). A diffusible cytotoxin of *Haemophilus ducreyi*. Proc. Natl. Acad. Sci. USA *94*, 4056–4061.

Cortes-Bratti, X., Karlsson, C., Lagergard, T., Thelestam, M., and Frisan, T. (2001). The *Haemophilus ducreyi* cytolethal distending toxin induces cell cycle arrest and apoptosis via the DNA damage checkpoint pathways. J. Biol. Chem. *276*, 5296–5302.

Cotter, S.E., Surana, N.K., Grass, S., and St Geme, J.W., 3rd (2006). Trimeric autotransporters require trimerization of the passenger domain for stability and adhesive activity. J. Bacteriol. *188*, 5400–5407.

Crosby, J.A., and Kachlany, S.C. (2007). TdeA, a TolC-like protein required for toxin and drug export in *Aggregatibacter (Actinobacillus) actinomycetemcomitans*. Gene *388*, 83–92.

de Bentzmann, S., Aurouze, M., Ball, G., and Filloux, A. (2006). FppA, a novel *Pseudomonas aeruginosa* prepilin peptidase involved in assembly of type IVb pili. J. Bacteriol. *188*, 4851–4860.

Delepelaire, P. (2004). Type I secretion in gram-negative bacteria. Biochim. Biophys. Acta *1694*, 149–161.

Deng, K., and Hansen, E.J. (2003). A CdtA-CdtC complex can block killing of HeLa cells by *Haemophilus ducreyi* cytolethal distending toxin. Infect. Immun. *71*, 6633–6640.

Diaz, R., Ghofaily, L.A., Patel, J., Balashova, N.V., Freitas, A.C., Labib, I., and Kachlany, S.C. (2006). Characterization of leukotoxin from a clinical strain of *Actinobacillus actinomycetemcomitans*. Microb. Pathog. *40*, 48–55.

Doughty, S.W., Ruffolo, C.G., and Adler, B. (2000). The type 4 fimbrial subunit gene of pasteurella multocida. Vet. Microbiol. *72*, 79–90.

Duong, F., and Wickner, W. (1998). Sec-dependent membrane protein biogenesis: SecYEG, preprotein hydrophobicity and translocation kinetics control the stop-transfer function. EMBO J. *17*, 696–705.

Elkins, C., Morrow, K.J., Jr., and Olsen, B. (2000). Serum resistance in *Haemophilus ducreyi* requires outer membrane protein DsrA. Infect. Immun. *68*, 1608–1619.

Elwell, C.A., and Dreyfus, L.A. (2000). DNase I homologous residues in CdtB are critical for cytolethal distending toxin-mediated cell cycle arrest. Mol. Microbiol. *37*, 952–963.

Fabris, A.S., DiRienzo, J.M., Wikstrom, M., and Mayer, M.P. (2002). Detection of cytolethal distending toxin activity and cdt genes in *Actinobacillus actinomycetemcomitans* isolates from geographically diverse populations. Oral. Microbiol. Immunol. *17*, 231–238.

Fekkes, P., and Driessen, A.J. (1999). Protein targeting to the bacterial cytoplasmic membrane. Microbiol. Mol. Biol. Rev. *63*, 161–173.

Fine, D.H., and Furgang, D. (2002). Lactoferrin iron levels affect attachment of *Actinobacillus actinomycetemcomitans* to buccal epithelial cells. J. Periodontol. *73*, 616–623.

Fine, D.H., Furgang, D., Schreiner, H.C., Goncharoff, P., Charlesworth, J., Ghazwan, G., Fitzgerald-Bocarsly, P., and Figurski, D.H. (1999). Phenotypic variation in *Actinobacillus actinomycetemcomitans* during laboratory growth: implications for virulence. Microbiology *145*, 1335–1347.

Fine, D.H., Velliyagounder, K., Furgang, D., and Kaplan, J.B. (2005). The *Actinobacillus actinomycetemcomitans* autotransporter adhesin Aae exhibits specificity for buccal epithelial cells from humans and old world primates. Infect. Immun. *73*, 1947–1953.

Fink, D.L., Cope, L.D., Hansen, E.J., and Geme, J.W., 3rd (2001). The Hemophilus influenzae Hap autotransporter is a chymotrypsin clan serine protease and undergoes autoproteolysis via an intermolecular mechanism. J. Biol. Chem. *276*, 39492–39500.

Fuller, T.E., Kennedy, M.J., and Lowery, D.E. (2000). Identification of *Pasteurella multocida* ;virulence genes in a septicemic mouse model using signature-tagged mutagenesis. Microb. Pathog. *29*, 25–38.

Grundy, F.J., Plaut, A., and Wright, A. (1987a). *Haemophilus influenzae* immunoglobulin A1 protease genes: cloning by plasmid integration-excision, comparative analyses, and localization of secretion determinants. J. Bacteriol. *169*, 4442–4450.

Grundy, F.J., Plaut, A.G., and Wright, A. (1987b). *Haemophilus influenzae* IgA1 proteases: determinants of cleavage site specificity and secretion. Adv. Exp. Med. Biol. *216B*, 1251–1260.

Guthmiller, J.M., Kolodrubetz, D., and Kraig, E. (1995). Mutational analysis of the putative leukotoxin transport genes in *Actinobacillus actinomycetemcomitans*. Microb. Pathog. *18*, 307–321.

Haase, E.M., Zmuda, J.L., and Scannapieco, F.A. (1999). Identification and molecular analysis of rough-colony-specific outer membrane proteins of *Actinobacillus actinomycetemcomitans*. Infect. Immun. *67*, 2901–2908.

Hansen, J.K., and Forest, K.T. (2006). Type IV pilin structures: insights on shared architecture, fiber assembly, receptor binding and type II secretion. J. Mol. Microbiol.Biotechnol. *11*, 192–207.

Harper, M., Boyce, J.D., Wilkie, I.W., and Adler, B. (2003). Signature-tagged mutagenesis of *Pasteurella multocida* ;identifies mutants displaying differential virulence characteristics in mice and chickens. Infect. Immun. *71*, 5440–5446.

Henderson, I.R., Navarro-Garcia, F., Desvaux, M., Fernandez, R.C., and Ala'Aldeen, D. (2004). Type V protein secretion pathway: the autotransporter story. Microbiol. Mol. Biol. Rev. *68*, 692–744.

Hendrixson, D.R., and St Geme, J.W., 3rd (1998). The *Haemophilus influenzae* Hap serine protease promotes adherence and microcolony formation, potentiated by a soluble host protein. Mol. Cell 2, 841–850.

Hendrixson, D.R., de la Morena, M.L., Stathopoulos, C., and St Geme, J.W., 3rd (1997). Structural determinants of processing and secretion of the *Haemophilus influenzae* hap protein. Mol. Microbiol. *26*, 505–518.

Highlander, S.K., Engler, M.J., and Weinstock, G.M. (1990). Secretion and expression of the *Pasteurella haemolytica* leukotoxin. J. Bacteriol. *172*, 2343–2350.

Hill, C., and Holland, I.B. (1967). Genetic basis of colicin E susceptibility in *Escherichia coli*. I. Isolation and properties of refractory mutants and the preliminary mapping of their mutations. J. Bacteriol. *94*, 677–686.

Hu, X., Nesic, D., and Stebbins, C.E. (2006). Comparative structure–function analysis of cytolethal distending toxins. Proteins *62*, 421–434.

Inouye, T., Ohta, H., Kokeguchi, S., Fukui, K., and Kato, K. (1990). Colonial variation and fimbriation of *Actinobacillus actinomycetemcomitans*. FEMS Microbiol. Lett. *57*, 13–17.

Inoue, T., Tanimoto, I., Ohta, H., Kato, K., Murayama, Y., and Fukui, K. (1998). Molecular characterization of low-molecular-weight component protein, Flp, in *Actinobacillus actinomycetemcomitans* fimbriae. Microbiol. Immunol. *42*, 253–258.

Inoue, T., Ohta, H., Tanimoto, I., Shingaki, R., and Fukui, K. (2000). Heterogeneous post-translational modification of *Actinobacillus actinomycetemcomitans* fimbrillin. Microbiol. Immunol. *44*, 715–718.

Jurcisek, J.A., Bookwalter, J.E., Baker, B.D., Fernandez, S., Novotny, L.A., Munson, R.S., Jr., and Bakaletz, L.O. (2007). The PilA protein of non-typeable *Haemophilus influenzae* plays a role in biofilm formation, adherence to epithelial cells and colonization of the mammalian upper respiratory tract. Mol. Microbiol. *65*, 1288–1299.

Kachlany, S.C., Fine, D.H., and Figurski, D.H. (2000a). Secretion of RTX leukotoxin by *Actinobacillus actinomycetemcomitans*. Infect. Immun. *68*, 6094–6100.

Kachlany, S.C., Planet, P.J., Bhattacharjee, M.K., Kollia, E., DeSalle, R., Fine, D.H., and Figurski, D.H. (2000b). Nonspecific adherence by *Actinobacillus actinomycetemcomitans* requires genes widespread in bacteria and archaea. J. Bacteriol. *182*, 6169–6176.

Kachlany, S.C., Planet, P.J., Desalle, R., Fine, D.H., Figurski, D.H., and Kaplan, J.B. (2001). flp-1, the first representative of a new pilin gene subfamily, is required for non-specific adherence of *Actinobacillus actinomycetemcomitans*. Mol. Microbiol. *40*, 542–554.

Kihara, A., and Ito, K. (1998). Translocation, folding, and stability of the HflKC complex with signal anchor topogenic sequences. J. Biol. Chem. *273*, 29770–29775.

Kilian, M., and Thomsen, B. (1983). Antigenic heterogeneity of immunoglobulin A1 proteases from encapsulated and non-encapsulated *Haemophilus influenzae*. Infect. Immun. *42*, 126–132.

Koronakis, V., Eswaran, J., and Hughes, C. (2004). Structure and function of TolC: the bacterial exit duct for proteins and drugs. Annu. Rev. Biochem. *73*, 467–489.

Kulkarni, K., Lewis, D.A., and Ison, C.A. (2003). Expression of the cytolethal distending toxin in a geographically diverse collection of *Haemophilus ducreyi* clinical isolates. Sex. Transm. Infect. *79*, 294–297.

Lally, E.T., Golub, E.E., Kieba, I.R., Taichman, N.S., Decker, S., Berthold, P., Gibson, C.W., Demuth, D.R., and Rosenbloom, J. (1991). Structure and function of the B and D genes of the *Actinobacillus actinomycetemcomitans* leukotoxin complex. Microb. Pathog. *11*, 111–121.

Lally, E.T., Hill, R.B., Kieba, I.R., and Korostoff, J. (1999). The interaction between RTX toxins and target cells. Trends Microbiol. *7*, 356–361.

Lara-Tejero, M., and Galan, J.E. (2000). A bacterial toxin that controls cell cycle progression as a deoxyribonuclease I-like protein. Science *290*, 354–357.

Lara-Tejero, M., and Galan, J.E. (2001). CdtA, CdtB, and CdtC form a tripartite complex that is required for cytolethal distending toxin activity. Infect. Immun. *69*, 4358–4365.

Lee, R.B., Hassane, D.C., Cottle, D.L., and Pickett, C.L. (2003). Interactions of *Campylobacter jejuni* cytolethal distending toxin subunits CdtA and CdtC with HeLa cells. Infect. Immun. *71*, 4883–4890.

Li, L., Matevski, D., Aspiras, M., Ellen, R.P., and Lepine, G. (2004). Two epithelial cell invasion-related loci of the oral pathogen *Actinobacillus actinomycetemcomitans*. Oral Microbiol. Immunol. *19*, 16–25.

Lomholt, H., Poulsen, K., and Kilian, M. (1995). Comparative characterization of the iga gene encoding IgA1 protease in Neisseria meningitidis, *Neisseria gonorrhoeae* and *Haemophilus influenzae*. Mol. Microbiol. *15*, 495–506.

Maeda, H., and Yamamoto, T. (1996). Pathogenic mechanisms induced by microbial proteases in microbial infections. Biol. Chem. Hoppe Seyler *377*, 217–226.

Manting, E.H., and Driessen, A.J. (2000). *Escherichia coli;* translocase: the unravelling of a molecular machine. Mol. Microbiol. *37*, 226–238.

Manting, E.H., van Der Does, C., Remigy, H., Engel, A., and Driessen, A.J. (2000). SecYEG assembles into a tetramer to form the active protein translocation channel. EMBO J. *19*, 852–861.

Mattick, J.S. (2002). Type IV pili and twitching motility. Annu. Rev. Microbiol. *56*, 289–314.

Mayer, M.P., Bueno, L.C., Hansen, E.J., and DiRienzo, J.M. (1999). Identification of a cytolethal distending toxin gene locus and features of a virulence-associated region in *Actinobacillus actinomycetemcomitans*. Infect. Immun. *67*, 1227–1237.

Meng, G., Surana, N.K., St Geme, J.W., 3rd, and Waksman, G. (2006). Structure of the outer membrane translocator domain of the *Haemophilus influenzae* Hia trimeric autotransporter. EMBO J. *25*, 2297–2304.

Mintz, K.P. (2004). Identification of an extracellular matrix protein adhesin, EmaA, which mediates the adhesion of *Actinobacillus actinomycetemcomitans* to collagen. Microbiology *150*, 2677–2688.

Miyoshi, S., and Shinoda, S. (2000). Microbial metalloproteases and pathogenesis. Microbes and infection/ Institut Pasteur 2, 91–98.

Nagel de Zwaig, R., and Luria, S.E. (1967). Genetics and physiology of colicin-tolerant mutants of *Escherichia coli;*. J. Bacteriol. *94*, 1112–1123.

Negrete-Abascal, E., Tenorio, V.R., Serrano, J.J., Garcia, C., and de la Garza, M. (1994). Secreted proteases from Actinobacillus pleuropneumoniae serotype 1 degrade porcine gelatin, hemoglobin and immunoglobulin A. Can. J. Vet. Res. *58*, 83–86.

Negrete-Abascal, E., Tenorio, V.R., Guerrero, A.L., Garcia, R.M., Reyes, M.E., and de la Garza, M. (1998). Purification and characterization of a protease from Actinobacillus pleuropneumoniae serotype 1, an antigen common to all the serotypes. Can. J. Vet. Res. *62*, 183–190.

Negrete-Abascal, E., Tenorio, V.R., and de la Garza, M. (1999). Secretion of proteases from *Pasteurella multocida ;*isolates. Curr. Microbiol. *38*, 64–67.

Negrete-Abascal, E., Garcia, R.M., Reyes, M.E., Godinez, D., and de la Garza, M. (2000). Membrane vesicles released by Actinobacillus pleuropneumoniae contain proteases and Apx toxins. FEMS Microbiol. Lett. *191*, 109–113.

Negrete-Abascal, E., Pacheco, S.V., Paniagua, G.L., Mendez, A.P., Caballero, J.I., Marquez, V.M., and Tenorio, V.R. (2004). Metalloproteases secreted by Actinobacillus suis. Curr. Microbiol. *49*, 55–58.

Nesic, D., Hsu, Y., and Stebbins, C.E. (2004). Assembly and function of a bacterial genotoxin. Nature *429*, 429–433.

Nika, J.R., Latimer, J.L., Ward, C.K., Blick, R.J., Wagner, N.J., Cope, L.D., Mahairas, G.G., Munson, R.S., Jr., and Hansen, E.J. (2002). *Haemophilus ducreyi* requires the flp gene cluster for microcolony formation in vitro. Infect. Immun. *70*, 2965–2975.

Nomura, M., and Witten, C. (1967). Interaction of colicins with bacterial cells. 3. Colicin-tolerant mutations in *Escherichia coli*;. J. Bacteriol. *94*, 1093–1111.

Nouwen, N., van der Laan, M., and Driessen, A.J. (2001). SecDFyajC is not required for the maintenance of the proton motive force. FEBS Lett. *508*, 103–106.

Ohara, M., Oswald, E., and Sugai, M. (2004). Cytolethal distending toxin: a bacterial bullet targeted to nucleus. J. Biochem. (Tokyo) *136*, 409–413.

Oliver, D.C., Huang, G., Nodel, E., Pleasance, S., and Fernandez, R.C. (2003). A conserved region within the Bordetella pertussis autotransporter BrkA is necessary for folding of its passenger domain. Mol. Microbiol. *47*, 1367–1383.

Peabody, C.R., Chung, Y.J., Yen, M.R., Vidal-Ingigliardi, D., Pugsley, A.P., and Saier, M.H., Jr. (2003). Type II protein secretion and its relationship to bacterial type IV pili and archaeal flagella. Microbiology *149*, 3051–3072.

Perez, B.A., Planet, P.J., Kachlany, S.C., Tomich, M., Fine, D.H., and Figurski, D.H. (2006). Genetic analysis of the requirement for flp-2, tadV, and rcpB in *Actinobacillus actinomycetemcomitans* biofilm formation. J. Bacteriol. *188*, 6361–6375.

Pickett, C.L., Cottle, D.L., Pesci, E.C., and Bikah, G. (1994). Cloning, sequencing, and expression of the *Escherichia coli;* cytolethal distending toxin genes. Infect. Immun. *62*, 1046–1051.

Planet, P.J., Kachlany, S.C., Fine, D.H., DeSalle, R., and Figurski, D.H. (2003). The Widespread Colonization Island of *Actinobacillus actinomycetemcomitans*. Nat. Genet. *34*, 193–198.

Plaut, A.G. (1983). The IgA1 proteases of pathogenic bacteria. Annu. Rev. Microbiol. *37*, 603–622.

Pohlner, J., Halter, R., Beyreuther, K., and Meyer, T.F. (1987). Gene structure and extracellular secretion of *Neisseria gonorrhoeae* IgA protease. Nature *325*, 458–462.

Pouedras, P., Andre, P.M., Donnio, P.Y., and Avril, J.L. (1992). Cleavage of immunoglobulin A1, A2 and G by proteases from clinical isolates of *Pasteurella multocida*. J. Med. Microbiol. *37*, 128–132.

Poulsen, K., Brandt, J., Hjorth, J.P., Thogersen, H.C., and Kilian, M. (1989). Cloning and sequencing of the immunoglobulin A1 protease gene (iga) of *Haemophilus influenzae* serotype b. Infect. Immun. *57*, 3097–3105.

Purdy, D., Buswell, C.M., Hodgson, A.E., McAlpine, K., Henderson, I., and Leach, S.A. (2000). Characterisation of cytolethal distending toxin (CDT) mutants of *Campylobacter jejuni*. J. Med. Microbiol. *49*, 473–479.

Purven, M., Falsen, E., and Lagergard, T. (1995). Cytotoxin production in 100 strains of *Haemophilus ducreyi* from different geographic locations. FEMS Microbiol. Lett. *129*, 221–224.

Rivero-Garcia, P.C., Cruz, C.V., Alonso, P.S., Vaca, S., and Negrete-Abascal, E. (2005). *Haemophilus paragallinarum* secretes metalloproteases. Can. J. Microbiol. *51*, 893–896.

Rosan, B., Slots, J., Lamont, R.J., Listgarten, M.A., and Nelson, G.M. (1988). *Actinobacillus actinomycetemcomitans* fimbriae. Oral Microbiol. Immunol. *3*, 58–63.

Rose, J.E., Meyer, D.H., and Fives-Taylor, P.M. (2003). Aae, an autotransporter involved in adhesion of *Actinobacillus actinomycetemcomitans* to epithelial cells. Infect. Immun. *71*, 2384–2393.

Ruffolo, C.G., Tennent, J.M., Michalski, W.P., and Adler, B. (1997). Identification, purification, and characterization of the type 4 fimbriae of *Pasteurella multocida*. Infect. Immun. *65*, 339–343.

Ruiz, T., Lenox, C., Radermacher, M., and Mintz, K.P. (2006). Novel surface structures are associated with the adhesion of *Actinobacillus actinomycetemcomitans* to collagen. Infect. Immun. *74*, 6163–6170.

Saier, M.H., Jr. (2006). Protein secretion and membrane insertion systems in gram-negative bacteria. J. Membrane biol. *214*, 75–90.

Schaller, A., Kuhn, R., Kuhnert, P., Nicolet, J., Anderson, T.J., MacInnes, J.I., Segers, R.P., and Frey, J. (1999). Characterization of apxIVA, a new RTX determinant of *Actinobacillus pleuropneumoniae*. Microbiology *145*, 2105–2116.

Schmidt, M.G., and Kiser, K.B. (1999). SecA: the ubiquitous component of preprotein translocase in prokaryotes. Microbes and infection/Institut Pasteur *1*, 993–1004.

Schreiner, H.C., Sinatra, K., Kaplan, J.B., Furgang, D., Kachlany, S.C., Planet, P.J., Perez, B.A., Figurski, D.H., and Fine, D.H. (2003). Tight-adherence genes of *Actinobacillus actinomycetemcomitans* are required for virulence in a rat model. Proc. Natl. Acad. Sci. USA *100*, 7295–7300.

Scott, D.A., and Kaper, J.B. (1994). Cloning and sequencing of the genes encoding *Escherichia coli;* cytolethal distending toxin. Infect. Immun. *62*, 244–251.

Shenker, B.J., McKay, T., Datar, S., Miller, M., Chowhan, R., and Demuth, D. (1999). *Actinobacillus actinomycetemcomitans* immunosuppressive protein is a member of the family of cytolethal distending toxins capable of causing a G2 arrest in human T cells. J. Immunol. *162*, 4773–4780.

Sijbrandi, R., Urbanus, M.L., ten Hagen-Jongman, C.M., Bernstein, H.D., Oudega, B., Otto, B.R., and Luirink, J. (2003). Signal recognition particle (SRP)-mediated targeting and Sec-dependent translocation of an extracellular *Escherichia coli;* protein. J. Biol. Chem. *278*, 4654–4659.

Siju, J., Kumar, A.A., Shivachandra, S.B., Chaudhuri, P., Srivastava, S.K., and Singh, V.P. (2007). Cloning and characterization of type 4 fimbrial gene (ptfA) of *Pasteurella multocida* ;serogroup B:2 (strain P52). Vet. Res. Commun. *31*, 397–404.

Skerker, J.M., and Shapiro, L. (2000). Identification and cell cycle control of a novel pilus system in Caulobacter crescentus. EMBO J. *19*, 3223–3234.

Soloaga, A., Veiga, M.P., Garcia-Segura, L.M., Ostolaza, H., Brasseur, R., and Goni, F.M. (1999). Insertion of *Escherichia coli;* alpha-haemolysin in lipid bilayers as a non-transmembrane integral protein: prediction and experiment. Mol. Microbiol. *31*, 1013–1024.

Spinola, S.M., Fortney, K.R., Katz, B.P., Latimer, J.L., Mock, J.R., Vakevainen, M., and Hansen, E.J. (2003). *Haemophilus ducreyi* requires an intact flp gene cluster for virulence in humans. Infect. Immun. *71*, 7178–7182.

St Geme, J.W., 3rd, and Cutter, D. (2000). The *Haemophilus influenzae* Hia adhesin is an autotransporter protein that remains uncleaved at the C terminus and fully cell associated. J. Bacteriol. *182*, 6005–6013.

St Geme, J.W., 3rd, de la Morena, M.L., and Falkow, S. (1994). A *Haemophilus influenzae* IgA protease-like protein promotes intimate interaction with human epithelial cells. Mol. Microbiol. *14*, 217–233.

Stevens, M.K., Latimer, J.L., Lumbley, S.R., Ward, C.K., Cope, L.D., Lagergard, T., and Hansen, E.J. (1999). Characterization of a *Haemophilus ducreyi* mutant deficient in expression of cytolethal distending toxin. Infect. Immun. *67*, 3900–3908.

Stevenson, A., Macdonald, J., and Roberts, M. (2003). Cloning and characterisation of type 4 fimbrial genes from *Actinobacillus pleuropneumoniae.* Vet. Microbiol. *92*, 121–134.

Strathdee, C.A., and Lo, R.Y. (1989a). Cloning, nucleotide sequence, and characterization of genes encoding the secretion function of the *Pasteurella* haemolytica leukotoxin determinant. J. Bacteriol. *171*, 916–928.

Strathdee, C.A., and Lo, R.Y. (1989b). Regulation of expression of the *Pasteurella* haemolytica leukotoxin determinant. J. Bacteriol. *171*, 5955–5962.

Sugai, M., Kawamoto, T., Peres, S.Y., Ueno, Y., Komatsuzawa, H., Fujiwara, T., Kurihara, H., Suginaka, H., and Oswald, E. (1998). The cell cycle-specific growth-inhibitory factor produced by *Actinobacillus actinomycetemcomitans* is a cytolethal distending toxin. Infect. Immun. *66*, 5008–5019.

Taichman, N.S., Simpson, D.L., Sakurada, S., Cranfield, M., DiRienzo, J., and Slots, J. (1987). Comparative studies on the biology of *Actinobacillus actinomycetemcomitans* leukotoxin in primates. Oral Microbiol. Immunol. *2*, 97–104.

Thelestam, M., and Frisan, T. (2004). Cytolethal distending toxins. Rev. Physiol. Biochem. Pharmacol. *152*, 111–133.

Tomasi, T.B., Jr., and Yurchak, A.M. (1972). The synthesis of secretory component by the human thymus. J. Immunol. *108*, 1132–1135.

Tomich, M., Fine, D.H., and Figurski, D.H. (2006). The TadV protein of *Actinobacillus actinomycetemcomitans* is a novel aspartic acid prepilin peptidase required for maturation of the Flp1 pilin and TadE and TadF pseudopilins. J. Bacteriol. *188*, 6899–6914.

Tomich, M., Planet, P.J., and Figurski, D.H. (2007). The tad locus: postcards from the widespread colonization island. Nat. Rev. Microbiol. *5*, 363–375.

Tsai, C.C., Shenker, B.J., DiRienzo, J.M., Malamud, D., and Taichman, N.S. (1984). Extraction and isolation of a leukotoxin from *Actinobacillus actinomycetemcomitans* with polymyxin B. Infect. Immun. *43*, 700–705.

Ueno, Y., Ohara, M., Kawamoto, T., Fujiwara, T., Komatsuzawa, H., Oswald, E., and Sugai, M. (2006). Biogenesis of the *Actinobacillus actinomycetemcomitans* cytolethal distending toxin holotoxin. Infect. Immun. *74*, 3480–3487.

Urbanus, M.L., Scotti, P.A., Froderberg, L., Saaf, A., de Gier, J.W., Brunner, J., Samuelson, J.C., Dalbey, R.E., Oudega, B., and Luirink, J. (2001). Sec-dependent membrane protein insertion: sequential interaction of nascent FtsQ with SecY and YidC. EMBO Rep. *2*, 524–529.

van der Laan, M., Houben, E.N., Nouwen, N., Luirink, J., and Driessen, A.J. (2001). Reconstitution of Sec-dependent membrane protein insertion: nascent FtsQ interacts with YidC in a SecYEG-dependent manner. EMBO Rep. *2*, 519–523.

van Rensburg, E., and du Preez, J.C. (2007). Effect of pH, temperature and nutrient limitations on growth and leukotoxin production by Mannheimia haemolytica in batch and continuous culture. J. Appl. Microbiol. *102*, 1273–1282.

Wang, Y., Shi, W., Chen, W., and Chen, C. (2003). Type IV pilus gene homologs pilABCD are required for natural transformation in *Actinobacillus actinomycetemcomitans.* Gene *312*, 249–255.

Watt, M.A., Mellors, A., and Lo, R.Y. (1997). Comparison of the recombinant and authentic forms of the *Pasteurella* haemolytica A1 glycoprotease. FEMS Microbiol. Lett. *147*, 37–43.

Welch, R.A. (2001). RTX toxin structure and function: a story of numerous anomalies and few analogies in toxin biology. Curr. Top. Microbiol. Immunol. *257*, 85–111.

Yamada, T., Komoto, J., Saiki, K., Konishi, K., and Takusagawa, F. (2006). Variation of loop sequence alters stability of cytolethal distending toxin (CDT): crystal structure of CDT from *Actinobacillus actinomycetemcomitans.* Protein Sci. *15*, 362–372.

Young, V.B., Knox, K.A., Pratt, J.S., Cortez, J.S., Mansfield, L.S., Rogers, A.B., Fox, J.G., and Schauer, D.B. (2004). In vitro and in vivo characterization of Helicobacter hepaticus cytolethal distending toxin mutants. Infect. Immun. *72*, 2521–2527.

Zecchinon, L., Fett, T., and Desmecht, D. (2005). How Mannheimia haemolytica defeats host defence through a kiss of death mechanism. Vet. Res. *36*, 133–156.

RTX Toxin-determined Virulence of *Pasteurellaceae*

7

Joachim Frey

Abstract

RTX toxins are bacterial pore forming toxins that are particularly abundant among pathogenic species of *Pasteurellaceae* where they play a major role in virulence. RTX toxins of several primary pathogens of the family of *Pasteurellaceae* are directly involved in causing necrotic lesions of the target organs. Many RTX toxins are mainly known as haemolysins due to their capacity to lyse erythrocytes *in vitro*, an effect that seems to be non-specific. It is now known for many RTX toxins that their specific targets are leukocytes, where RTX toxins bind to the corresponding β-subunit (CD18) of β_2 integrin and then cause a cytotoxic effect. For several RTX toxins the binding to CD18 was shown to be host specific and seems to be the basis determining the host range of a given RTX toxin. Observations on very closely related species of the *Pasteurellaceae* family with different RTX toxins indicate that these latter contribute to a significant part to the host specificity of the pathogen itself. RTX toxins induce a strong immunological response generating neutralizing antibodies. They therefore constitute important antigens in modern subunit vaccines.

Introduction

Virulence of bacterial pathogens is a measure of their capacity to cause an infectious disease. In contrast to other epidemiological perceptions such as 'incidence', 'morbidity' or 'mortality' that are dependent on the pathogen, the status of the receptive host and environmental factors, virulence is solely defined by the pathogen. Bacterial virulence is generally multi-factorial and depends on a number of virulence factors that determine the ability of the pathogenic bacterium to adhere, colonize, escape the host's immune defence, invade and finally cause damage to the host's tissue. Damage to the host tissue and resulting disease can occur either indirectly by the bacterial proliferation, by over-induction of the host's immune defence mechanisms or directly by aggressins such as toxins. Toxins often are cause of severe, acute forms of disease. In pathogenic species of the family of *Pasteurellaceae*, several virulence attributes that are accountable for the various steps in infection such as adhesins, capsule, iron acquisition systems and various toxins have been investigated and addressed in other chapters (Bossé *et al.*, 2002; Harper *et al.*, 2006; Fine *et al.*, 2006). Secreted protein toxins have been found to play a particular role in virulence determination of *Pasteurellaceae* species. Background information on the secretion and the molecular mode of action of protein toxins of *Pasteurellaceae* is given in Chapter 6 of this book. The current review deals with the role that RTX-protein toxins play in the severity of disease in the susceptible individual. Furthermore the role of inactivated toxins and toxin derivatives in inducing protective immunity and as components in vaccines are discussed.

RTX toxins a major class of virulence factors in *Pasteurellaceae*

Bacterial RTX toxins, named by their repeats in the structural toxin (Strathdee and Lo, 1989), represent a homogenous family of pore forming toxins that are particularly wide spread in species of the family *Pasteurellaceae*. RTX toxins are pore forming protein toxins with a molecular mass in the range of 100–300 kDa that are post-translationally activated by acylation via a specific activator protein that is encoded on the same operon as the structural toxin protein (Ludwig *et al.*, 1996; Stanley *et al.*, 1998; Strathdee and Lo, 1989). The repeated structure, which gave RTX toxins their name, is composed by up to 47 copies of iterative glycine-rich nonapeptides that are generally on the C-terminal half of the protein. These repeats bind Ca^{2+}, which is involved in the binding of RTX toxins to erythrocytes and other target cells, a process that also involves the acylated amino-acid residues. Many RTX toxins show a pronounced haemolytic phenotype under *in vitro* conditions. However, leukocytes were found to be the main targets of RTX toxins where they bind to β_2-integrins and then induce signalling cascades leading to apoptosis at low concentrations and necrosis and cell lysis at higher concentrations of the toxin. These events finally result in inflammation, tissue lesions and consequently to disease (Sun *et al.*, 1999a; Clinkenbeard and Upton, 1991; Confer *et al.*, 1990; Zecchinon, 2007; Dileepan *et al.*, 2007; Thumbikat *et al.*, 2005; Lally *et al.*, 1999). Specific binding of various RTX toxins to different β_2-integrins was shown to account for the pronounced host cell specificity of several RTX toxins, a mechanism that seems finally to determine together with surface adhesions, the host specificity of the pathogen (Dileepan *et al.*, 2005; Dileepan *et al.*, 2007; Zecchinon *et al.*, 2004a). Genetic analysis of RTX toxins of various species of *Pasteurellaceae* and of a few other Gram-negative bacteria provided evidence of horizontal transfer of genes encoding RTX toxins and led to speculation that RTX toxins might have evolved from *Pasteurellaceae* and had conferred to certain species host specific pathogenicity (Frey and Kuhnert, 2002).

Leukotoxin-determined virulence of *Mannheimia haemolytica*

Mannheimia haemolytica is the key agent of mannheimiosis or pasteurellosis, a severe pneumonia in cattle, sheep and goats, which involves several predisposing factors such as other bacteria (e.g. *Arcanobacterium pyogenes*, *Mycoplasma bovis*), viruses (e.g. parainfluenza virus, bovine respiratory syncytial virus, bovine herpesvirus), environmental factors (ammoniac, dust, sudden temperature changes) or also stress (transport, crowding). Predisposing factors seem to modify the epithelium of the upper respiratory tract thus allowing *M. haemolytica* to leave the normal habitat of the nasopharynx and invade the lungs, where it is responsible for the resulting severe pneumonia, which causes high morbidity and mortality, in particular in cattle (Vogel and Parrot, 1994; Edwards, 1996). Besides several virulence factors such as capsule, lipopolysaccharide, outer membrane proteins and neuraminidase, leukotoxin (Lkt) a pore forming, weakly haemolytic protein toxin belonging to the family of RTX toxins (Burrows *et al.*, 1993; Strathdee and Lo, 1989) appears to be the main virulence factor of *M. haemolytica* (Zecchinon *et al.*, 2005). Hence a deletion mutant of the Leukotoxin gene *lktA* in *M. haemolytica* leads to a strain that is unable to cause pulmonary lesions in spite of its capacity to colonize the upper respiratory tract of calves as does the wild type parent strain (Tatum *et al.*, 1998). Under *in vitro* conditions, necrosis of ruminant neutrophils can be induced with purified leukotoxin (Cudd *et al.*, 2001; Czuprynski *et al.*, 1991) giving further evidence for the cytotoxic effect of leukotoxin during infections with *M. haemolytica*. Leukotoxin is cytotoxic only for leukocytes of ruminants and not other animals, indicating that the specificity is responsible for the restricted host range of *M. haemolytica* to cattle, sheep and goats with younger animals being particularly receptive for the disease (Shewen and Wilkie, 1982; Sun *et al.*, 1999b; Highlander, 2001). The overall structure of leukotoxin is well conserved. However, the study of the leukotoxin genes of several bovine and ovine isolates of *M. haemolytica* revealed a complex mosaic structure of the leukotoxin, which indicates that it evolved from a series of inter- and intra species horizontal gene transfers between major lineages of *M. haemolytica* (Davies *et al.*, 2002; Larsen *et al.*, 2007a; Larsen *et al.*, 2007b). While variations in the structure of the leukotoxin have an impact on the strength of cytotoxicity, they seem not to affect the host specificity of this toxin for ruminant leukocytes (Davies and Baillie, 2003). Host specificity of leukotoxin is due to specific interactions of the toxin with the β-subunit CD18 of the β_2 integrins of bovine, ovine and caprine leukocytes (Dassanayake *et al.*, 2007; Zecchinon *et al.*, 2006; Deshpande *et al.*, 2002). In these ruminant species CD18 is conserved but is distinct from CD18 of other species (Dassanayake *et al.*, 2007; Zecchinon *et al.*, 2004a; Zecchinon *et al.*, 2004b). In a study using murine mastocytoma cells P815 that were stably transfected with the gene for ovine CD18 Dassanayake and collaborators showed that the cells became susceptible to the *M. haemolytica* leukotoxin Lkt, while non-transfected cells were not (Dassanayake *et al.*, 2007). Another study showed that a human leukocyte cell line that is devoid of β_2 integrin and that was transfected with the genes for the bovine $\alpha_L\beta_2$ integrin ($CD11a_{bovine}/CD18_{bovine}$) was susceptible to the cytotoxic effect of Lkt, while the same cell line transfected with the $CD11a_{bovine}/CD18_{human}$ genes to produce a hybrid bovine/human $\alpha_L\beta_2$ integrin was not susceptible to leukotoxin (Zecchinon, 2007; Dileepan *et al.*, 2005). Hence leukotoxin of *M. haemolytica* exerts its cytotoxicity to bovine leukocytes via a specific interaction with a CD18 moiety and thus confers *M. haemolytica* a species-specific pathogenic phenotype whose strength seems to depend both on the modular composition and the quantity of Lkt expressed as well as on other virulence attributes.

RTX toxin determined virulence of *Actinobacillus pleuropneumoniae*

Among the different species of the genus *Actinobacillus*, the most prominent pathogen is *Actinobacillus pleuropneumoniae*, a severe, contagious pulmonary disease of pigs. *A. pleuropneumoniae* still causes severe economic losses in industrialized production worldwide in spite of drastic control measures by diagnostic surveillance and vaccination that have been introduced during the last 20 years and that have significantly improved the epidemic situation. Porcine pleuropneumonia can be either acute, extensive and fibrino-haemorrhagic causing very high mortality rapidly after infection, or can be chronic, localized, necrotizing and associated with pleuritis causing low mortality but high morbidity (Shope, 1964; Taylor, 1999; Bossé *et al.*, 2002). Serological subtyping of *A. pleuropneumoniae* strains based on the surface located lipopolysaccharide and the capsular polysaccharide into 15 different serotypes together with extensive epidemiological studies allowed to discriminate highly pathogenic serotypes such as serotypes 1, 5 and also 9 and 11 from less virulent serotypes (Taylor, 1999; Nicolet, 1992). Experimental models of infection using pigs or mice with the serotype reference strains corroborated the epidemiological studies and showed serotypes 1, 5, 9 and 11 to be particularly virulent (Komal and Mittal, 1990; Nicolet, 1992). The serotyping method, although based on indirect virulence markers, allowed many countries to eradicate the most severe forms of porcine pleuropneumonia.

The discovery of four different RTX toxins ApxI, ApxII, ApxIII and ApxIV in the various serotypes of *A. pleuropneumoniae* shed a new light on the molecular mechanisms of pathogenicity of this pathogen (Frey, 1995; Schaller *et al.*, 1999; Chang *et al.*, 1993; Jansen *et al.*, 1992a). Initial characterization of these toxins revealed that ApxI is strongly haemolytic and cytotoxic, ApxII is weakly haemolytic and weakly cytotoxic, while ApxIII is non-haemolytic and strongly cytotoxic. Phenotypic and genotypic analyses of large numbers of strains of all serotypes that had been isolated from different geographic areas showed, with very few exceptions, that all serotypes had their characteristic pattern of secreted Apx toxins (Beck *et al.*, 1994; Frey *et al.*, 1993; Kamp *et al.*, 1994; Jansen *et al.*, 1992b). Hence, while all *A. pleuropneumoniae* secrete ApxIV, but only during infection, the strongly pathogenic serotypes 1, 5, 9 and 11 secrete in addition toxins ApxI and ApxII, the group of serotypes 2, 4, 6, 8 and 15 that generally cause infections with lower mortality by high morbidity secrete ApxII and ApxIII, while the remaining serotypes each only secrete one additional Apx toxin, serotype 10 and 14 ApxI, serotype 7, 12 and 13 ApxII and serotype 3 ApxIII. These observations lead to the conclusions that *A. pleuropneumoniae* strains respectively serotypes that secrete two Apx toxins in addition to ApxIV show increased virulence and that those among them secreting ApxI belong to the group of particularly virulent *A. pleuropneumoniae* causing high mortality in infected pig herds (Frey, 1995; Frey, 1994).

While, it was obvious that the presence of capsule, lipopolysaccharide and iron acquisition mechanisms are essential for the bacterium to survive in the host and cause disease (Inzana, 1990; Ward and Inzana, 1994; Rioux *et al.*, 1999; Gerlach *et al.*, 1992) epidemiological and experimental evidence showed that the different RTX toxins play a primary role in virulence of *A. pleuropneumoniae* and contribute to most of the pathological consequences of porcine pleuropneumonia (Frey, 1994; Frey, 1995; Rycroft *et al.*, 1991).

The direct impact of RTX toxins in virulence was confirmed by experimental infections of pigs using molecular genetically defined mutants and *trans*-complemented mutant strains of *A. pleuropneumoniae*. A non-haemolytic serotype 5 mutant lacking the ApxI toxin genes *apxICA* and the type I secretion system genes *apxIBD* that hence is unable to secrete neither ApxI nor ApxII (in spite of the residual *apxIICA* genes) was shown to be non pathogenic in pigs and also in mice (which serve as an experimental animal model to study virulence), even at doses that are much higher than the LD_{50} of the parent strain (Reimer *et al.*, 1995). Partial complementation of this deletion mutant with type I secretion genes *apxIBD*, cloned on a broad host-range plasmid, restored secretion of ApxII and resulted in a recombinant strain that caused pleuropneumonia in experimentally infected pigs, but at a higher dose of infection and with less mortality than the wild type strain. This strain still remained non-virulent in a mouse infection model. Complementation of the same deletion mutant with the cloned ApxI toxin genes *apxICA* and type I secretion genes *apxIBD* restored production and secretion of both ApxI and ApxII and led to reappearance of full virulence with high mortality level in pigs and mice even greater than those induced by the wild type (Reimer *et al.*, 1995). Similar results were obtained using transposon mutants of a serotype 1 strain, which showed that insertional inactivation of the *apxIBD* genes resulting in a non-haemolytic $ApxI^-/ApxII^-$ phenotype are non-pathogenic, while insertional inactivation of the *apxIA* gene resulting in a weakly haemolytic $ApxI^-/ApxII^+$ phenotype retains some virulence in pigs (Tascon *et al.*, 1994). A more recent study using knock-out mutants derived from the highly virulent serotype 1 *A. pleuropneumoniae* type strain 4074 revealed that derivatives of this strain mutated in both *apxIA* and *apxIIA*, which have an $ApxI^-/ApxII^-$ phenotype were unable to induce pathological lung lesions in pigs and hence were non-virulent. Isogenic mutants of strain 4074 that were deleted in either *apxIA* (phenotype: $ApxI^-/ApxII^+$) or in *apxIIA* ($ApxI^+/ApxII^-$) both showed a reduced virulence compared to the parent strain (Boekema *et al.*, 2004). These results indicate that both ApxI and ApxII of *A. pleuropneumoniae* serotype 1 are necessary for full virulence.

The direct role of Apx toxins in the pathogenesis of porcine pleuropneumonia was studied by endobronchial inoculation of recombinant ApxI, ApxII and ApxIII, as well as with culture supernatants from *A. pleuropneumoniae* serotype 8 and 9 containing toxins ApxII plus ApxIII and ApxI plus ApxII respectively. The studies showed that ApxI, ApxIII and to a lesser extent ApxII are able to induce fibro-hemorrhagic pneumonia and purulent pneumonia with both native and recombinant Apx toxins. Necrosis, however, was only induced by recombinant ApxI or by a combination of recombinant ApxI plus ApxII (Kamp *et al.*, 1997). This study is an important contribution to the understanding of the impact of Apx toxins in triggering the development of clinical symptoms and lung lesions typical of porcine pleuropneumonia. Moreover, the recombinant Apx toxins used in this study revealed that these symptoms and lung lesions were induced in the absence of *A. pleuropneumoniae* lipopolysaccharide (LPS), since the Apx toxins were produced in *Escherichia coli* K12-derived safety strains (Kamp *et al.*, 1997). In as how far *E. coli* K12 LPS would subsidize the missing host strain LPS or whether Apx toxins do not request LPS as a cofactor to be a functional toxin in contrast to what was suggested for other RTX toxins (Czuprynski and Welch, 1995; Li and Clinkenbeard, 1999) has, however not been investigated yet.

While the haemolytic strength of the three different major Apx toxins ApxI, ApxII and ApxIII can be explained by their pore forming activity and the pore sizes made in artificial lipid bilayers (Maier *et al.*, 1996), the differences in pathologies induced in lung tissue is expected to be due to further, yet unknown attributes of these RTX toxins. The 200–300 kDa large RTX protein, named ApxIV (Schaller *et al.*, 1999) exhibits structural and functional features of RTX toxins, in particular the glycine-rich nonapeptide repeats with Ca^{2+}-binding capacity and shows high similarity to the calcium-dependent processing and cross-linking 'clip and link' FrpA protein (Osicka *et al.*, 2004). It is present in all *A. pleuropneumoniae* serotypes and is specific to the species (Schaller *et al.*, 1999). ApxIV is not expressed *in vivo* using current media and culture conditions, but induces high levels of serum antibodies during infection (Schaller *et al.*, 1999; Dreyfus *et al.*, 2004), which are useful in serodiagnosis of porcine pleuropneumonia (Dreyfus *et al.*, 2004). Unfortunately several attempts to create knockout mutants of the ApxIV gene of *A. pleuropneumoniae* in the author's laboratory and elsewhere were unsuccessful and hence did not allow assessing the role of ApxIV in virulence. Most recently an *A. pleuropneumoniae* serotype 7 strain (AP76) was found to carry insertion element ISAP*1* in the *apxIV* gene, thus silencing this gene by insertional inactivation (Tegetmeyer *et al.*, 2008).

The molecular mechanism of Apx toxins has been attributed to cytotoxicity against porcine leukocytes, neutrophils and phagocytes (Dom *et al.*, 1992; Bossé *et al.*, 2002; Kamp *et al.*, 1991). Knockout mutants of the strongly virulent *A. pleuropneumoniae* seroptype 1 strain 4074 that are devoid of RTX toxins did not activate or kill porcine neutrophils (Jansen *et al.*, 1995). The respective receptors for the Apx toxins in porcine neutrophils, however, have not yet been identified.

RTX toxins from '*Actinobacillus porcitonsillarum*', *Actinobacillus suis* and *[Actinobacillus] rossii*

While endobronchial inoculation with purified recombinant ApxI, ApxII and ApxIII leads to pleuropneumonic symptoms in pigs showing their direct role in causing these symptoms, it is important to note that these toxins, when expressed by other porcine *Actinobacillus* species, do not render them pathogenic or only weakly pathogenic. Hence, '*Actinobacillus porcitonsillarum*' is a non-pathogenic bacterium that is frequently found on the tonsills of healthy pigs (Gottschalk *et al.*, 2003; Tonpitak *et al.*, 2007). '*A. porcitonsillarum*' actively secretes ApxII via its endogenous *apxIIBD* encoded type I secretion system (Kuhnert *et al.*, 2005) but does not induce pathologies in pigs (Gottschalk *et al.*, 2003). Other *Actinobacillus* species, in which Apx toxins seem to cause only mild or no pathologies are *Actinobacillus suis* and [*Actinobacillus*] *rossii* (Kamp *et al.*, 1994; Mayor *et al.*, 2006). *A. suis* strains secrete both ApxI and ApxII (in *A. suis* they are also are named AshI and AshII) but do not cause pleuropneumonia in pigs (Burrows and Lo, 1992). *A. suis* seems solely to cause moderate respiratory distress in early weaned pigs with minor pathologies in the respiratory tract (VanOstaaijen *et al.*, 1997). It is therefore tempting to speculate that the strong tissue damage caused by the classical RTX toxins ApxI, ApxII or ApxIII as seen in *A. pleuropneumoniae* infections might request an interaction or synergy of the species specific ApxIV and with other factors such as capsule, iron uptake and LPS that help *A. pleuropneumoniae* to propagate and persist in the host and that are not present in *A. suis* and [*A.*] *rossii*.

Aqx, an RTX toxin of the equine pathogen *Actinobacillus equuli*

Actinobacillus equuli is the etiological agent of septicaemia in neonatal foals also known as 'sleepy foal disease' which is spread worldwide (Rycroft *et al.*, 1998; Rycroft and Garside, 2000). Two subspecies have been formerly described namely *A. equuli* subsp. *equuli* and *A. equuli* subsp. *haemolyticus* (Christensen *et al.*, 2002). The haemolytic *A. equuli* subsp. *haemolyticus* strains all secrete a 110-kDa RTX toxin named Aqx. Toxin Aqx is encoded on a classical four gene RTX operon *aqxCABD* which contains besides the structural toxin gene also the activator and type I secretion genes (Berthoud *et al.*, 2002). Aqx shows 61% amino-acid sequence similarity and 65% DNA sequence similarity to Lkt of *M. haemolytica*, which is its phylogenetically most closely related RTX toxin (Kuhnert *et al.*, 2003a). While there was previously some confusion on the correct identification of *A. equuli*, it could be demonstrated that erroneously identified haemolytic *A. suis*, isolated from horses indeed were *A. equ-*

uli containing the *aqx* toxin genes (Kuhnert *et al.*, 2003b). In contrast, *aqx* carrying *Actinobacillus* species were never isolated from pigs (Kuhnert *et al.*, 2003b). Aqx toxin was shown to have a specific cytotoxic activity towards equine lymphocytes and was scarcely toxic to porcine lymphocytes, in contrast to the haemolytic activity that is equally pronounced on equine, porcine and ovine erythrocytes (Kuhnert *et al.*, 2003b). A serological analysis of antibodies to Aqx in horses and foals revealed that adult horses generally showed antibodies towards Aqx and hence must have had contact with haemolytic *A. equuli* (Berthoud *et al.*, 2004). This corresponds to the large distribution of *A. equuli* in healthy mares and explains the neutralizing activity of sera of mares towards haemolysin of *A. equuli* (Rycroft *et al.*, 1998). While also foals generally do show Aqx neutralizing antibodies, serum from foals taken before an intake of colostrum did not react with Aqx (Berthoud *et al.*, 2004). In a sequential study it was shown that a foal that was seronegative for anti-Aqx antibodies after birth had seroconverted one hour after taking in colostrums in a time-dependent manner. This study indicated that adult horses, in which *A. equuli* is part of the normal flora in the oral cavity and alimentary tract of horses, do have specific antibodies against Aqx, which seems to neutralize its toxic activity, while newborn foals need to acquire it via the maternal colostrums. It is therefore assumed that pathologies caused by *A. equuli* typically in newborn foals might be due to a failure of acquisition of neutralizing anti-Aqx antibodies from colostrum (Berthoud *et al.*, 2004).

Specificity of Apx and Aqx toxins for porcine respectively equine lymphocytes

Although the role of the different Apx toxins in the pathogenesis and epidemiology of *A. pleuropneumoniae* has been studied extensively, there is currently only little knowledge on the host cell specificity of these toxins and no knowledge on the receptor binding sites. One of the problems seems to be that much emphasis was given to the haemolytic activity, a parameter that is easily to measure and can be neutralized with antibodies directed against the given Apx toxin. However, relatively few studies with Apx toxins, in particular ApxI and ApxII, which are found in *A. pleuropneumoniae* as well as in *A. suis*, were performed with leukocytes. In a comparative study of the haemolytic ApxI and ApxII toxins from the porcine pathogen *A. suis* with the haemolytic RTX toxin Aqx from the equine pathogen *A. equuli*, it was shown that the strength of the haemolytic activity for each of the toxins was the same on erythrocytes of porcine equine or even ovine origin. In contrast, when testing the cytotoxic activity, the toxins ApxI and ApxII from porcine pathogens showed a much higher toxicity toward porcine lymphocytes compared to equine lymphocytes while Aqx, the RTX toxin from *A. equuli* was specific to equine lymphocytes and only showed weak cytotoxic activity towards porcine lymphocytes (Kuhnert *et al.*, 2003b). This is of particular interest, since *A. suis* (the porcine pathogen) and *A. equuli* (the equine pathogen) are taxonomically and phylogenetically very closely related and often are confounded in routine diagnostics. The target specificity of ApxI and ApxII for porcine and Aqx for equine leukocytes in a very similar bacterial background indicates that these toxins seem to be involved in the specificity of the pathogens host range (Kuhnert *et al.*, 2003b). Hence while erythrocytes are most valuable as indicator cells for the detection of RTX toxins in diagnostic procedures in form of blood agar medium, as well as to study several lytic aspects of RTX toxins, erythrocytes seem not to be the specific targets of many RTX toxins including ApxI, ApxII and Aqx. Currently, the specific ligand on RTX toxins for the target host cell has not been investigated in detail.

Leukotoxin-determined virulence of *Aggregatibacter actinomycetemcomitans*

Aggregatibacter actinomycetemcomitans is the causative organism of localized aggressive periodontitis, a rapidly progressing degenerative disease of the gingival and periodontal ligaments primarily in adolescent humans (Zambon, 1985). It also can be implicated in subacute infective endocarditis in humans (Das *et al.*, 1997). However, *Agg. actinomycetemcomitans* is also detected with considerable frequency in subjects without clinical evidence of periodontal diseases (Gmür and Guggenheim, 1994). *Agg. actinomycetemcomitans* produces several virulence factors including an exotoxic leukotoxin Ltx, which is a member of the RTX family of toxins and is assumed to be a major virulence factor. Epidemiological evidence revealed that highly leukotoxic strains of *Agg. actinomycetemcomitans* are more associated with aggressive forms of periodontitis in humans than the low- or non- leukotoxin producing strains (Haraszthy *et al.*, 2000; Haubek and Westergaard, 2004; Brogan *et al.*, 1994; Spitznagel *et al.*, 1991; Mombelli *et al.*, 1999). Expression of Ltx seems to be transcriptionally regulated and strains of *Agg. actinomycetemcomitans* that cause aggressive forms of periodontitis are characterized by particular promoter structures (Bueno *et al.*, 1998). Highly toxigenic strains are characterized by a 530-bp deletion in the promoter region of the leukotoxin operon that results in a 10–20 times higher leukotoxin production (Brogan *et al.*, 1994). Strains showing this particular strong promoter structure were found at high occurrence in patients with localized juvenile periodontitis in contrast to healthy patients or adults with periodontitis, which carried mostly *Agg. actinomycetemcomitans* with the less active *ltx* promoter (Zambon *et al.*, 1996) thus bringing in agreement the genotypic with the phenotypic observations on the primordial impact of leukotoxin production in disease of infected patients.

Leukotoxin of *Agg. actinomycetemcomitans* was shown in antecedent studies to specifically interact with leukocytes from humans and primates (Mangan *et al.*, 1991; Taichman *et al.*, 1980; Taichman *et al.*, 1987). Lally and collaborators had provided compelling evidence of a β_2 integrin family, human LFA-1 (lymphocyte function-associated antigen-1), as a receptor for *Agg. actinomycetemcomitans* leukotoxin Ltx (Lally *et al.*, 1997). Based on experiments performed with recombinant chimeric β_2 integrins expressed in a cell line that is resistant to Ltx, human CD18 was identified as the receptor for Ltx (Dileepan *et al.*, 2007). The domain on CD18 containing three cysteine-rich tandem repeats, named I-EGF-2, I-EGF-3 and I-EGF-4 which have epidermal growth factor (EGF)-like structures and are located on the extracellular region of human CD18 were identified most recently as a specific receptor domain for interacting with Ltx (Dileepan *et al.*, 2007). This receptor is expected to interact with a specific ligand on Ltx. Based on studies with chimeric RTX toxins made of leukotoxin and Ltx of *Agg. actinomycetemcomitans*, it is expected that the glycine-rich nonapeptide repeats are representing the respective ligand on the bacterial RTX toxin (Dileepan *et al.*, 2007; Lally *et al.*, 1994). While the studies on *Agg. actinomycetemcomitans* Ltx and *M. haemolytica* Lkt give currently the most detailed insight on the molecular interaction and specificities of RTX toxins, these studies will serve to unravel the binding sites and specificity determinants of other medically important RTX toxins and will be useful to develop specific therapeutic agents that could block the toxic effects of RTX toxins.

RTX toxins in vaccines

The final goal of research on molecular mechanisms of microbial pathogenicity is to realize, by applying the acquired knowledge, the development of novel therapeutics and prophylactics in order to cure, prevent or possibly eradicate infectious diseases. RTX toxin revealed to be interesting targets as antigens in potential vaccines against infectious diseases caused by species of the *Pasteurellaceae* family, since they showed to introduce specific neutralizing antibodies in immunized individuals (Berthoud *et al.*, 2002; Frey and Nicolet, 1990; Frey *et al.*, 1992; Frey and Nicolet, 1988; Lo *et al.*, 1985).

Monoclonal and polyclonal antibodies directed against leukotoxin Lkt of *M. haemolytica* showed a strong neutralising activity against the cytotoxic effect of Lkt (Conlon *et al.*, 1991; Gerbig, Jr. *et al.*, 1992; Lo *et al.*, 1985). In early experiments using leukotoxin obtained from supernatant of pathogenic *M. haemolytica* A1 or non-pathogenic serotype 11 to immunize cattle, it was shown that the immunized animals included antitoxic activity in the serum but they were not protected against mannheimiosis. In order to induce protective immunity in cattle, serospecific agglutinating activity as obtained by vaccination with *M. haemolytica* A1 whole cell bacteria plus antitoxic activity against Lkt was necessary (Shewen and Wilkie, 1988). Similar results were obtained when cattle were vaccinated using recombinant leukotoxin. Recombinant leukotoxin strongly enhanced the efficacy of vaccines containing bacterins in decreasing clinical signs and pneumonic lesions in vaccinated animals after challenge (Shewen and Wilkie, 1988). Lkt alone, however, did not protect from disease (Shewen and Wilkie, 1988). Furthermore, vaccination with recombinant Lkt alone did not protect animals against disease (Conlon *et al.*, 1991). In order to study the feasibility of edible vaccines for cattle, hydrophobic transmembrane domains of Lkt were expressed in transgenic white clover where it was shown to be stable in plant tissue after dying and to induce neutralizing antibodies in rabbits (Lee *et al.*, 2001).

Vaccination with serotype specific vaccines, based on washed, inactivated bacteria, against porcine pleuropneumonia that is caused by the different serotypes of *A. pleuropneumoniae* requested precise knowledge of the epidemiological situation of porcine pleuropneumonia in areas where these vaccines were used, since the vaccines did not confer cross-protection against different serotypes. The RTX toxins ApxI, ApxII and ApxIII or recombinant derivatives of them were shown to induce neutralizing antibodies and protective immunity in the mouse or in pigs that were vaccinated with detoxified individual Apx toxin or combination of all three (Haga *et al.*, 1997; Tumamao *et al.*, 2004b; Xu *et al.*, 2006). Subunit vaccines based solely on ApxI, ApxII and ApxIII were shown to protect pigs of the most prevalent serotypes 1–12 primarily from mortality but not from pneumonic lesions (Kobisch and van den Bosch, 1992; van den Bosch *et al.*, 1992). However, supplementing the vaccine based on ApxI, ApxII and ApxIII with a 42-kDa outer membrane protein of *A. pleuropneumoniae* led to the formulation of a successful vaccine Porcilis® APP (Intervet), that confers high protection against all 12 major serotypes 1–12 of *A. pleuropneumoniae* under experimental conditions (Kobisch and van den Bosch, 1992; Tumamao *et al.*, 2004b; van den Bosch and Frey, 2003; van den Bosch *et al.*, 1990) as well as in field conditions (Jungbloot, 2005; Martelli *et al.*, 1996; Pommier *et al.*, 1996; Valks *et al.*, 1996). Interestingly, the vaccine failed to fully protect against *A. pleuropneumoniae* serovar 15 that is specifically present on the Australian continent. This failure was attributed to significant differences between the ApxIII toxin of serovar 15 and the ApxIII toxin from the other serotypes (2, 3, 4, 6 and 8) that is included in Porcilis® APP (Tumamao *et al.*, 2004a). This study further confirms the important role of anti-Apx antibodies in protective immunity against porcine pleuropneumonia. An approach to develop live vaccines against porcine pleuropneumonia consists in deleting the activator genes *apxIC* and *apxIIC* to obtain attenuated mutants that secrete inactive, but immunologically protective ApxI and ApxII which protect pigs against pleuropneumonia caused by a subset of *A. pleuropneumoniae* serotypes (Lin *et al.*, 2007; Prideaux *et al.*, 1999). Recently it was shown that mice that were fed with powdered transgenic plants expressing ApxII induced neutralizing anti-ApxII antibodies and were protected from injection of a lethal dose of *A. pleuropneumoniae* serotype 2 (Lee *et al.*, 2006). This approach might conduct to the development of future edible vaccines for pigs.

Conclusions

Many studies have highlighted the predominant role of RTX toxins of various pathogenic species of the family of *Pasteurellaceae* in causing infectious diseases by their cytotoxic and especially leukotoxic activities. Moreover, specific binding capacities of RTX toxins to β_2 integrins of the specific hosts of the respective pathogens revealed that RTX toxins are to some extent at the bases of host specificity of several pathogens of *Pasteurellaceae*. The profound knowledge on epidemiology and molecular mechanisms of RTX toxins in *Pasteurellaceae* has led to the development of valuable novel vaccines that are either already on the market or are expected to be introduced in the future.

References

Beck, M., Vandenbosch, J.F., Jongenelen, I.M.C.A., Loeffen, P.L.W., Nielsen, R., Nicolet, J., and Frey, J. (1994). RTX toxin genotypes and phenotypes in *Actinobacillus pleuropneumoniae* field strains. J. Clin. Microbiol. *32*, 2749–2754.

Berthoud, H., Frey, J., and Kuhnert, P. (2002). Characterization of Aqx and its operon: the hemolytic RTX determinant of *Actinobacillus equuli*. Vet. Microbiol. *87*, 159–174.

Berthoud, H., Frey, J., Sternberg, S., Straub, R., and Kuhnert, P. (2004). Antibodies to Aqx toxin of *Actinobacillus equuli* in horses and foals. Vet. Rec. *155*, 231–233.

Boekema, B.K., Kamp, E.M., Smits, M.A., Smith, H.E., and Stockhofe-Zurwieden, N. (2004). Both ApxI and ApxII of *Actinobacillus pleuropneumoniae* serotype 1 are necessary for full virulence. Vet. Microbiol. *100*, 17–23.

Bossé, J.T., Janson, H., Sheehan, B.J., Beddek, A.J., Rycroft, A.N., Kroll, J.S., and Langford, P.R. (2002). *Actinobacillus pleuropneumoniae*: pathobiology and pathogenesis of infection. Microbes Infect. *4,* 225–235.

Brogan, J.M., Lally, E.T., Poulsen, K., Kilian, M., and Demuth, D.R. (1994). Regulation of *Actinobacillus actinomycetemcomitans* leukotoxin expression: analysis of the promoter regions of leukotoxic and minimally leukotoxic strains. Infect. Immun. *62,* 501–508.

Bueno, L.C., Mayer, M.P., and DiRienzo, J.M. (1998). Relationship between conversion of localized juvenile periodontitis-susceptible children from health to disease and *Actinobacillus actinomycetemcomitans* leukotoxin promoter structure. J. Periodontol. *69,* 998–1007.

Burrows, L.L. and Lo, R.Y. (1992). Molecular characterization of an RTX toxin determinant from *Actinobacillus suis.* Infect. Immun. *60,* 2166–2173.

Burrows, L.L., Olahwinfield, E., and Lo, R.Y.C. (1993). Molecular analysis of the leukotoxin determinants from *Pasteurella haemolytica* serotypes 1 to 16. Infect. Immun. *61,* 5001–5007.

Chang, Y.F., Shi, J., Ma, D., Shin, S., and Lein, D.H. (1993). Molecular analysis of the *Actinobacillus pleuropneumoniae* RTX toxin-III gene cluster. DNA Cell Biol. *12,* 351–362.

Christensen, H., Bisgaard, M., and Olsen, J.E. (2002). Reclassification of equine isolates previously reported as *Actinobacillus equuli,* variants of *A. equuli, A. suis* or taxon 11 of Bisgaard and proposal of two subspecies of *A. equuli: A. equuli* subsp. *equuli* and *A. equuli* subsp. *haemolyticus.* Int. J Syst. Evol. Microbiol. 52, 1569–1576.

Clinkenbeard, K.D. and Upton, M.L. (1991). Lysis of bovine platelets by *Pasteurella haemolytica* leukotoxin. Amer. J. Vet. Res. *52,* 453–457.

Confer, A.W., Panciera, R.J., Clinkenbeard, K.D., and Mosier, D.A. (1990). Molecular aspects of virulence of *Pasteurella haemolytica.* Can. J. Vet. Res. *54 Suppl,* S48-S52.

Conlon, J.A., Shewen, P.E., and Lo, R.Y. (1991). Efficacy of recombinant leukotoxin in protection against pneumonic challenge with live *Pasteurella haemolytica* A1. Infect. Immun. *59,* 587–591.

Cudd, L.A., Ownby, C.L., Clarke, C.R., Sun, Y., and Clinkenbeard, K.D. (2001). Effects of *Mannheimia haemolytica* leukotoxin on apoptosis and oncosis of bovine neutrophils. Am. J. Vet. Res. *62,* 136–141.

Czuprynski, C.J., Noel, E.J., Ortiz Carranza, O., and Srikumaran, S. (1991). Activation of bovine neutrophils by partially purified *Pasteurella haemolytica* leukotoxin. Infect. Immun. *59,* 3126–3133.

Czuprynski, C.J. and Welch, R.A. (1995). Biological effects of RTX toxins: The possible role of lipopolysaccharide. Trends Microbiol. *3,* 480–483.

Das, M., Badley, A.D., Cockerill, F.R., Steckelberg, J.M., and Wilson, W.R. (1997). Infective endocarditis caused by HACEK microorganisms. Annu. Rev. Med. *48,* 25–33.

Dassanayake, R.P., Shanthalingam, S., Davis, W.C., and Srikumaran, S. (2007). *Mannheimia haemolytica* leukotoxin-induced cytolysis of ovine (*Ovis aries*) leukocytes is mediated by CD18, the β subunit of β2-integrins. Microb. Pathog. *42,* 167–173.

Davies, R.L. and Baillie, S. (2003). Cytotoxic activity of *Mannheimia haemolytica* strains in relation to diversity of the leukotoxin structural gene *lktA.* Vet. Microbiol. *92,* 263–279.

Davies, R.L., Campbell, S., and Whittam, T.S. (2002). Mosaic structure and molecular evolution of the leukotoxin operon (*lktCABD*) in *Mannheimia (Pasteurella) haemolytica, Mannheimia glucosida,* and *Pasteurella trehalosi.* J. Bacteriol. *184,* 266–277.

Deshpande, M.S., Ambagala, T.C., Ambagala, A.P., Kehrli, M.E., Jr., and Srikumaran, S. (2002). Bovine CD18 is necessary and sufficient to mediate *Mannheimia (Pasteurella) haemolytica* leukotoxin-induced cytolysis. Infect. Immun. *70,* 5058–5064.

Dileepan, T., Kachlany, S.C., Balashova, N.V., Patel, J., and Maheswaran, S.K. (2007). Human CD18 is the functional receptor for *Aggregatibacter actinomycetemcomitans* leukotoxin. Infect. Immun. *doi:10.1128/IAI.00314–07.*

Dileepan, T., Kannan, M.S., Walcheck, B., Thumbikat, P., and Maheswaran, S.K. (2005). Mapping of the binding site for *Mannheimia haemolytica* leukotoxin within bovine CD18. Infect. Immun. *73,* 5233–5237.

Dom, P., Haesebrouck, F., and De Baetselier, P. (1992). Stimulation and suppression of the oxygenation activity of porcine pulmonary alveolar macrophages by *Actinobacillus pleuropneumoniae* and its metabolites. Amer. J. Vet. Res. *53,* 1113–1118.

Dreyfus, A., Schaller, A., Nivollet, S., Segers, R.P., Kobisch, M., Mieli, L., Soerensen, V., Hüssy, D., Miserez, R., Zimmermann, W., Inderbitzin, F., and Frey, J. (2004). Use of recombinant ApxIV in serodiagnosis of *Actinobacillus pleuropneumoniae* infections, development and prevalidation of the ApxIV ELISA. Vet. Microbiol. *99,* 227–238.

Edwards, A.J. (1996). Respiratory diseases of feedlot cattle in the central USA. Bovine Pract. *30,* 5–7.

Fine, D.H., Kaplan, J.B., Kachlany, S.C., and Schreiner, H.C. (2006). How we got attached to *Actinobacillus actinomycetemcomitans*: a model for infectious diseases. Periodontol. 2000 *42,* 114–157.

Frey, J. (1994). RTX-toxins in *Actinobacillus pleuropneumoniae* and their potential role in virulence. In Molecular Mechanisms of Bacterial Virulence, C.I.Kado and J.H.Crosa, eds. (Dordrecht, Boston, London: Kluwer Academic Publishers), pp. 325–340.

Frey, J. (1995). Virulence in *Actinobacillus pleuropneumoniae* and RTX toxins. Trends Microbiol. *3,* 257–261.

Frey, J., Beck, M., Stucki, U., and Nicolet, J. (1993). Analysis of hemolysin operons in *Actinobacillus pleuropneumoniae.* Gene *123,* 51–58.

Frey, J. and Kuhnert, P. (2002). RTX toxins in *Pasteurellaceae.* Int. J. Med. Microbiol. 292, 149–158.

Frey, J. and Nicolet, J. (1988). Regulation of hemolysin expression in *Actinobacillus pleuropneumoniae* serotype 1 by Ca^{2+}. Infect. Immun. *56,* 2570–2575.

Frey, J. and Nicolet, J. (1990). Hemolysin patterns of *Actinobacillus pleuropneumoniae.* J. Clin. Microbiol. *28,* 232–236.

Frey, J., van den Bosch, H., Segers, R., and Nicolet, J. (1992). Identification of a second hemolysin (HlyII) in *Actinobacillus pleuropneumoniae* serotype 1 and expression of the gene in *Escherichia coli.* Infect. Immun. *60,* 1671–1676.

Gerbig, D.G., Jr., Cameron, M.R., Struck, D.K., and Moore, R.N. (1992). Characterization of a neutralizing monoclonal antibody to *Pasteurella haemolytica* leukotoxin. Infect. Immun. *60*, 1734–1739.

Gerlach, G.F., Klashinsky, S., Anderson, C., Potter, A.A., and Willson, P.J. (1992). Characterization of two genes encoding distinct transferrin- binding proteins in different *Actinobacillus pleuropneumoniae* isolates. Infect. Immun. *60*, 3253–3261.

Gmür, R. and Guggenheim, B. (1994). Interdental supragingival plaque. A natural habitat of *Actinobacillus actinomycetemcomitans, Bacteroides forsythus, Campylobacter rectus,* and *Prevotella nigrescens*. J. Dent. Res. *73*, 1421–1428.

Gottschalk, M., Broes, A., Mittal, K.R., Kobisch, M., Kuhnert, P., Lebrun, A., and Frey, J. (2003). Non-pathogenic *Actinobacillus* isolates antigenically and biochemically similar to *Actinobacillus pleuropneumoniae*: a novel species? Vet. Microbiol. *92*, 87–101.

Haga, Y., Ogino, S., Ohashi, S., Ajito, T., Hashimoto, K., and Sawada, T. (1997). Protective efficacy of an affinity-purified hemolysin vaccine against experimental swine pleuropneumonia. J. Vet. Med. Sci. *59*, 115–120.

Haraszthy, V.I., Hariharan, G., Tinoco, E.M., Cortelli, J.R., Lally, E.T., Davis, E., and Zambon, J.J. (2000). Evidence for the role of highly leukotoxic *Actinobacillus actinomycetemcomitans* in the pathogenesis of localized juvenile and other forms of early-onset periodontitis. J. Periodontol. *71*, 912–922.

Harper, M., Boyce, J.D., and Adler, B. (2006). *Pasteurella multocida* pathogenesis: 125 years after Pasteur. FEMS Microbiol. Lett. *265*, 1–10.

Haubek, D. and Westergaard, J. (2004). Detection of a highly toxic clone of *Actinobacillus actinomycetemcomitans* (JP2) in a Moroccan immigrant family with multiple cases of localized aggressive periodontitis. Int. J. Paediatr. Dent. *14*, 41–48.

Highlander, S.K. (2001). Molecular genetic analysis of virulence in *Mannheimia (Pasteurella) haemolytica*. Front. Biosci. *6*, D1128-D1150.

Inzana, T.J. (1990). Capsules and virulence in the HAP group of bacteria. Can. J. Vet. Res. *54 Suppl*, S22-S27.

Jansen, R., Briaire, J., Kamp, E.M., and Smits, M.A. (1992a). Comparison of the cytolysin II genetic determinants of *Actinobacillus pleuropneumoniae* serotypes. Infect. Immun. *60*, 630–636.

Jansen, R., Briaire, J., Kamp, E.M., and Smits, M.A. The cytolysin genes of *Actinobacillus pleuropneumoniae*. Proceedings of the International Pig Veterinary Society Congress 12, 197. 1992b.

Jansen, R., Briaire, J., Smith, H.E., Dom, P., Haesebrouck, F., Kamp, E.M., Gielkens, A.L.J., and Smits, M.A. (1995). Knockout mutants of *Actinobacillus pleuropneumoniae* serotype 1 that are devoid of RTX toxins do not activate or kill porcine neutrophils. Infect. Immun. *63*, 27–37.

Jungbloot, J. (2005). [APP: A pathogen with many faces]. Die landwirtschaftliche Zeitung *11*, 140–142.

Kamp, E.M., Popma, J.K., Anakotta, J., and Smits, M.A. (1991). Identification of hemolytic and cytotoxic proteins of *Actinobacillus pleuropneumoniae* by use of monoclonal antibodies. Infect. Immun. *59*, 3079–3085.

Kamp, E.M., Stockhofezurwieden, N., vanLeengoed, L.G., and Smits, M.A. (1997). Endobronchial inoculation with Apx toxins of *Actinobacillus pleuropneumoniae* leads to pleuropneumonia in pigs. Infect. Immun. *65*, 4350–4354.

Kamp, E.M., Vermeulen, T.M.M., Smits, M.A., and Haagsma, J. (1994). Production of Apx toxins by field strains of *Actinobacillus pleuropneumoniae* and *Actinobacillus suis*. Infect. Immun. *62*, 4063–4065.

Kobisch, M. and van den Bosch, J.F. Efficacy of an *Actinobacillus pleuropneumoniae* subunit vaccine. Proceedings of the International Pig Veterinary Society Congress 12, 216. 1992.

Komal, J.P. and Mittal, K.R. (1990). Studies on the interactions of two different serotypes of *Actinobacillus pleuropneumoniae*. Comp. Immunol. Microbiol. Infect. Dis. *13*, 25–34.

Kuhnert, P., Berthoud, H., Christensen, H., Bisgaard, M., and Frey, J. (2003a). Phylogenetic relationship of equine *Actinobacillus* species and distribution of RTX toxin genes among clusters. Vet. Res. *34*, 353–359.

Kuhnert, P., Berthoud, H., Straub, R., and Frey, J. (2003b). Host cell specific activity of RTX toxins from haemolytic *Actinobacillus equuli* and *Actinobacillus suis*. Vet. Microbiol. *92*, 161–167.

Kuhnert, P., Schlatter, Y., and Frey, J. (2005). Characterization of the type I secretion system of the RTX toxin ApxII in '*Actinobacillus porcitonsillarum*'. Vet. Microbiol. *107*, 225–232.

Lally, E.T., Golub, E.E., and Kieba, I.R. (1994). Identification and immunological characterization of the domain of *Actinobacillus actinomycetemcomitans* leukotoxin that determines its specificity for human target cells. J. Biol. Chem. *269*, 31289–31295.

Lally, E.T., Hill, R.B., Kieba, I.R., and Korostoff, J. (1999). The interaction between RTX toxins and target cells. Trends Microbiol. *7*, 356–361.

Lally, E.T., Kieba, I.R., Sato, A., Green, C.L., Rosenbloom, J., Korostoff, J., Wang, J.F., Shenker, B.J., Ortlepp, S., Robinson, M.K., and Billings, P.C. (1997). RTX toxins recognize a beta 2 integrin on the surface of human target cells. J. Biol. Chem. *272*, 30463–30469.

Larsen, J., Kuhnert, P., Frey, J., Christensen, H., Bisgaard, M., and Olsen, J.E. (2007a). Analysis of gene order data supports vertical inheritance of the leukotoxin operon and genome rearrangements in the 5′ flanking region in genus *Mannheimia*. BMC Evol. Biol. *7*, 184.

Larsen, J., Pedersen, A.G., Christensen, H., Bisgaard, M., Angen, Ø., Ahrens, P., and Olsen, J.E. (2007b). Evidence for vertical inheritance and loss of the leukotoxin operon in genus *Mannheimia*. J. Mol. Evol. *64*, 423–437.

Lee, K.Y., Kim, D.H., Kang, T.J., Kim, J., Chung, G.H., Yoo, H.S., Arntzen, C.J., Yang, M.S., and Jang, Y.S. (2006). Induction of protective immune responses against the challenge of *Actinobacillus pleuropneumoniae* by the oral administration of transgenic tobacco plant expressing ApxIIA toxin from the bacteria. FEMS Immunol. Med. Microbiol. *48*, 381–389.

Lee, R.W.H., Strommer, J., Hodgins, D., Shewen, P.E., Niu, Y., and Lo, R.Y.C. (2001). Towards development of an edible vaccine against bovine pneumonic pasteurellosis using transgenic white clover expressing a *Mannheimia haemolytica* A1 leukotoxin 50 fusion protein. Infect. Immun. *69*, 5786–5793.

Li, J. and Clinkenbeard, K.D. (1999). Lipopolysaccharide complexes with *Pasteurella haemolytica* leukotoxin. Infect. Immun. *67*, 2920–2927.

Lin, L., Bei, W., Sha, Y., Liu, J., Guo, Y., Liu, W., Tu, S., He, Q., and Chen, H. (2007). Construction and immunogencity of a Δ*apxIC*/Δ*apxIIC* double mutant of *Actinobacillus pleuropneumoniae* serovar 1. FEMS Microbiol. Lett. *274*, 55–62.

Lo, R.Y.C., Shewen, P.E., Strathdee, A., and Greer, C.N. (1985). Cloning and expression of the leukotoxin gene of *Pasteurella haemolytica* A1 in *Escherichia coli* K-12. Infect. Immun. *50*, 667–671.

Ludwig, A., Garcia, F., Bauer, S., Jarchau, T., Benz, R., Hoppe, J., and Goebel, W. (1996). Analysis of the in vivo activation of hemolysin (HlyA) from *Escherichia coli*. J. Bacteriol. *178*, 5422–5430.

Maier, E., Reinhard, N., Benz, R., and Frey, J. (1996). Channel-forming activity and channel size of the RTX toxins ApxI, ApxII, and ApxIII of *Actinobacillus pleuropneumoniae*. Infect. Immun. *64*, 4415–4423.

Mangan, D.F., Taichman, N.S., Lally, E.T., and Wahl, S.M. (1991). Lethal effects of *Actinobacillus actinomycetemcomitans* leukotoxin on human T lymphocytes. Infect. Immun. *59*, 3267–3272.

Martelli, P., Guadagnini, P.F., Foccoli, E., and Ballarini, G. (1996). Efficacy of an *Actinobacillus pleuropneumoniae* subunit vaccine in the control of pleuropneumonia: a field trial. In Proceedings/International Pig Veterinary Society (IPVS) Congress, Italy, International Pig Veterinary Society (IPVS) Congress, Italy), p. 214.

Mayor, D., Korczak, B.M., Christensen, H., Bisgaard, M., Frey, J., and Kuhnert, P. (2006). Distribution of RTX toxin genes in strains of [*Actinobacillus*] *rossii* and [*Pasteurella*] *mairii*. Vet. Microbiol. *116*, 194–201.

Mombelli, A., Gmür, R., Lang, N.P., Corbert, E., and Frey, J. (1999). *Actinobacillus actinomycetemcomitans* in Chinese adults. Serotype distribution and analysis of the leukotoxin gene promoter locus. J. Clin. Periodontol. *26*, 505–510.

Nicolet, J. (1992). *Actinobacillus pleuropneumoniae*. In Diseases of Swine, A.D. Leman, B.E. Straw, W.L. Mengeling, S. D'Allaire, and D.J. Taylor, eds. (Ames, Iowa: Iowa State University Press), pp. 401–408.

Osicka, R., Prochazkova, K., Sulc, M., Linhartova, I., Havlicek, V., and Sebo, P. (2004). A novel 'clip-and-link' activity of repeat in toxin (RTX) proteins from gram-negative pathogens. Covalent protein cross-linking by an Asp-Lys isopeptide bond upon calcium-dependent processing at an Asp-Pro bond. J. Biol. Chem. *279*, 24944–24956.

Pommier, P., Ridremont, B., Wessel-Roberts, S., and Keïta, A. (1996). Field study into efficacy of a new *Actinobacillus pleuropneumoniae* subunit vaccine. In Proceedings/International Pig Veterinary Society (IPVS) Congress, Italy, International Pig Veterinary Society (IPVS) Congress, Italy), p. 206.

Prideaux, C.T., Lenghaus, C., Krywult, J., and Hodgson, A.L.M. (1999). Vaccination and protection of pigs against pleuropneumonia with a vaccine strain of *Actinobacillus pleuropneumoniae* produced by site-specific mutagenesis of the ApxII operon. Infect. Immun. *67*, 1962–1966.

Reimer, D., Frey, J., Jansen, R., Veit, H.P., and Inzana, T.J. (1995). Molecular investigation of the role of ApxI and ApxII in the virulence of *Actinobacillus pleuropneumoniae* serotype 5. Microb. Pathog. *18*, 197–209.

Rioux, S., Galarneau, C., Harel, J., Frey, J., Nicolet, J., Kobisch, M., Dubreuil, J.D., and Jacques, M. (1999). Isolation and characterization of mini-Tn*10* lipopolysaccharide mutants of *Actinobacillus pleuropneumoniae* serotype 1. Can. J. Microbiol. *45*, 1017–1026.

Rycroft, A.N. and Garside, L.H. (2000). *Actinobacillus* species and their role in animal disease. Vet. J. *159*, 18–36.

Rycroft, A.N., Williams, D., Cullen, J.M., and Macdonald, J. (1991). The cytotoxin of *Actinobacillus pleuropneumoniae* (pleurotoxin) is distinct from the haemolysin and is associated with a 120 kDa polypeptide. J. Gen. Microbiol. *137*, 561–568.

Rycroft, A.N., Woldeselassie, A., Gordon, P.J., and Bjornson, A. (1998). Serum antibody in equine neonatal septicaemia due to *Actinobacillus equuli*. Vet. Rec. *143*, 254–255.

Schaller, A., Kuhn, R., Kuhnert, P., Nicolet, J., Anderson, T.J., MacInnes, J.I., Segers, R.P.A.M., and Frey, J. (1999). Characterization of *apxIVA*, a new RTX determinant of *Actinobacillus pleuropneumoniae*. Microbiology *145*, 2105–2116.

Shewen, P.E. and Wilkie, B.N. (1982). Cytotoxin of *Pasteurella haemolytica* acting on bovine leukocytes. Infect. Immun. *35*, 91–94.

Shewen, P.E. and Wilkie, B.N. (1988). Vaccination of calves with leukotoxic culture supernatant from *Pasteurella haemolytica*. Can. J. Vet. Res. *52*, 30–36.

Shope, R.E. (1964). Porcine contagious pleuropneumonia. Experimental transmission, etiology and pathology. J. Exp. Med. *119*, 357–368.

Spitznagel, J., Kraig, E., and Kolodrubetz, D. (1991). Regulation of leukotoxin in leukotoxic and nonleukotoxic strains of *Actinobacillus actinomycetemcomitans*. Infect. Immun. *59*, 1394–1401.

Stanley, P., Koronakis, V., and Hughes, C. (1998). Acylation of *Escherichia coli* hemolysin: A unique protein lipidation mechanism underlying toxin function. Microbiol. Mol. Biol. Rev. *62*, 309–333.

Strathdee, C.A. and Lo, R.Y. (1989). Cloning, nucleotide sequence, and characterization of genes encoding the secretion function of the *Pasteurella haemolytica* leukotoxin determinant. J. Bacteriol. *171*, 916–928.

Sun, Y., Clinkenbeard, K.D., Clarke, C., Cudd, L., Highlander, S.K., and Dabo, S.M. (1999a). *Pasteurella haemolytica* leukotoxin induced apoptosis of bovine lymphocytes involves DNA fragmentation. Vet. Microbiol. *65*, 153–166.

Sun, Y., Clinkenbeard, K.D., Cudd, L.A., Clarke, C.R., and Clinkenbeard, P.A. (1999b). Correlation of *Pasteurella haemolytica* leukotoxin binding with susceptibility to intoxication of lymphoid cells from various species. Infect. Immun. *67*, 6264–6269.

Taichman, N.S., Dean, R.T., and Sanderson, C.J. (1980). Biochemical and morphological characterization of the killing of human monocytes by a leukotoxin derived from *Actinobacillus actinomycetemcomitans*. Infect. Immun. *28*, 258–268.

Taichman, N.S., Simpson, D.L., Sakurada, S., Cranfield, M., DiRienzo, J., and Slots, J. (1987). Comparative studies on the biology of *Actinobacillus actinomycetemcomitans* leukotoxin in primates. Oral Microbiol. Immunol. *2*, 97–104.

Tascon, R.I., Vazquez-Boland, J.A., Gutierrez-Martin, C.B., Rodriguez-Barbosa, I., and Rodriguez-Ferri, E.F. (1994). The RTX haemolysins ApxI and ApxII are major virulence factors of the swine pathogen *Actinobacillus pleuropneumoniae*: Evidence from mutational analysis. Mol. Microbiol. *14*, 207–216.

Tatum, F.M., Briggs, R.E., Sreevatsan, S.S., Zehr, E.S., Hsuan, S.L., Whiteley, L.O., Ames, T.R., and Maheswaran, S.K. (1998). Construction of an isogenic leukotoxin deletion mutant of *Pasteurella haemolytica* serotype 1: characterization and virulence. Microb. Pathog. *24*, 37–46.

Taylor, D.J. (1999). *Actinobacillus pleuropneumoniae*. In Diseases of Swine, B.E.Straw, S.D'Allaire, W.L.Mengeling, and D.J.Taylor, eds. (Ames, Iowa: Iowa State University Press), pp. 343–354.

Tegetmeyer, H.E., Jones, S.C., Langford, P.R., Baltes, N. (2008). ISApl*1*, a novel insertion element of *Actinobacillus pleuropneumoniae*, prevents ApxIV-based serological detection of serotype 7 strain AP76. Vet. Microbiol. *128*, 342–353.

Thumbikat, P., Dileepan, T., Kannan, M.S., and Maheswaran, S.K. (2005). Mechanisms underlying *Mannheimia haemolytica* leukotoxin-induced oncosis and apoptosis of bovine alveolar macrophages. Microb. Pathog. *38*, 161–172.

Tonpitak, W., Rohde, J., and Gerlach, G.-F. (2007). Prevalence of '*Actinobacillus porcitonsillarum*' in porcine tonsils and development of a diagnosis duplex PCR differentiating between '*Actinobacillus porcitonsillarum*' and *Actinobacillus pleuropneumoniae*. Vet. Microbiol. *122*, 157–165.

Tumamao, J.Q., Bowles, R.E., van den Bosch, H., Klaasen, H.L., Fenwick, B.W., and Blackall, P.J. (2004a). An evaluation of the role of antibodies to *Actinobacillus pleuropneumoniae* serovar 1 and 15 in the protection provided by sub-unit and live streptomycin-dependent pleuropneumonia vaccines. Aust. Vet. J. *82*, 773–780.

Tumamao, J.Q., Bowles, R.E., van den Bosch, H., Klaasen, H.L., Fenwick, B.W., Storie, G.J., and Blackall, P.J. (2004b). Comparison of the efficacy of a subunit and a live streptomycin-dependent porcine pleuropneumonia vaccine. Aust. Vet. J. *82*, 370–374.

Valks, M.M.H., Nell, T., and van den Bosch, J.F. (1996). A clinical field trial in finishing pigs to evaluate the efficacy of a new APP subunit vaccine. In Proceedings/ International Pig Veterinary Society (IPVS) Congress, Italy, p. 208.

van den Bosch, H. and Frey, J. (2003). Interference of outer membrane protein PalA with protective immunity against *Actinobacillus pleuropneumoniae* infections in vaccinated pigs. Vaccine *21*, 3601–3607.

van den Bosch, J.F., Jongenelen, I.M.C.A., Pubben, A.N.B., van Vugt, F.G.A., and Segers, R.P.A.M. Protection induced by a trivalend *A. pleuropneumoniae* subunit vaccine. Proceedings of the International Pig Veterinary Society Congress 12, 194. 1992.

van den Bosch, J.F., Pennings, A.M.M.A., Cuiijpers, M.E.C.M., Pubben, A.N.B., van Vugt, F.G.A., and van der Linden, M.F.I. Heterologous protection induced by an *A. pleuropneumoniae* subunit vaccine. Proceedings of the International Pig Veterinary Society Congress 11, 11. 1990.

VanOstaaijen, J., Frey, J., Rosendal, S., and MacInnes, J.I. (1997). *Actinobacillus suis* strains isolated from healthy and diseased swine are clonal and carry *apxICABD*(var. suis) and *apxIICA*(var. suis) toxin genes. J. Clin. Microbiol. *35*, 1131–1137.

Vogel, G.J. and Parrot, C. (1994). Mortality survey in feedyards: The incidence of death from digestive, respiratory and other causes in feedyards on the Great Plains. Comp. Cont. Educ. Pract. Vet. *16*, 227–234.

Ward, C.K. and Inzana, T.J. (1994). Resistance of *Actinobacillus pleuropneumoniae* to bactericidal antibody and complement is mediated by capsular polysaccharide and blocking antibody specific for lipopolysaccharide. J. Immunol. *153*, 2110–2121.

Xu, F., Chen, X., Shi, A., Yang, B., Wang, J., Li, Y., Guo, X., Blackall, P.J., and Yang, H. (2006). Characterization and immunogenicity of an *apxIA* mutant of *Actinobacillus pleuropneumoniae*. Vet. Microbiol. *118*, 230–239.

Zambon, J.J. (1985). *Actinobacillus actinomycetemcomitans* in human periodontal disease. J. Clin. Periodontol. *12*, 1–20.

Zambon, J.J., Haraszthy, G.G., Hariharan, G., Lally, E.T., and Demuth, D.R. (1996). The microbiology of early-onset periodontitis: association of highly toxic *Actinobacillus actinomycetemcomitans* strains with localized juvenile periodontitis. J. Periodontol. *67*, 282–290.

Zecchinon, L. Dissection moléculaire de l'interaction leucotoxine-β2-intégrine LFA-1 (CD11a/CD18), responsable de la spécificité d'espèce de *Mannheimia haemolytica* envers les ruminants. PhD Thesis Liège, Belgium. 2007.

Zecchinon, L., Fett, T., Baise, E., and Desmecht, D. (2004a). Characterization of the caprine (*Capra hircus*) beta-2 integrin CD18-encoding cDNA and identification of mutations potentially responsible for the ruminant-specific virulence of *Mannheimia haemolytica*. Mol. Membr. Biol. *21*, 289–295.

Zecchinon, L., Fett, T., Baise, E., and Desmecht, D. (2004b). Molecular cloning and characterisation of the CD18 partner in ovine (*Ovis aries*) β2-integrins. Gene *334*, 47–52.

Zecchinon, L., Fett, T., and Desmecht, D. (2005). How *Mannheimia haemolytica* defeats host defence through a kiss of death mechanism. Vet. Res. *36*, 133–156.

Zecchinon, L., Fett, T., Vanden Bergh, P., and Desmecht, D. (2006). LFA-1 and associated diseases: The dark side of a receptor. Clin. Applied Immunol. Rev. *6*, 201–216.

Outer Membrane Proteins and Iron Uptake of *Actinobacillus pleuropneumoniae*

8

Jacqueline W. Chung, Mario Jacques and James W. Coulton

Abstract

Essential for the integrity and selective permeability of the membrane, outer membrane (OM) proteins of Gram-negative bacteria have critical roles in adaptation and infection within a host niche. With prominent roles in the pathogenesis of several bacterial pathogens, OM proteins have growing appeal as novel targets for anti-infectives and therapeutics. Members of the *Pasteurellaceae* family represent important human and animal pathogens that include *Haemophilus, Actinobacillus, Pasteurella* and, more recently, the newly added *Mannheimia* genera of organisms. Characterization of cell surface proteins has highlighted several redundant iron acquisition receptors for transferrin, siderophores, and haem/haem-containing proteins in *Pasteurellaceae.* In addition, the identification of several immunogenic lipoproteins and OM proteins has driven research for an effective cross-protective vaccine for these organisms. This chapter will review OM proteins and iron uptake systems of the swine pathogen, *Actinobacillus pleuropneumoniae,* causative agent of porcine pleuropneumonia, with reference to homologues in other members of *Pasteurellaceae.*

Introduction

Gram-negative bacteria possess a cell envelope composed of two membranes separated by the periplasmic space and a peptidoglycan layer. The cytoplasmic or inner membrane (IM) is a phospholipid bilayer containing active transport systems. The outer membrane (OM) of Gram-negative bacteria is an asymmetrical bilayer composed of an inner leaflet of phospholipids, an outer leaflet of lipopolysaccharide (LPS), and OM proteins that constitute approximately 50% of the OM mass (Lin *et al.*, 2002). OM proteins include both lipoproteins that are anchored by a lipid covalently linked to the mature protein and integral proteins that typically display β-barrel structural motifs. Proteins that are destined for the OM are first synthesized in the cytoplasm with an N-terminal signal sequence that targets them for Sec-mediated translocation across the IM. Following translocation across the IM, lipoproteins that are lipidated at the N-terminal cysteine and integral OM proteins are processed by signal peptidases and routed to the OM by different pathways. Lipoproteins interact with periplasmic chaperone LolA of the Lol system and are delivered to the OM assembly site. Based on studies with *Escherichia coli* as a model system, most lipoproteins face the periplasm; however, surface-exposed lipoproteins have been reported in other bacteria (Bos and Tommassen, 2004; Narita *et al.*, 2004). The delivery of β-barrel OM proteins to the OM is less understood although transport is reported to involve periplasmic chaperones Skp, DegP and SurA and the YaeT/YfgL/YfiO/NlpB complex at the OM serves as the site of OM protein assembly (Ruiz *et al.*, 2006).

Because they are in direct contact with the environment, OM proteins are actively engaged in activities involved with cell-cell interactions, ion transport, and cell signalling. The functional roles of most bacterial OM proteins are associated with the virulence of several Gram-negative bacterial pathogens. OM proteins are key players that facilitate adherence, colonization and persistence in the host. Several bacterial adaptive responses to the host environment can be attributed to changes at the OM whereby the expression of OM proteins mediate iron uptake, resistance to antimicrobial peptides, serum or other anti-infectives, and resistance to bile (Lin *et al.*, 2002). Given their growing clinical importance, research in the OM proteome of important pathogens provide a resource of novel targets for the design of antimicrobial drugs and vaccines.

Actinobacillus pleuropneumoniae causes porcine pleuropneumonia, a highly infectious respiratory disease that contributes to major economic losses in the swine industry. The disease, transmitted by aerosol or by direct contact with infected pigs, may result in rapid death or in severe pathology characterized by hemorrhagic and necrotic lesions in the lung (Bossé *et al.*, 2002). Exposure to the organism may lead to chronic infection such that animals fail to thrive; alternatively, they survive as asymptomatic carriers that transmit the disease to healthy herds (Sebunya and Saunders, 1983; Rycroft and Garside, 2000; Hodgetts *et al.*, 2004). Based on capsular antigens, 15 serotypes have been described for *A. pleuropneumoniae*; serotypes 1, 5 and 7 are the predominant serotypes in North America, whereas serotype 2 is prevalent in Europe (Dubreuil *et al.*, 2000; Blackall *et al.*, 2002; Jacques, 2004). Although several virulence factors have been described for *A. pleuropneumoniae* including LPS, receptors for iron acquisition, capsule and Apx toxins (Apx I–IV), pathogenesis of this organism remains poorly understood (Haesebrouck *et al.*, 1997; Bossé *et al.*, 2002; Jacques, 2004). All serotypes are capable of causing disease but some serotypes are more virulent than others, attributed to the amount of capsule and to different combinations of Apx toxins (Jacques *et al.*, 1988; Rosendal and MacInnes, 1990; Frey, 1995; Bossé *et al.*, 2002). Vaccines for controlling infection by *A. pleuropneumoniae* include inactivated whole-cell bacterins or subunit vaccines containing Apx toxins; however they provide only partial protection, have little impact on morbidity or may be serotype-specific (Rossi-Campos *et al.*, 1992; Chiers *et al.*, 1998; Van Overbeke *et al.*, 2001; van den Bosch and Frey, 2003; Haesebrouck *et al.*, 2004). With the availability of genome sequences of clinically important *A. pleuropneumoniae* serotypes 1, 5b, and 7, and other members of *Pasteurellaceae*, genome-based strategies identify conserved antigens or proteins, and provide insights into pathogenesis of these organisms to allow the development of live attenuated vaccines.

OM proteins of *Actinobacillus pleuropneumoniae*

In silico analyses of OM proteins

The published genome of *Haemophilus influenzae* represented the first release of a completed genome sequence of a free-living organism (Fleischmann *et al.*, 1995). A rapid expansion of the bacterial genomic database followed, elucidating the biology of over 500 bacteria to date. With availability of whole-genome sequences, the protein complement of subcellular constituents is feasible through genomic mining, using bioinformatic programs trained to identify genes of interest. Apart from *H. influenzae*, complete genome sequences of other *Pasteurellaceae* members include *A. pleuropneumoniae*, *Aggregatibacter actinomycetemcomitans*, [*Haemophilus*] *ducreyi*, *Histophilus somni*, '*Mannheimia succiniciproducens*', *Pasteurella multocida*, and a draft genome sequence of *Mannheimia haemolytica* are now available (Gioia *et al.*, 2006; see also Chapter 3). Genome-based strategies, such as reverse vaccinology, can be used to computationally identify potential antigens of important pathogens prior to experimental verification (Rappuoli, 2001). Exploiting the genome sequence of *A. pleuropneumoniae* serotype 5b, Chung *et al.* (2007) performed bioinformatic analyses to predict genes encoding OM proteins, based upon subcellular localization tools that detected signal peptide sequences and β-barrel motifs. Three predictor programs (Proteome Analyst, PSORT-b, BOMP) identified integral membrane proteins, predominantly of the β-barrel structural motif while two other predictor programs (Lipo, and LipoP) detected lipoproteins based on lipoprotein signal peptide sequences. The outcome from these five genome scanning programs provided a consensus prediction list of 93 OM proteins accounting for 45 integral membrane proteins (Table 8.1) and 48 lipoproteins (Table

8.2) present in *A. pleuropneumoniae*. This list postulated proteins localized at the cell surface and subsequently guided an experimental search for OM proteins.

Lipoproteins

One of the two major protein groups present at the OM, lipoproteins are distinguished by an N-terminal lipid tail that anchors these proteins to the OM. Apart from its characteristic signal sequence flanking the N-terminal cysteine and the lipid moiety attached, lipoproteins do not share common structural features. Lipoproteins exhibit broad functional diversity ranging from maintenance of the cell surface structure to substrate transport to cell signalling. Compared with integral OM proteins, the structural and biochemical functions of individual lipoproteins are not as well characterized; specific protein domains cannot be assigned to functional roles. Indeed, recent undertakings in creating a comprehensive database of bacterial lipoproteins (DOLOP; http://www.mrc-1mb.cam.ac.uk/genomes/dolop/) demonstrate that several lipoproteins are uncharacterized (Babu *et al.*, 2006). All bacteria share a common set of lipoproteins, but have also evolved a unique set of lipoproteins to meet their needs. In particular, a variety of roles have been ascribed for lipoproteins of pathogenic bacteria. These lipid-modified proteins have been reported to have roles in host-pathogen interactions, translocation of virulence factors and activating inflammatory responses (Babu *et al.*, 2006). Owing to their highly immunogenic nature and association in pathogenesis, lipoproteins of *Pasteurellaceae* species have been studied as attractive candidates for vaccine development. The *in silico* analysis of *A. pleuropneumoniae* serotype 5 genome showed that there were at least 48 lipoproteins localized at the OM (Table 8.2) (Chung *et al.*, 2007). Several of them have not yet been identified and most are matched to hypothetical proteins. Some lipoproteins had homology to the OM lipoprotein LolB, an opacity-associated protein (OapB), lipoprotein VacJ and a copper homeostasis protein. Lipoproteins that have been cloned in *A. pleuropneumoniae* include peptidoglycan-associated lipoprotein (Pal), outer membrane lipoprotein A (OmlA) and tranferrin binding lipoprotein B (TbpB) (Gerlach *et al.*, 1992a,b, 1993; Gonzalez *et al.*, 1995; Frey *et al.*, 1996). All three lipoproteins have been used individually in their recombinant forms as vaccine subunits combined with Apx toxins. A detailed description of TbpB will be provided in the section devoted to OM proteins of iron uptake.

PalA (peptidoglycan-associated lipoprotein A)

PAL proteins are described as integral components of the OM of several Gram-negative bacteria. PAL proteins are members of the OmpA protein; a subset of the PAL family are lipoproteins that are tightly but non-covalently attached to the peptidoglycan. These PAL proteins are strongly immunogenic and are highly conserved within a given bacterial species but also show great similarities between different species. In *E. coli*, PAL is associated with two complex forms: one composed of OM proteins TolA, TolQ and TolR, and another involving periplasmic protein TolB (Bouveret *et al.*, 1999). Of the OM proteins identified in *A. pleuropneumoniae*, PalA is among the most immunodominant. First detected by sera from pigs that were naturally or experimentally infected by *A. pleuropneumoniae*, PalA was cloned from serotype 2 and identified as a 14-kDa OM protein with homology to the peptidoglycan-associated lipoprotein of *E. coli* (Frey *et al.*, 1996). Studies by Frey *et al.* (1996) demonstrated, by serological and PCR analysis, that PalA was conserved in 12 serotypes of *A. pleuropneumoniae*, the total number at the time. They also detected reactivity with a polyclonal anti-PalA antibody to a 14-kDa protein present in *A. suis*, *A. equuli*, and *A. lignieresii*. Sequence analysis showed that PalA was homologous to the family of PAL proteins of Gram-negative bacteria with the greatest similarity to immunogenic PAL antigens of other *Pasteurellaceae* members: the P6 protein of *H. influenzae* (82% amino acid similarity) and a 16-kDa protein of *P. multocida* (97% amino acid similarity). In particular, several studies have reported OM protein P6 of *H. influenzae* as a promising vaccine. Subsequently van den Bosch and coworkers evaluated the PalA of *A. pleuropneumoniae* as a potential vaccine antigen and assessed its capacity to induce protective immunity in pigs that were infected with serotype 1 (van den Bosch and Frey, 2003). PalA was tested alone and in combination with

Table 8.1 Prediction list of OM proteins from *A. pleuropneumoniae* serotype 5b

Gene ID[1]	Protein homology (Swiss-Prot)	% identity	Pfam family
ap0007	OPP28_HAEIN outer membrane protein P2 precursor (OMP P2)	23	
ap0052	None found		
ap0114	YADA1_YEREN adhesin YadA precursor	33	
ap0115	OMPA_RICRI outer membrane protein A precursor (190-kDa antigen) (cell surface antigen) (rOmpA)	21	YadA-like C-terminal domain
ap0220	COME_HAEIN competence protein E precursor (DNA transformation protein ComE)	52	Bacterial type II and III secretion system protein
ap0266	TBB2_NEIMB transferrin-binding protein 2 precursor (TBP-2)	36	Transferrin binding protein-like solute binding protein
ap0267	TBP1_HAEIN probable transferrin-binding protein 1 precursor	41	TonB-dependent receptor
ap0268	TBP12_NEIMB transferrin-binding protein 1 precursor	54	
ap0280	YE62_HAEIN hypothetical protein HI1462	55	Outer membrane efflux protein
ap0300	HEMR_YEREN haemin receptor precursor	32	Plug domain
ap0301	HXUC2_HAEIN haem/haemopexin utilization protein C precursor	23	TonB-dependent receptor
ap0334	PAL_PASMU outer membrane protein P6 precursor (OMP P6) (P6-like) (peptidoglycan-associated lipoprotein)	64	OmpA family
ap0453	OM26_HAEIN outer membrane protein 26 precursor	28	OmpH-like protein
ap0454	D153_HAEIN protective surface antigen D15 precursor (80-kDa D15 antigen) (D-15-Ag) (outer membrane protein D15)	56	Surface antigen
ap0495	YADA1_YEREN adhesin YadA precursor	30	Bacterial haemagglutinins, invasins
ap0514	OMP52_HAEIN outer membrane protein P5 precursor (OMP P5)	37	OmpA family
ap0621	Y4XJ_RHISN hypothetical 44.3-kDa protein y4xJ	27	Bacterial type II and III secretion system protein
ap0632	Y1467_HAEIN hypothetical ABC transporter ATP-binding protein HI1467	66	TonB-dependent receptor
ap0719	OPP28_HAEIN outer membrane protein P2 precursor (OMP P2)	38	Porin
ap0928	Y698_HAEIN protein HI0698 precursor	49	Surface antigen
ap0940	SRPC_PSEPU solvent efflux pump outer membrane protein SrpC precursor	22	Outer membrane efflux protein
ap1018	PLPA_PASHA outer membrane lipoprotein 1 precursor (PLP1)	66	Lipoprotein
ap1030	BTUB_PHOLL vitamin B12 transporter BtuB precursor (Cobalamin receptor)	28	
ap1046	OMP47_PASMU 47-kDa outer membrane protein precursor	44	Outer membrane protein transport protein

ap1076	HLYA_PROMI haemolysin precursor	23	
ap1077	FHAC_BORPE haemolysin activator-like protein FhaC precursor	28	
ap1079	OSTA_HAEDU organic solvent tolerance protein precursor	70	Organic solvent tolerance protein
ap1175	HGP4_HAEIN probable haemoglobin and haemoglobin–haptoglobin binding protein 4 precursor	65	TonB-dependent receptor
ap1176	HGP4_HAEIN probable haemoglobin and haemoglobin-haptoglobin binding protein 4 precursor	57	Plug domain
ap1215	OMPW_VIBCH outer membrane protein W precursor	46	OmpW family
ap1382	LAMB1_YERPE maltoporin 1 precursor (maltose-inducible porin 1)	61	LamB porin
ap1453	HXUC2_HAEIN haem/haemopexin utilization protein C precursor	24	TonB-dependent receptor
ap1581	OMP51_HAEIN outer membrane protein P5 precursor (OMP P5)	49	
ap1736	TBP1_HAEIN probable transferrin-binding protein 1 precursor	40	TonB-dependent receptor
ap1737	TBP2_NEIMB transferrin-binding protein 2 precursor (TBP-2)	34	Lipoprotein
ap1752	BEXD_HAEIN capsule polysaccharide export protein BexD precursor	71	Polysaccharide biosynthesis/export protein
ap1880	FKBB_ECOLI FKBP-type 22-kDa peptidyl-prolyl cis-trans isomerase (PPIase)	57	
ap1996	OM24_PASMU 24-kDa outer membrane protein precursor	46	Membrane bound lytic transglycosylase interacting protein
ap2035	OMP52_HAEIN outer membrane protein P5 precursor (OMP P5)	54	OmpA family
ap2108	TOM70_MOUSE mitochondrial precursor proteins import receptor (translocase of outer membrane TOM70)	22	Tetratricopeptide repeat
ap2142	HGBA_HAEDU haemoglobin and haemoglobin-haptoglobin binding protein precursor	35	
ap2143	HPUB_NEIMAh-haptoglobin utilization protein B precursor	26	
ap2145	HHUA_HAEIN haemoglobin–haptoglobin binding protein A precursor (haemoglobin–haptoglobin utilization protein A)	35	
ap2196	Y974_HAEIN hypothetical protein HI0974	40	
ap2211	FHUA_ECOLI ferrichrome-iron receptor precursor (ferric hydroxamate uptake)	28	TonB-dependent receptor

Chung *et al*. 'Outer membrane proteome of *Actinobacillus pleuropneumoniae*: LC-MS/MS analyses validate *in silico* predictions.' Proteomics. 2007: *7*; 1854–1865.

[1]Annotation of the *A. pleuropneumoniae* genome was ongoing at the time Table 8.1 was generated, therefore Gene IDs represent old locus tags of the updated and annotated *A. pleuropneumoniae* L20 genome, publicly available through the National Research Council genome browser (http://informatics.bio.nrc.ca/genomics-db/mainmenu.do).

Table 8.2 Prediction list of lipoproteins from *A. pleuropneumoniae* serotype 5b

Gene ID[1]	Protein homology (Swiss-Prot)	% identity
ap0032	HBPA_HAEIN haem-binding protein A precursor (haemin-binding lipoprotein)	23
ap0039	None found	
ap0040	PCP_HAEIN outer membrane lipoprotein pcp precursor (15-kDa lipoprotein)	67
ap0132	Y449_HAEIN hypothetical protein HI0449 precursor	43
ap0142	Y960_HAEIN hypothetical protein HI0960 precursor	32
ap0175	APBE_HAEIN thiamine biosynthesis lipoprotein apbE precursor	55
ap0249	Y1099_HAEIN hypothetical protein HI1099	39
ap0258	CUTF_ECOLI copper homeostasis protein cutF precursor (Lipoprotein nlpE)	25
ap0266	TBB2_NEIMB transferrin-binding protein 2 precursor (TBP-2)	36
ap0363	YI05_PASMU hypothetical protein PM1805 precursor	26
ap0365	YI05_PASMU hypothetical protein PM1805 precursor	24
ap0391	None found	
ap0407	POTD1_HAEIN spermidine/putrescine-binding periplasmic protein 1 precursor (SPBP)	79
ap0417	GLQP_HAEIN glycerophosphoryl diester phosphodiesterase precursor (Glycerophosphodiester phosphodiesterase) (surface-exposed lipoprotein D) (protein D) (immunoglobulin D-binding protein) (IgD-binding protein	79
ap0429	HEL_HAEIN lipoprotein E precursor (outer membrane protein P4) (OMP P4)	65
ap0476	SMPA_HAEIN small protein A homologue precursor	53
ap0615	None found	
ap0627	YI97_PASMU hypothetical lipoprotein PM1897 precursor	46
ap0673	Y983_HAEIN hypothetical protein HI0983	25
ap0681	YD14_HAEIN hypothetical lipoprotein HI1314 precursor	65
ap0712	MLTB_ECOLI membrane-bound lytic murein transglycosylase B precursor (murein hydrolase B) (35-kDa soluble lytic transglycosylase)	26
ap0753	YB92_HAEIN hypothetical protein HI1192 precursor	57
ap0865	LOLB_HAEDU outer-membrane lipoprotein lolB precursor	66
ap0914	MLTA_ECOLI membrane-bound lytic murein transglycosylase A precursor (murein hydrolase A) (Mlt38)	45
ap0977	Y922_HAEIN Hhypothetical lipoprotein HI0922 precursor	42
ap1032	None found	
ap1191	YGIW_ECOLI protein ygiW precursor	41
ap1252	Y470_HAEDU hypothetical UPF0169 lipoprotein HD0470 precursor	70
ap1287	Y246_HAEIN protein HI0246 precursor	41
ap1426	Y366_HAEIN hypothetical protein HI0366 precursor	46
ap1451	None found	
ap1519	YG55_HAEIN hypothetical protein HI1655	44
ap1562	OAPB_HAEIN opacity-associated proteins oapB	32
ap1619	None found	
ap1640	None found	
ap1690	LPPL_PSEAE	48
ap1745	Y966_HAEIN hypothetical protein HI0966	67
ap1804	None found	

Table 8.2 Prediction list of lipoproteins from *A. pleuropneumoniae* serotype 5b

Gene ID[1]	Protein homology (Swiss-Prot)	% identity
ap1920	MLTC_HAEDU membrane-bound lytic murein transglycosylase C precursor (Murein hydrolase C)	78
ap1927	OMLA_ACTPL outer membrane lipoprotein A precursor	62
ap2059	Y966_HAEIN hypothetical protein HI0966	23
ap2084	YK67_YEAST hypothetical 35.8-kDa protein in PRP16-SRP40 intergenic region	28
ap2100	PEPO_LACHE neutral endopeptidase (Endopeptidase O)	32
ap2105	VACJ_HAEIN lipoprotein vacJ homologue precursor	56
ap2116	LPPB_HAESO outer membrane antigenic lipoprotein B precursor	39
ap2117	LPPB_HAESO outer membrane antigenic lipoprotein B precursor	38
ap2118	LPPB_HAEIN outer membrane antigenic lipoprotein B precursor	46
ap2147	Y973_HAEIN hypothetical protein HI0973	27

[1]Annotation of the *A. pleuropneumoniae* genome was ongoing at the time Table 8.2 was generated, therefore Gene IDs represent old locus tags of the updated and annotated *A. pleuropneumoniae* L20 genome, publicly available through the National Research Council genome browser (http://informatics.bio.nrc.ca/genomics-db/mainmenu.do).

the Apx toxins, well-characterized protective antigens of *A. pleuropneumoniae*. Animals that were immunized with PalA developed antibody titres against the lipoprotein but showed more severe symptoms, had a higher mortality rate and succumbed more rapidly. In addition, PalA antibodies diminished the protective effects of anti-Apx toxin antibodies. These studies highlighted risks in selecting immunogens, not all of which will induce the desired protective effect.

Despite sharing a high degree of similiarity with PalA of *A. pleuropneumoniae*, it is unclear whether a similar outcome would manifest with the OM protein P6 of *H. influenzae*. *H. influenzae* strains include typeable and non-typeable isolates based upon capsular serotypes a to f; strains that do not bind to antibodies to one of six capsular types are classified as non-typeable. The majority of bacteremic illnesses are caused by *H. influenzae* type b, which establishes infection in the upper respiratory tract and is the leading cause of bacterial meningitis. In contrast, nontypeable strains are associated with localized respiratory tract infections such as acute otitis media, sinusitis, and bronchitis (Ecevit *et al.*, 2004). The P6 protein is conserved antigenically in both typeable and non-typeable strains of *H. influenzae*. Its appeal as a vaccine candidate is demonstrated in reports showing that P6 is surface exposed, provides protective immunity in animal model of *H. influenzae* infection and induces protective immune responses in humans (Munson, Jr. and Granoff, 1985; Nelson *et al.*, 1988; Yamanaka and Faden, 1993; Hotomi *et al.*, 1999). More recent studies have described P6 as a potent immunomodulator of human macrophages with the capacity to selectively induce cytokine production of IL-10, TNF-α, and IL-8 in significant levels, thereby contributing to macrophage-associated inflammation (Berenson *et al.*, 2005). The functional role of P6 was recently defined by Murphy *et al.* (2006) with P6 mutants that were generated by replacing the gene encoding P6 with a chloramphenicol resistance cassette (Murphy *et al.*, 2006). The P6 mutant revealed morphological changes reflected in colony size, vesicle formation and stability. In addition, the mutant was hypersensitive to antibiotics and complement-mediated killing by human serum. These experiments showed that P6 plays an integral role in maintaining OM structure and integrity. PAL homologues have also been identified and found to be conserved in *Pasteurellaceae* members, [*H.*] *ducreyi*, *P. multocida* and *Agg. actinomycetemcomitans*. [*H.*] *ducreyi*, the causative agent of genital ulcer disease, expresses a surface localized

18-kDa PAL homologue that has 71% similarity to the P6 protein of *H. influenzae* (Spinola *et al.*, 1996). DNA sequence analysis of the 16-kDa OM lipoprotein from *P. multocida* showed 81% identity with the P6 gene (Kasten *et al.*, 1995). The gene encoding this P6-like protein was present in all 16 somatic serotypes of *P. multocida* but only weakly hybridized with DNA from *Actinobacillus* species. Interestingly, vaccination with the recombinant P6-like protein did not provide protection against avian cholera caused by *P. multocida*. The PAL homologue in *Agg. actinomycetemcomitans* (AaPAL) was identified by immunoproteomic analyses that included MS determination of the sequence (Paul-Satyaseela *et al.*, 2006). *Agg. actinomycetemcomitans* is an invasive periodontopathogen associated with severe periodontal disease in young individuals and adults. Infection by *Agg. actinomycetemcomitans* develops into chronic inflammation that results in deterioration of tooth-supporting tissues. Like other PAL proteins, AaPAL was strongly immunoreactive.

OmlA

Gerlach and coworkers cloned the gene encoding a protective lipoprotein designated as outer membrane lipoprotein A (OmlA) from *A. pleuropneumoniae* serotype 1; the recombinant protein protected vaccinated pigs from death after aerosol challenge (Gerlach *et al.*, 1993). Immunoblot analysis with serum against OmlA protein detected homologues from serotype 2, 8, 9, 11, and 12. In contrast, southern blots detected the presence of the gene in other serotypes, suggesting antigenic variability. OmlA was also cloned from serotypes 5 and 5a, which had 57% and 61% amino acid sequence identity respectively to OmlA of serotype 1 (Bunka *et al.*, 1995; Ito *et al.*, 1995). Immunization with OmlA from serotype 5 induced a significant antibody response in pigs and lowered their mortality rate (Bunka *et al.*, 1995). Analysis by southern blot and immunoblots revealed the presence of OmlA homologues in serotypes 5a, 5b, and 10, further confirming antigenic diversity of OmlA among *A. pleuropneumoniae* strains. Ito *et al.* also cloned *omlA* gene from serotype 7; the amino acid sequence displayed 64.5% and 71.6% identity to the *omlA* genes of serotype 1 and 5 respectively (Ito *et al.*, 1998). Homologues to the *omlA* gene of serotype 7 were detected in serotype 3, 4, and 6 while immunoblot analyses confirmed the expression of yet a third antigenically distinct OmlA protein in *A. pleuropneumoniae*. Although OmlA from serotypes 1 and 5 is constitutively expressed, a recent study by Jacobsen *et al.* (2005b) revealed the induction of OmlA expression in *A. pleuropneumoniae* serotype 7 grown in the presence of biological fluids such as bronchoalveolar lavage fluid (BALF) from pigs. Furthermore, others have applied the genetic diversity of the *omlA* gene as a diagnostic tool to distinguish between serotypes and aid in epidemiological studies (Osaki *et al.*, 1997; Gram and Ahrens, 1998; Cho and Chae, 2003). Using PCR-RFLP analysis with primers generated from the conserved genetic regions of *omlA* from serotype 1 and 5a, Osaki *et al.* showed that 12 serotypes of *A. pleuropneumoniae* could be divided into five possible groups of allelic variation (Osaki *et al.*, 1997). Since the *omlA* gene is species-specific for *A. pleuropneumoniae*, PCR assays based on *omlA* have been developed for routine typing and detection of *A. pleuropneumoniae* in subclinical infections. Although the function of OmlA is not yet known, a homologue of the protein (PlpE) is present in *M. haemolytica* (Pandher *et al.*, 1998). Anti-PlpE antibodies in bovine immune serum demonstrated complement-mediated killing, contributing to host immune defences against *M. haemolytica*.

Integral OM proteins

In contrast to IM proteins that have α-helical transmembrane domains, integral OM proteins are characterized by β-barrel structures. Transmembrane domains of OM proteins span the OM with β-strands of alternating polar and non-polar amino acids that adopt a barrel-like conformation that functions as transporters for specific molecules across the OM barrier (Santoni *et al.*, 2000). Alternatively, integral OM proteins may also function as enzymes and adhesins. This class of proteins is largely responsible for OM integrity and selective permeability of the membrane. In *A. pleuropneumoniae* serotype 5b, 45 integral OM proteins are predicted at the cell surface (Table 8.1); (Chung *et al.*, 2007). Those that have been characterized in *A. pleuropneu-*

moniae include maltose-inducible OM protein, iron-regulated proteins, and adhesins, including autotransporter adhesins.

Maltose-inducible OM protein

Several early studies reported OM profiles of *A. pleuropneumoniae* revealing major and minor OM proteins that produced patterns distinguishing between serotypes; however many of these proteins were not identified. Rapp *et al.* (1986) documented the effects of different factors such as medium composition, colony type, time of harvest, and both *in vitro* and *in vivo* passage, all of which did not affect the OM protein profiles of *A. pleuropneumoniae* strains. The most revealing studies showed OM proteins that were expressed under nutrient deprivation, particularly conditions that resembled the *in vivo* environment. Gram-negative bacteria respond to stress and to certain environmental factors by changing the protein composition of the OM, thus adapting or promoting survival. This may be reflected in the induction or repression of certain OM proteins as well as changes in the abundance levels. Deneer and Potter (1989a) reported *A. pleuropneumoniae* grown under iron restriction that resulted in the induction of several iron-regulated OM proteins. Some of these are among the best-characterized integral OM proteins of *A. pleuropneumoniae* and are further discussed in later sections. Following this study, Deneer and coworkers investigated OM protein profiles elaborated under the presence of maltose (Deneer and Potter, 1989b). *A. pleuropneumoniae* serotype 1 expressed a 42-kDa OM protein in abundance when grown in PPLO media with 0.4% maltose and exhibited physical properties of porins that suggested the OM protein functioned as a porin in maltose transport. Compared with iron-inducible OM proteins (IROMP) that Deneer *et al.* had previously described, induction of the 42-kDa OM protein was not as rapid and was present in small amounts in bacteria grown without maltose. This maltose-inducible OM protein was also assayed for phage lambda binding since the maltose-inducible LamB protein in *E. coli* functions not only for maltose and maltodextrin transport but serves as a receptor for lambda and several other bacteriophages. Among seven serotypes (1–7) that were tested, the 42-kDa OM protein was present in all but serotype 4. Not all isolates representing the other six serotypes expressed the OM protein in response to maltose, suggesting that the 42-kDa OM protein did not appear to be required for virulence. However, immunoblot results with convalescent antisera against serotype 1 detected the 42-kDa OM protein, indicating that this protein was expressed *in vivo*.

OM proteins expressed under β-NAD restriction

Using the V-factor, β-nicotinamide adenine dinucleotide (β-NAD), divides *A. pleuropneumoniae* serotypes into two biotypes: biotype I is β-NAD-dependent for growth whereas biotype II is β-NAD-independent. The majority of *A. pleuropneumoniae* strains are β-NAD-dependent and therefore warranted studies to investigate effects of β-NAD deprivation on OM protein profiles. *A. pleuropneumoniae* grown under reduced β-NAD concentrations represented *in vivo* conditions in the lumen of lung alveoli. O'Reilly *et al.* (1991) reported the enhanced production of three OM proteins under β-NAD restriction with molecular weights of 17, 31 and 69 kDa. However, they concluded that the abundance of these proteins were more reflective of biomass levels obtained under these growth conditions and could not define the function of these OM proteins. In contrast, Van Overbeke *et al.* reported the production of a 55-kDa OM protein under β-NAD restriction that appeared to confer enhanced adherence to porcine alveolar epithelial cells (Van Overbeke *et al.*, 2002). The initial steps in *A. pleuropneumoniae* colonization are still poorly understood, although adherence to host cells is recognized as an essential prerequisite for the pathogenesis of most bacterial pathogens, including *A. pleuropneumoniae*. *A. pleuropneumoniae* adherence to host cells has been demonstrated in several studies examining tissue sections and using *in vitro* models. Characterized adhesins of *A. pleuropneumoniae* include LPS and fimbriae, but few OM proteins had been described as potential adhesins (Bélanger *et al.*, 1990; Dom *et al.*, 1994; Jacques and Paradis, 1998; Zhang *et al.*, 2000). N-terminal sequencing of the 55-kDa protein did not match any known proteins, thereby confirming identification of a novel OM protein (Van Overbeke *et al.*, 2002). Increased adhesion scores indicating the number of bacteria interacting with epithelial cells were observed with *A. pleuropneumoniae* serotypes 5a, 9 and 10 grown under β-NAD restriction. This observation correlated with the production of fimbriae and the 55-kDa OM protein. The involvement of these proteins was further demonstrated by proteolytic enzyme treatment to β-NAD-restricted cells that resulted in reduced adhesion. Effects of β-NAD restriction were applied towards a new bacterin vaccine. Bacterins were generated from *A. pleuropneumoniae* serotype 10 grown under β-NAD-replete and β-NAD-limiting conditions and then evaluated on their ability to confer protection against the infection with the same serotype (Van Overbeke *et al.*, 2003). The β-NAD-restricted bacterin provided partial protection that was modestly better than the β-NAD-replete bacterin; the percentage of severe lung lesions and affected lung tissue was significantly lower in animals immunized with the β-NAD-restricted bacterin, but all vaccinated animals retained high bacterial titres in lungs. However, immunoblot analysis with sera from vaccinated pigs did not reveal the presence of antibodies against the 55-kDa OM protein, suggesting low immunogenicity.

Autotransporter adhesins/collagen adhesins

Recently, *A. pleuropneumoniae* adherence to swine lung collagen was documented with the identification of a 60-kDa OM protein (Enriquez-Verdugo *et al.*, 2004). Described as a collagen adhesin, the 60-kDa OM protein demonstrated strong interaction with four types of pig lung-collagens and antibodies against the 60-kDa protein inhibited binding between collagen and whole cells. Collagen makes up 60% of connective tissue in the lung but is not readily exposed unless there is tissue damage. Whether collagen adhesion plays a critical role in the early stages of *A. pleuropneumoniae* infection is uncertain. Contact with collagen may be facilitated by proteolytic enzymes produced by *A. pleuropneumoniae* to degrade mucus and gain access to the epithelium during initial stages of colonization. Alternatively, collagen adhesion by *A. pleuropneumoniae* may be more pronounced during later stages of infection when collagen is more abundant with lesion formation, thereby promoting the spread of *A. pleuropneumoniae* in the lung. For many pathogens such as meningitis-associated *E. coli* and *Yersinia* spp., collagen adhesion is an essential step towards colonization (Pouttu *et al.*, 1999; Heise and Dersch, 2006). *Yersinia*, in particular, expresses YadA protein, the prototype of a class of OM autotransporter adhesins that mediate interaction with host cells by binding to macromolecules constituting the extracellular matrix (ECM). Unlike typical autotransporter proteins, YadA adhesins remain cell associated and have a 'lollipop' conformation composed of a knob-like head domain, a coiled-coil pillar stalk, and a C-terminal domain anchored to the membrane. In addition to binding several collagen types, YadA operates as an important virulence factor protecting bacteria against defensins, and conferring resistance to serum complement and phagocytosis (Heise and Dersch, 2006). Interestingly, YadA homologues were listed among the 45 integral OM proteins of *A. pleuropneumoniae* serotype 5 that were predicted by Chung *et al.* (2007), one of which may correspond with the 60-kDa collagen adhesin. Although the gene encoding the 60-kDa collagen adhesin of *A. pleuropneumoniae* has not yet been sequenced for identification, it appears to differ from the 55-kDa OM adhesin protein described by Van Overbeke *et al.* (2002) in that the collagen adhesin is expressed under β-NAD-replete conditions and, unlike the 55-kDa OM adhesin, had a blocked N-terminus. Several other *Pasteurellaceae* members such as *H. influenzae*, *[H.] ducreyi*, *Agg. actinomycetemcomitans* and *P. multocida* have demonstrated adherence to ECM proteins (Abeck *et al.*, 1992; Jacques and Paradis, 1998; Mintz and Fives-Taylor, 1999; Fink *et al.*, 2002; Dabo *et al.*, 2003). A YadA homologue was recently identified in *Agg. actinomycetemcomitans* (Mintz, 2004). Infection by *Agg. actinomycetemcomitans* is established within the connective tissue of the oral cavity, where contact with ECM components is likely. A transposon mutagenesis study identified a gene encoding EmaA, an OM protein that mediated collagen adhesion. Based on its amino acid sequence, EmaA showed conserved structural features related to YadA that included a secretion signal, a head domain of degenerate repeats, a conserved neck region, a stalk domain of probable coiled-coil structure and a conserved C-terminal anchor domain.

In vivo studies have provided additional evidence of the presence of this class of adhesins in *A. pleuropneumoniae*. A homologue to the autotransporter adhesin, *Haemophilus* surface fibrils (Hsf) of encapsulated and a subset of non-typeable *H. influenzae*, was identified as one of the genes expressed by *A. pleuropneumoniae* in necrotic porcine lung tissue at day 7 (Baltes and Gerlach, 2004). Similarly to Chung *et al.* (2007), *in silico* analyses by Baltes and Gerlach (2004) revealed several putative autotransporter adhesin genes. Like YadA, Hsf represents cell surface proteins that form trimeric structures comprised of head-stalk-anchor architecture (Koretke *et al.*, 2006). Non-typeable *H. influenzae* also expresses the *H. influenzae* adhesin (Hia), another trimeric autotransporter adhesin that shares 72% amino acid identity with Hsf (St Geme III *et al.*, 1996). Both Hsf and Hia are large OM proteins with molecular weights of 240 kDa and 115 kDa, respectively. Both proteins mediate adherence to a variety of human epithelial cells (St Geme III *et al.*, 1996; St Geme III and Cutter, 2000). A recent study elucidated the architecture and adhesive activity of Hsf through structural modelling (Cotter *et al.*, 2005). Two binding regions of Hsf, determined by amino acid sequence analysis and adherence to host cells, were homologous to binding domains within Hia, and harboured acidic binding pockets; disruption of these pockets diminished adhesive activity. Furthermore, Hallstrom *et al.* (2006) described Hsf as a vitronectin-binding protein. Vitronectin, present in the ECM and plasma, prevents the membrane attack complex formed by the complement system. Hsf, expressed by *E. coli*, tightly adhered to soluble and immobilized vitronectin whereas Hsf mutants of *H. influenzae* showed reduced binding. Incubation with human serum reduced the viability of Hsf mutants compared to wild-type, demonstrating that Hsf contributed to serum resistance of *H. influenzae*.

OM proteins identified by *in vivo* studies

Several genetic technologies have been developed to identify genes involved in the virulence and survival of important pathogens in a living host. These include approaches such as *in vivo* expression technology (IVET), signature-tagged mutagenesis (STM) and selective capture of transcribed sequences (SCOTS), which have all been applied to identify genes expressed by *A. pleuropneumoniae in vivo* and provide insight into its pathogenesis. Furthermore, STM studies provided a collection of mutants with potential as live-attenuated vaccines. STM technology applies classic transposon mutagenesis developed for the large-scale analysis of multiple transposon-insertion mutants that are screened simultaneously in one animal. Each mutant is tagged with a barcode or unique DNA sequence carried on the inserted transposon that enables identification of a mutant within a pool. Virulence genes are discovered through the negative selection of mutants; STM mutants that are attenuated or not recovered indicate disruption of genes that are critical for survival in a given host. While several novel virulence genes have been identified with this approach, many of them are classically defined as housekeeping genes that are essential for growth *in vivo*. In *A. pleuropneumoniae*, components of the cell surface as well as genes involved in metabolism, transport of substrates, stress responses, and regulation were identified in attenuated mutants that were screened in porcine infection models and recovered 20 or 24 hours post infection (Fuller *et al.*, 2000; Sheehan *et al.*, 2003). STM disrupted genes encoding two putative lipoproteins with homology to LppB of *Hist. somni* and NlpI of *E. coli* (Sheehan *et al.*, 2003). NlpI is a 34-kDa lipoprotein that contributes to the adhesion and invasion of intestinal epithelial cells by an *E. coli* isolate of Crohn's disease (Barnich *et al.*, 2004). In addition, NlpI mutants have altered cell morphology and are osmosensitive (Ohara *et al.*, 1999).

A. pleuropneumoniae produces capsular polysaccharides that are serotype-specific and serve as a protective barrier to host immune cells and effectors. Recovery of three attenuated STM mutants revealed insertions in *cpxB*, *cpxC*, and *cpxD*, genes of the operon *cpxDCBA* involved in export of capsular polysaccharides (Sheehan *et al.*, 2003). Characterization of the *cpx* operon showed homology to genes encoding capsule export in *H. influenzae*, *bexDCBA* (Ward and Inzana, 1997). The gene encoding OM protein, BexD, is homologous to *cpxD*, which likely encodes a putative OM lipoprotein that facilitates capsular polysaccharide transport across the OM.

Disruption of genes encoding AopA, and homologues of major OM proteins such as OmpA2 of [*H.*] *ducreyi*, PomA of *M. haemolytica*, the P2 porin and OmpP5 of *H. influenzae* also affected survival of *A. pleuropneumoniae* in the porcine host (Fuller *et al.*, 2000; Sheehan *et al.*, 2003). AopA is an immunogenic, 48-kDa OM protein that is present in at least 12 serotypes of *A. pleuropneumoniae*, but not detected by anti-AopA antibodies in related Gram-negative bacteria (Cruz *et al.*, 1996). In contrast, OmpA proteins are conserved in almost all Gram-negative bacteria and have multifunctional roles that include stabilizing the OM, serving as receptors for phage and colicin, mediating F-dependent conjugation, and contributing to serum resistance of *E. coli* K1 (Datta *et al.*, 1977; Chai and Foulds, 1978; Sonntag *et al.*, 1978; Prasadarao *et al.*, 2002). In addition, OmpA proteins are immunogenic and can illicit antibodies with opsonic, bactericidal or protective activities (Mahasreshti *et al.*, 1997). PomA is an OmpA-like OM protein that is heat-modifiable and recognized by antibodies from cattle vaccinated or experimentally infected with *M. haemolytica* (Zeng *et al.*, 1999).

OmpP5 of *H. influenzae* is also related to OmpA proteins; antibodies that were raised against purified OmpP5 recognized the OmpA protein of *E. coli* (van Alphen *et al.*, 1983; Duim *et al.*, 1997). In non-typeable *H. influenzae*, OmpP5 is expressed as a fibrin subunit that is assembled as a filamentous structure at the cell surface and contributes to virulence and attachment to epithelial cells and mucin (Hardy *et al.*, 2003). Genomic mining of *A. pleuropneumoniae* serotype 5b identified at least three homologues of OmpP5 from *H. influenzae* that have yet to be determined for their functionality as adhesins (Chung *et al.*, 2007).

In contrast to STM, SCOTS identifies genes that are up-regulated but not necessarily vital for *in vivo* survival. Bacterial transcripts are isolated from infected tissues and are captured with biotinylated genomic bacterial DNA. Selective enrichment of sequences transcribed *in vivo* is achieved with PCR and subtractive hybridization with transcripts prepared from bacteria grown *in vitro*. Baltes and Gerlach (2004) reported differential expression of 46 genes that were up-regulated in necrotic porcine lung tissue at day 7 post-infection, but not expressed under culture conditions. The identification of an Hsf homologue in *A. pleuropneumoniae* highlighted a novel class of putative non-pilus adhesins in *A. pleuropneumoniae*. Other OM proteins identified by SCOTS analysis included receptors for iron acquisition, such as haemoglobin-binding protein HgbA, and the lipoprotein subunit of the transferrin receptor, TbpB. Previously identified by STM, the *in vivo* expression of PomA was also verified by SCOTS.

Iron uptake systems

Introduction

Iron, required by almost all bacteria, is essential for diverse cell functions that include DNA replication and repair, respiration, ATP generation, and metabolism. Under physiological conditions, exposed to oxygen and at neutral pH, iron converts from its ferrous (FeII) to ferric (FeIII) state which is extremely insoluble. Availability of this element is further diminished in the animal host where free iron is tightly complexed with iron-binding glycoproteins transferrin and lactoferrin, and haem-containing proteins. This sequestration serves as a host defence mechanism to withhold essential bacterial nutrients; levels of free iron (10^{-18} to 10^{-24} M) are well below the levels required for microbial growth (10^{-6} to 10^{-8}) (Jacques, 2004). To overcome these limitations, bacteria have developed efficient strategies to exploit iron reserves in the host, mechanisms that are now recognized as virulence factors for several pathogens. Bacteria acquire iron from the host by two main mechanisms: production of siderophores and their cognate receptors; and receptor-mediated uptake of iron from host proteins bound at the surface (Cornelissen, 2003).

Both systems serve to localize iron complexes at the cell surface for their transport across the OM (Faraldo-Gomez and Sansom, 2003). OM receptors that mediate iron uptake are members of a family of TonB-dependent transporters. In association with IM proteins ExbB and ExbD, TonB spans the periplasm, transducing energy derived from the proton motive force (PMF) at the IM to ligand-bound receptors for the uptake of substrates. Following binding of ligand, conformational changes signal the ligand-loaded status of the receptor. TonB undergoes its own conformational changes that enable it to mechanically interact with OM receptors at the Ton box, an amino acid sequence near the N-terminus at the periplasmic face of the OM protein. The Ton box is essential for the energization of the receptor and passage of substrate into the cell.

OM receptors for iron uptake represent the best characterized OM proteins of *A. pleuropneumoniae* and other members of *Pasteurellaceae*. Iron acquisition systems, particularly those involved in haem uptake, are highly redundant in *H. influenzae* and *P. multocida*, demonstrating an adaptation in utilizing iron from blood sources. *A. pleuropneumoniae* is also capable of utilizing haem proteins such as haemoglobin (Hb), as well as host transferrin as sole sources of iron for growth *in vitro*. These findings evolved from several reports that identified iron-repressible OM proteins in *A. pleuropneumoniae*, using chemical iron-chelators to produce iron-restricted conditions in culture (Deneer and Potter, 1989a; Niven *et al.*, 1989; Ricard *et al.*, 1991). Like other mucosal pathogens, *A. pleuropneumoniae* and other members of *Pasteurellaceae* rely upon direct receptor-mediated uptake of host proteins as iron sources (Cornelissen, 2003). While *A. pleuropneumoniae* is capable of using exogenous siderophores from other microbes, several studies report that *A. pleuropneumoniae* itself does not appear to produce any detectable siderophores (Deneer and Potter, 1989a; Niven *et al.*, 1989; D'Silva *et al.*, 1995). However, Diarra *et al.* have described the production of a putative iron chelator by *A. pleuropneumoniae*, for which the cognate receptor and uptake mechanism remain to be determined (Diarra *et al.*, 1996; Bossé *et al.*, 2002).

Beyond OM receptors, other participating components for iron uptake have been identified in *A. pleuropneumoniae*. The *afuABC* operon in *A. pleuropneumoniae* is a proposed periplasmic binding protein-dependent transport system, involved in the passage of iron into the cytoplasm (Chin *et al.*, 1996). The operon encodes proteins of an ATP-binding cassette (ABC) transport system with homology to the FbpABC proteins of *Neisseria*, HitABC proteins of *H. influenzae*, and YfuABC proteins of *Yersinia*, all involved in the transport of iron across the IM. For iron transport across the OM, *A. pleuropneumoniae* requires a functional TonB system, of which there are two. The *tonB1* system, *tonB1-exbB1-exbD1*, described by Tonpitak *et al.* is dedicated to iron uptake by transferrin-binding and bears resemblance to Ton systems in *Neisseria* and *Pseudomonas* (Tonpitak *et al.*, 2000; Beddek *et al.*, 2004). In contrast, the gene order of the *tonB2* system, *exbB2-exbD2-tonB2*, matches that in other *Pasteurellaceae* members suggesting that this system is indigenous to *A. pleuropneumoniae* (Beddek *et al.*, 2004). Moreover, TonB2 is critical for *in vitro* growth using haem, porcine Hb or the hydroxamate siderophore ferrichrome as the sole source of iron. Both *tonB1* and *tonB2* are up-regulated under iron-starvation, though recent details suggest that *exbB1* of the *tonB1* system and associated genes for transferrin-binding are specifically regulated by the ferric uptake regulator Fur protein (Jacobsen *et al.*, 2005a). In addition, Fur represses the expression of the *afuABC* operon in *A. pleuropneumoniae* (Hsu *et al.*, 2003). Present in several pathogens, Fur regulates the expression of several genes associated with iron acquisition and virulence. Iron itself acts as a cofactor for Fur, which together will block gene transcription by binding to a consensus sequence, termed the Fur box, located in the promoter region. When iron is scarce, the Fur protein loses its cofactor and disengages from the Fur box, freeing the promoter for polymerase and gene expression.

Current studies using microarray and *in silico* analyses reveal that *A. pleuropneumoniae* possesses several unidentified auxillary systems to adapt to iron shortage in the host. In this section, we describe receptors for transferrin, for the hydroxamate siderophore ferrichrome, and for

Hb, all of which have been cloned and characterized in *A. pleuropneumoniae* (Fig. 8.1 and Table 8.3).

Transferrin-binding proteins (Tbps)

Transferrin is a serum glycoprotein divided into two lobes (N and C) that serve as binding sites for iron (Ratledge and Dover, 2000; Cornelissen, 2003). Mammalian hosts produce transferrin for delivery of iron to tissues. Bacteria are able to exploit this iron source by expressing specific receptors for transferrin that are present in *Pasteurellaceae* and *Neisseriaceae*, the latter contributing to most of our knowledge on transferrin-binding proteins (Tbps). Expressed under iron restriction, Tbps exhibit host specificity such that *A. pleuropneumoniae* can only use porcine transferrin and not transferrin from any other animal species (Gonzalez *et al.*, 1990; D'Silva *et al.*, 1995). Two Tbps of dissimilar size make up the cognate receptor; however, in *A. pleuropneumoniae*, more than one nomenclature exists for these two proteins as a result of isolated works from several independent laboratories. Interestingly, genome analysis of *A. pleuropneumoniae* serotype 5b shows that there may be more than one set of Tbps as more than one homologue to both receptor components has been predicted by bioinformatic programs (Chung *et al.*, 2007). TbpA, also identified as Tbp1 and TfbB, is a 100-kDa integral protein that forms a transmembrane channel for iron transport and has strong similarity to the corresponding Tbp1 in *Neisseria* and to TonB-dependent OM receptors of *E. coli* (Gonzalez *et al.*, 1995; Medrano *et al.*, 1997; Wilke *et al.*, 1997). Litt *et al.* (2000) demonstrated interaction between TonB of *N. meningitidis* and the TonB box motif present in TbpA from *A. pleuropneumoniae*, demonstrating TonB interaction and conserved sites between TonBs and TbpAs for cross-species recognition.

TbpB, also known as Tbp2 and TfbA, represents a 60-kDa surface-exposed lipoprotein, anchored to the OM of *A. pleuropneumoniae* (Gerlach *et al.*, 1992a; Gerlach *et al.*, 1992b; Gonzalez *et al.*, 1995). Among different serotypes of *A. pleuropneumoniae*, the *tbpB* gene exists in at least three different isoforms that distinguish se-

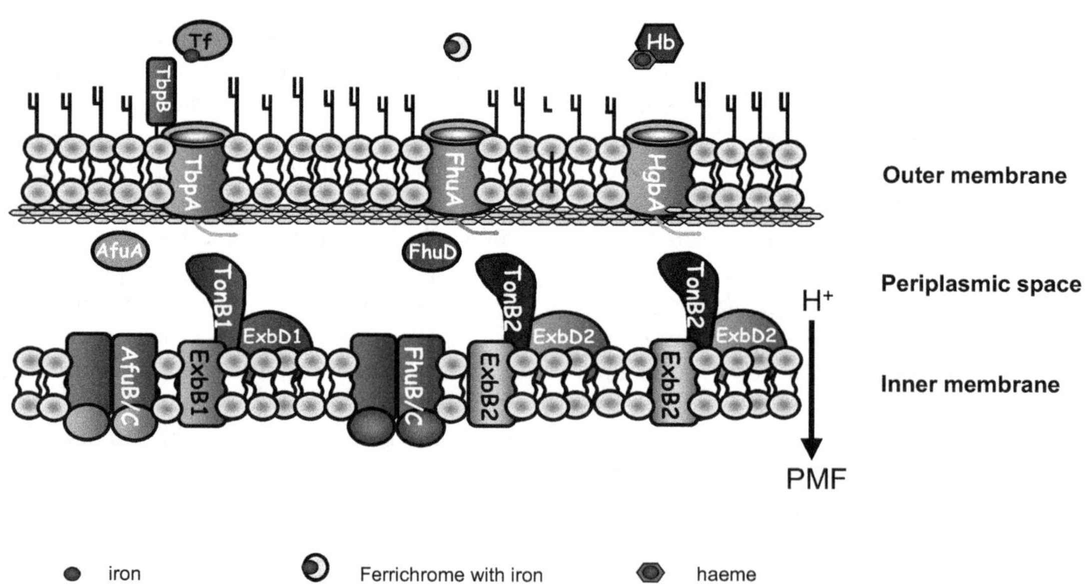

Figure 8.1 Representation of iron-uptake systems identified in *A. pleuropneumoniae.* The AfuABC proteins are homologous to the FbpABC proteins of *Neisseria*, responsible for transporting iron, from transferrin (Tf), into the cell. Tf is bound to TbpBA that is energized by the TonB1 system, one of two Ton systems that transduces energy from the proton motive force (PMF) at the inner membrane. Ferrichrome is taken up by FhuA, bound by FhuD and delivered to FhuBC for iron transport into the cell. Haemoglobin (Hb) interacts with HgbA, whereupon haem is removed and crosses the outer membrane. Ferrichrome and haem uptake are energized by the TonB2 system. A colour version of this figure is located in the plate section at the back of the book.

Table 8.3 Iron uptake systems characterized in *Pasteurellaceae*

Iron uptake mechanism	OM protein receptors of *A. pleuropneumoniae*	OM proteins/homologues in other *Pasteurellaceae* species	References
Transferrin binding	TbpBA	Tbps in *H. influenzae*	Loosmore *et al.* (1996)
		M. haemolytica	Ogunnariwo *et al.* (1997)
		P. multocida	Ogunnariwo and Schryvers (2001)
		A. suis	Bahrami *et al.* (2003)
Siderophore binding	FhuA	FhuA in *[H.] parasuis*	Mikael *et al.* (2003) del Rio *et al.* (2006)
Haemoglobin binding	HgbA	Hup in *H. influenzae*	Morton *et al.* (2004)
		HupA in *[H.] ducreyi*	Srikumar *et al.* (2004)
		HgbA in *H. influenzae*	Srikumar *et al.* (2004)
		[H.] ducreyi	Afonina *et al.* (2006)
		P. multocida	Bosch *et al.* (2002)
		Agg. actinomycetemcomitans	Srikumar *et al.* (2004)
		A. suis	Bahrami and Niven (2005)
Haemoglobin–haptoglobin binding		Hgps/Hgbs in *H. influenzae*	Morton *et al.* (1999) Cope *et al.* (2000)
		Hup in *H. influenzae*	Morton *et al.* (2004)
Free haem, haem–haemopexin, haem–albumin binding		HxuC in *H. influenzae*	Cope *et al.* (1995) Cope *et al.* (2001) Morton *et al.* (2007)
		Hup in *H. influenzae*	Morton *et al.* (2004)

rotypes 2, 3, 4, 7, 8, 9, 10 and 11 from serotypes 1, 6 and 12, with serotype 5a and 5b separated from the rest (Gerlach *et al.*, 1992b). Gerlach *et al.* (1992a) demonstrated that TbpB can also bind to haem and discriminate between apo- and holo-transferrin, suggesting recognition of the 'closed' iron-bound conformation of transferrin. Evidence also suggests that TbpB preferentially binds to holo-transferrin, important for efficient discrimination between iron-depleted transferrin and iron-laden transferrin in the host. Studies by Strutzberg *et al.* (1995) described three different domains of 13 or 14 amino acids in length that were responsible for transferrin-binding and that mapped to the variable N-terminal half of TbpB (Fuller *et al.*, 1998). Although TbpB is capable of inducing a strong, serotype-specific immune response, the N-terminal half is not recognized by convalescent-phase serum (Strutzberg *et al.*, 1995). In contrast, the conserved C-terminal half of the protein contains an epitope that is detected by both convalescent-phase serum and serum from immunized pigs, and may serve to divert antibodies away from obstructing the binding site (Strutzberg *et al.*, 1995). The mechanism of iron removal from transferrin and subsequent passage through TbpA is still unclear although several observations suggest that TbpA and TbpB function together for optimal binding and utilization of transferrin (Wilke *et al.*, 1997; Litt *et al.*, 2000; Fuller *et al.*, 1998). Affinity isolation of TbpB from *A. pleuropneumoniae* with *N. meningitidis* TbpA-transferrin complexes demonstrates interactions that are functionally conserved between TbpB and TbpA (Fuller *et al.*, 1998). This observation was also extended to *H. influenzae*, in which TbpA of *N. meningitidis* facilitated isolation of TbpB from *H. influenzae* (Fuller *et al.*, 1998).

Genetic arrangement of *tbp* genes is consistent in *Neisseria*, *H. influenzae*, *M. haemolytica* and *A. pleuropneumoniae*, where the *tbpB* gene precedes *tbpA*, and both are coordinately transcribed (Fig. 8.2). In contrast, *tbpA* of *P. multocida* is downstream of a leucyl-tRNA synthetase gene and an IS element (Ogunnariwo and Schryvers,

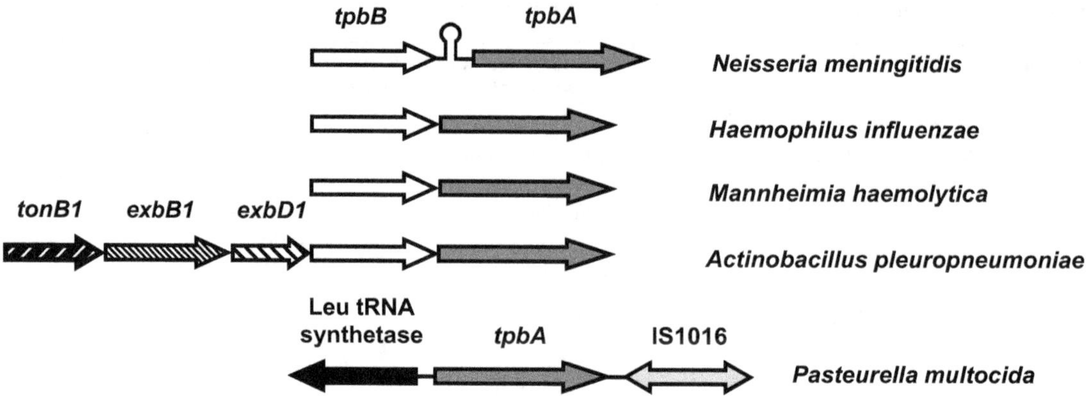

Figure 8.2 Genetic arrangement of *tbp* genes from *N. meningitidis* and members of *Pasteurellaceae*. The *tbp* genes of *N. meningitidis* are separated by a potential stem-loop structure. Genes of the TonB1 system are genetically linked to *tbpBA* of *A. pleuropneumoniae*. The *tbpA* gene of *P. multocida* is flanked by a gene encoding a leucine tRNA synthetase and an IS element. Genes are not drawn to scale. Adapted from Cornelissen (2003).

2001). *P. multocida* does not appear to have a linked *tbpB* gene, suggesting that TbpA can sufficiently function as a single transferrin receptor in this organism. In *Neisseria* and *H. influenzae*, a putative promoter region is located immediately upstream of *tbpB*, whereas *A. pleuropneumoniae* has one of two *tonB* systems, immediately upstream of *tbpBA*. Tonpitak *et al.* (2000) identified the *tonB1* system, *tonB1-exbB1-exbD1*, and demonstrated by RT-PCR analysis that the *exbBD* genes and *tbpB* are transcriptionally linked. The same linkage of a *ton* system to *tbpBA* transcribed as one unit was identified in related swine pathogen, *A. suis* (Bahrami *et al.*, 2003). Possession of more than one *tonB* system has been described in other organisms, in which individual TonBs are assigned to specific functions (Beddek *et al.*, 2004). Due to its linkage to *tbpBA*, the *tonB1* system of *A. pleuropneumoniae* may serve as energy couplers devoted to transferrin utilization. This hypothesis is supported by a nonpolar *exbB1* deletion mutant, in which the expression of *exbB1-exbD1* genes are required for transferrin uptake (Tonpitak *et al.*, 2000). The second *tonB* system of *A. pleuropneumoniae*, *tonB2*, is not positionally associated with any iron uptake genes, but is also required for transferrin uptake (Beddek *et al.*, 2004). Moreover, the *tonB2* system, but not *tonB1*, is essential for the acquisition of haem, Hb and ferrichrome. Analyses of these two systems revealed that TonB1 cannot substitute functionally for TonB2 although TonB2 may interact with Exb proteins encoded in the *tonB1* locus, thereby allowing transferrin utilization. The requirement of both Ton systems for transferrin uptake may explain how these two systems evolved. Beddek *et al.* (2004) proposed that *tonB1-exbB-exbD* was first acquired in association with transferrin uptake but a mutation event in *tonB1* may have driven the surrogate role of TonB2 in this pathway. The *exbB1-exbD1* linkage to *tbpBA* may confer a biological advantage for transferrin uptake although TonB1 itself does not appear to have a significant role in this process. *In vivo* studies show that functional *exbB1* is essential for colonization and persistence of *A. pleuropneumoniae* in the porcine lung, demonstrating the importance of Tbp expression for pathogenesis (Baltes *et al.*, 2001). This is further supported by the avirulent phenotypes of individual *tbpB* and *tbpA* mutants as well as a *tbpBA* mutant. All these genetically engineered strains were eradicated in infected animals before an immune response could be detected or manifested into clinical disease (Baltes *et al.*, 2002). However TonB2 is more important for virulence than TonB1 in *A. pleuropneumoniae*. In a porcine model of acute infection a *tonB2* mutant was highly attenuated whereas the *tonB1* mutant was attenuated only at a lower infectious dose (Beddek *et al.*, 2004). Based on these studies TonB2 may contribute to *A. pleuropneumoniae* virulence at acute stages of infection prior to rapid death when bacterial titres are high and iron assimilation from sources other than transferrin is sufficient. In contrast, transferrin uptake may be required when more bacterial replication is required *in vivo*, thereby engaging proteins encoded in *tonB1* system for efficient expression of Tbps.

Regulation of *tbpB* expression may, in part, contribute to *A. pleuropneumoniae* persistence in the porcine host. As a vaccine candidate, TbpB is highly immunogenic and is capable of inducing a protective immune response (Gerlach *et al.*, 1992a). The downregulation of immune-targeted proteins is one mechanism by which pathogens may persist in their host and avoid eradication. Over the course of disease, Hennig *et al.* (1999) observed decreased expression of TbpB in *A. pleuropneumoniae*, suggesting modulation of this protective antigen during disease progression. In addition, *in vitro* experiments with BALF from infected pigs showed that TbpB is regulated by an alternative, iron-independent mechanism. Although this observation may suggest regulatory components derived from host factors, previous evidence suggests that transcription of TbpB is regulated by Fur, which has been cloned and identified in *A. pleuropneumoniae* (Hsu *et al.*, 2003; Jacobsen *et al.*, 2005a). A *fur* mutant resulted in the constitutive expression of *tbpB* and *exbB* and growth deficiencies *in vitro* (Jacobsen *et al.*, 2005a). Moreover, the *fur* mutant showed reduced virulence and could not persist in the porcine host. These findings support the hypothesis that downregulation of immunogenic *tbpB* is important for persistence whereas constitutive expression of *tbpB* renders *A. pleuropneumoniae* more susceptible to the host immune response and subsequent clearance.

Siderophore ferrichrome uptake: ferric hydroxamate uptake (Fhu) receptor

Siderophores, high-affinity iron chelators, are produced by several bacteria that express OM receptors for their uptake. Hydroxamates and catecholates are the two broad classes of siderophores based on the chemical group involved in chelating iron (Faraldo-Gomez and Sansom, 2003). Determination of the crystal structures of three TonB-dependent receptors from *E. coli* – FhuA, FepA, and FecA – have elucidated interactions between this class of OM proteins with the Ton system, and has revealed structural features related to the uptake of siderophores ferrichrome, enterobactin, and ferric citrate, respectively. These features include a β-barrel architecture formed by 22 transmembrane β-strands, long extracellular loops for ligand recognition and binding, and an N-terminal globular domain, referred to as a plug or cork, that occludes the interior of the β-barrel from the periplasmic face. Proximal to the N-terminus is the conserved Ton box that physically interacts with TonB complex for energy. Current models propose that following substrate binding, the extracellular loops close around the substrate to prevent release into the medium. Substrate binding triggers a cascade of conformational changes at the periplasmic end that signal the requirement for TonB and preparation for passage of substrate across the OM.

Utilization of siderophores produced by other microbial species has been documented in *E. coli*, *Salmonella* Typhimurium, and *Neisseria*, indicating that these bacteria express receptors that can recognize and scavenge iron chelators that they do not secrete. *A. pleuropneumoniae* is able to use exogenous hydroxamate and catecholate siderophores, among which the hydroxamate ferrichrome and bis-catechol-based siderophores were particularly apt in promoting growth for *A. pleuropneumoniae in vitro* (Diarra *et al.*, 1996). Ferrichrome is produced by fungi *Aspergillus*, *Penicillium* and *Ustilago* and is used as an iron source by many organisms (Neilands, 1984; Baltes *et al.*, 2003). Genes of the *fhu* operon, *fhuACDB*, in *E. coli* encode the receptor for ferrichrome plus accessory periplasmic and ATP-transporter components, respectively. FhuA, the TonB-dependent receptor for ferrichrome, also functions as a receptor for phages, bacterial toxin colicin M and antibiotics albomycin and rifamycin CGP4832 (Mikael *et al.*, 2002). Mikael *et al.* (2002) cloned the homologues of the *E. coli fhu* genes in *A. pleuropneumoniae* with *fhuA* transcribed as the last gene in the operon arrangement, *fhuCDBA*. The *fhuA* gene of *A. pleuropneumoniae* encodes the 77-kDa OM receptor and shares 35% identity and 48% similarity with the nucleotide sequence of *fhuA* in *E. coli*. FhuD is a 35.6-kDa periplasmic protein that translocates ferric hydroxamate from the OM to the IM. FhuC (28.5 kDa) and FhuB (69.4 kDa) are IM-associated components of an ABC transporter for internalization of ferric hydroxamate.

Based on sequence alignments with FhuA of *E. coli*, Mikael *et al.* (2002) proposed a 3D homology model of FhuA in *A. pleuropneumoniae* that showed 11 extracellular loops and 10 periplasmic turns contributing to a 22 β-strand structure. The FhuA model of *A. pleuropneumoniae* had the same overall fold as the solved crystal structure of *E. coli* FhuA with significant deviations at the extracellular and periplasmic-loop regions. In particular two extracellular loops (L3 and L4) involved in ligand recognition and uptake were longer in length compared with corresponding loops in *E. coli*. Whether these structural variations indicate functional differences between the FhuAs of *E. coli* and *A. pleuropneumoniae* has not been experimentally confirmed, although *A. pleuropneumoniae* is not susceptible to antibiotics albomycin and rifamycin CGP 4832, nor to colicin M, all of which are FhuA-specific ligands in *E. coli*.

PCR analysis shows that *fhuA* is conserved in all 15 serotypes of *A. pleuropneumoniae*, with at least 12 serotypes containing genes of the entire *fhu* operon (Mikael *et al.*, 2002; Shakarji *et al.*, 2006). Comparison and alignment of the predicted amino acid sequence of FhuA from all 15 serotypes established phylogenetic relationships that were defined by dendrogram analysis (Fig. 8.3). Based on sequence relatedness, FhuA divides *A. pleuropneumoniae* into one branch containing serotypes 1, 11, 9, 3, 8, 13, 14, and 15; and into another branch containing serotypes 2, 4, 7, 5, 10, 12, and 6 (Shakarji *et al.*, 2006). These data support evidence of clonal populations within *A. pleuropneumoniae* and correlate with group clusters based on structural similarities of the O-antigen that confer serological cross-reactivity. For example, cross-reactivity is shared between serotypes 1, 9 and 11; 3 and 8; and 4 and 7. Interestingly, dendrogram analysis grouped the two biotypes of *A. pleuropneumoniae* together rather than on a separate branch. Although differentiated by their requirement for β-NAD, biotype II serotypes 13 and 14 have FhuA sequences that show a common ancestry among biotype I (Shakarji *et al.*, 2006).

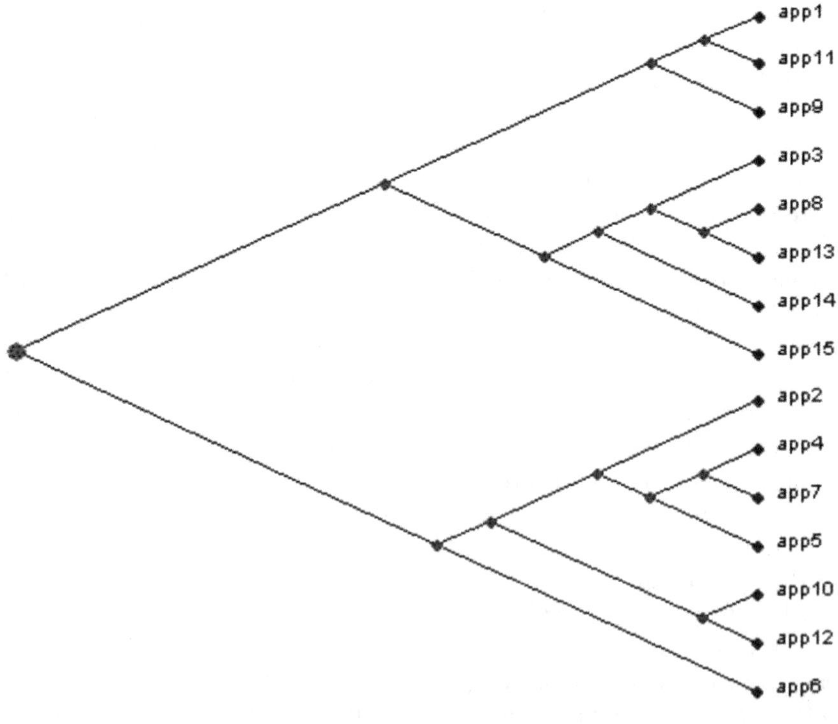

Figure 8.3 Dendrogram analysis showing sequence relatedness of FhuA from reference strains representing *A. pleuropneumoniae* serotypes 1 to 15. Reproduced with permission from Shakarji *et al.*, Can. J. Microbiol. 2006: *52*, 391–396.

Unlike most iron acquisition systems such as Tbps that are up-regulated in response to iron starvation, regulation of *fhuA* appears to be iron-independent in *A. pleuropneumoniae*. Despite the presence of putative Fur boxes in regions upstream of *fhuA* and *fhuC*, RT-PCR analyses showed that *fhuA* transcript levels did not increase in response to low iron levels or in the presence of ferrichrome (Mikael *et al.*, 2002; Mikael *et al.*, 2003). Similarly, transcript levels of *fhuC* were unaffected by iron restriction, indicating that the *fhu* operon in *A. pleuropneumoniae* is not regulated by iron levels in the environment. It appears that *fhuA* is constitutively expressed and is unlike any other iron uptake system known to date. Mutations in *fhuA* confirm its functional role as the receptor for ferrichrome, but also show that FhuA does not contribute or affect *A. pleuropneumoniae* virulence (Mikael *et al.*, 2003; Baltes *et al.*, 2003; Shakarji *et al.*, 2006). Animals that were challenged with either *fhuA* mutants or parent controls experienced the same morbidity and disease outcome. Considering that *A. pleuropneumoniae* can use several alternative iron sources from the host, it is not surprising that ferrichrome uptake is unnecessary *in vivo*. The ability to use ferrichrome may be more important to adapt to habitats where siderophore-producing microbes coexist. Other swine pathogens of *Pasteurellaceae*, such as *A. suis* and [*H.*] *parasuis*, can also use ferrichrome as a sole iron source (Mikael *et al.*, 2003). Furthermore, the ability to use catechol-based siderophores indicates that *A. pleuropneumoniae* possesses other receptors that are specific for this class of siderophores (Diarra *et al.*, 1996; Shakarji *et al.*, 2006).

Haemoglobin and haem uptake

Among the potential sources of iron in the host, haem-containing proteins are the most abundant. Haem is reduced, ferrous or Fe(II) iron protoporphyrin IX whereas the oxidized, ferric form is formally referred to as haemin. However the term 'haem' is generally used to refer to the molecule in either oxidation state. Haem is predominantly located intracellularly and thus unattainable for extracellular pathogens that require the release of haem-containing proteins into the environment (Genco and Dixon, 2001). Haem is available usually after tissue damage, resulting in release of intracellular materials. In serum, haem is sequestered by proteins, haemopexin, haemoglobin (Hb), albumin, and lipoproteins, with haemopexin having the greatest affinity for haem (Genco and Dixon, 2001). Lysed erythrocytes yield Hb, the source of most circulating haem that then becomes sequestered by serum protein haptoglobin. Haem acquisition by bacteria requires specific OM receptors that bind to haem and haem proteins, followed by transport of haem into the cell by the TonB energy-dependent mechanism.

Utilization of Hb as a source of iron is well documented among members of *Pasteurellaceae*, with *H. influenzae* containing the most described variety of haem-uptake systems. The use of haem by *A. pleuropneumoniae* is documented by several studies that demonstrate *A. pleuropneumoniae* growth *in vitro* with pig Hb as the sole iron source (Archambault *et al.*, 1999; Archambault *et al.*, 2003). Negrete-Abascal *et al.* (1994) also reported proteases in culture supernatant that degrade porcine IgA and Hb. The presence of extracellular proteases may be important for the removal of haem from its protein carrier, Hb, once bound to its cognate receptor, thereby allowing for haem transport across the OM. LPS binding to pig Hb was demonstrated in a study that used immunogold labeling and EM to visualize the binding of pig Hb to *A. pleuropneumoniae* (Bélanger *et al.*, 1995). LPS may serve as a docking station for Hb prior to its uptake by specific OM receptors. To identify these receptors, Archambault *et al.* (1999) used flow cytometry to compare the Hb-binding activity between *A. pleuropneumoniae* grown under iron-restricted and iron-replete conditions. Unlike transferrin-binding, Hb-binding activity was not host specific as determined by *A. pleuropneumoniae* binding to Hb from pig, sheep, human, goat, bovine, rabbit, or horse. Iron restriction promoted expression of Hb receptors, particularly a 75-kDa iron-regulated protein that Archambault *et al.* (2003) subsequently detected in OM preparations that were subject to affinity purification with bovine haemin or Hb immobilized on agarose. Verifying the growth of 12 *A. pleuropneumoniae* serotypes on pig Hb as an iron source, this study isolated OM proteins of 75 kDa and 104 kDa that were not observed in

A. pleuropneumoniae grown under iron-replete conditions. MALDI-TOF analysis of the 75 kDa revealed peptide sequences with homologies to iron-regulated OM proteins, transporter proteins, and TonB-dependent receptors. Although associated with haemin and Hb-binding, the gene encoding the 75-kDa OM protein has not been cloned or assigned an identity, but may correspond to a 76-kDa protein, previously described by Deneer and Potter (1989a), that bound to Congo red and haemin.

A 105-kDa OM protein with Hb-binding activity was recently identified as Hb-binding protein A (HgbA) of *A. pleuropneumoniae* and may correlate with the previously described 104-kDa OM protein (Srikumar *et al.*, 2004). Srikumar *et al.* also applied Hb-agarose affinity purification to isolate HgbA from OM vesicles of *A. pleuropneumoniae* and obtained peptide sequences that were used to design primers. Sequence analysis of the gene encoding HgbA revealed an ORF of 2838 bp that translated into 946 amino acids, and a putative Fur-binding sequence that was 60 bp upstream from the proposed start codon. Upstream of *hgbA* was an ORF with sequence homology to HugZ in *Plesiomonas shigelloides*, a protein that appears necessary to prevent haem toxicity (Henderson *et al.*, 2001; Bosch *et al.*, 2002). At the primary sequence level, a signal sequence was detected in the first 23 aa and a Ton box, characteristic of TonB-dependent receptors, was predicted at the N-terminus. Another conserved motif of TonB-dependent receptors, termed the TonB boxC, was predicted at the C-terminus. NPNL and FRAP amino acid motifs, specific for haem/Hb transporters, were also detected within HgbA (Srikumar *et al.*, 2004). Upregulation of *hgbA* under iron restriction was confirmed with RT-PCR results that showed an increase in transcript levels of *hgbA* in response to decreasing iron levels in culture medium. Further characterization with an *hbgA* mutant showed Hb-binding activity but it was incapable of using Hb as a sole source of iron for growth, indicating that deleted internal sequences from residues Asp592 to Pro807 contributed to Hb import into the cell. In particular, the deleted sequence contained the FRAP motif, a region that is essential for haem uptake by the HemR receptor of *Y. enterocolitica* (Bracken *et al.*, 1999; Srikumar *et al.*, 2004).

HgbA showed similarities with other Hb-binding OM proteins in *Pasteurellaceae* such as HgbA of *H. influenzae*, HgbA of *Agg. actinomycetemcomitans* and HupA of [*H.*] *ducreyi*. However, based on mass spectrometric sequences of internal peptide sequences, HgbA from *A. pleuropneumoniae* is most closely related to the HgbA protein from *P. multocida* (Bosch *et al.*, 2002). The genetic arrangement of *hgbA* in *P. multocida* is also preceded by two ORF (PM0298 and PM0299) with significant identity to the *hugX* and *hugZ* genes of *P. shigelloides*; together, these two ORF and *hgbA* constitute a single transcriptional unit. Inactivation of PM0298 and PM0299 affected the recovery of *P. multocida* mutants, indicating that these genes are essential for cell viability (Bosch *et al.*, 2002). Adaptation of the recombinase-based *in vivo* expression technology (RIVET) demonstrated the rapid induction of the PM0298-PM0299-*hgbA* transcriptional unit in an infected mouse, consistent with the low iron response of iron-regulated genes encoding Hb receptors (Bosch *et al.*, 2002).

To construct a 3D homology model of HgbA, Pawelek and Coulton (2004) applied comparative modelling and a hidden Markov model, using known X-ray crystal structures of TonB-dependent OM receptors, BtuB, FepA, FecA, and FhuA, as a template. The HgbA model shares highest structural similarity to BtuB and proposes a 22-stranded β-barrel domain containing a globular N-terminal cork domain and 11 predicted extracellular loops (Fig. 8.4). Ten of the 11 loops showed fold similarity to known structures of loop regions involved in haem binding, iron binding, and protein-protein interactions. Specifically, recognition and binding to Hb or haem proteins were assigned to HgbA loops 2 and 7. Loop 2 shared structural homology with a loop in bovine endothelial nitric oxide synthase that is nearby a haem-binding pocket while loop 7 contains a conserved histidine residue in a motif associated with haem/Hb binding.

A recent study by Shakarji *et al.* (2006) assessed the roles of HgbA and FhuA as virulence determinant in a piglet model of intranasal infection. Both proteins were also identified in

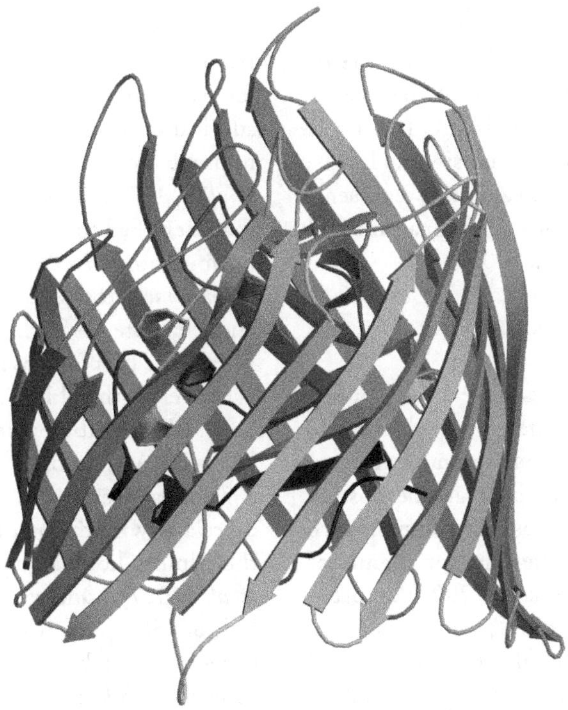

Figure 8.4 Homology-based model of HgbA of *A. pleuropneumoniae*. Flat arrows represent β-strands, lines represent loops and flat coils positioned within the barrel represent helices. Reprinted from Journal of Molecular Graphics and Modelling, Pawelek and Coulton, *23*, 211–221, 2004, with permission from Elsevier. A colour version of this figure is located in the plate section at the back of the book.

recently described serotypes of *A. pleuropneumoniae*, updating their conservation to all 15 serotypes. Compared with the parent strain, infection with the *hgbA* mutant resulted in a lower mortality rate, less severe symptoms, fewer lung lesions, and reduced colonization in the respiratory tract. These observations indicate that HgbA may play a role in *A. pleuropneumoniae* virulence. Similarly, *hgbA* mutants of [*H.*] *ducreyi* are attenuated in establishing infection in human hosts, even at 10× the infectious dose of the parental strain (Afonina *et al.*, 2006). A homologue of HgbA from [*H.*] *ducreyi* was also among the unidentified haem-related OM proteins that were predicted through *in silico* analysis of the *A. pleuropneumoniae* serotype 5b genome (Chung *et al.*, 2007). The other homologues included HemR of *Y. enterolitica*, HpuB of *N. meningitidis*, and HxuC2, HhuA, and HbpA of *H. influenzae*. In *P. multocida*, at least nine proteins have been functionally assigned as haem/haemoglobin-binding receptors (Bosch *et al.*, 2004). The redundant nature of haem acquisition systems in *Pasteurellaceae* organisms is best represented by *H. influenzae*, from which several haem-related receptors have been characterized. In addition to Hb, *H. influenzae* can use Hb–haptoglobin, haem–haemopexin, and haem–albumin complexes as iron sources for *in vitro* growth (Morton *et al.*, 2007). Binding to Hb and Hb-haptoglobin is mediated by proteins designated as the Hgps by one research group and HgbA, HgbB, HgbC by another independent study (Morton *et al.*, 1999; Cope *et al.*, 2000). Genes encoding these proteins contain CCAA repeats that vary in length, contributing to frame-shift mutations and the introduction of stop codons that result in phase variability of these proteins.

Products of the *hxuCBA* gene cluster have demonstrated interaction with the haem–haemopexin complex as well as Hb, the haem–albumin complex and low levels of free haem (Cope *et al.*, 1995; Cope *et al.*, 2001; Morton *et al.*, 2004). HxuA is a 100-kDa secreted protein that acts as a haemophore for haem-haemopexin capture and a shuttle to HxuC, the cognate OM receptor. Amino acid sequence analysis of HxuC shows characteristics of other TonB-dependent receptors involved in ligand-binding and transport. Based on mutation studies, both HxuA and HxuB are necessary for use of the haem–haemopexin complex whereas HxuC has a role in the uptake of free haem and is essential for the use of haem–albumin complexes, thus representing a versatile OM receptor with high affinity for haem uptake (Morton *et al.*, 2007).

Microarray studies: response to iron limitation

Whole-genome microarrays have been applied to understand *Pasteurellaceae* genome differences and pathogenesis at the transcriptome level, particularly in response to infection in the host or conditions that mimic the host environment with iron restriction as one of the common themes of study. Recently, Deslandes *et al.* (2007) reported the first microarray study of *A. pleuropneumoniae* and described the differential expression for 210 genes between iron-replete and iron-limiting conditions. Of the 118 genes that were down-regulated under iron restriction, the majority are involved with energy metabolism in association with the electron transport respiratory chain, of which several enzymes use haem molecules as cofactors, have Fe-S clusters or are activated by Fe^{2+}. Microarray studies by Paustian *et al.* (2001) described a similar effect with *P. multocida* in response to iron limitation with decreased expression of genes involved in energy metabolism and electron transport, and demonstrated a shift towards anaerobic metabolism. Interestingly, OmpW, an OM protein linked to broad adaptive functions under environmental stresses, was also significantly repressed in *A. pleuropneumoniae* (Deslandes *et al.*, 2007). Not surprisingly, genes involved in iron-transport were significantly represented among the 92 genes that were up-regulated to counter iron starvation. These included previously described genes of the *tonB1* system linked with *tbpA*, *hgbA* and the *hugZ* haem utilization gene upstream of *hgbA*, and *tonB2* gene. Consistent with observations by Mikael *et al.* (2003), expression of the *fhu* operon did not change under iron limitation.

Significant among the up-regulated genes were unidentified iron acquisition systems of *A. pleuropneumoniae* (Deslandes *et al.*, 2007). They also described a gene cluster that shared homology with the HmbR Hb receptor from *N. meningitidis* that demonstrates high affinity for Hb and that contributes to *N. meningitidis* survival in an infant rat infection model. These findings are similar to homologues that Chung *et al.* (2007) described in their *in silico* analyses that identified the presence of several unreported haem-related proteins in the OM of *A. pleuropneumoniae* serotype 5b. HgbA is the only cloned and characterized Hb receptor of *A. pleuropneumoniae* (Srikumar *et al.*, 2004). Mutations in *hgbA* verify that HgbA is the only receptor responsible for Hb uptake in *A. pleuropneumoniae*. However Archambault *et al.* (2003) had reported a 75-kDa haem- and Hb-binding protein that correlates with the predicted molecular weight of the HmbR homologue described by Deslandes *et al.* (2007). One reason that may explain the lack of Hb uptake from putative OM receptors other than HgbA is the absence of the *hemO* gene in *A. pleuropneumoniae* (Deslandes *et al.*, 2007). In *N. meningitidis*, *hmbR* gene is downstream of the *hemO* gene that encodes a haem oxygenase, essential for haem utilization (Zhu *et al.*, 2000). Alternatively, HgbA may be specific for Hb whereas these putative receptors share homology with proteins that bind to Hb, but recognize Hb complexed to other host proteins for specific haem uptake. Indeed, several of the homologues reported by Chung and coworkers were receptors that recognized Hb-haptoglobin or haem-haemopexin (Chung *et al.*, 2007).

Homologues to *yfeABCD*, a periplasmic binding protein-dependent transport system, were also up-regulated in *A. pleuropneumoniae* under iron restriction (Deslandes *et al.*, 2007). Together with the *yfeE* operon, this system is Fur-responsive and is responsible for chelated iron utilization in *E. coli* mutant lacking siderophore enterobactin. Iron-dependent induction of *yfe* genes has also been documented in microarray studies of *P. multocida*, *H. parasuis* and *H. influenzae*, and *M. haemolytica* (Paustian *et al.*, 2001; Melnikow *et al.*, 2005; Whitby *et al.*, 2006; Roehrig *et al.*, 2007). Whilst all genes of the *yfeABCD* operon are present in *A. pleuropneumoniae*, they are not clustered. Instead *yfeAB* is near two ORFs both annotated as Omp64, while *yfeCD* is located 160 bp downstream of the second ORF. Omp64 shares homology with CopB protein that is involved in iron acquisition from transferrin and lactoferrin by *Moraxella catarrhalis* (Aebi *et al.*, 1996). These data suggest components of novel iron acquisition in *A. pleuropneumoniae* comprised of Omp64 as the OM representative and YfeBCD as the IM counterparts.

Three other ORFs of unknown function were up-regulated but had homologies with periplasmic lipoproteins involved in iron transport, iron-dependent peroxidase, and high affinity Fe^{2+}/Pb^{2+} permeases (Deslandes *et al.*, 2007). The lipoprotein, OmlA, and three hypothetical OM proteins also had enhanced expression under iron-restricted conditions. Their functions and that of other hypothetical genes that were up-regulated, may be elucidated in the context of iron restriction, survival in the host environment and adaptation to stress.

Vaccines

Animal pathogens of *Pasteurellaceae* contribute to major economic losses in the livestock industry worldwide. In addition to losses attributed to animals that succumb to infection are the steep costs for the control and treatment of disease outbreaks. With pressures to reduce the use of antibiotics in agricultural livestock, vaccination against bacterial pathogens has emerged as a safer and more cost-effective approach for disease control. To date, a fully cross-protective and safe vaccine against all *A. pleuropneumoniae* serotypes does not exist. Pigs surviving natural or experimental infection with *A. pleuropneumoniae* develop immunity that protects them from future infections with homologous and heterologous serotypes (Nielsen, 1984; Rycroft and Garside, 2000). However, the antigens that confer such complete protection are still unknown. Investigation for such potential vaccine candidates involves consideration of criteria such as accessibility to host immune effectors, induction of humoral immunity, antigenic conservation, and *in vivo* expression, as determined by immunodetection with sera (Chakravarti *et al.*, 2000). *A. pleuropneumoniae* vaccines based on whole cell bacteria contain surface features that include capsule polysaccharides, LPS and OM proteins, but still fail to prevent disease or the development of lung lesions (Rycroft and Garside, 2000). Reasons that may explain the limitations of bacterins are the absence of Apx toxin production, and other virulence factors, some of which may be compromised by the inactivation process; preparation of bacterins by heat-killing, irradiation or chemical treatment may reduce or alter the antigenicity of the vaccine, spoiling relevant immunogenic epitopes (Huter *et al.*, 2000; Haesebrouck *et al.*, 2004). Furthermore, the expression of relevant antigens may be lost under the growth conditions used to generate whole-cell bacterins. This observation was demonstrated with *A. pleuropneumoniae* grown under β-NAD-replete and β-NAD-restricted conditions, in which the latter promoted the expression of fimbriae and a 55-kDa OM protein adhesin that enhanced adhesion of *A. pleuropneumoniae* to host cells (Van Overbeke *et al.*, 2002). Bacterins were produced by both growth conditions and then evaluated on their protective efficacy (Van Overbeke *et al.*, 2003). Although both bacterins induced partial protection against severe infection, animals immunized with the β-NAD-restricted bacterin had fewer lung lesions and the percentage of affected lung tissue was significantly lower than in animals immunized with β-NAD-replete bacterin (Van Overbeke *et al.*, 2003).

Because they can actively replicate *in vivo*, live attenuated vaccines are more immunogenic than killed bacterins; however the former presents underlying risks of safety and reversion to virulence (Ulmer *et al.*, 2006). Furthermore, construction of live vaccines requires a molecular understanding of pathogenicity of the bacterium to disrupt genes and antigens that facilitate its survival and render damage in the host. Molecular-based strategies such as STM have not only identified putative virulence genes of *A. pleuropneumoniae*, but have also provided a potential collection of live-attenuated mutants as vaccines. Fuller *et al.* (2000) evaluated seven STM mutants as live vaccines and demonstrated protection against homologous challenge. Recently, Maas and coworkers described the development of a sixfold mutant that evolved from a live negative marker vaccine of *A. pleuropneumoniae* for differentiation of infected and vaccinated animals (DIVA) (Maas *et al.*, 2006a, b). The DIVA vaccine is based on the absence of one immunogenic protein in the vaccine strain that can be used serologically to discriminate from wild-type strains, enabling routine maintenance of pathogen-free herds (Maas *et al.*, 2006a). The highly attenuated sixfold mutant had deletions in three genes for anaerobic respiration that promoted persistence, a deletion in the *fur* gene for the constitutive expression of immunogenic proteins, and deletions in previously described virulence genes, *apx II* and *ureC* that encode Apx toxin and urease respectively (Baltes *et al.*, 2001; Maas *et al.*, 2006a,b). The mutant was still able to colonize animals but without any clinical symptoms or tissue damage. Upon single aerosol application, animals were also protected from clinical symptoms with heterologous challenge.

As an alternative, subunit vaccines containing a few defined antigens have been favoured over the bacterin vaccines of *A. pleuropneumoniae*. Previous studies demonstrated that immunization with one immunogenic protein is not enough to prevent infection or lung damage by *A. pleuropneumoniae* (Gerlach *et al.*, 1993). Apx toxins, essential for *A. pleuropneumoniae* pathogenesis, can only induce partial protection on their own, although subunit vaccines containing Apx toxins with other antigens such as OM proteins, decrease clinical symptoms, enhance performance of animals and provide better cross-protection (Madsen *et al.*, 1995; Haesebrouck *et al.*, 1997, 2004). Antibodies to OM proteins, in particular, have been reported as potent opsonins for enhanced phagocytosis of *A. pleuropneumoniae* by porcine polymorphonuclear leukocytes (Thwaits and Kadis, 1991). More often, the identification of several OM proteins of *A. pleuropneumoniae* have described their antigenic conservation among serotypes and their efficacy as protective antigens. Studies that identified OM lipoprotein OmlA from serotypes 1 and 5, also demonstrated that the recombinant protein protected immunized pigs from death in homologous challenge (Gerlach *et al.*, 1993; Bunka *et al.*, 1995). Vaccines with both Apx toxin and Tbps appear to be the most promising subunit combination, performing better than bacterin vaccines, and sustaining minimal clinical signs, minor lung damage, and less mortality (Rossi-Campos *et al.*, 1992; Van Overbeke *et al.*, 2001).

Other approaches to cell-surface based vaccines include enrichment of protective antigens from cell-free culture supernatants (CFS). Goethe *et al.* (2000) described a method for a non-recombinant subunit vaccine that applied detergent sodium deoxycholate to extract OM lipoproteins without disrupting the integrity of the OM of *A. pleuropneumoniae* in broth cultures. Using lipoproteins OmlA, TbpB and integral OM protein TbpA as markers, the CFS were enriched in lipoproteins but free of periplasmic and cytoplasmic contaminants as well as integral OM proteins. Although individual antigenic components were not identified, pigs that were immunized with the CFS from sodium deoxycholate extract cultures demonstrated a strong humoral response that provided protection against *A. pleuropneumoniae* with mild clinical symptoms and less lung damage (Goethe *et al.*, 2000). Bacterial ghosts are another vaccine concept focused on preserving the antigenic properties of cell surface antigens. Expression of the bacteriophage phiX174 lysis gene E in Gram-negative bacteria results in the leakage of cytoplasmic contents, leaving behind an empty cell envelope or membranous ghost that represents the functional and antigenic structures of live bacteria. Pigs vaccinated with *A. pleuropneumoniae* ghosts or formalin-inactivated *A. pleuropneumoniae* were completely protected against challenge with the homologous strain. Moreover, bacteria could not be recovered from ghost-vaccinated pigs, eliminating the potential of asymptomatic carrier animals (Hensel *et al.*, 2000; Huter *et al.*, 2000). Comparisons between the immune sera from convalescent pigs and pigs that were immunized with *A. pleuropneumoniae* ghosts or formalin-inactivated bacteria showed that the ghost sera contain a unique set of antibodies that prevented colonization of *A. pleuropneumoniae*. In particular, a 100-kDa protein was more strongly recognized by ghost sera than convalescent sera, and was also detected by ghost sera of homologous and heterologous *A. pleuropneumoniae* serotypes, indicating cross-protective potential (Huter *et al.*, 2000). Interestingly, the ghost sera could not detect OM lipoproteins OmlA or Tbps that may have been subject to the preparation and growth conditions used to generate bacterial ghosts.

Current strategies for vaccine design now integrate the power of bioinformatics with available genome sequences. Using predictor programs, a genome-based selection of potential antigens provides a comprehensive inventory that can be further scrutinized prior to extensive experimental trials. This approach can be extended to comparing sequences between related bacteria to study genetic diversity and identify conserved antigens that would provide cross-protection. Whilst several predicted antigens would still require testing for their protective efficacies in relevant animal models, it cannot be denied that bioinformatic tools and high-throughput proteomics have expedited the discovery process for potential vaccine candidates (Chakravarti *et al.*, 2000). The success of this reverse vaccinology approach has been demonstrated with novel antigen discovery for *N. meningitidis*; genome-mining followed by expression and serological studies identified new antigens different from those detected by conventional approaches and exposed underrepresented and promising protein candidates localized at the bacterial cell surface (Pizza *et al.*, 2000; Rappuoli, 2001). Such an approach has been applied to identify putative vaccine candidates in *A. pleuropneumoniae* as well as *H. influenzae* (Chakravarti *et al.*, 2000; Chung *et al.*, 2007). Microarray studies with transcriptional profiling and comparative genomics can further distinguish conserved antigens from this dataset that are up-regulated during infection and provide cross-protection against all serotypes. Such collaborative efforts are predicted to refine the search for potential candidates for vaccine development against *A. pleuropneumoniae* and related pathogens.

Conclusions

Characterization of OM proteins of *A. pleuropneumoniae* and other *Pasteurellaceae* species demonstrate their importance in adaptation and survival in the host environment. The majority of these OM proteins are associated with iron uptake, a prominent mechanism in the lifestyle of these pathogens. The availability of genome sequences, advances in membrane proteomics and in structural biology, and complementary microarray studies will close the gap of unidentified OM proteins, elucidate their biological roles, and reveal those proteins that are essential for *in vivo* infection. These studies will ultimately provide insight into novel targets for the design of antimicrobial drugs and vaccines against pathogens of *Pasteurellaceae*.

Acknowledgements

Natural Sciences and Engineering Research Council (NSERC) of Canada awarded Strategic grant STPGP 306730-04 to J. W. Coulton and M. Jacques and also Discovery grant 003428 to M. Jacques. We acknowledge support from the NSERC-supported Research Network on Bacterial Pathogens of Swine and from the Centre de Recherche en Infectologie porcine (CRIP) financed by the Fonds de recherche sur la nature et technologies, Québec. We thank J.A. Kashul for editing the manuscript, and D.M. Carter for providing a template for Fig. 8.1.

References

Abeck, D., Johnson, A.P., and Mensing, H. (1992). Binding of *Haemophilus ducreyi* to extracellular matrix proteins. Microb. Pathog. *13*, 81–84.

Aebi, C., Stone, B., Beucher, M., Cope, L.D., Maciver, I., Thomas, S.E., McCracken, G.H., Jr., Sparling, P.F., and Hansen, E.J. (1996). Expression of the CopB outer membrane protein by *Moraxella catarrhalis* is regulated by iron and affects iron acquisition from transferrin and lactoferrin. Infect. Immun. *64*, 2024–2030.

Afonina, G., Leduc, I., Nepluev, I., Jeter, C., Routh, P., Almond, G., Orndorff, P.E., Hobbs, M., and Elkins, C. (2006). Immunization with the *Haemophilus ducreyi* hemoglobin receptor HgbA protects against infection in the swine model of chancroid. Infect. Immun. *74*, 2224–2232.

Archambault, M., Labrie, J., Rioux, C.R., Dumas, F., Thibault, P., Elkins, C., and Jacques, M. (2003). Identification and preliminary characterization of a 75-kDa hemin- and hemoglobin-binding outer membrane protein of *Actinobacillus pleuropneumoniae* serotype 1. Can. J. Vet. Res. *67*, 271–277.

Archambault, M., Rioux, S., and Jacques, M. (1999). Evaluation of the hemoglobin-binding activity of *Actinobacillus pleuropneumoniae* using fluorescein-labeled pig hemoglobin and flow cytometry. FEMS Microbiol. Lett. *173*, 17–25.

Babu, M.M., Priya, M.L., Selvan, A.T., Madera, M., Gough, J., Aravind, L., and Sankaran, K. (2006). A database of bacterial lipoproteins (DOLOP) with functional assignments to predicted lipoproteins. J. Bacteriol. *188*, 2761–2773.

Bahrami, F., and Niven, D.F. (2005). Iron acquisition by *Actinobacillus suis*: identification and characterization of a single-component haemoglobin receptor and encoding gene. Microb. Pathog. *39*, 45–51.

Bahrami, F., Ekins, A., and Niven, D.F. (2003). Iron acquisition by *Actinobacillus suis*: identification and characterization of transferrin receptor proteins and encoding genes. Vet. Microbiol. *94*, 79–92.

Baltes, N., and Gerlach, G.F. (2004). Identification of genes transcribed by *Actinobacillus pleuropneumoniae* in necrotic porcine lung tissue by using selective capture of transcribed sequences. Infect. Immun. *72*, 6711–6716.

Baltes, N., Hennig-Pauka, I., and Gerlach, G.F. (2002). Both transferrin binding proteins are virulence factors in *Actinobacillus pleuropneumoniae* serotype 7 infection. FEMS Microbiol. Lett. *209*, 283–287.

Baltes, N., Tonpitak, W., Gerlach, G.F., Hennig-Pauka, I., Hoffmann-Moujahid, A., Ganter, M., and Rothkötter, H.J. (2001). *Actinobacillus pleuropneumoniae* iron transport and urease activity: effects on bacterial virulence and host immune response. Infect. Immun. *69*, 472–478.

Baltes, N., Tonpitak, W., Hennig-Pauka, I., Gruber, A.D., and Gerlach, G.F. (2003). *Actinobacillus pleuropneumoniae* serotype 7 siderophore receptor FhuA is not required for virulence. FEMS Microbiol. Lett. *220*, 41–48.

Barnich, N., Bringer, M.A., Claret, L., and Darfeuille-Michaud, A. (2004). Involvement of lipoprotein NlpI in the virulence of adherent invasive *Escherichia coli* strain LF82 isolated from a patient with Crohn's disease. Infect. Immun. *72*, 2484–2493.

Beddek, A.J., Sheehan, B.J., Bossé, J.T., Rycroft, A.N., Kroll, J.S., and Langford, P.R. (2004). Two TonB systems in *Actinobacillus pleuropneumoniae*: their roles in iron acquisition and virulence. Infect. Immun. *72*, 701–708.

Bélanger, M., Bégin, C., and Jacques, M. (1995). Lipopolysaccharides of *Actinobacillus pleuropneumoniae* bind pig hemoglobin. Infect. Immun. *63*, 656–662.

Bélanger, M., Dubreuil, D., Harel, J., Girard, C., and Jacques, M. (1990). Role of lipopolysaccharides in adherence of *Actinobacillus pleuropneumoniae* to porcine tracheal rings. Infect. Immun. *58*, 3523–3530.

Berenson, C.S., Murphy, T.F., Wrona, C.T., and Sethi, S. (2005). Outer membrane protein P6 of nontypeable *Haemophilus influenzae* is a potent and selective inducer of human macrophage proinflammatory cytokines. Infect. Immun. *73*, 2728–2735.

Blackall, P.J., Klaasen, H.L., van den Bosch, H., Kuhnert, P., and Frey, J. (2002). Proposal of a new serovar of *Actinobacillus pleuropneumoniae*: serovar 15. Vet. Microbiol. *84*, 47–52.

Bos, M.P. and Tommassen, J. (2004). Biogenesis of the Gram-negative bacterial outer membrane. Curr. Opin. Microbiol. *7*, 610–616.

Bosch, M., Garrido, M.E., Llagostera, M., Pérez de Rozas, A.M., Badiola, I., and Barbé, J. (2002). Characterization of the *Pasteurella multocida hgbA* gene encoding a hemoglobin-binding protein. Infect. Immun. *70*, 5955–5964.

Bosch, M., Garrido, M.E., Pérez de Rozas, A.M., Badiola, I., Barbé, J., and Llagostera, M. (2004). *Pasteurella multocida* contains multiple immunogenic haemin- and haemoglobin-binding proteins. Vet. Microbiol. *99*, 103–112.

Bossé, J.T., Janson, H., Sheehan, B.J., Beddek, A.J., Rycroft, A.N., Kroll, J.S., and Langford, P.R. (2002). *Actinobacillus pleuropneumoniae*: pathobiology and pathogenesis of infection. Microbes. Infect. *4*, 225–235.

Bouveret, E., Bénédetti, H., Rigal, A., Loret, E., and Lazdunski, C. (1999). In vitro characterization of peptidoglycan-associated lipoprotein (PAL)-peptidoglycan and PAL–TolB interactions. J. Bacteriol. *181*, 6306–6311.

Boyce, J.D., Cullen, P.A., Nguyen, V., Wilkie, I., and Adler, B. (2006). Analysis of the *Pasteurella multocida* outer membrane sub-proteome and its response to the *in vivo* environment of the natural host. Proteomics. *6*, 870–880.

Bracken, C.S., Baer, M.T., Abdur-Rashid, A., Helms, W., and Stojiljkovic, I. (1999). Use of heme-protein complexes by the *Yersinia enterocolitica* HemR receptor: histidine residues are essential for receptor function. J. Bacteriol. *181*, 6063–6072.

Bunka, S., Christensen, C., Potter, A.A., Willson, P.J., and Gerlach, G.F. (1995). Cloning and characterization of a protective outer membrane lipoprotein of *Actinobacillus pleuropneumoniae* serotype 5. Infect. Immun. *63*, 2797–2800.

Chai, T.J., and Foulds, J. (1978). Two bacteriophages which utilize a new *Escherichia coli* major outer membrane protein as part of their receptor. J. Bacteriol. *135*, 164–170.

Chakravarti, D.N., Fiske, M.J., Fletcher, L.D., and Zagursky, R.J. (2000). Application of genomics and proteomics for identification of bacterial gene products as potential vaccine candidates. Vaccine *19*, 601–612.

Chiers, K., Van Overbeke, I., De Laender, P., Ducatelle, R., Carel, S., and Haesebrouck, F. (1998). Effects of endobronchial challenge with *Actinobacillus pleuropneumoniae* serotype 9 of pigs vaccinated with inactivated vaccines containing the Apx toxins. Vet. Q. *20*, 65–69.

Chin, N., Frey, J., Chang, C.F., and Chang, Y.F. (1996). Identification of a locus involved in the utilization of iron by *Actinobacillus pleuropneumoniae*. FEMS Microbiol. Lett. *143*, 1–6.

Cho, W.S. and Chae, C. (2003). Differentiation of twelve *Actinobacillus pleuropneumoniae* serotypes by outer membrane lipoprotein gene-based restriction fragment length polymorphism. J. Vet. Med. B Infect. Dis. Vet. Public Health *50*, 90–94.

Chung, J.W., Ng-Thow-Hing, C., Budman, L.I., Gibbs, B.F., Nash, J.H., Jacques, M., and Coulton, J.W. (2007). Outer membrane proteome of *Actinobacillus pleuropneumoniae*: LC-MS/MS analyses validate in silico predictions. Proteomics. *7*, 1854–1865.

Cope, L.D., Hrkal, Z., and Hansen, E.J. (2000). Detection of phase variation in expression of proteins involved in hemoglobin and hemoglobin-haptoglobin binding by nontypeable *Haemophilus influenzae*. Infect. Immun. *68*, 4092–4101.

Cope, L.D., Love, R.P., Guinn, S.E., Gilep, A., Usanov, S., Estabrook, R.W., Hrkal, Z., and Hansen, E.J. (2001). Involvement of HxuC outer membrane protein in utilization of hemoglobin by *Haemophilus influenzae*. Infect. Immun. *69*, 2353–2363.

Cope, L.D., Yogev, R., Muller-Eberhard, U., and Hansen, E.J. (1995). A gene cluster involved in the utilization of both free heme and heme:hemopexin by *Haemophilus influenzae* type b. J. Bacteriol. *177*, 2644–2653.

Cornelissen, C.N. (2003). Transferrin-iron uptake by Gram-negative bacteria. Front Biosci. *8*, d836-d847.

Cotter, S.E., Yeo, H.J., Juehne, T., and St Geme, J.W., III (2005). Architecture and adhesive activity of the *Haemophilus influenzae* Hsf adhesin. J. Bacteriol. *187*, 4656–4664.

Cruz, W.T., Nedialkov, Y.A., Thacker, B.J., and Mulks, M.H. (1996). Molecular characterization of a common 48-kilodalton outer membrane protein of *Actinobacillus pleuropneumoniae*. Infect. Immun. *64*, 83–90.

D'Silva, C.G., Archibald, F.S., and Niven, D.F. (1995). Comparative study of iron acquisition by biotype 1 and biotype 2 strains of *Actinobacillus pleuropneumoniae*. Vet. Microbiol. *44*, 11–23.

Dabo, S.M., Confer, A.W., and Quijano-Blas, R.A. (2003). Molecular and immunological characterization of *Pasteurella multocida* serotype A:3 OmpA: evidence of its role in *P. multocida* interaction with extracellular matrix molecules. Microb. Pathog. *35*, 147–157.

Datta, D.B., Arden, B., and Henning, U. (1977). Major proteins of the *Escherichia coli* outer cell envelope membrane as bacteriophage receptors. J. Bacteriol. *131*, 821–829.

del Rio M.L., Navas, J., Martin, A.J., Gutiérrez, C.B., Rodriguez-Barbosa J.L., and Rodriguez-Ferri E.F. (2006). Molecular characterization of *Haemophilus parasuis* ferric hydroxamate uptake (*fhu*) genes and constitutive expression of the FhuA receptor. Vet. Res. *37*, 49–59.

Deneer, H.G., and Potter, A.A. (1989a). Effect of iron restriction on the outer membrane proteins of *Actinobacillus* (*Haemophilus*) *pleuropneumoniae*. Infect. Immun. *57*, 798–804.

Deneer, H.G., and Potter, A.A. (1989b). Identification of a maltose-inducible major outer membrane protein in *Actinobacillus* (*Haemophilus*) *pleuropneumoniae*. Microb. Pathog. *6*, 425–432.

Deslandes, V., Nash, J.H., Harel, J., Coulton, J.W., and Jacques, M. (2007). Transcriptional profiling of *Actinobacillus pleuropneumoniae* under iron-restricted conditions. BMC. Genomics *8*, 72.

Diarra, M.S., Dolence, J.A., Dolence, E.K., Darwish, I., Miller, M.J., Malouin, F., and Jacques, M. (1996). Growth of *Actinobacillus pleuropneumoniae* is promoted by exogenous hydroxamate and catechol siderophores. Appl. Environ. Microbiol. *62*, 853–859.

Dom, P., Haesebrouck, F., Ducatelle, R., and Charlier, G. (1994). *In vivo* association of *Actinobacillus pleuropneumoniae* serotype 2 with the respiratory epithelium of pigs. Infect. Immun. *62*, 1262–1267.

Dubreuil, J.D., Jacques, M., Mittal, K.R., and Gottschalk, M. (2000). *Actinobacillus pleuropneumoniae* surface polysaccharides: their role in diagnosis and immunogenicity. Anim. Health Res. Rev. *1*, 73–93.

Duim, B., Bowler, L.D., Eijk, P.P., Jansen, H.M., Dankert, J., and van Alphen, L. (1997). Molecular variation in the major outer membrane protein P5 gene of non-encapsulated *Haemophilus influenzae* during chronic infections. Infect. Immun. *65*, 1351–1356.

Ecevit, I.Z., McCrea, K.W., Pettigrew, M.M., Sen, A., Marrs, C.F., and Gilsdorf, J.R. (2004). Prevalence of the *hifBC*, *hmw1A*, *hmw2A*, *hmwC*, and *hia* Genes in *Haemophilus influenzae* Isolates. J. Clin. Microbiol. *42*, 3065–3072.

Enríquez-Verdugo, I., Guerrero, A.L., Serrano, J.J., Godínez, D., Rosales, J.L., Tenorio, V., and de la Garza, M. (2004). Adherence of *Actinobacillus pleuropneumoniae* to swine-lung collagen. Microbiology *150*, 2391–2400.

Faraldo-Gómez, J.D. and Sansom, M.S. (2003). Acquisition of siderophores in gram-negative bacteria. Nat. Rev. Mol. Cell Biol. *4*, 105–116.

Fink, D.L., Green, B.A., and St Geme, J.W., III (2002). The *Haemophilus influenzae* Hap autotransporter binds to fibronectin, laminin, and collagen IV. Infect. Immun. *70*, 4902–4907.

Fleischmann, R.D., Adams, M.D., White, O., Clayton, R.A., Kirkness, E.F., Kerlavage, A.R., Bult, C.J., Tomb, J.F., Dougherty, B.A., and Merrick, J.M. (1995). Whole-genome random sequencing and assembly of *Haemophilus influenzae* Rd. Science *269*, 496–512.

Frey, J. (1995). Virulence in *Actinobacillus pleuropneumoniae* and RTX toxins. Trends Microbiol. *3*, 257–261.

Frey, J., Kuhnert, P., Villiger, L., and Nicolet, J. (1996). Cloning and characterization of an *Actinobacillus pleuropneumoniae* outer membrane protein belonging to the family of PAL lipoproteins. Res. Microbiol. *147*, 351–361.

Fuller, C.A., Yu, R., Irwin, S.W., and Schryvers, A.B. (1998). Biochemical evidence for a conserved interaction between bacterial transferrin binding protein A and transferrin binding protein B. Microb. Pathog. *24*, 75–87.

Fuller, T.E., Martin, S., Teel, J.F., Alaniz, G.R., Kennedy, M.J., and Lowery, D.E. (2000). Identification of *Actinobacillus pleuropneumoniae* virulence genes using signature-tagged mutagenesis in a swine infection model. Microb. Pathog. *29*, 39–51.

Genco, C.A., and Dixon, D.W. (2001). Emerging strategies in microbial haem capture. Mol. Microbiol. *39*, 1–11.

Gerlach, G.F., Anderson, C., Klashinsky, S., Rossi-Campos, A., Potter, A.A., and Willson, P.J. (1993). Molecular characterization of a protective outer membrane lipoprotein (OmlA) from *Actinobacillus pleuropneumoniae* serotype 1. Infect. Immun. *61*, 565–572.

Gerlach, G.F., Anderson, C., Potter, A.A., Klashinsky, S., and Willson, P.J. (1992a). Cloning and expression of a transferrin-binding protein from *Actinobacillus pleuropneumoniae*. Infect. Immun. *60*, 892–898.

Gerlach, G.F., Klashinsky, S., Anderson, C., Potter, A.A., and Willson, P.J. (1992b). Characterization of two genes encoding distinct transferrin-binding proteins in different *Actinobacillus pleuropneumoniae* isolates. Infect. Immun. *60*, 3253–3261.

Gioia, J., Qin, X., Jiang, H., Clinkenbeard, K., Lo, R., Liu, Y., Fox, G.E., Yerrapragada, S., McLeod, M.P., McNeill, T.Z., Hemphill, L., Sodergren, E., Wang,

Q., Muzny, D.M., Homsi, F.J., Weinstock, G.M., and Highlander, S.K. (2006). The genome sequence of *Mannheimia haemolytica* A1: insights into virulence, natural competence, and *Pasteurellaceae* phylogeny. J. Bacteriol. *188*, 7257–7266.

Goethe, R., Gonzáles, O.F., Lindner, T., and Gerlach, G.F. (2000). A novel strategy for protective *Actinobacillus pleuropneumoniae* subunit vaccines: detergent extraction of cultures induced by iron restriction. Vaccine *19*, 966–975.

Gonzalez, G.C., Caamano, D.L., and Schryvers, A.B. (1990). Identification and characterization of a porcine-specific transferrin receptor in *Actinobacillus pleuropneumoniae*. Mol. Microbiol. *4*, 1173–1179.

Gonzalez, G.C., Yu, R.H., Rosteck, P.R., Jr., and Schryvers, A.B. (1995). Sequence, genetic analysis, and expression of *Actinobacillus pleuropneumoniae* transferrin receptor genes. Microbiology *141 (Pt 10)*, 2405–2416.

Gram, T. and Ahrens, P. (1998). Improved diagnostic PCR assay for *Actinobacillus pleuropneumoniae* based on the nucleotide sequence of an outer membrane lipoprotein. J. Clin. Microbiol. *36*, 443–448.

Haesebrouck, F., Chiers, K., Van Overbeke, I., and Ducatelle, R. (1997). *Actinobacillus pleuropneumoniae* infections in pigs: the role of virulence factors in pathogenesis and protection. Vet. Microbiol. *58*, 239–249.

Haesebrouck, F., Pasmans, F., Chiers, K., Maes, D., Ducatelle, R., and Decostere, A. (2004). Efficacy of vaccines against bacterial diseases in swine: what can we expect? Vet. Microbiol. *100*, 255–268.

Hallström, T., Trajkovska, E., Forsgren, A., and Riesbeck, K. (2006). *Haemophilus influenzae* surface fibrils contribute to serum resistance by interacting with vitronectin. J. Immunol. *177*, 430–436.

Hardy, G.G., Tudor, S.M., and St Geme, J.W., III (2003). The pathogenesis of disease due to nontypeable *Haemophilus influenzae*. Methods Mol. Med. *71*, 1–28.

Heise, T. and Dersch, P. (2006). Identification of a domain in *Yersinia* virulence factor YadA that is crucial for extracellular matrix-specific cell adhesion and uptake. Proc. Natl. Acad. Sci. USA *103*, 3375–3380.

Henderson, D.P., Wyckoff, E.E., Rashidi, C.E., Verlei, H., and Oldham, A.L. (2001). Characterization of the *Plesiomonas shigelloides* genes encoding the heme iron utilization system. J. Bacteriol. *183*, 2715–2723.

Hennig, I., Teutenberg-Riedel, B., and Gerlach, G.F. (1999). Downregulation of a protective *Actinobacillus pleuropneumoniae* antigen during the course of infection. Microb. Pathog. *26*, 53–63.

Hensel, A., Huter, V., Katinger, A., Raza, P., Strnistschie, C., Roesler, U., Brand, E., and Lubitz, W. (2000). Intramuscular immunization with genetically inactivated (ghosts) *Actinobacillus pleuropneumoniae* serotype 9 protects pigs against homologous aerosol challenge and prevents carrier state. Vaccine *18*, 2945–2955.

Hodgetts, A., Bossé, J.T., Kroll, J.S., and Langford, P.R. (2004). Analysis of differential protein expression in *Actinobacillus pleuropneumoniae* by Surface Enhanced Laser Desorption Ionisation--ProteinChip (SELDI) technology. Vet. Microbiol. *99*, 215–225.

Hotomi, M., Yamanaka, N., Saito, T., Shimada, J., Suzumoto, M., Suetake, M., and Faden, H. (1999). Antibody responses to the outer membrane protein P6 of non-typeable *Haemophilus influenzae* and pneumococcal capsular polysaccharides in otitis-prone children. Acta Otolaryngol. *119*, 703–707.

Hsu, Y.M., Chin, N., Chang, C.F., and Chang, Y.F. (2003). Cloning and characterization of the *Actinobacillus pleuropneumoniae fur* gene and its role in regulation of ApxI and AfuABC expression. DNA Seq. *14*, 169–181.

Huter, V., Hensel, A., Brand, E., and Lubitz, W. (2000). Improved protection against lung colonization by *Actinobacillus pleuropneumoniae* ghosts: characterization of a genetically inactivated vaccine. J. Biotechnol. *83*, 161–172.

Ito, H., Osaki, M., Uchida, I., Ohya, T., and Sekizaki, T. (1998). Demonstration of the third antigenically distinct outer membrane lipoprotein (OmlA) in *Actinobacillus pleuropneumoniae* serotype 7. FEMS Microbiol. Lett. *167*, 303–308.

Ito, H., Uchida, I., Sekizaki, T., Ooishi, E., Kawai, T., Okabe, T., Taneno, A., and Terakado, N. (1995). Molecular cloning of an *Actinobacillus pleuropneumoniae* outer membrane lipoprotein (OmlA) from serotype 5a. Microb. Pathog. *18*, 29–36.

Jacobsen, I., Gerstenberger, J., Gruber, A.D., Bossé, J.T., Langford, P.R., Hennig-Pauka, I., Meens, J., and Gerlach, G.F. (2005a). Deletion of the ferric uptake regulator Fur impairs the *in vitro* growth and virulence of *Actinobacillus pleuropneumoniae*. Infect. Immun. *73*, 3740–3744.

Jacobsen, I.D., Meens, J., Baltes, N., and Gerlach, G.F. (2005b). Differential expression of non-cytoplasmic *Actinobacillus pleuropneumoniae* proteins induced by addition of bronchoalveolar lavage fluid. Vet. Microbiol. *109*, 245–256.

Jacques, M. (2004). Surface polysaccharides and iron-uptake systems of *Actinobacillus pleuropneumoniae*. Can. J. Vet. Res. *68*, 81–85.

Jacques, M., Foiry, B., Higgins, R., and Mittal, K.R. (1988). Electron microscopic examination of capsular material from various serotypes of *Actinobacillus pleuropneumoniae*. J. Bacteriol. *170*, 3314–3318.

Jacques, M. and Paradis, S.E. (1998). Adhesin-receptor interactions in *Pasteurellaceae*. FEMS Microbiol. Rev. *22*, 45–59.

Kasten, R.W., Hansen, L.M., Hinojoza, J., Bieber, D., Ruehl, W.W., and Hirsh, D.C. (1995). *Pasteurella multocida* produces a protein with homology to the P6 outer membrane protein of *Haemophilus influenzae*. Infect. Immun. *63*, 989–993.

Koretke, K.K., Szczesny, P., Gruber, M., and Lupas, A.N. (2006). Model structure of the prototypical non-fimbrial adhesin YadA of *Yersinia enterocolitica*. J. Struct. Biol. *155*, 154–161.

Lin, J., Huang, S., and Zhang, Q. (2002). Outer membrane proteins: key players for bacterial adaptation in host niches. Microbes. Infect. *4*, 325–331.

Litt, D.J., Palmer, H.M., and Borriello, S.P. (2000). *Neisseria meningitidis* expressing transferrin binding proteins of *Actinobacillus pleuropneumoniae* can utilize porcine transferrin for growth. Infect. Immun. *68*, 550–557.

Loosmore, S.M., Yang, Y.P., Coleman, D.C., Shortreed, J.M., England, D.M., Harkness, R.E., Chong, P.S., and Klein, M.H. (1996). Cloning and expression of the *Haemophilus influenzae* transferrin receptor genes. Mol. Microbiol. *19*, 575–586.

Maas, A., Jacobsen, I.D., Meens, J., and Gerlach, G.F. (2006a). Use of an *Actinobacillus pleuropneumoniae* multiple mutant as a vaccine that allows differentiation of vaccinated and infected animals. Infect. Immun. *74*, 4124–4132.

Maas, A., Meens, J., Baltes, N., Hennig-Pauka, I., and Gerlach, G.F. (2006b). Development of a DIVA subunit vaccine against *Actinobacillus pleuropneumoniae* infection. Vaccine *24*, 7226–7237.

Madsen, M.E., Carnahan, K.G., and Thwaits, R.N. (1995). Evaluation of pig lungs following an experimental challenge with *Actinobacillus pleuropneumoniae* serotype 1 and 5 in pigs inoculated with either hemolysin protein and/or outer membrane proteins. FEMS Microbiol. Lett. *131*, 329–335.

Mahasreshti, P.J., Murphy, G.L., Wyckoff, J.H., III, Farmer, S., Hancock, R.E., and Confer, A.W. (1997). Purification and partial characterization of the OmpA family of proteins of *Pasteurella haemolytica*. Infect. Immun. *65*, 211–218.

Medrano, A., Querol, E., and Daban, M. (1997). Cloning and sequencing of the gene encoding the outer-membrane protein Tbp1 from *Actinobacillus pleuropneumoniae*. Expression of Tbp1 and Tbp2. Behring Inst. Mitt. 410–423.

Melnikow, E., Dornan, S., Sargent, C., Duszenko, M., Evans, G., Gunkel, N., Selzer, P.M., and Ullrich, H.J. (2005). Microarray analysis of *Haemophilus parasuis* gene expression under *in vitro* growth conditions mimicking the in vivo environment. Vet. Microbiol. *110*, 255–263.

Mikael, L.G., Pawelek, P.D., Labrie, J., Sirois, M., Coulton, J.W., and Jacques, M. (2002). Molecular cloning and characterization of the ferric hydroxamate uptake (*fhu*) operon in *Actinobacillus pleuropneumoniae*. Microbiology *148*, 2869–2882.

Mikael, L.G., Srikumar, R., Coulton, J.W., and Jacques, M. (2003). *fhuA* of *Actinobacillus pleuropneumoniae* encodes a ferrichrome receptor but is not regulated by iron. Infect. Immun. *71*, 2911–2915.

Mintz, K.P. (2004). Identification of an extracellular matrix protein adhesin, EmaA, which mediates the adhesion of *Actinobacillus actinomycetemcomitans* to collagen. Microbiology *150*, 2677–2688.

Mintz, K.P. and Fives-Taylor, P.M. (1999). Binding of the periodontal pathogen *Actinobacillus actinomycetemcomitans* to extracellular matrix proteins. Oral Microbiol. Immunol. *14*, 109–116.

Molloy, M.P., Herbert, B.R., Slade, M.B., Rabilloud, T., Nouwens, A.S., Williams, K.L., and Gooley, A.A. (2000). Proteomic analysis of the *Escherichia coli* outer membrane. Eur. J. Biochem. *267*, 2871–2881.

Morton, D.J., Seale, T.W., Madore, L.L., VanWagoner, T.M., Whitby, P.W., and Stull, T.L. (2007). The haem-haemopexin utilization gene cluster (*hxuCBA*) as a virulence factor of *Haemophilus influenzae*. Microbiology *153*, 215–224.

Morton, D.J., Smith, A., Ren, Z., Madore, L.L., VanWagoner, T.M., Seale, T.W., Whitby, P.W., and Stull, T.L. (2004). Identification of a haem-utilization protein (Hup) in *Haemophilus influenzae*. Microbiology *150*, 3923–3933.

Morton, D.J., Whitby, P.W., Jin, H., Ren, Z., and Stull, T.L. (1999). Effect of multiple mutations in the hemoglobin- and hemoglobin-haptoglobin-binding proteins, HgpA, HgpB, and HgpC, of *Haemophilus influenzae* type b. Infect. Immun. *67*, 2729–2739.

Munson, R.S., Jr. and Granoff, D.M. (1985). Purification and partial characterization of outer membrane proteins P5 and P6 from *Haemophilus influenzae* type b. Infect. Immun. *49*, 544–549.

Murphy, T.F., Kirkham, C., and Lesse, A.J. (2006). Construction of a mutant and characterization of the role of the vaccine antigen P6 in outer membrane integrity of nontypeable *Haemophilus influenzae*. Infect. Immun. *74*, 5169–5176.

Narita, S., Matsuyama, S., and Tokuda, H. (2004). Lipoprotein trafficking in *Escherichia coli*. Arch. Microbiol. *182*, 1–6.

Negrete-Abascal, E., Tenorio, V.R., Serrano, J.J., Garcia, C., and de la Garza, M. (1994). Secreted proteases from *Actinobacillus pleuropneumoniae* serotype 1 degrade porcine gelatin, hemoglobin and immunoglobulin A. Can. J. Vet. Res. *58*, 83–86.

Neilands, J.B. (1984). Siderophores of bacteria and fungi. Microbiol. Sci. *1*, 9–14.

Nelson, M.B., Murphy, T.F., Vankeulen, H., Rekosh, D., and Apicella, M.A. (1988). Studies on P6, an important outer-membrane protein antigen of *Haemophilus influenzae*. Rev. Infect. Dis. *10 Suppl 2*, S331-S336.

Nielsen, R. (1984). *Haemophilus pleuropneumoniae* serotypes--cross protection experiments. Nord. Vet. Med. *36*, 221–234.

Niven, D.F., Donga, J., and Archibald, F.S. (1989). Responses of *Haemophilus pleuropneumoniae* to iron restriction: changes in the outer membrane protein profile and the removal of iron from porcine transferrin. Mol. Microbiol. 3, 1083–1089.

O'Reilly, T., Niven, D.F., and Brown, M.R. (1991). Phenotypic variation in the outer membrane protein composition of *Actinobacillus* (*Haemophilus*) *pleuropneumoniae*: non-specific effect of exogenous pyridine nucleotide supply. Vet. Microbiol. *29*, 159–172.

Ogunnariwo, J.A. and Schryvers, A.B. (2001). Characterization of a novel transferrin receptor in bovine strains of *Pasteurella multocida*. J. Bacteriol. *183*, 890–896.

Ogunnariwo, J.A., and Woo, T.K., Lo, R.Y., Gonzalez G.C., Schryvers, A.B. (1997). Characterization of the *Pasteurella haemolytica* transferrin receptor genes and the recombinant receptor proteins. Microb. Pathog. *23*, 273–284.

Ohara, M., Wu, H.C., Sankaran, K., and Rick, P.D. (1999). Identification and characterization of a new lipoprotein, NlpI, in *Escherichia coli* K-12. J. Bacteriol. *181*, 4318–4325.

Osaki, M., Sato, Y., Tomura, H., Ito, H., and Sekizaki, T. (1997). Genetic diversity of the genes encoding the outer membrane lipoprotein (*omlA*) of *Actinobacillus pleuropneumoniae*. J. Vet. Med. Sci. *59*, 213–215.

Pandher, K., Confer, A.W., and Murphy, G.L. (1998). Genetic and immunologic analyses of PlpE, a lipoprotein important in complement-mediated killing of

Pasteurella haemolytica serotype 1. Infect. Immun. *66*, 5613–5619.

Paul-Satyaseela, M., Karched, M., Bian, Z., Ihalin, R., Borén, T., Arnqvist, A., Chen, C., and Asikainen, S. (2006). Immunoproteomics of *Actinobacillus actinomycetemcomitans* outer-membrane proteins reveal a highly immunoreactive peptidoglycan-associated lipoprotein. J. Med. Microbiol. *55*, 931–942.

Paustian, M.L., May, B.J., and Kapur, V. (2001). *Pasteurella multocida* gene expression in response to iron limitation. Infect. Immun. *69*, 4109–4115.

Pawelek, P.D. and Coulton, J.W. (2004). Hemoglobin-binding protein HgbA in the outer membrane of *Actinobacillus pleuropneumoniae*: homology modelling reveals regions of potential interactions with hemoglobin and heme. J. Mol. Graph. Model. *23*, 211–221.

Pizza, M., Scarlato, V., Masignani, V., Giuliani, M.M., Arico, B., Comanducci, M., Jennings, G.T., Baldi, L., Bartolini, E., Capecchi, B., Galeotti, C.L., Luzzi, E., Manetti, R., Marchetti, E., Mora, M., Nuti, S., Ratti, G., Santini, L., Savino, S., Scarselli, M., Storni, E., Zuo, P., Broeker, M., Hundt, E., Knapp, B., Blair, E., Mason, T., Tettelin, H., Hood, D.W., Jeffries, A.C., Saunders, N.J., Granoff, D.M., Venter, J.C., Moxon, E.R., Grandi, G., and Rappuoli, R. (2000). Identification of vaccine candidates against serogroup B *meningococcus* by whole-genome sequencing. Science *287*, 1816–1820.

Pouttu, R., Puustinen, T., Virkola, R., Hacker, J., Klemm, P., and Korhonen, T.K. (1999). Amino acid residue Ala-62 in the FimH fimbrial adhesin is critical for the adhesiveness of meningitis-associated *Escherichia coli* to collagens. Mol. Microbiol. *31*, 1747–1757.

Prasadarao, N.V., Blom, A.M., Villoutreix, B.O., and Linsangan, L.C. (2002). A novel interaction of outer membrane protein A with C4b binding protein mediates serum resistance of *Escherichia coli* K1. J. Immunol. *169*, 6352–6360.

Rapp, V.J., Munson, R.S., Jr., and Ross, R.F. (1986). Outer membrane protein profiles of *Haemophilus pleuropneumoniae*. Infect. Immun. *52*, 414–420.

Rappuoli, R. (2001). Reverse vaccinology, a genome-based approach to vaccine development. Vaccine *19*, 2688–2691.

Ratledge, C., and Dover, L.G. (2000). Iron metabolism in pathogenic bacteria. Annu. Rev. Microbiol. *54*, 881–941.

Ricard, M.A., Archibald, F.S., and Niven, D.F. (1991). Isolation and identification of a putative porcine transferrin receptor from *Actinobacillus pleuropneumoniae* biotype 1. J. Gen. Microbiol. *137*, 2733–2740.

Roehrig, S.C., Tran, H.Q., Spehr, V., Gunkel, N., Selzer, P.M., and Ullrich, H.J. (2007). The response of *Mannheimia haemolytica* to iron limitation: implications for the acquisition of iron in the bovine lung. Vet. Microbiol. *121*, 316–329.

Rosendal, S., and MacInnes, J.I. (1990). Characterization of an attenuated strain of *Actinobacillus pleuropneumoniae*, serotype 1. Am. J. Vet. Res. *51*, 711–717.

Rossi-Campos, A., Anderson, C., Gerlach, G.F., Klashinsky, S., Potter, A.A., and Willson, P.J. (1992). Immunization of pigs against *Actinobacillus pleuropneumoniae* with two recombinant protein preparations. Vaccine *10*, 512–518.

Ruiz, N., Kahne, D., and Silhavy, T.J. (2006). Advances in understanding bacterial outer-membrane biogenesis. Nat. Rev. Microbiol. *4*, 57–66.

Rycroft, A.N., and Garside, L.H. (2000). *Actinobacillus* species and their role in animal disease. Vet. J. *159*, 18–36.

Rycroft, A.N., and Taylor, D.J. (1987). Preparation and characterisation of envelope proteins from *Haemophilus pleuropneumoniae*. Vet. Microbiol. *15*, 303–314.

Santoni, V., Molloy, M., and Rabilloud, T. (2000). Membrane proteins and proteomics: un amour impossible? Electrophoresis *21*, 1054–1070.

Sebunya, T.N., and Saunders, J.R. (1983). *Haemophilus pleuropneumoniae* infection in swine: a review. J. Am. Vet. Med. Assoc. *182*, 1331–1337.

Shakarji, L., Mikael, L.G., Srikumar, R., Kobisch, M., Coulton, J.W., and Jacques, M. (2006). Fhua and HgbA, outer membrane proteins of *Actinobacillus pleuropneumoniae*: their role as virulence determinants. Can. J. Microbiol. *52*, 391–396.

Sheehan, B.J., Bossé, J.T., Beddek, A.J., Rycroft, A.N., Kroll, J.S., and Langford, P.R. (2003). Identification of *Actinobacillus pleuropneumoniae* genes important for survival during infection in its natural host. Infect. Immun. *71*, 3960–3970.

Sonntag, I., Schwarz, H., Hirota, Y., and Henning, U. (1978). Cell envelope and shape of *Escherichia coli*: multiple mutants missing the outer membrane lipoprotein and other major outer membrane proteins. J. Bacteriol. *136*, 280–285.

Spinola, S.M., Hiltke, T.J., Fortney, K., and Shanks, K.L. (1996). The conserved 18, 000-molecular-weight outer membrane protein of *Haemophilus ducreyi* has homology to PAL. Infect. Immun. *64*, 1950–1955.

Srikumar, R., Mikael, L.G., Pawelek, P.D., Khamessan, A., Gibbs, B.F., Jacques, M., and Coulton, J.W. (2004). Molecular cloning of haemoglobin-binding protein HgbA in the outer membrane of *Actinobacillus pleuropneumoniae*. Microbiology *150*, 1723–1734.

St Geme, J.W., III, and Cutter, D. (2000). The *Haemophilus influenzae* Hia adhesin is an autotransporter protein that remains uncleaved at the C terminus and fully cell associated. J. Bacteriol. *182*, 6005–6013.

St Geme, J.W., III, Cutter, D., and Barenkamp, S.J. (1996). Characterization of the genetic locus encoding *Haemophilus influenzae* type b surface fibrils. J. Bacteriol. *178*, 6281–6287.

Strutzberg, K., von Olleschik, L., Franz, B., Pyne, C., Schmidt, M.A., and Gerlach, G.F. (1995). Mapping of functional regions on the transferrin-binding protein (TfbA) of *Actinobacillus pleuropneumoniae*. Infect. Immun. *63*, 3846–3850.

Tabatabai, L.B., and Zehr, E.S. (2004). Identification of five outer membrane-associated proteins among cross-protective factor proteins of *Pasteurella multocida*. Infect. Immun. *72*, 1195–1198.

Thwaits, R.N., and Kadis, S. (1991). Immunogenicity of *Actinobacillus pleuropneumoniae* outer membrane proteins and enhancement of phagocytosis by antibodies to the proteins. Infect. Immun. *59*, 544–549.

Tonpitak, W., Thiede, S., Oswald, W., Baltes, N., and Gerlach, G.F. (2000). *Actinobacillus pleuropneumoniae* iron transport: a set of *exbBD* genes is transcriptionally linked to the *tbpB* gene and required for utilization of transferrin-bound iron. Infect. Immun. *68*, 1164–1170.

Ulmer, J.B., Valley, U., and Rappuoli, R. (2006). Vaccine manufacturing: challenges and solutions. Nat. Biotechnol. *24*, 1377–1383.

van den Bosch, H. and Frey, J. (2003). Interference of outer membrane protein PalA with protective immunity against *Actinobacillus pleuropneumoniae* infections in vaccinated pigs. Vaccine *21*, 3601–3607.

van Alphen, L., Riemens, T., Poolman, J., and Zanen, H.C. (1983). Characteristics of major outer membrane proteins of *Haemophilus influenzae*. J. Bacteriol. *155*, 878–885.

Van Overbeke, I., Chiers, K., Charlier, G., Vandenberghe, I., Van Beeumen, J., Ducatelle, R., and Haesebrouck, F. (2002). Characterization of the *in vitro* adhesion of *Actinobacillus pleuropneumoniae* to swine alveolar epithelial cells. Vet. Microbiol. *88*, 59–74.

Van Overbeke, I., Chiers, K., Donne, E., Ducatelle, R., and Haesebrouck, F. (2003). Effect of endobronchial challenge with *Actinobacillus pleuropneumoniae* serotype 10 of pigs vaccinated with bacterins consisting of *A. pleuropneumoniae* serotype 10 grown under NAD-rich and NAD-restricted conditions. J. Vet. Med. B Infect. Dis. Vet. Public Health *50*, 289–293.

Van Overbeke, I., Chiers, K., Ducatelle, R., and Haesebrouck, F. (2001). Effect of endobronchial challenge with *Actinobacillus pleuropneumoniae* serotype 9 of pigs vaccinated with a vaccine containing Apx toxins and transferrin-binding proteins. J. Vet. Med. B Infect. Dis. Vet. Public Health *48*, 15–20.

Ward, C.K. and Inzana, T.J. (1997). Identification and characterization of a DNA region involved in the export of capsular polysaccharide by *Actinobacillus pleuropneumoniae* serotype 5a. Infect. Immun. *65*, 2491–2496.

Whitby, P.W., VanWagoner, T.M., Seale, T.W., Morton, D.J., and Stull, T.L. (2006). Transcriptional profile of *Haemophilus influenzae*: effects of iron and heme. J. Bacteriol. *188*, 5640–5645.

Wilke, M., Franz, B., and Gerlach, G.F. (1997). Characterization of a large transferrin-binding protein from *Actinobacillus pleuropneumoniae* serotype 7. Zentralbl. Veterinarmed. B *44*, 73–86.

Yamanaka, N. and Faden, H. (1993). Antibody response to outer membrane protein of nontypeable *Haemophilus influenzae* in otitis-prone children. J. Pediatr. *122*, 212–218.

Zeng, H., Pandher, K., and Murphy, G.L. (1999). Molecular cloning of the *Pasteurella haemolytica pomA* gene and identification of bovine antibodies against PomA surface domains. Infect. Immun. *67*, 4968–4973.

Zhang, Y., Tennent, J.M., Ingham, A., Beddome, G., Prideaux, C., and Michalski, W.P. (2000). Identification of type 4 fimbriae in *Actinobacillus pleuropneumoniae*. FEMS Microbiol. Lett. *189*, 15–18.

Zhu, W., Hunt, D.J., Richardson, A.R., and Stojiljkovic, I. (2000). Use of heme compounds as iron sources by pathogenic *neisseriae* requires the product of the *hemO* gene. J. Bacteriol. *182*, 439–447.

Lipopolysaccharides, Biofilms and Quorum Sensing in *Pasteurellaceae*

9

Thomas J. Inzana, W. Edward Swords, Indra Sandal and Shivakumara Siddaramappa

Abstract

The endotoxin of the *Pasteurellaceae* is an essential component of the bacterial cell surface, as it is for all Gram-negative bacteria. However, the *Pasteurellaceae* are unusual in that some members make an endotoxin that is a lipopolysaccharide (LPS) while others make a lipooligosaccharide (LOS). The LOS can be a relatively simple antigen, or a deceptively complex molecule that can be decorated with a variety of components that influence host response and interactions, contain structures that mimic host antigens, and phase vary in the expression of these antigenic epitopes. In this chapter, we will give a brief summary of what is known regarding LPS/LOS structure and genetics for different members of the *Pasteurellaceae*, and discuss how the glycose and other associated components affect persistence and virulence *in vivo*. Furthermore, we will discuss the formation of biofilms, which are produced by many members of this group, and how these surface-attached biofilm communities may promote bacterial persistence *in vivo*, even in the face of immune effectors and antimicrobial treatment. Finally, we will describe a related field of bacterial quorum sensing, which is an area of much recent work for these and other bacterial species.

Introduction

Over the past 30 years a tremendous amount of work has been done by a number of different groups on the genetics and biochemistry of the lipopolysaccharides (LPS) and lipooligosaccharides (LOS), which are the endotoxin components, of the *Pasteurellaceae*. These molecules are one of the major surface components that interact directly with a variety of host environments the bacteria may be present in. There is also a growing appreciation for the breadth of the role(s) of these surface structures in the infectious process. While most species of the *Pasteurellaceae* colonize host mucosal surfaces, many are capable of disseminating and infecting a variety of systemic sites, such as the lungs, nervous system, joints, fetus, and other internal organs. The LPS/LOS plays an important role in both colonization and the capability to disseminate and cause systemic disease. The LPS/LOS contributes to the disease process directly through inducing inflammation, and by protecting the bacterium against host defences, such as resistance to complement-mediated killing and to phagocytosis.

The formation of biofilms by environmental bacteria has long been recognized, and the importance of biofilms in diseases and medical devices by bacteria such as *Pseudomonas aeruginosa*, *Staphylococcus aureus* and others is well established. However, it is now being recognized that most bacterial species can form a biofilm under at least some growth or environmental conditions, including species in the *Pasteurellaceae*. The formation of a biofilm by these bacteria may have a profound influence on persistence of the bacteria in the host and influence disease outcome. Unfortunately, biofilms have only been investigated in a few species of the family *Pasteurellaceae*, and this is an area requiring extensive research.

Quorum sensing (QS) is a simple form of communication that occurs between bacteria within a population that reach a particular density. Mediated by small soluble molecules, QS can influence the capability of a bacterial population to form a biofilm, cause disease, be motile, and produce a variety of structural components and proteins. Thus, QS is an important adaptation mechanism for bacteria living in a particular environment. Although all of the *Pasteurellaceae* live on mucosal surfaces of human or animal hosts, a functional QS system is present in all species that have been studied. However, the role of this QS system to the lifestyle of the organism or its role in the disease process is unknown in most of the species, and not well understood in any of them. Therefore, QS is another field that could yield tremendous insight following intensive investigation.

Lipopolysaccharide/lipooligosaccharide structure

Bacterial LPS are glycolipids found on the surface of Gram-negative bacteria, where they are an integral component of the bacterial outer membrane. As such, they are at the interface between the microbe and the environment and/or host. LPS was first recognized as the 'endotoxin' responsible for the fulminant host inflammatory response to Gram-negative bacteria. The lipid portion of endotoxin is the so-called lipid A moiety that confers the 'endotoxic' properties upon LPS by means of its recognition by the Toll-like receptor 4/CD14/MD-2 complex found on many host cells. Lipid A is the major lipid component of the outer leaflet of the outer membrane, and in the *Pasteurellaceae* is similar in structure to lipid A from enteric bacteria, and is responsible for the biological activity of endotoxin, such as induction of cytokines and inflammation (Inzana *et al.*, 1988; Syrogiannopoulos *et al.*, 1988; Yoo *et al.*, 1995). Lipid A typically consists of a diglucosamine structure to which fatty acyl groups are attached. The most common acylation pattern among Gram-negative bacteria is hexa-acylated lipid A, which is the 'molecular pattern' recognized by Toll-like receptor 4 (Chow *et al.*, 1999; Hirschfeld *et al.*, 2000; Hoshino *et al.*, 1999; Poltorak *et al.*, 2000). In some bacterial species, the number and length of the acyl groups on the lipid A moiety can be differentially altered, which can have profound implications for the endotoxic potency of the LPS, and as a correlate, for the recognition of the bacterium by the host innate immune system (Ernst, 1999). Alteration of the number and chain length of acyl groups on the lipid A can also drastically change the susceptibility of the organism to antimicrobial compounds, including defensins (Gunn, 1996; Guo, 1998). While there is at present no evidence for temporal regulation of lipid A acylation among members of the *Pasteurellaceae*, work with *Haemophilus influenzae htrB* mutants that express a mixture of tetra- and penta-acylated lipid A species (Nichols *et al.*, 1997) has shown a defect in persistence in a tissue culture model systems (Swords *et al.*, 2002), as well as animal infection models (DeMaria, 1997). Of note, this mutant was also highly susceptible to defensin killing (Starner, 2002). The LOS of *Histophilus somni*, through the biologically active lipid A, has been shown to be responsible for apoptosis of bovine endothelial cells through the release of purinergic compounds [e.g. adenosine 5′-triphosphate (ATP)] that interact through the P2X family of purinergic receptors and is dependent upon caspase activation (Sylte *et al.*, 2005); apoptosis is also enhanced by factors such as TNF-α, and oxygen and nitrogen intermediates (Sylte *et al.*, 2003, 2004, 2006). Overall, the biological activity of *Hist. somni* lipid A, as measured by *Limulus* amoebocyte lysate assay, Shwartzman dermal reaction, and mouse lethality, is generally less than that of *Escherichia coli* lipid A, but not significantly (Inzana *et al.*, 1988).

LPS differs from LOS in regard to the presence or absence of a polymeric O-antigen, respectively, and in the structure of the core oligosaccharide. LPS appears as a ladder of distinct bands that span the length of a polyacrylamide gel following separation of LPS molecules by sodium dodecyl sulfate-polyacrylamide gel electrophoresis (SDS-PAGE; Fig. 9.1, lane 5). Each higher molecular size band represents an LPS species with one additional O-antigen unit, which can differ from one to about four glycoses. In contrast, LOS appears as a heterogeneous mixture of closely separated bands of low molecular size by SDS-PAGE (Fig. 9.1 lanes 1–4). In general, different genera of the *Pasteurellaceae* appear to

Figure 9.1 Electrophoretic profiles of LOS and LPS of various genera following SDS-PAGE. Lanes: 1, *H. influenzae* (LOS); 2, *Hist. somni* (LOS); 3, *N. gonorrhoeae* (LOS); 4, *P. multocida* (LOS); 5, *A. pleuropneumoniae* serotype 7 (LPS). Reproduced by permission of Wiley InterScience.

produce either an LPS or an LOS. Most species in the genera *Actinobacillus, Aggregatibacter,* and *Mannheimia* produce an LPS, whereas most species in the genera *Haemophilus, Histophilus,* and *Pasteurella* produce an LOS (unpublished observations). Exceptions are some *Haemophilus parainfluenzae* strains that express full-length LPS, although the role (if any) of an O-antigen in these strains in regard to persistence and/or virulence has not been established (Roberts *et al.*, 1986), and serotypes 2 and 8 of *Mannheimia haemolytica* appear to be devoid of O-antigen (Lacroix *et al.*, 1993). The structures of the O-antigens of most of the *Actinobacillus pleuropneumoniae* serotypes have been determined (Altman *et al.*, 1990), and vary in glycose chain length depending on the serotype. For instance, the O-antigen of *A. pleuropneumoniae* serotype 7 is a repeating tetrasaccharide unit, which consists of β-D-galactose linked (1–4) to β-D-*N*-acetylgalactosamine, which is linked (1–4) to α-L-rhamnose and (3–1) to β-D-galactose. In contrast, the O-antigen of serotype 5 consists of a linear unbranched homopolymer of (1–6)-linked β-D-galactose residues (Perry *et al.*, 1990). As a result *A. pleuropneumoniae* serotype 7 LPS appears as a typical ladder by SDS-PAGE (Fig. 9.1), while most of the LPS from serotype 5 is of low molecular size, resembling a micoheterogeneous LOS (Inzana *et al.*, unpublished data). A similar variation in SDS-PAGE profiles, demonstrating either a uniform ladder or a faint ladder and predominantly low molecular size LPS, has been shown for strains of *M. haemolytica* (Lacroix *et al.*, 1993). Of interest is that the structure of the O-antigen of *A. pleuropneumoniae* serotype 5 is similar to that of *M. haemolytica* serotypes 4 and 10, which consists of only D-galactose residues of the disaccharide repeating unit [α-D-galactose(1–3)β-D-galactose(1–)*n* (Richards and Leitch, 1989). In contrast, the O-antigens of *M. haemolytica* serotypes 1, 6 and 9 consist of β-D-galactose(1–3)β-D-galactose(1–3)β-D-*N*-acetylgalactosamine, and that of serotypes 3 and 15 β-D-*N*-acetylglucosamine(1–4)α-L-rhamnose. The O-antigen of *Aggregatibacter actinomycetemcomitans* consists of a polymer of trisaccharide repeating units composed of D-fucose, L-rhamnose, and D-*N*-acetylgalactosamine residues (Perry *et al.*, 1996). These LPS structures also have a typical ladder-like banding pattern by SDS-PAGE (Lacroix *et al.*, 1993).

The LPS/LOS inner core regions that have been structurally characterized among the *Pasteurellaceae* are typical of those of other Gram-negative bacterial species. The linker region for *H. influenzae* is a phosphorylated 3-deoxy-D-*manno*-2-octulosonic acid (Kdo) rather than the di-Kdo structure found in many endotoxins, and the Kdo-kinase that adds the phosphate to this structure is essential for virulence (White *et al.*, 1997; White *et al.*, 1999), (Fig. 9.2C). Infection studies have shown that a fully assembled inner

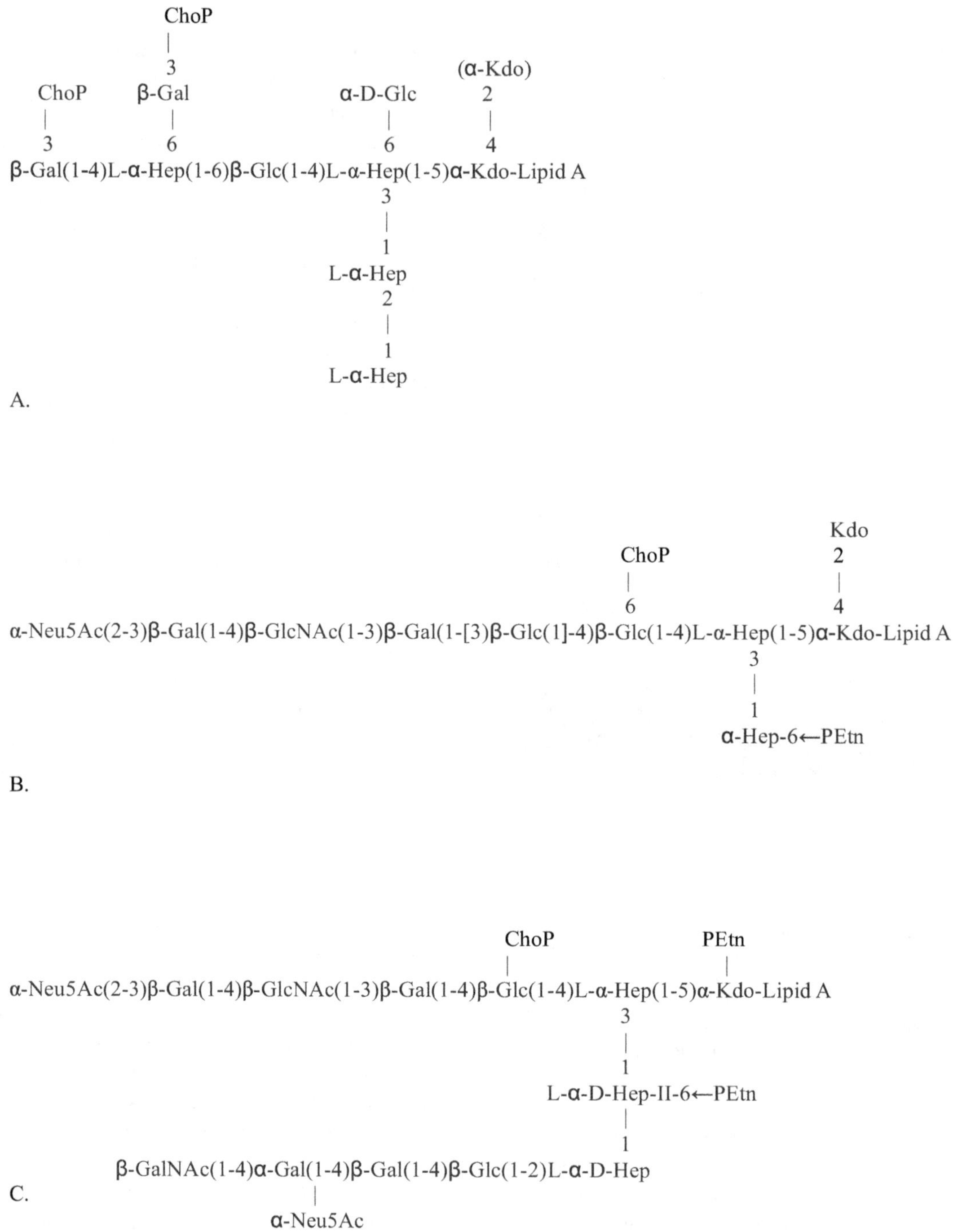

Figure 9.2 LOS structures from (A) *P. multocida* strain VP161 (St Michael *et al.*, 2005b), (B) *Hist. somni* strain 738 (sialylated) (Cox *et al.*, 1998) and (C) *H. influenzae* strain RM118 (sialylated) (Hood *et al.*, 2004). Abbreviations: Glc, glucose; Gal, galactose; GlcNAc, *N*-acetyglucosamine; GalNAc, *N*-acetylgalactosamine; Hep, heptose; Kdo, 3-deoxy-D-*manno*-2-octulosonic acid; ChoP, phosphorylcholine; PEtn, phosphatidylethanolamine; Neu5Ac, *N*-acetylneuraminic acid.

core region is essential to the virulence of *H. influenzae* in the chinchilla model for otitis media (DeMaria, 1997). Additional structural work has shown that the inner core may be substituted with glycine and ethanolamine (Li *et al.*, 2001). It is of particular interest that the structure of the inner core of the LOS of *Pasteurella multocida* strain 70 (St Michael *et al.*, 2005c) is similar to the LPS core structures for *M. haemolytica* (Brisson, 2002), *A. pleuropneumoniae* (Michael *et al.*, 2004), [*Haemophilus*] *ducreyi* (Schweda *et al.*, 1994), and *H. influenzae* (Cox *et al.*, 2001). In each case the core oligosaccharide contains a triheptosyl unit with identical linkages attached to Kdo. Outer core oligosaccharide chains can extend from each heptose, or may be substituted with phosphoethanolamine (PEtn). Variation in oligosaccharide extensions from each heptose between strains and as a result of phase variation (see below) have been well characterized in *H. influenzae* and accounts for extensive structural and antigenic heterogeneity (Schweda *et al.*, 2007). Monoclonal antibodies raised to the inner core of a *M. haemolytica* mutant are bactericidal and protective in a mouse model against challenge with wild type *M. haemolytica* (Logan *et al.*, 2006). Of note, *P. multocida* strains, such as VP161, express LOS molecules that contain the phospho-Kdo linker, but can simultaneously express glycoforms containing di-Kdo structures (Harper *et al.*, 2007) (Fig. 9.2A). Infection studies in chickens revealed that for this organism the di-Kdo glycoforms were essential to virulence (Harper *et al.*, 2007). The inner core of *Hist. somni* also consists of the di-Kdo structure (Cox *et al.*, 1998; Cox *et al.*, 2003; St Michael *et al.*, 2004; St Michael *et al.*, 2005a, 2006). Furthermore, the entire LOS structure of *Hist. somni* strain 2336 (St Michael *et al.*, 2005a) is remarkably similar to that of *Neisseria gonorrhoeae* strain F62 (Yamasaki *et al.*, 1991), which also forms a micro-heterogeneous LOS (Fig. 9.1). The inner cores of the endotoxins of other members of the *Pasteurellaceae* have not been as well characterized.

For species expressing LOS, the outer carbohydrate oligosaccharide chains are quite diverse in structure, but overall conserved in composition. Their structure, synthesis, and assembly have been dissected in detail, particularly that of *H. influenzae* (Apicella *et al.*, 1985; Campagnari *et al.*, 1987; Harvey, 2001; Hood *et al.*, 2001b; Hood *et al.*, 1996; Jennings *et al.*, 1995; Kwaik *et al.*, 1991, 1992; Mansson *et al.*, 2002; Mansson *et al.*, 2003; McLaughlin *et al.*, 1992; Phillips *et al.*, 1993; Phillips *et al.*, 2000; Risberg *et al.*, 1999; Schweda *et al.*, 2007). Their outer core glycose composition consists of glucose, galactose, and the amino sugars *N*-acetylglucosamine or *N*-acetylgalactosamine. In addition, these structures may contain a number of different substitutions that include phosphorylcholine (ChoP), PEtn, sialic acid, and acetyl, glycine, and phosphate groups (Schweda *et al.*, 2000). Although lacking the O-antigen of LPS, the outer core oligosaccharide of LOS may be much more complex, mimic oligosaccharides of host cell surface glycolipids, and be subject to phase variation. Phase variation is the reversible gain or loss of specific glycoses in the oligosaccharide resulting in extension or truncation of the oligosaccharide, respectively. The molecular basis for LOS phase variation has been extensively investigated in *H. influenzae* and has been described in several excellent reviews (Hood *et al.*, 1999b; Richards *et al.*, 2001; Schweda *et al.*, 2007). A similar mechanism of LOS phase variation has been described in *Hist. somni* (Inzana *et al.*, 1992; Inzana *et al.*, 1997). Briefly, a variable number of tandem repeats (VNTR) of 1, 2 or 4 nucleotides present in the open reading frame (ORF) either immediately upstream of the start codon or within the ORF act as a translational switch to turn on or off expression of the gene. During replication, slipped strand mispairing of the replicating strand can result in the loss or gain of one or more VNTR units, resulting in the start codon being placed in or out of frame with the ORF. This translational switch can cause the glycosyl transferase that adds the specific glycose to the outer core to be non-functional when the gene is out of frame. The VNTR that are located immediately downstream of the start codons are not required for translational expression of the gene, and appear to act only as a molecular switch (High *et al.*, 1996). Examples of the variation in LOS structure that can result from phase variation have been reported for *Hist. somni* strain 2336 and its clonal phase variant 738 (Cox *et al.*, 1998; St Michael *et al.*, 2005a). These variants differ in the linkage of their terminal galactose-

N-acetylglucosamine residue [(1–4) in 2336 and (1–3) in 738] off heptose I and decoration of heptose II with galactose in strain 2336, which is missing in 738 (Fig. 9.2B).

In addition to the common glycoses, modifying components of the LOS may also be subject to phase variation, such as ChoP, PEtn, and neuraminic (sialic) acid. Of interest is that although LOS phase variation occurs in *H. influenzae* and *Hist. somni* LOS, the same VNTR are either in different genes, or homologous genes contain different VNTR. For instance, the gene that codes for choline kinase, required for ChoP synthesis, in *H. influenzae* (*lic1A*) contains repeats of 5′-CAAT-3′, whereas the homologous gene in *Hist. somni* contains 5′-AACC-3′ (S. Elswaifi *et al.* unpublished). The gene for galactosyl transferase (*lob1*), which is involved in LOS synthesis in *Hist. somni*, contains the VNTR 5′-CAAT-3′ immediately downstream of potential start codons. However, this gene has 61% homology to *lex2B* from *H. influenzae*, which lacks VNTR, but appears to be cotranscribed with *lex2A*, which contains the VNTR 5′-GCAA-3′ in the same position as the VNTR in the *lob1* ORF of *Hist. somni* (Griffin *et al.*, 2003; Inzana *et al.*, 1997; McQuiston *et al.*, 2000). Likewise *lob2A* (an *N*-acetylglucosamine transferase) from *Hist. somni* contains the VNTR 5′-GA-3′ about two-thirds of the way through the open reading frame and this gene has 56% homology to *lic2B* of *H. influenzae* (Wu *et al.*, 2000). The gene *lic2B* lacks VNTR, but the product of *lic2B* is homologous to Lic2A, which is encoded by a gene containing the VNTR 5′-CAAT-3′ just downstream of potential start codons (High *et al.*, 1993). Therefore, this similar, though different, genetic mechanism may be the result of convergent or divergent evolution. It is likely that phase variation enables the bacteria to adapt to different host sites or resist adaptive host immunity. In cattle infected with *Hist. somni* phase variation to a new LOS phenotype is correlated with development of antibodies to the previous phenotypes, and development of antibodies to all recognized phenotypes is correlated with clearance of bacteria from the lungs (Inzana *et al.*, 1992). However, phase variation has not been described in other *Haemophilus* species or other species of the *Pasteurellaceae*. Although the LOS of [*Haemophilus*] *parasuis* is similar in electrophoretic profile to that of other *Haemophilus* species, no evidence of LOS phase variation could be detected following subculture of individual colonies for 2 weeks, as determined by SDS-PAGE (Inzana, 2004). With the advent of ready and broadscale genome sequencing, the capability of a species to undergo phase variation can be predicted by the presence of VNTR in ORFs with homology to genes that encode for glycosyl transferases (Hood *et al.*, 1996).

Phosphorylcholine is now recognized as an LOS component in several members of the *Pasteurellaceae* including *H. influenzae* (Weiser *et al.*, 1997), *Hist. somni* (Howard, 2000), and *P. multocida*. For *H. influenzae*, choline is scavenged from host cells or mucus via the GlpQ phospholipase (Fan, 2001); a surface enzyme with similar phospholipase activity has been reported from *H. parainfluenzae*. The addition of ChoP to LOS glycoforms is determined by the *lic1* operon, which encodes factors that mediate the uptake of choline from exogenous sources, and phosphorylation and assimilation of ChoP into a discrete set of LOS glycoforms. The *H. influenzae lic1* operon undergoes phase-variation via a slip-strand mechanism among 5′-CAAT-3′ repeats within the *licA* gene (Weiser *et al.*, 1989a,b), as described above. In *H. influenzae*, synthesis of ChoP-positive glycoforms is also controlled at the level of synthesis of the acceptor LOS glycoforms via an oxygen-sensitive regulatory mechanism (Wong and Akerley, 2005).

The role of ChoP *in vivo* appears to be multifactorial. For *H. influenzae*, it is clear that variants expressing ChoP predominate during persistent infection (Tong *et al.*, 2000; Weiser and Pan, 1998; Weiser *et al.*, 1998; Weiser *et al.*, 1990). As for pneumococci, ChoP mediates adherence to the platelet-activating factor receptor on host cells (Gould and Weiser, 2001; Gould and Weiser, 2002; Swords *et al.*, 2000; Swords *et al.*, 2001). ChoP can also confer resistance to bacterial killing by antimicrobial peptides (Lysenko *et al.*, 2000b) and diminishes the bioactivity of purified LOS (West-Barnette *et al.*, 2006). Because ChoP positive variants predominate within biofilm communities (Hong *et al.*, 2007a; Hong *et al.*, 2007b; West-Barnette *et al.*, 2006), this modification seems to represent a way whereby *H. influenzae* simultaneously resists and evades host immunity. The role of ChoP and other surface modifications in biofilm formation and maturation will be discussed in a later section. For at least some types of bacterial variants, the expression of ChoP targets the bacteria for both host innate and adaptive responses. These include the binding of C-reactive protein, which can initiate bacterial killing by the alternative complement pathway, and the binding of anti-ChoP antibody to the bacterial surface (Weiser *et al.*, 1998). It should be noted that available evidence indicates that at least some structural presentations of ChoP on the *H. influenzae* surface are inaccessible for either C-reactive protein or antibody binding (Lysenko *et al.*, 2000a). As mentioned, a comparable system of ChoP expression is present in *Hist. somni* (Howard, 2000), except that the VNTR in the *Hist. somni lic1A* homologue is 5′-AACC-3′ (Elswaifi *et al.*, unpublished data). For *Hist. somni*, ChoP positive LOS forms also mediate platelet activation (Kuckleburg *et al.*, 2007), and promote colonization of mucosal surfaces, whereas ChoP negative variants predominate in systemic sites (Elswaifi *et al.*, unpublished data). Therefore, the bacteria must lose ChoP in order to efficiently disseminate.

In addition to phase variation, the structure of the LOS outer core oligosaccharide can mimic host cell oligosaccharides. The lacto-*N*-neotetraose component (galactose-*N*-acetylglucosamine-galactose-glucose) of the LOS of *H. influenzae* and *Hist. somni* are identical in structure to the terminal oligosaccharide of cell surface glycosphingolipids (Mandrell *et al.*, 1992b; Preston *et al.*, 1996). ChoP and/or sialic acid are also host components that are expressed on the LOS of several genera of the *Pasteurellaceae* as well. Thus, LOS mimicry of host antigenic structures is an important mechanism for evading host immunity.

Many members of the *Pasteurellaceae* express sialylated LOS forms. These include *H. influenzae* (Mandrell *et al.*, 1992), *Hist. somni* (Inzana *et al.*, 2002), [*H.*] *ducreyi* (Bozue *et al.*, 1999), and possibly others. Of interest is when *H. influenzae* strain RM118 is grown is the presence of sialic acid the tetrasaccharide α-neuraminic acid(2–3) β-galactose(1–4)β-*N*-acetylglucosamine(1–3) β-galactose is added to heptose 1 (Fig. 9.2C) *en bloc* via an isoprenoid lipid carrier, which is similar to the method used to attach O-antigen to the LPS of bacteria producing LPS O-side chains. Thus, some groups propose the endotoxin of *H. influenzae* is an LPS (Hood *et al.*, 2004). In *H. influenzae*, and likely *Hist. somni*, sialic acid is catabolized via a TRAP-type membrane transporter and added to a discrete set of LOS glycoforms. Several distinct sialyl transferase types have been identified in *H. influenzae* (Fox *et al.*, 2006; Hood *et al.*, 2001a; Jones *et al.*, 2002), and at least two in *Hist. somni* (M.D. Howard *et al.*, unpublished). One *Hist. somni* sialyl transferase preferably sialylates the galactose(1–4)*N*-acetylneuraminic acid configuration while the other preferentially sialylates the galactose(1–3)*N*-acetylneuraminic acid configuration. Thus, efficient sialylation is assured regardless of the linkage conferred by phase variation. Sialic acid has been shown to confer resistance to complement-mediated killing for both *H. influenzae* and *Hist. somni* (Hood *et al.*, 1999a; Inzana *et al.*, 2002), at least in part through binding of complement factor H. Factor H bound to the cell surface acts as a cofactor with factor I to mediate cleavage of C3b to its inactive form C3bi (Pangburn and Muller-Eberhard, 1978), dissociates factor Bb from the C3Bb complex (Pangburn *et al.*, 1977), and decreases deposition of C4 and mannan-binding lectin (Gulati *et al.*, 2002). Each of these processes results in inactivation or incomplete activation of the complement cascade. Work by Bouchet and colleagues established that *H. influenzae* actively sialylates its surface *in vivo*, and that sialylation-defective mutants are unable to persist within the chinchilla middle ear (Bouchet *et al.*, 2003). However, comparable studies with [*H.*] *ducreyi* have shown that sialylated LOS glycoforms are not required for the formation of the pustules associated with chancroid disease (Young *et al.*, 1999). Sialylation of *Hist. somni* LOS also results in reduced binding of specific antibodies (Inzana *et al.*, 2002), and less efficient uptake by bovine polymorphonuclear leukocytes (Inzana *et al.*, unpublished data).

Biofilm formation by the *Pasteurellaceae*

The definition of a biofilm is, at present, quite broad and may be applied to essentially any surface-attached microbial community, especially if an observable matrix material is present. One defining characteristic of a biofilm *in vivo* is enhanced resistance of the bacteria to clearance by extraneous bactericidal factors, whether they are host immune components or other antimicrobial factors. For *H. influenzae*, the presence of biofilm communities had long been postulated within the context of chronic otitis media infections. Even as recently as ten years ago, it was thought that cases of otitis media in which viable bacteria were not recovered from middle-ear fluids represented 'sterile' cases of otitis media or, alternatively, incidents of viral infection. Early indirect evidence for a wider bacterial aetiology featuring biofilm communities included detection of bacterial RNA and bacterial components within recovered effusion fluids by molecular techniques (Dingman *et al.*, 1998; Rayner *et al.*, 1998). Later, actual biofilm communities were observed on tympanostomy tubes from patients with chronic otitis media, and a biofilm infection was reconstituted in the chinchilla model of otitis media. More recently, biofilms formed by *H. influenzae* (as well as other airway opportunists) have been detected in tissues from patients with a range of upper respiratory infections, including adenotonsillitis (Galli *et al.*, 2007), otitis media (Hall-Stoodley *et al.*, 2006), and rhinosinusitis (Sanderson *et al.*, 2006).

In vitro studies on biofilm formation by *H. influenzae* by Murphy and Kirkham showed that biofilm formation in a static assay system is a common phenotype among strains recovered from patients with chronic obstructive pulmonary disease (COPD), but that biofilm formation does not absolutely correlate with any specific patient outcome or duration of infection (Murphy and Kirkham, 2002). Later, Swords *et al.* and Greiner *et al.* both demonstrated that sialylation is essential to biofilm formation in static and continuous-flow biofilm assay systems, as well as *H. influenzae* persistence *in vivo* (Greiner *et al.*, 2004; Jurcisek *et al.*, 2005; Swords *et al.*, 2004). In a later study, *H. influenzae* mutants unable to take up sialic acid from culture medium were shown to have a severe defect for survival within a biofilm (Allen *et al.*, 2005). In addition to changes in sialylation, work from other groups has shown that *H. influenzae* biofilms may express a unique set of proteins as compared with planktonic cultures (Gallaher *et al.*, 2006; Murphy *et al.*, 2005; Webster *et al.*, 2006). For example, Murphy and colleagues have shown that statically grown *H. influenzae* biofilms express high levels of a peroxiredoxin, and these results are consistent with measurements of this protein in sputum samples from patients with COPD (Murphy *et al.*, 2005). Other *H. influenzae* phenotypes associated with biofilm formation/maturation include expression of type IV pili (Jurcisek *et al.*, 2007), presence of extracellular DNA (Jurcisek and Bakaletz, 2007), and expression of ChoP positive LOS (Hong *et al.*, 2007a,b; West-Barnette *et al.*, 2006). In the latter studies, ChoP positive variants localize within the centre of biofilm communities and correlate absolutely with diminished inflammatory responses to *H. influenzae* bacteria or purified LOS. ChoP expression is positively correlated with the formation of mature biofilm communities *in vitro* and within the chinchilla middle-ear chamber (Hong *et al.*, 2007a,b).

Hist. somni also forms biofilm communities *in vitro* and *in vivo*. A unique feature of *Hist. somni* that distinguishes it from other members of the *Pasteurellaceae* is the substantial difference in the amount and architecture of biofilm formed by pathogenic strains compared to commensal strains. For instance, pneumonia isolate strain 2336 produces a mature biofilm that consists of thick, homogenous mound-shaped microcolonies encased in an amorphous extracellular matrix with profound water channels. In contrast, urogenital isolate strain 129Pt forms a biofilm of cell clusters that are tower shaped or distinct filamentous structures intertwined with each other by strands of extracellular matrix (Sandal *et al.*, 2007). Nonetheless, all strains of *Hist. somni* tested can form some biofilm based on the ring assay for microbial deposit on polyvinyl chloride (PVC) wells. Of particular interest is that *Hist. somni* forms a biofilm within 48 h under most types of growth conditions in broth, except during rapid shaking (~200 rpm). *Hist. somni* also displays a distinct, phenotypic life cycle that progresses through stages of attachment, growth, maturation, and detachment (Sandal *et al.*, 2007). Although *Hist. somni* in a biofilm is more resistant to most antibiotics than planktonic cells, the biofilm cells are more sensitive to tetracycline and some aminoglycosides than planktonic forms (Inzana et al., unpublished data). These results are reproducible, but not clearly understood. Since all members of the *Pasteurellaceae* are inhabitants of mucosal surfaces of mammals, the formation of a biofilm may be crucial to their persistence *in vivo*, and formation of a more robust biofilm may provide a selective advantage for strains that cause systemic disease. Recently we have shown that an exopolysaccharide (EPS) is a component of the *Hist. somni* biofilm matrix, and that this EPS is only produced under stress conditions, such as low oxygen tension, high NaCl concentration, stationary phase of growth, or conditions that favour biofilm formation. Of particular interest is that this EPS is composed of mannose and galactose (3:1), and appears highly similar in structure to yeast mannans (Inzana *et al.*, unpublished data).

Experimental respiratory challenge of calves with *Hist. somni* strain 2336 results in pneumonia, arthritis, and cardiac disease within 7 days post infection. Lesions from heart and lung tissue from an infected calf diagnosed with myocarditis and pneumonia were found to contain substantial biofilms, as determined by transmission electron microscopy (TEM), cryo-ultrathin TEM, freeze-fracture scanning electron microscopy (SEM), and fluorescence *in situ* hybridization (FISH) (Sandal *et al.*, unpublished data). *Hist. somni* and possibly other coinfecting pathogens in lesions may be protected from vaccine-induced immunity and antibiotic treatment in the biofilm. Therefore, strategies to treat or prevent *Hist. somni* diseases should address biofilm formation, and not just the planktonic bacteria. Furthermore, biofilm formation by *Hist. somni* in the bovine host may serve as a model for the study of systemic biofilm infections in humans.

The porcine pathogen [*H.*] *parasuis* is also capable of forming biofilms, but for this bacterium biofilm formation *in vitro* is not associated with virulence, but appears to be more common among less virulent strains (Jin *et al.*, 2006). Jin and coworkers used an intranasal pig infection model to show that biofilm formation depends on the recovery site, and is maintained when bacteria are passaged in the upper respiratory tract. [*H.*] *parasuis* isolated from the nasal cavities of experimentally infected pigs maintains the biofilm phenotype, which is lost from bacteria recovered from the lungs and brain. Therefore, biofilm formation in [*H.*] *parasuis* appears to be associated with colonization and is not compatible with dissemination (Jin *et al.*, 2006). Nonetheless, about 85% of [*H.*] *parasuis* strains forms a biofilm on polystyrene surfaces, as well as on air-liquid interfaces, and unlike *A. pleuropneumoniae*, the biofilm phenotype is maintained by passage on either agar or in broth (Jin *et al.*, 2006).

Agg. actinomycetemcomitans is a dental/periodontal pathogen within the *Pasteurellaceae* family that forms biofilm communities on tooth surfaces (Kaplan *et al.*, 2003a). Biofilm formation has been shown to play a significant role in the colonization and pathogenesis of *Agg. actinomycetemcomitans* (Kaplan *et al.*, 2003a; Kaplan *et al.* 2003b). The EPS in the matrix of the *Agg. actinomycetemcomitans* biofilm is poly-*N*-acetyl-glucosamine (PGA), production of which is conferred by a genetic locus with homology to regions with similar functions in *Staphylococcus epidermidis*, *E. coli*, and in the *Bordetella* species. PGA plays an important role in attachment and intercellular adhesion of the bacteria to abiotic surfaces (Agladze *et al.*, 2005; Heilmann *et al.*, 1996; Wang *et al.*, 2004), to protection from host innate defences including phagocytosis and antimicrobial peptides (Vuong *et al.*, 2004), and virulence (Kropec *et al.*, 2005). *Agg. actinomycetemcomitans* also produces an enzyme called dispersin, which cleaves the matrix PGA and disperses the biofilm (Izano *et al.*, 2007c; Kaplan *et al.*, 2003b, 2004). Elegant work by Jeffrey Kaplan and colleagues has shown that dispersin expression is increased late in biofilm maturation and correlates with the dissolution of a biofilm community (Kaplan *et al.*, 2003b, 2004). In addition, PGA mutants are biofilm-deficient (Izano *et al.*, 2007b).

The related veterinary pathogen *A. pleuropneumoniae* also forms biofilms (Kaplan and Mulks, 2005), and has homologues of the PGA matrix and dispersin genes that play similar roles (Izano *et al.*, 2007a; Kaplan *et al.*, 2004). The biofilm matrix may be a virulence factor in swine pleuropneumonia and a potential vaccine candidate (Kaplan *et al.*, 2004). However, only two reference strains, L20 (serotype 5B) and 56513 (serotype 11), and about 50–60% of field isolates of *A. pleuropneumoniae* form biofilms in polystyrene microtitre plates as well as in glass tubes at the air–liquid interface (Kaplan and Mulks, 2005). Cells within the *A. pleuropneumoniae* biofilm are also more resistant to antibiotics than are planktonic cells, and following pretreatment of the biofilm with dispersin B the cells become more susceptible to antibiotics (Izano *et al.*, 2007a). One interesting characteristic of the *A. pleuropneumoniae in vitro* biofilm is that single colonies loose their biofilm phenotype when transferred from agar to broth (Kaplan and Mulks, 2005). This change from a biofilm-positive to a biofilm-negative phenotype is correlated with a change in the subsequent colony morphology on agar. When broth cultures are plated on agar two distinct colony types form. One colony type is large, smooth textured, opaque and named 'smooth'; the other is very small, rough-textured, clear, dark, and termed 'rough'. Rough colonies are biofilm-positive and smooth colonies are biofilm-negative. The conversion from a biofilm-positive to biofilm-negative phenotype is irreversible (Kaplan and Mulks, 2005). A similar type of change in biofilm phenotype in broth culture is observed in *Agg. actinomycetemcomitans* (Fine *et al.*, 1999). A recent study also demonstrated that an *A. pleuropneumoniae arcA* mutant is attenuated in a porcine infection model, reduced in long term survival on lung epithelium, and deficient in autoaggregation and biofilm formation (Buettner *et al.*, 2007). ArcA is part of a regulatory circuit controlling responses to oxygen-depleted environments, and prior work has shown that other mutations affecting anaerobic metabolism are attenuating (Baltes *et al.*, 2005; Jacobsen *et al.*, 2005). Because the bacteria within biofilm communities are generally thought to be within an anaerobic environment, these data collectively suggest that *A. pleuropneumoniae* survival and metabolism within a biofilm is essential to persistence *in vivo*. Another DNA region that may be important in biofilm formation and enhance colonization or pathogenesis of *A. pleuropneumoniae* is the *flp-rcp-tad* locus, which consists of 14 genes and has a homologous locus in *Agg. actinomycetemcomitans* (Kachlany *et al.*, 2001). A mutation in *Agg. actinomycetemcomitans flp-1*, which encodes the major pilin protein subunit, abolishes biofilm formation *in vitro* and the mutant fails to colonize the mouth of a rat model of periodontal disease (Schreiner *et al.*, 2003). Similar results were obtained for *P. multocida* septicaemia (Fuller *et al.*, 2000). In addition, Ross (2006) showed that *P. multocida* grows as a biofilm in the porcine lung during 'systemic pasteurellosis'. A comprehensive strain survey of these and other pathogens using the static, high-throughput Calgary Biofilm Device showed that both of these species formed biofilms that are more resistant to killing by antibiotics than planktonic cells (Olson *et al.*, 2002). Fimbriae or pili have been shown to be responsible for *P. multocida* twitching motility (Fuller *et al.*, 2000), which is required for the initial steps of biofilm formation in many pathogens, including *Agg. actinomycetemcomitans* (Perez *et al.*, 2006), [*H.*] *ducreyi* (Spinola *et al.*, 2003) and *H. influenzae* (Murphy and Kirkham, 2002). It has yet to be determined if there is a relationship between pili and biofilm formation by *M. haemolytica* or *Hist. somni*.

Definition of biofilm phenotypes/genotypes in *Pasteurellaceae* species as well as understanding the role(s) of biofilms in opportunistic infections, are important emerging areas of research. The ability to use the bacterium's natural host for experimental infections provides an opportunity to obtain new information regarding animal infections and biofilms in general.

Quorum sensing in the *Pasteurellaceae*

Bacterial quorum sensing or signalling (QS) results from the release of soluble signal mediators by bacterial communities, which elicit a coordinated response when a particular cellular quorum is attained. These small molecules, known as autoinducers, are produced by QS bacteria and have been shown to influence the regulation of genes whose functions are more efficient at high cell densities. Density-dependent gene regulation has been implicated in swarming motility, biofilm formation, EPS production, and type III secretion (Keller and Surette, 2006; Kendall and Sperandio, 2007). This mechanism is used by some bacteria to sense and respond to members of their own species, and also to other bacterial species occupying the same ecological niche. In view of its characterization among diverse species, QS appears to be a conserved function during bacterial evolution (Federle and Bassler, 2003; Fuqua and Greenberg, 1998). Acylated homoserine lactone (AHL) molecules [also known as autoinducer-1 (AI-1) signals] mediate QS only in Gram-negative bacteria, and small peptide molecules [also known as autoinducing peptides (AIPs)] mediate QS only in Gram-positive bacteria (Fuqua *et al.*, 1994; Sturme *et al.*, 2002). Due to the species-specificity of the AI-1 and AIP molecules, the QS they mediate is sometimes referred to as 'intraspecies' or 'species-specific' communication (Federle and Bassler, 2003; Fuqua *et al.*, 1994).

A furanosyl borate diester autoinducer signal molecule has been shown to regulate QS- dependent bioluminescence in the Gram-negative marine bacterium *Vibrio harveyi* (Chen *et al.*, 2002). Since an AI-1 molecule can also regulate QS dependent bioluminescence in this species, the furanosyl borate diester has been referred to as autoinducer-2 (AI-2). The product of *luxS* (S-ribosylhomocysteinase) has been shown to be an intermediary enzyme in the biosynthesis of the AI-2 molecule. In *V. harveyi*, LuxP, which is a periplasmic protein, acts as a receptor for AI-2 molecules and transmits the signal to LuxQ, which is a sensor histidine kinase. A transcriptional regulator (LuxR) regulates gene expression in response to the AI-2 type QS signals (Chen *et al.*, 2002; Federle and Bassler, 2003). Homologues of these and other genes involved in QS have been identified and characterized in several members of the genus *Vibrio*. Using *V. harveyi* mutants as reporter strains that do not respond to the AHL (AI-1) signal molecules, it has been shown that culture supernatants from several *luxS*-containing bacteria can induce luminescence via the AI-2 system (Federle and Bassler, 2003). These initial observations lead to the proposal of the 'universal signal' hypothesis (Chen *et al.*, 2002). Like many eubacterial species, most members of the *Pasteurellaceae* have LuxS homologues and should presumably activate the AI-2 QS network. While there remains much work to be done in this area with the *Pasteurellaceae*, some studies have shed light on important quorum signalling effects on biofilm formation and/or virulence.

Fig. 9.3 shows the characteristic pattern of luminescence induction in the *V. harveyi* reporter strain BB170 by culture supernatants from different *Pasteurellaceae* species. The broad variation in the response indicates that AI-2 is a 'family of molecules' rather than a single 'universal signal' molecule. Several other studies also suggest that structurally diverse AI-2 signals could be produced by different *luxS*-containing bacteria (Bassler *et al.*, 1997; Forsyth and Cover, 2000; Greenberg *et al.*, 1979; Miller *et al.*, 2004). Furthermore, the *luxS* gene from the organism whose culture supernatants induced the weakest response, *Hist. somni*, was able to complement production of AI-2 by *E. coli* DH5α (which has a defective *luxS*) to a high degree, confirming that *Hist. somni luxS* is functional, but that its AI-2 is not effectively recognized by *V. harveyi* (Siddaramappa *et al.*, unpublished data).

Agg. actinomycetemcomitans has a *luxS* homologue that confers QS events both within and between species (Fong *et al.*, 2001). AI-2 signalling in this species increases leukotoxin and iron transport protein expression (Fong *et al.*, 2003). Mutation of *luxS* in *Agg. actinomycetemcomitans* has been shown to reduce planktonic growth under iron-limited aerobic culture conditions as well as biofilm formation (Fong *et al.*, 2003; Shao *et al.*, 2007). Furthermore, two putative periplasmic proteins (LsrB and RbsB) have been proposed to function as receptors for AI-2 in *Agg. actinomycetemcomitans* and a *lsrB-rbsB* double mutant has also been shown to be biofilm deficient (Shao *et al.*, 2007). There are indications that a comparable AI-2 network is present in *A. pleuropneumoniae* (Nielsen and Boye, 2005).

Malott and Lo (2002) surveyed strains within various species of the *Pasteurellaceae* from animals for the presence of AI-2 activity using the *V. harveyi* biosensor system. They found significant levels of AI-2 signal within culture supernatants of *M. haemolytica* and *P. multocida*, in addition to the aforementioned *A. pleuropneumoniae*. Compared to its wild-type parent, a

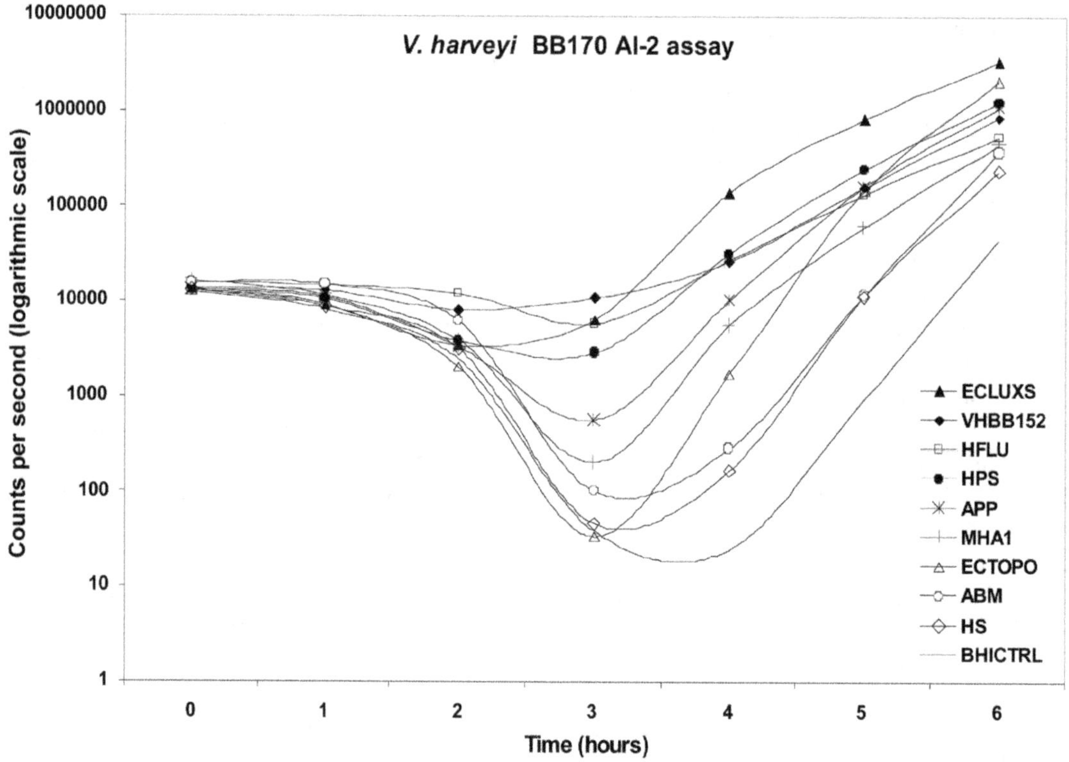

Figure 9.3 Luminescence induction in the *V. harveyi* BB170 reporter strain by culture supernatants from controls and different *Pasteurellaceae* species. VHBB152; *V. harveyi* strain BB152 (positive control), ABM; Autoinducer Bioassay Medium (negative control), BHICTRL; growth medium only (negative control), ECTOPO; *E. coli* DH5α carrying the TOPO vector only (negative control), ECLUXS; *E. coli* DH5α containing pCR2.1TOPO*luxS* (*luxS* from *Hist. somni*), HFLU; *H. influenzae*, HPS; *[H.] parasuis*, APP; *A. pleuropneumoniae*, MHA1; *M. haemolytica* A1, HS; *Hist. somni*.

M. haemolytica luxS mutant has been observed to grow faster under iron-restricted conditions, synthesize more capsular polysaccharide, and adhere better to polarized bovine tracheal cells (van der Vinne *et al.*, 2005). Daines and colleagues showed that many *H. influenzae* strains produce an AI-2 signal and that inactivation of *luxS* decreases the amount of AI-2 in culture filtrates (Daines *et al.*, 2005). However, *H. influenzae luxS* mutants form biofilms *in vitro*, and are more invasive to human epithelial cells. Furthermore, in the chinchilla model for otitis media these mutants caused a more fulminant otitis media disease, suggesting that biofilm production in the middle ear may actually modify the virulence of *H. influenzae*.

Conclusions and future trends

Certainly, much has been learned about the interconnecting topics of LPS/LOS structures, biofilm formation and persistence, and quorum signalling in the *Pasteurellaceae* family. Important topics for continuing development include using current knowledge about key metabolic requirements of bacteria within biofilm communities to identify targets for new approaches to control, treat, or even prevent these infections. There is also a general need for continued work on LPS/LOS genetics and biochemistry, especially in some of the less-studied but closely associated species. For example, while we now have a broad appreciation of the major glycoform populations and their biological relevance in *H. influenzae*, considerably less is known about the same topics for *H. parainfluenzae* and *Haemophilus haemolyticus*, which (especially in the latter case) recent work tells us are persisting in the same host niches in opportunistic infections (McCrea *et al.*, 2007; Mukundan *et al.*, 2007). We also have little understanding of the factors that come into play in the interstrain and interspecies dynamics in polymicrobial infections in either human or animal pathogens. For most of the diseases under study, hosts are simultaneously colonized by multiple strains and species that can have profoundly different abilities to persist *in vivo*, with equally diverse consequences for the host. A better understanding of how quorum signalling affects metabolism and biofilm formation *in vitro* and *in vivo* will be an important step in understanding these population dynamics, since most of the pathogens within *Pasteurellaceae* share the same AI-2 quorum signalling network of genes.

Biofilm formation by the animal pathogens within *Pasteurellaceae* is an important and extremely understudied topic. Recent work with *Hist. somni* indicates that the ability to form biofilms may be a key determinant in virulence vs. commensalism, and thus understanding how to prevent or treat these biofilms may be a key to solving the problem of opportunistic infections such as shipping fever in cattle or Glasser's disease and pleuropneumonia in swine. Because these problems have a large economic impact, they would be a vital area for additional research.

Acknowledgements

The authors would like to thank their colleagues in advance for their understanding that this chapter was intended as a window into the field rather than an exhaustive and all-inclusive review. Work in our laboratories is supported by grants from the NIH and USDA.

References

Agladze, K., Wang, X., and Romeo, T. (2005). Spatial periodicity of *Escherichia coli* K-12 biofilm microstructure initiates during a reversible, polar attachment phase of development and requires the polysaccharide adhesin PGA. J. Bacteriol. *187*, 8237–8246.

Allen, S., Zaleski, A., Johnston, J.W., Gibson, B.W., and Apicella, M.A. (2005). Novel sialic acid transporter of *Haemophilus influenzae*. Infect. Immun. *73*, 5291–5300.

Altman, E., Griffith, D.W., and Perry, M.B. (1990). Structural studies of the O-chains of the lipopolysaccharides produced by strains of *Actinobacillus (Haemophilus) pleuropneumoniae* serotype 5. Biochem. Cell Biol. *68*, 1268–1271.

Apicella, M.A., Dudas, K.C., Campagnari, A., Rice, P., Mylotte, J.M., and Murphy, T.F. (1985). Antigenic heterogeneity of lipid A of *Haemophilus influenzae*. Infect. Immun. *50*, 9–14.

Baltes, N., N'diaye, M., Jacobsen, I., Maas, A., Buettner, F., and Gerlach, G.F. (2005). Deletion of the anaerobic regulator HlyX causes reduced colonization and persistence of *Actinobacillus pleuropneumoniae* in the porcine respiratory tract. Infect. Immun. *73*, 4614–4619.

Bassler, B.L., Greenberg, E.P., and Stevens, A.M. (1997). Cross-species induction of luminescence in the quorum-sensing bacterium *Vibrio harveyi*. J. Bacteriol. *179*, 4043–4045.

Bouchet, V., Hood, D.W., Li, J., Brisson, J.R., Randle, G., Martin, A., Li, Z., Goldstein, R., and Schweda, E.K. (2003). Host-derived sialic acid is incorporated into *Haemophilus influenzae* lipopolysaccharide and is a major virulence factor in experimental otitis media. Proc. Natl. Acad. Sci. USA *100*, 8898–8903.

Bozue, J.A., Tullius, M.V., Wang, J., Gibson, B.W., Munson, R.S. Jr. (1999). *Haemophilus ducreyi* produces a novel sialyltransferase. Identification of the sialyltransferase gene and construction of mutants deficient in the production of the sialic acid-containing glycoform of the lipooligosaccharide. J. Biol. Chem. *274*, 4106–4114.

Brisson, J.-R., Crawford, E., Uhrin, D., Khieu. N.H., Perry, M.B., Severn, W.B., and Richards, J.C. (2002). The core oligosaccharide component from *Mannheimia (Pasteurella) haemolytica* serotype Al lipopolysaccharide contains l-*glycero*-d-*manno* and d-*glycero*-d-*manno*-heptoses. Analysis of the structure and confirmation by high resolution NMR spectroscopy. Can. J. Chem. *80*, 949–963.

Buettner, F., Maas, A., and Gerlach, G.F. (2007). An *Actinobacillus pleuropneumoniae arcA* deletion mutant is attenuated and deficient in biofilm formation. Vet. Microbiol. Epub ahead of print.

Campagnari, A.A., Gupta, M.R., Dudas, K.C., Murphy, T.F., and Apicella, M.A. (1987). Antigenic diversity of lipooligosaccharides of nontypeable *Haemophilus influenzae*. Infect. Immun. *55*, 882–887.

Chen, X., Schauder, S., Potier, N., Van Dorsselaer, A., Pelczer, I., Bassler, B.L., and Hughson, F.M. (2002). Structural identification of a bacterial quorum-sensing signal containing boron. Nature *415*, 545–549.

Chow, J.C., Young, D.W., Golenbock, D.T., Christ, W.J., and Gusovsky, F. (1999). Toll-like receptor-4 mediates lipopolysaccharide-induced signal transduction. J. Biol. Chem. *274*, 10689–10692.

Cox, A.D., Howard, M.D., Brisson, J.-R., Van Der Zwan, M., Thibault, P., Perry, M.B., and Inzana, T.J. (1998). Structural analysis of the phase-variable lipooligosaccharide from *Haemophilus somnus* strain 738. Eur. J. Biochem. *253*, 507–516.

Cox, A.D., Howard, M.D., and Inzana, T.J. (2003). Structural analysis of the lipooligosaccharide from the commensal *Haemophilus somnus* strain 1P. Carbohydr. Res. *338*, 1223–1228.

Cox, A.D., Masoud, H., Thibault, P., Brisson, J.R., van der Zwan, M., Perry, M.B., and Richards, J.C. (2001). Structural analysis of the lipopolysaccharide from the nontypeable *Haemophilus influenzae* strain SB 33. Eur. J. Biochem. *268*, 5278–5286.

Daines, D.A., Bothwell, M., Furrer, J., Unrath, W., Nelson, K., Jarisch, J., Melrose, N., Greiner, L., Apicella, M., and Smith, A.L. (2005). *Haemophilus influenzae* luxS mutants form a biofilm and have increased virulence. Microb. Pathogen. *39*, 87–96.

DeMaria, T.F., Apicella, M.A., Nichols, W.A., and Leake, E.R. (1997). Evaluation of the virulence of nontypeable *Haemophilus influenzae* lipopolysaccharide *htrB* and *rfaD* mutants in the chinchilla model of otitis media. Infect. Immun. *65*, 4431–4435.

Dingman, J.R., Rayner, M.G., Mishra, S., Zhang, Y., Ehrlich, M.D., Post, J.C., and Ehrlich, G.D. (1998). Correlation between presence of viable bacteria and presence of endotoxin in middle-ear effusions. J. Clin. Microbiol. *36*, 3417–3419.

Ernst, R.K., Yi, E.C., Guo, L., Lim, K.B., Burns, J.L., Hackett, M., Miller, S.I. (1999). Specific lipopolysaccharide found in cystic fibrosis airway *Pseudomonas aeruginosa*. Science *286*, 1561–1565.

Fan, X., H. Goldfine, E. Lysenko, E., J.N. Weiser (2001). The transfer of choline from the host to the bacterial cell surface requires *glpQ* in *Haemophilus influenzae*. Mol. Microbiol. *41*, 1029–1036.

Federle, M.J., and Bassler, B.L. (2003). Interspecies communication in bacteria. J. Clin. Invest. *112*, 1291–1299.

Fine, D. Furgang, H., D., Schreiner, H.C., Goncharoff, P., Charlesworth, J., Ghazwan, G., Fitzgerald-Bocarsly, P., and Figurski, D.H. (1999). Phenotypic variation in *Actinobacillus actinomycetemcomitans* during laboratory growth: implications for virulence. Microbiol. *145*, 1335–1347.

Fong, K.P., Chung, W.O., Lamont, R.J., and Demuth, D.R. (2001). Intra- and interspecies regulation of gene expression by *Actinobacillus actinomycetemcomitans* LuxS. Infect. Immun. *69*, 7625–7634.

Fong, K.P., Gao, L., and Demuth, D.R. (2003). luxS and arcB control aerobic growth of *Actinobacillus actinomycetemcomitans* under iron limitation. Infect. Immun. *71*, 298–308.

Forsyth, M.H., and Cover, T.L. (2000). Intercellular communication in *Helicobacter pylori*: *luxS* is essential for the production of an extracellular signaling molecule. Infect. Immun. *68*, 3193–3199.

Fox, K.L., Cox, A.D., Gilbert, M., Wakarchuk, W., Li, J., Makepeace, K., Richards, J., Moxon, E.R., and Hood, D.W. (2006). Identification of a bifunctional lipopolysaccharide sialyltransferase in *Haemophilus influenzae*: incorporation of disialic acid. J. Biol. Chem. *281*, 40024–40032.

Fuller, T.E., Kennedy, M.J., and Lowery, D.E. (2000). Identification of *Pasteurella multocida* virulence genes in a septicemic mouse model using signature-tagged mutagenesis. Microb. Pathogen. *29*, 25–38.

Fuqua, C., and Greenberg, E.P. (1998). Self perception in bacteria: quorum sensing with acylated homoserine lactones. Curr. Opin. Microbiol. *1*, 183–189.

Fuqua, W.C., Winans, S.C., and Greenberg, E.P. (1994). Quorum sensing in bacteria: the LuxR-LuxI family of cell density-responsive transcriptional regulators. J. Bacteriol. *176*, 269–275.

Gallaher, T.K., Wu, S., Webster, P., and Aguilera, R. (2006). Identification of biofilm proteins in nontypeable *Haemophilus influenzae*. BMC Microbiol. *6*, 65.

Galli, J., Calo, L., Ardito, F., Imperiale, M., Bassotti, E., Fadda, G., and Paludetti, G. (2007). Biofilm formation by *Haemophilus influenzae* isolated from adeno-tonsil tissue samples, and its role in recurrent adenotonsillitis. Acta Otolaryngol. *27*, 134–138.

Gould, J.M., and Weiser, J.N. (2001). Expression of C-reactive protein in the human respiratory tract. Infect. Immun. *69*, 1747–1754.

Gould, J.M., and Weiser, J.N. (2002). The inhibitory effect of C-reactive protein on bacterial phosphorylcholine platelet-activating factor receptor-mediated adherence is blocked by surfactant. J. Infect. Dis. *186*, 361–371.

Greenberg, E.P., Hastings, J.W., and Ulitzer, S. (1979). Induction of luciferase synthesis in *Beneckea harveyi* by other marine bacteria. Arch. Microbiol. *120*, 87–91.

Greiner, L., Watanabe, H., Phillips, N.J., Shao, J., Morgan, A., Zaleski, A., Gibson, B.W., and Apicella, M.A. (2004). Nontypeable *Haemophilus influenzae* strain 2019 produces a biofilm containing N-acetylneuraminic acid that may mimic sialylated O-linked glycans. Infect. Immun. *72*, 4249–4260.

Griffin, R., Cox, A.D., Makepeace, K., Richards, J.C., Moxon, E.R., and Hood, D.W. (2003). The role of *lex2* in lipopolysaccharide biosynthesis in *Haemophilus influenzae* strains RM7004 and RM153. Microbiol. *149*, 3165–3175.

Gulati, S., Sastry, K., Jensenius, J.C., Rice, P.A., and Ram, S. (2002). Regulation of the mannan-binding lectin pathway of complement on *Neisseria gonorrhoeae* by C1-inhibitor and alpha 2-macroglobulin. J. Immunol. *168*, 4078–4086.

Gunn, J.S., and Miller, S.I. (1996). PhoP-PhoQ activates transcription of *pmrAB*, encoding a two-component regulatory system involved in *Salmonella typhimurium* antimicrobial peptide resistance. J. Bacteriol. *178*, 6857–6864.

Guo, L., Lim, K.B., Poduje, C.M., Daniel, M., Gunn, J.S., Hackett, M., and Miller, S.I. (1998). Lipid A acylation and bacterial resistance against vertebrate antimicrobial peptides. Cell *95*, 189–198.

Hall-Stoodley, L., Hu, F.Z., Gieseke, A., Nistico, L., Nguyen, D., Hayes, J.D., Forbes, M., Greenberg, D.P., Dice, B., Burrows, A., *et al.* (2006). Direct detection of bacterial biofilms on the middle ear mucosa of children with chronic otitis media. JAMA *296*, 202–211.

Harper, M., Boyce, J.D., Cox, A.D., St. Michael, F., Wilkie, I.W., Blackall, P.J., and Adler, B. (2007). *Pasteurella multocida* expresses two lipopolysaccharide glycoforms simultaneously, but only a single form is required for virulence: identification of two acceptor-specific heptosyl I transferases. Infect. Immun. *75*, 3885–3893.

Harvey, H.A., W.E. Swords, and M.A. Apicella (2001). The mimicry of human glycolipids and glycosphingolipids by the lipooligosaccharides of pathogenic *Neisseria* and *Haemophilus*. J. Autoimmun. *16*, 257–262.

Heilmann, C., Schweitzer, O., Gerke, C., Vanittanakom, N., Mack, D., and Gotz, F. (1996). Molecular basis of intercellular adhesion in the biofilm-forming *Staphylococcus epidermidis*. Mol. Microbiol. *20*, 1083–1091.

High, N.J., Deadman, M.E., and Moxon, E.R. (1993). The role of repetitive DNA motif (5'-CAAT-3') in the variable expression of the *Haemophilus influenzae* lipopolysaccharide epitope αGal(1–4)βGal. Mol. Microbiol. *9*, 1275–1282.

High, N.J., Jennings, M.P., and Moxon, E.R. (1996). Tandem repeats of the tetramer 5'-CAAT-3' present in *lic2A* are required for phase variation but not lipopolysaccharide biosynthesis in *Haemophilus influenzae*. Mol. Microbiol. *20*, 165–174.

Hirschfeld, M., Ma, Y., Weis, J.H., Vogel, S.N., and Weis, J.J. (2000). Cutting edge: repurification of lipopolysaccharide eliminates signaling through both human and murine toll-like receptor 2. J. Immunol. *165*, 618–622.

Hong, W., Mason, K., Jurcisek, J.A., Novotny, L.A., Bakaletz, L.O., and Swords, W.E. (2007a). Phosphorylcholine decreases early inflammation and promotes the establishment of stable biofilm communities of nontypeable *Haemophilus influenzae* strain 86–028NP in a chinchilla model of otitis media. Infect. Immun. *75*, 958–965.

Hong, W., Pang, B., West-Barnette, S., and Swords, W.E. (2007b). Phosphorylcholine expression by nontypeable *Haemophilus influenzae* correlates with maturation of biofilm communities in vitro and in vivo. J. Bacteriol. *189*, 8300–8307.

Hood, D.W., Cox, A.D., Gilbert, M., Makepeace, K., Walsh, S., Deadman, M.E., Cody, A., Martin, A., Mansson, M., Schweda, E.K., *et al.* (2001a). Identification of a lipopolysaccharide alpha-2, 3-sialyltransferase from *Haemophilus influenzae*. Mol. Microbiol. *39*, 341–350.

Hood, D.W., Cox, A.D., Wakarchuk, W.W., Schur, M., Schweda, E.K., Walsh, S.L., Deadman, M.E., Martin, A., Moxon, E.R., and Richards, J.C. (2001b). Genetic basis for expression of the major globotetraose-containing lipopolysaccharide from *H. influenzae* strain Rd (RM118). Glycobiol. *11*, 957–967.

Hood, D.W., Deadman, M.E., Allen, T., Masoud, H., Martin, A., Brisson, J.R., Fleischmann, R., Venter, J.C., Richards, J.C., and Moxon, E.R. (1996). Use of the complete genome sequence information of *Haemophilus influenzae* strain Rd to investigate lipopolysaccharide biosynthesis. Mol. Microbiol. *22*, 951–965.

Hood, D.W., Deadman, M.E., Jennings, M.P., Bisercic, M., Fleischmann, R.D., Venter, J.C., and Moxon, E.R. (1996). DNA repeats identify novel virulence genes in *Haemophilus influenzae*. Proc. Natl. Acad. Sci. USA *93*, 11121–11125.

Hood, D.W., Makepeace, K., Deadman, M.E., Rest, R.F., Thibault, P., Martin, A., Richards, J.C., and Moxon, E.R. (1999a). Sialic acid in the lipopolysaccharide of *Haemophilus influenzae*: strain distribution, influence on serum resistance and structural characterization. Mol. Microbiol. *33*, 679–692.

Hood, D.W., Randle, G., Cox, A.D., Makepeace, K., Li, J., Schweda, E.K., Richards, J.C., and Moxon, E.R. (2004). Biosynthesis of cryptic lipopolysaccharide glycoforms in *Haemophilus influenzae* involves a mechanism similar to that required for O-antigen synthesis. J. Bacteriol. *186*, 7429–7439.

Hood, D.W., Richards, J.C., and Moxon, E.R. (1999b). *Haemophilus influenzae* lipopolysaccharide. Biochem. Soc. Trans. *27*, 493–498.

Hoshino, K., Takeuchi, O., Kawai, T., Sanjo, H., Ogawa, T., Takeda, Y., Takeda, K., and Akira, S. (1999). Cutting edge: Toll-like receptor 4 (TLR4)-deficient mice are hyporesponsive to lipopolysaccharide: evidence for TLR4 as the Lps gene product. J. Immunol. *162*, 3749–3752.

Howard, M.D., Cox, A.D., Weiser, J.N., Schurig, G.G., Inzana, T.J. (2000). Antigenic diversity of *Haemophilus somnus* lipooligosaccharide:phase-variable accessibility of the phosphorylcholine epitope. J. Clin. Microbiol. *38*, 4412–4419.

Inzana, T.J., and L.B. Corbeil. (2004). *Haemophilus*. In Pathogenesis of Bacterial Infections of Animals, in C. Gyles, J. Prescott, and G. Songer, eds. (Ames, Iowa, Iowa State University Press), pp. pp.243–258.

Inzana, T.J., Glindemann, G., Cox, A.D., Wakarchuk, W., and Howard, M.D. (2002). Incorporation of N-acetylneuraminic acid into *Haemophilus somnus* lipooligosaccharide (LOS): enhancement of resistance to serum and reduction of LOS antibody binding. Infect. Immun. *70*, 4870–4879.

Inzana, T.J., Gogolewski, R.P., and Corbeil, L.B. (1992). Phenotypic phase variation in *Haemophilus somnus* lipooligosaccharide during bovine pneumonia and after in vitro passage. Infect. Immun. *60*, 2943–2951.

Inzana, T.J., Hensley, J., McQuiston, J., Lesse, A.J., Campagnari, A.A., Boyle, S.M., and Apicella, M.A. (1997). Phase variation and conservation of lipooligosaccharide epitopes in *Haemophilus somnus*. Infect. Immun. *65*, 4675–4681.

Inzana, T.J., Iritani, B., Gogolewski, R.P., Kania, S.A., and Corbeil, L.B. (1988). Purification and characterization of lipooligosaccharides from four strains of '*Haemophilus somnus*'. Infect. Immun. *56*, 2830–2837.

Izano, E.A., Sadovskaya, I., Vinogradov, E., Mulks, M., Velliyagounder, K., Ragunath, C., Kher, W.B., Ramasubbu, N., Jabbouri, S., Perry, M.B., *et al.* (2007a). Poly-N-acetylglucosamine mediates biofilm formation and antibiotic resistance in *Actinobacillus pleuropneumoniae*. Microb. Pathog. *43*, 1–9.

Izano, E.A., Sadovskaya, I., Vinogradov, E., Mulks, M.H., Velliyagounder, K., Ragunath, C., Kher, W.B., Ramasubbu, N., Jabbouri, S., Perry, M.B., *et al.* (2007b). Poly-N-acetylglucosamine mediates biofilm formation and antibiotic resistance in *Actinobacillus pleuropneumoniae*. Microb. Pathog. *43*, 1–9.

Izano, E.A., Wang, H., Ragunath, C., Ramasubbu, N., and Kaplan, J.B. (2007c). Detachment and killing of *Aggregatobacter actinomycetemcomitans* biofilms by dispersin B and SDS. J. Dent. Res. *86*, 618–622.

Jacobsen, I., Hennig-Pauka, I., Baltes, N., Trost, M., and Gerlach, G.F. (2005). Enzymes involved in anaerobic respiration appear to play a role in *Actinobacillus pleuropneumoniae* virulence. Infect. Immun. *73*, 226–234.

Jennings, M.P., Hood, D.W., Peak, I.R., Virji, M., and Moxon, E.R. (1995). Molecular analysis of a locus for the biosynthesis and phase-variable expression of the lacto-N-neotetraose terminal lipopolysaccharide structure in *Neisseria meningitidis*. Mol. Microbiol. *18*, 729–740.

Jin, H., Zhou, R., Kang, M., Luo, R., Cai, X., and Chen, H. (2006). Biofilm formation by field isolates and reference strains of *Haemophilus parasuis*. Vet. Microbiol. *118*, 117–123.

Jones, P.A., Samuels, N.A., Phillips, N.J., Munson, R.S., Bozue, J.A., Arseneau, J.A., Nichols, W.A., Zaleski, A., Gibson, B.W., and Apicella, M.A. (2002). *Haemophilus influenzae* type B strain A2 has multiple sialyltransferases involved in lipooligosaccharide sialylation. J. Biol. Chem. *277*, 14598–14611.

Jurcisek, J.A., and Bakaletz, L.O. (2007). Biofilms formed by nontypeable *Haemophilus influenzae* in vivo contain both double-stranded DNA and type IV pilin protein. J. Bacteriol. *189*, 3868–3875.

Jurcisek, J.A., Bookwalter, J., Baker, B., Fernandez, S., Novotny, L.A., Munson, R.S., Jr., and Bakaletz, L.O. (2007). The PilA protein of nontypeable *Haemophilus influenzae* plays a role in biofilm formation, adherence to epithelial cells and colonization of the mammalian upper respiratory tract. Mol. Microbiol. *65*, 1288–1299.

Jurcisek, J.A., Greiner, L., Watanabe, H., Zaleski, A., Apicella, M.A., and Bakaletz, L.O. (2005). Role of sialic acid and complex carbohydrate biosynthesis in biofilm formation by nontypeable *Haemophilus influenzae* in the chinchilla middle ear. Infect. Immun. *73*, 3210–3218.

Kachlany, S.C., Planet, P.J., DeSalle, R., Fine, D.H., and Figurski, D.H. (2001). Genes for tight adherence of *Actinobacillus actinomycetemcomitans*: from plaque to plague to pond scum. Trends Microbiol. *9*, 429–437.

Kaplan, J.B., Meyenhofer, M.F., and Fine, D.H. (2003a). Biofilm growth and detachment of *Actinobacillus actinomycetemcomitans*. J. Bacteriol. *185*, 1399–1404.

Kaplan, J.B., and Mulks, M. (2005). Biofilm formation is prevalent among field isolates of *Actinobacillus pleuropneumoniae*. Vet. Microbiol. *108*, 89–94.

Kaplan, J.B., Ragunath, C., Ramasubbu, N., and Fine, D.H. (2003b). Detachment of *Actinobacillus actinomycetemcomitans* biofilm cells by an endogenous beta-hexosaminidase activity. J. Bacteriol. *185*, 4693–4698.

Kaplan, J.B., Velliyagounder, K., Ragunath, C., Rohde, H., Mack, D., Knobloch, J.K., and Ramasubbu, N. (2004). Genes involved in the synthesis and degradation of matrix polysaccharide in *Actinobacillus actinomycetemcomitans* and *Actinobacillus pleuropneumoniae* biofilms. J. Bacteriol. *186*, 8213–8220.

Keller, L., and Surette, M.G. (2006). Communication in bacteria: an ecological and evolutionary perspective. Nature Rev. *4*, 249–258.

Kendall, M.M., and Sperandio, V. (2007). Quorum sensing by enteric pathogens. Curr. Opin. Gastroenterol. *23*, 10–15.

Kropec, A., Maira-Litran, T., Jefferson, K.K., Grout, M., Cramton, S.E., Gotz, F., Goldmann, D.A., and Pier, G.B. (2005). Poly-*N*-acetylglucosamine production in *Staphylococcus aureus* is essential for virulence in murine models of systemic infection. Infect. Immun. *73*, 6868–6876.

Kuckleburg, C.J., Elswaifi, S.F., Inzana, T.J., and Czuprynski, C.J. (2007). Expression of phosphorylcholine by *Histophilus somni* induces bovine platelet aggregation. Infect. Immun. *75*, 1045–1049.

Kwaik, Y.A., McLaughlin, R.E., Apicella, M.A., and Spinola, S.M. (1991). Analysis of *Haemophilus influenzae* type b lipooligosaccharide-synthesis genes that assemble or expose a 2-Keto-3-deoxyoctulosonic acid epitope. Mol. Microbiol. *5*, 2475–2480.

Kwaik, Y.A., McLaughlin, R.E., Apicella, M.A., and Spinola, S.M. (1992). Cloning and analysis of lipooligosaccharide synthesis genes of *Haemophilus influenzae* type b that assemble or expose a 2-keto-3-deoxyoctulosonic acid epitope in *Escherichia coli*. J. Infect. Dis. *165*, 195–196.

Lacroix, R.P., Duncan, J.R., Jenkins, R.P., Leitch, R.A., Perry, J.A., and Richards, J.C. (1993). Structural and serological specificities of *Pasteurella haemolytica* lipopolysaccharides. Infect. Immun. *61*, 170–181.

Li, J., Bauer, S.H., Mansson, M., Moxon, E.R., Richards, J.C., and Schweda, E.K. (2001). Glycine is a common substituent of the inner core in *Haemophilus influenzae* lipopolysaccharide. Glycobiol. *11*, 1009–1015.

Logan, S.M., Chen, W., Aubry, A., Gidney, M.A., Lacelle, S., St Michael, F., Kuolee, R., Higgins, M., Neufeld, S., and Cox, A.D. (2006). Production of a D-glycero-D-manno-heptosyltransferase mutant of *Mannheimia haemolytica* displaying a veterinary pathogen specific conserved LPS structure; development and functionality of antibodies to this LPS structure. Vet. Microbiol. *116*, 175–186.

Lysenko, E., Richards, J.C., Cox, A.D., Stewart, A., Martin, A., Kapoor, M., and Weiser, J.N. (2000a). The position of phosphorylcholine on the lipopolysaccharide of *Haemophilus influenzae* affects binding and sensitivity to C-reactive protein-mediated killing. Mol. Microbiol. *35*, 234–245.

Lysenko, E.S., Gould, J., Bals, R., Wilson, J.M., and Weiser, J.N. (2000b). Bacterial phosphorylcholine decreases susceptibility to the antimicrobial peptide LL-37/hCAP18 expressed in the upper respiratory tract. Infect. Immun. *68*, 1664–1671.

Malott, R.J., and Lo, R.Y. (2002). Studies on the production of quorum-sensing signal molecules in *Mannheimia haemolytica* A1 and other Pasteurellaceae species. FEMS Microbiol. Lett. *206*, 25–30.

Mandrell, R.E., McLaughlin, R., Kwaik, Y.A., Lesse, A., Yamasaki, R., Gibson, B., Spinola, S.A., and Apicella, M.A. (1992). Lipooligosaccharides (LOS) of some *Haemophilus* species mimic human glycosphingolipids, and some LOS are sialyated. Infect. Immun. *60*, 1322–1328.

Mansson, M., Hood, D.W., Li, J., Richards, J.C., Moxon, E.R., and Schweda, E.K. (2002). Structural analysis of the lipopolysaccharide from nontypeable *Haemophilus influenzae* strain 1003. Eur. J. Biochem. *269*, 808–818.

Mansson, M., Hood, D.W., Moxon, E.R., and Schweda, E.K. (2003). Structural diversity in lipopolysaccharide expression in nontypeable *Haemophilus influenzae*. Eur. J. Biochem. *270*, 610–624.

McCrea, K.W., Xie, J., Lacross, N., Patel, M., Mukundan, D., Murphy, T.F., Marrs, C.F., and Gilsdorf, J.R. (2008). Relationships of non-typeable *Haemophilus influenzae* to hemolytic and non-hemolytic *Haemophilus haemolyticus* strains. J. Clin. Microbiol. *46*, 406–416.

McLaughlin, R., Spinola, S.M., and Apicella, M.A. (1992). Generation of lipooligosaccharide mutants of *Haemophilus influenzae* type b. J. Bacteriol. *174*, 6455–6459.

McQuiston, J.H., McQuiston, J.R., Cox, A.D., Wu, Y., Boyle, S.M., and Inzana, T.J. (2000). Characterization of a DNA region containing 5′-CAAT-3′ DNA sequences involved in lipooligosaccharide biosynthesis in *Haemophilus somnus*. Microb. Pathog. *28*, 301–312.

Michael, F.S., Brisson, J.R., Larocque, S., Monteiro, M., Li, J., Jacques, M., Perry, M.B., and Cox, A.D. (2004). Structural analysis of the lipopolysaccharide derived core oligosaccharides of *Actinobacillus pleuropneumoniae* serotypes 1, 2, 5a and the genome strain 5b. Carbohyd. Res. *339*, 1973–1984.

Miller, S.T., Xavier, K.B., Campagna, S.R., Taga, M.E., Semmelhack, M.F., Bassler, B.L., and Hughson, F.M. (2004). *Salmonella typhimurium* recognizes a chemically distinct form of the bacterial quorum-sensing signal AI-2. Mol. Cell *15*, 677–687.

Mukundan, D., Ecevit, Z., Patel, M., Marrs, C.F., and Gilsdorf, J.R. (2007). Pharyngeal colonization dynamics of *Haemophilus influenzae* and *Haemophilus haemolyticus* in healthy adult carriers. J. Clin. Microbiol. *45*, 3207–3217.

Murphy, T.F., and Kirkham, C. (2002). Biofilm formation by nontypeable *Haemophilus influenzae*: strain variability, outer membrane antigen expression and role of pili. BMC Microbiol. *2*, 7.

Murphy, T.F., Kirkham, C., Sethi, S., and Lesse, A. (2005). Expression of a peroxiredoxin-glutaredoxin by *Haemophilus influenzae* in bioflims and during human respiratory tract infection. FEMS Immunol. Med. Microbiol. *44*, 81–89.

Nichols, W.A., Raetz, C.R.H., Clementz, T., Smith, A.L., Hanson, J.A., Ketterer, M.R., Sunshine, M., and Apicella, M.A. (1997). *htrB* of *Haemophilus influenzae*: determination of biochemical activity and effects on virulence and lipooligosaccharide toxicity. J Endotoxin Res. *4*, 163–172.

Nielsen, K.K., and Boye, M. (2005). Real-time quantitative reverse transcription-PCR analysis of expression stability of *Actinobacillus pleuropneumoniae* housekeeping genes during in vitro growth under iron-depleted conditions. Appl. Environ. Microbiol. *2005*, 71.

Olson, M.E., Ceri, H., Morck, D., Buret, A., and Read, R. (2002). Biofilm bacteria: formation and comparative susceptibility to antibiotics. Can. J. Vet. Res. *66*, 86–92.

Pangburn, M.K., and Muller-Eberhard, H.J. (1978). Complement C3 convertase: cell surface restriction of beta1H control and generation of restriction on neuraminidase-treated cells. Proc. Natl. Acad. Sci. USA *75*, 2416–2420.

Pangburn, M.K., Schreiber, R.D., and Muller-Eberhard, H.J. (1977). Human complement C3b inactivator: isolation, characterization, and demonstration of an absolute requirement for the serum protein beta1H for cleavage of C3b and C4b in solution. J. Exp. Med. *146*, 257–270.

Perez, B.A., Planet, P.J., Kachlany, S.C., Tomich, M., Fine, D.H., and Figurski, D.H. (2006). Genetic analysis of the requirement for *flp-2*, *tadV*, and *rcpB* in *Actinobacillus actinomycetemcomitans* biofilm formation. J. Bacteriol. *188*, 6361–6375.

Perry, M.B., Altman, E., Brisson, J.-R., Beynon, L.M., and Richards, J.C. (1990). Structural characteristics of the antigenic capsular polysaccharides and lipopolysaccharides involved in the serological classification of *Actinobacillus pleuropneumoniae* strains. Natl. Res. Coun. Can. Ottawa, Ontario.

Perry, M.B., MacLean, L.L., Gmur, R., and Wilson, M.E. (1996). Characterization of the O-polysaccharide structure of lipopolysaccharide from *Actinobacillus actinomycetemcomitans* serotype b. Infect. Immun. *64*, 1215–1219.

Phillips, N.J., Apicella, M.A., Griffiss, J.M., and Gibson, B.W. (1993). Structural studies of the lipooligosaccharides from *Haemophilus influenzae* type b strain A2. Biochem. *32*, 2003–2012.

Phillips, N.J., Miller, T.J., Engstrom, J.J., Melaugh, W., McLaughlin, R., Apicella, M.A., and Gibson, B.W. (2000). Characterization of chimeric lipopolysaccharides from *Escherichia coli* strain JM109 transformed with lipooligosaccharide synthesis genes (*lsg*) from *Haemophilus influenzae*. J. Biol. Chem. *275*, 4747–4758.

Poltorak, A., Ricciardi-Castagnoli, P., Citterio, S., and Beutler, B. (2000). Physical contact between lipopolysaccharide and toll-like receptor 4 revealed by genetic complementation. Proc. Natl. Acad. Sci. USA *97*, 2163–2167.

Preston, A., Mandrell, R.E., Gibson, B.W., and Apicella, M.A. (1996). The lipopolysaccharides of pathogenic gram-negative bacteria. Crit. Rreviews Microbiol. *22*, 139–180.

Rayner, M.G., Zhang, Y., Gorry, M.C., Chen, Y., Post, J.C., and Ehrlich, G.D. (1998). Evidence of bacterial metabolic activity in culture-negative otitis media with effusion. JAMA *279*, 296–299.

Richards, J.C., Cox, A.D., Schweda, E.K., Martin, A., Hood, D.W., and Moxon, E.R. (2001). Structure and functional genomics of lipopolysaccharide expression in *Haemophilus influenzae*. Adv. Exp. Med. Biol. *491*, 515–524.

Richards, J.C., and Leitch, R.A. (1989). Elucidation of the structure of the *Pasteurella haemolytica* serotype T10 lipopolysaccharide O-antigen by n.m.r. spectroscopy. Carbohyd. Res. *186*, 275–286.

Risberg, A., Masoud, H., Martin, A., Richards, J.C., Moxon, E.R., and Schweda, E.K. (1999). Structural analysis of the lipopolysaccharide oligosaccharide epitopes expressed by a capsule-deficient strain of *Haemophilus influenzae* Rd. Eur. J. Biochem. *261*, 171–180.

Roberts, M.C., Mintz, C.S., and Morse, S.A. (1986). Characterization of *Haemophilus parainfluenzae* strains with low-Mr or ladder-like lipopolysaccharides. J. Gen. Microbiol. *132*, 611–616.

Ross, R.F. (2006). *Pasteurella multocida* and its role in porcine pneumonia. Anim. Health Res. Rev. *7*, 13–29.

Sandal, I., Hong, W., Swords, W.E., and Inzana, T.J. (2007). Characterization and comparison of biofilm development by pathogenic and commensal isolates of *Histophilus somni*. J. Bacteriol. *189*, 8179–8185.

Sanderson, A.R., Leid, J., and Hunsaker, D. (2006). Bacterial biofilms on the sinus mucosa of human subjects with chronic rhinosinusitis. Laryngoscope *116*, 1121–1126.

Schreiner, H.C., Sinatra, K., Kaplan, J.B. Furgang, D., Kachlany, S.C., Planet, P.J., Perez, B.A., Figurski, D.H., and Fine, D.H. (2003). Tight-adherence genes of *Actinobacillus actinomycetemcomitans* are required for virulence in a rat model. Proc. Natl. Acad. Sci. USA *100*, 7295–7300.

Schweda, E.K., Brisson, J.R., Alvelius, G., Martin, A., Weiser, J.N., Hood, D.W., Moxon, E.R., and Richards, J.C. (2000). Characterization of the phosphocholine-substituted oligosaccharide in lipopolysaccharides of type b *Haemophilus influenzae*. Eur. J. Biochem. *267*, 3902–3913.

Schweda, E.K., Richards, J.C., Hood, D.W., and Moxon, E.R. (2007). Expression and structural diversity of the lipopolysaccharide of *Haemophilus influenzae*: implication in virulence. Int. J. Med. Microbiol. *297*, 297–306.

Schweda, E.K., Sundstrom, A.C., Eriksson, L.M., Jonasson, J.A., and Lindberg, A.A. (1994). Structural studies of the cell envelope lipopolysaccharides from *Haemophilus ducreyi* strains ITM 2665 and ITM 4747. J. Biol. Chem. *269*, 12040–12048.

Shao, H., Lamont, R.J., and Demuth, D.R. (2007). Autoinducer 2 is required for biofilm growth of *Aggregatibacter (Actinobacillus) actinomycetemcomitans*. Infect. Immun. *75*, 4211–4218.

Spinola, S.M., Fortney, K.R., Katz, B.P., Latimer, J.L., Mock, J.R., Vakevainen, M., and Hansen, E.J. (2003). *Haemophilus ducreyi* requires an intact flp gene cluster for virulence in humans. Infect. Immun. *71*, 7178–7182.

St Michael, F., Howard, M.D., Li, J., Duncan, A.J., Inzana, T.J., and Cox, A.D. (2004). Structural analysis of the lipooligosaccharide from the commensal *Haemophilus somnus* genome strain 129Pt. Carbohyd. Res. *339*, 529–535.

St Michael, F., Inzana, T.J., and Cox, A.D. (2006). Structural analysis of the lipooligosaccharide-derived oligosaccharide of *Histophilus somni* (*Haemophilus somnus*) strain 8025. Carbohyd. Res. *341*, 281–284.

St Michael, F., Li, J., Howard, M.D., Duncan, A.J., Inzana, T.J., and Cox, A.D. (2005a). Structural analysis of the oligosaccharide of *Histophilus somni* (*Haemophilus somnus*) strain 2336 and identification of several lipooligosaccharide biosynthesis gene homologues. Carbohyd. Res. *340*, 665–672.

St Michael, F., Li, J., Vinogradov, E., Larocque, S., Harper, M., and Cox, A.D. (2005b). Structural analysis of the lipopolysaccharide of *Pasteurella multocida* strain VP161: identification of both Kdo-P and Kdo-Kdo species in the lipopolysaccharide. Carbohyd. Res. *340*, 59–68.

St Michael, F., Vinogradov, E., Li, J., and Cox, A.D. (2005c). Structural analysis of the lipopolysaccharide from *Pasteurella multocida* genome strain Pm70 and identification of the putative lipopolysaccharide glycosyltransferases. Glycobiol. *15*, 323–333.

Starner, T.D., W.E. Swords, M.A. Apicella, and P.B. McCray., Jr. (2002). Lipid A acylation is a determinant of resistance of nontypeable *Haemophilus influenzae* to killing by beta-defensins. Infect. Immun. *70*, 5287–5289.

Sturme, M.H., Kleerebezem, M., Nakayama, J., Akkermans, A.D., Vaugha, E.E., and de Vos, W.M. (2002). Cell to cell communication by autoinducing peptides in gram-positive bacteria. Antonie van Leeuwenhoek *81*, 233–243.

Swords, W.E., Buscher, B., Ver Steeg, K., Nichols, W., Preston, A., Weiser, J.N., Gibson, B., and Apicella, M.A. (2000). Nontypeable *Haemophilus influenzae* adhere to and invade human bronchial epithelial cells by an interaction of lipooligosaccharide with the PAF receptor. Mol. Microbiol. *37*, 13–27.

Swords, W.E., Chance, D.L., Cohn, L.A., Shao, J., Apicella, M.A., and Smith, A.L. (2002). Acylation of the lipooligosaccharide of *Haemophilus influenzae* and colonization: an *htrB* mutation diminishes the colonization of human airway epithelial cells. Infect. Immun. *70*, 4661–4668.

Swords, W.E., M.R. Ketterer, J. Shao, C.A. Campbell, J.N. Weiser, and Apicella, M.A. (2001). Binding of the nontypeable *Haemophilus influenzae* lipooligosaccharide to the PAF receptor initiates host cell signaling. Cell Microbiol. *8*, 525–536.

Swords, W.E., Moore, M.L., Godzicki, L., Bukofzer, G., Mitten, M.J., and VonCannon, J. (2004). Sialylation of lipooligosaccharides promotes biofilm formation by nontypeable *Haemophilus influenzae*. Infect. Immun. *72*, 106–113.

Sylte, M.J., Inzana, T.J., and Czuprynski, C.J. (2004). Reactive oxygen and nitrogen intermediates contribute to *Haemophilus somnus* lipooligosaccharide-mediated apoptosis of bovine endothelial cells. Vet. Immunol. Immunopathol. *97*, 207–217.

Sylte, M.J., Kuckleburg, C.J., Inzana, T.J., Bertics, P.J., and Czuprynski, C.J. (2005). Stimulation of P2X receptors enhances lipooligosaccharide-mediated apoptosis of endothelial cells. J Leukoc. Biol. *77*, 665–672.

Sylte, M.J., Kuckleburg, C.J., Leite, F.P., Inzana, T.J., and Czuprynski, C.J. (2006). Tumor necrosis factor-alpha enhances *Haemophilus somnus* lipooligosaccharide-induced apoptosis of bovine endothelial cells. Vet. Immunol. Immunopathol. *110*, 303–309.

Sylte, M.J., Leite, F.P., Kuckleburg, C.J., Inzana, T.J., and Czuprynski, C.J. (2003). Caspase activation during *Haemophilus somnus* lipooligosaccharide-mediated apoptosis of bovine endothelial cells. Microb. Pathog. *35*, 285–291.

Syrogiannopoulos, G.A., Hansen, E.J., Erwin, A.L., Munford, R.S., Rutledge, J., Reisch, J.S., and McCracken, G.H., Jr. (1988). *Haemophilus influenzae* type b lipooligosaccharide induces meningeal inflammation. J. Infect. Dis. *157*, 237–244.

Tong, H.H., Blue, L.E., James, M.A., Chen, Y.P., and DeMaria, T.F. (2000). Evaluation of phase variation of nontypeable *Haemophilus influenzae* lipooligosaccharide during nasopharyngeal colonization and development of otitis media in the chinchilla model. Infect. Immun. *68*, 4593–4597.

van der Vinne, A.N., Lo, R.Y., and Shewen, P.E. (2005). Construction and analysis of a *Mannheimia haemolytica* A1 *luxS* mutant. Vet. Microbiol. *110*, 53–66.

Vuong, C., Voyich, J.M., Fischer, E.R., Braughton, K.R., Whitney, A.R., DeLeo, F.R., and Otto, M. (2004). Polysaccharide intercellular adhesin (PIA) protects *Staphylococcus epidermidis* against major components of the human innate immune system. Cell. Microbiol. *6*, 269–275.

Wang, X., Preston, J.F., 3rd, and Romeo, T. (2004). The *pgaABCD* locus of *Escherichia coli* promotes the synthesis of a polysaccharide adhesin required for biofilm formation. J. Bacteriol. *186*, 2724–2734.

Webster, P., Wu, S., Gomez, G., Apicella, M.A., Plaut A.G., and St. Geme, J.W., 3rd. (2006). Distribution of bacterial proteins in biofilms formed by nontypeable *Haemophilus influenzae*. J. Histochem. Cytochem. *54*, 829–842.

Weiser, J.N., Lindberg, A.A., Manning, E.J., Hansen, E.J., and Moxon, E.R. (1989a). Identification of a chromosomal locus for expression of lipopolysaccharide epitopes in *Haemophilus influenzae*. Infect. Immun. *57*, 3045–3052.

Weiser, J.N., Love, J.M., and Moxon, E.R. (1989b). The molecular mechanism of phase variation of *Haemophilus influenzae* lipopolysaccharide. Cell *59*, 657–665.

Weiser, J.N., and Pan, N. (1998). Adaptation of *Haemophilus influenzae* to acquired and innate humoral immunity based on phase variation of lipopolysaccharide. Mol. Microbiol. *30*, 767–775.

Weiser, J.N., Pan, N., McGowan, K.L., Musher, D., Martin, A., and Richards, J. (1998). Phosphorylcholine on the lipopolysaccharide of *Haemophilus influenzae* contributes to persistence in the respiratory tract and sensitivity to serum killing mediated by C-reactive protein. J. Exp. Med. *187*, 631–640.

Weiser, J.N., Shchepetov, M., and Chong, S.T. (1997). Decoration of lipopolysaccharide with phosphorylcholine: a phase-variable characteristic of *Haemophilus influenzae*. Infect. Immun. *65*, 943–950.

Weiser, J.N., Williams, A., and Moxon, E.R. (1990). Phase-variable lipopolysaccharide structures enhance the invasive capacity of *Haemophilus influenzae*. Infect. Immun. *58*, 3455–3457.

West-Barnette, S., Rockel, A., and Swords, W.E. (2006). Biofilm growth increases phosphorylcholine content and decreases potency of nontypeable *Haemophilus influenzae* endotoxins. Infect. Immun. *74*, 1828–1836.

White, K.A., Kaltashov, I.A., Cotter, R.J., and Raetz, C.R.H. (1997). A mono-functional 3-deoxy-D-manno-octulosonic acid (Kdo) transferase and a Kdo kinase in extracts of *Haemophilus influenzae*. J. Biol. Chem. *272*, 16555–16563.

White, K.A., Lin, S., Cotter, R.J., and Raetz, C.R. (1999). A *Haemophilus influenzae* gene that encodes a membrane bound 3-deoxy-D-manno-octulosonic acid (Kdo) kinase. Possible involvement of kdo phosphorylation in bacterial virulence. J. Biol. Chem. *274*, 31391–31400.

Wong, S.M., and Akerley, B.J. (2005). Environmental and genetic regulation of the phosphorylcholine epitope of *Haemophilus influenzae* lipooligosaccharide. Mol. Microbiol. *55*, 724–738.

Wu, Y., McQuiston, J.H., Cox, A., Pack, T.D., and Inzana, T.J. (2000). Molecular cloning and mutagenesis of a DNA locus involved in lipooligosaccharide biosynthesis in *Haemophilus somnus*. Infect. Immun. *68*, 310–319.

Yamasaki, R., Nasholds, W., Schneider, H., and Apicella, M.A. (1991.). Epitope expression and partial structural characterization of F62 lipooligosaccharide (LOS) of *Neisseria gonorrhoeae*. Mol. Immunol. *28*, 1233–1242.

Yoo, H.S., Maheswaran, S.K., Lin, G., Townsend, E.L., and Ames, T.R. (1995). Induction of inflammatory cytokines in bovine alveolar macrophages following stimulation with *Pasteurella haemolytica* lipopolysaccharide. Infect. Immun. *63*, 381–388.

Young, R., Fortney, K., Haley, J.C., Hood, A.F., Campagnari, A.A., Wang, J., Bozue, J.A., Munson, R.S., Jr., and Spinola, S. (1999). Expression of sialylated or paragloboside-like lipooligosaccharides are not required for pustule formation by *Haemophilus ducreyi* in human volunteers. Infect. Immun. *67*, 6335–6340.

Mechanisms of Antimicrobial Resistance in *Pasteurellaceae* 10

Stefan Schwarz

Abstract

Members of the family *Pasteurellaceae* cause a wide variety of diseases in humans and animals. Antimicrobial agents represent the most powerful tools to control such infections. Acquisition of resistance genes as well as development of resistance-mediating mutations, however, strongly reduce the efficacy of the antimicrobial agents. In this chapter, a review of the current knowledge of the genetic basis of resistance to antimicrobial agents is given with particular reference to resistances to β-lactam antibiotics, tetracyclines, aminoglycosides/aminocyclitols, phenicols, sulphonamides, trimethoprim, macrolides and quinolones. Moreover, the role of plasmids and transposons in the spread of the resistance genes among *Pasteurellaceae* and members of other bacterial families is highlighted to provide insight into the possibilities of horizontal dissemination, co-selection and persistence of antimicrobial resistance genes.

Introduction

Bacteria of the family *Pasteurellaceae* represent important pathogens in both, humans and animals. Among humans, *Haemophilus influenzae, Aggregatibacter actinomycetemcomitans*, and *Pasteurella multocida* are the most relevant pathogens. While *H. influenzae* is associated with a wide variety of diseases including childhood pneumonia, meningitis, bacteraemia, community-acquired pneumonia, acute otitis media, acute sinusitis, and acute exacerbations of chronic bronchitis (Tristam *et al.*, 2007), *Agg. actinomycetemcomitans* is mostly associated with periodontal diseases. *P. multocida* infections in humans are mainly the result of dog and cat bites. The most relevant animal pathogens among the *Pasteurellaceae* are *P. multocida, Mannheimia haemolytica, Actinobacillus pleuropneumoniae,* [*Haemophilus*] *parasuis*, and *Histophilus somni*. *P. multocida* is involved in a variety of diseases including haemorrhagic septicaemia in cattle and water buffaloes, pneumonia in cattle, sheep and pigs, atrophic rhinitis in pigs, fowl cholera in poultry, snuffles in rabbits, and mastitis in sheep (Kehrenberg *et al.*, 2006b). *M. haemolytica* is often involved in respiratory tract infections in cattle whereas *A. pleuropneumoniae* is the causative agent of a haemorrhagic pleuropneumonia in pigs. [*H.*] *parasuis* causes the 'Glässer's disease' in pigs, which is characterized by high fevers, polyserositis, polysynovitis, respiratory distress and meningitis, while *Hist. somni* is the etiological agent of thromboembolic meningoencephalitis in cattle (Kehrenberg *et al.*, 2006b). Besides these major pathogens, various other members of the family *Pasteurellaceae* have been reported to be less frequently associated with diseases in humans and animals (Kehrenberg *et al.*, 2006b).

Antimicrobial agents are commonly applied to treat or prevent infections caused by the aforementioned pathogens. However, as most other bacterial pathogens, members of the family *Pasteurellaceae* also respond to the selective pressure imposed by the use of antimicrobial agents by developing or acquiring resistance genes or resistance mediating mutations. In this regard, mobile genetic elements, such as plasmids or transposons that carry resistance genes, play an important role in the dissemination of antimicrobial resistance with members of the family *Pasteurellaceae* acting as donors and/or acceptors. Several reviews have been published during recent years which focused on the genetic basis of antimicrobial resistance in bacteria of the family *Pasteurellaceae* (Kehrenberg *et al.*, 2001b, 2006b; Tristam *et al.*, 2007). The present chapter provides an update of the current knowledge on genes and mutations conferring antimicrobial resistance among *Pasteurellaceae*. Moreover, a brief description of the most recent data on the susceptibility status of selected members of the family *Pasteurellaceae* is provided.

Susceptibility testing and susceptibility status of *Pasteurellaceae*

Although prudent use guidelines request the identification of the causative pathogen and the determination of its *in vitro* susceptibility prior to the onset of an antimicrobial therapy, the mostly acute character of the diseases – and in veterinary medicine, the rapid spread of the causative organisms within animal herds – often require an immediate therapeutic intervention in which the initial choice of the antimicrobial agents may be revised on the basis of the results of the diagnostic tests. In this regard, two major aspects need to be taken into consideration: (a) the correct performance of *in vitro* susceptibility tests and (b) representative data on the susceptibility status of members of the *Pasteurellaceae*.

Susceptibility testing

The results of *in vitro* susceptibility testing shall ensure a most efficacious antimicrobial therapy by excluding antimicrobial agents to which the causative bacterial pathogen shows resistance or reduced susceptibility under *in vitro* conditions. The correct performance of the tests is essential in order to most accurately predict the clinical response of the bacterium. Several members of the family *Pasteurellaceae* are classified as 'fastidious organisms' since they require specific growth conditions, e.g. supplementation of the media with components essential for the growth of the respective bacteria. In this regard, performance of *in vitro* susceptibility testing should follow a standardized and internationally accepted guideline. Such guidelines are available for human pathogens of the family *Pasteurellaceae* – namely *H. influenzae* – from various organizations, including among others the Clinical and Laboratory Standards Institute (CLSI), the British Society for Antimicrobial Chemotherapy (BSAC), the Deutsches Institut für Normung (DIN) from Germany, and the Comité de l'Antibiogramme de la Societé Française de Microbiologie (SFM). A variety of different test media for *H. influenzae* have been proposed by the different guidelines, including *Haemophilus* test medium (HTM), Mueller-Hinton broth, IsoSensitest broth or IsoSensitest broth-based HTM, all of which, however, are supplemented by β-NAD and contain an iron source such as haemoglobin, haematin, or haemin (Tristam *et al.*, 2007). As pointed out by Tristam *et al.* (2007), proper quality controls are essential since the commercially available *Haemophilus* test medium has a short shelf-life (Jacobs *et al.*, 2002).

For veterinary pathogens of the family *Pasteurellaceae*, specific recommendations concerning the performance of *in vitro* susceptibility testing by disk diffusion, agar dilution, or broth dilution are available only from the CLSI. The forthcoming CLSI document M31-A3 (2008) specifies the particular requirements for testing of *Hist. somni*, and *A. pleuropneumoniae*. For disk diffusion, a chocolate Mueller-Hinton agar is recommended while for broth dilution methods the veterinary fastidious medium (VFM) should be used. Detailed information on the composition of both media is given in the CLSI document M31-A3. In contrast, few specific recommendations concerning susceptibility testing of *P. multocida* and *M. haemolytica* have been defined by CLSI so far. Thus, susceptibility testing of these organisms follows mainly the recommendations given for other fast-growing aerobic bacteria. Currently, there are no CLSI-approved specifications available for the susceptibility testing of [*H.*] *parasuis*.

The CLSI documents also contain clinical breakpoints on the basis of which isolates are classified as susceptible, intermediate or resistant. In contrast to epidemiological cut-off values, clinical breakpoints may predict the clinical outcome when the antimicrobial agent in question is dosed and administered as recommended (Bywater *et al.*, 2006). While clinical breakpoints for human-pathogenic *H. influenzae* or *Haemophilus* spp. are available from CLSI, BSAC, CA-SFM, and DIN for numerous antimicrobial agents, veterinary-specific clinical breakpoints applicable to bovine or porcine *P. multocida, M. haemolytica, Hist. somni*, and/or *A. pleuropneumoniae* but also to not further specified respiratory tract pathogens from cattle or swine are currently available only for a relatively small number of antimicrobial agents, including amoxicillin, ceftiofur, enrofloxacin, florfenicol, gentamicin, spectinomycin, tetracycline, tiamulin, tilmicosin, and tulathromycin (CLSI, 2008; Schwarz *et al.*, 2008).

Susceptibility status of selected pathogens of the family *Pasteurellaceae*

As an important human pathogen, antimicrobial resistance of *H. influenzae* has been monitored over the years in human medicine at various levels. Such activities included studies on a multinational level, such as the SENTRY Antimicrobial Surveillance Program (Fedler *et al.*, 2006), the Alexander Project (Jacobs *et al.*, 2003; Felmingham *et al.*, 2005), or European surveillance studies (Fluit *et al.*, 2005; Jansen *et al.*, 2006), but also studies on a national level (Jacobs *et al.*, 2004; Castanheira *et al.*, 2006; Oin *et al.*, 2007; Sener *et al.*, 2007), or even on a hospital level (Fukutomi *et al.*, 2004; Daoud *et al.*, 2006). Moreover, there have been drug-specific studies such as the MYSTIC (Meropenem Yearly Susceptibility Test Information Collection) (Jones *et al.*, 2005) or PROTEKT (Prospective Resistant Organism Tracking and Epidemiology for the Ketolide Telithromycin) (Marchese and Schito, 2007) which also determined susceptibility data of human *H. influenzae*. A summary of susceptibility data (MIC_{50}, MIC_{90}, per cent susceptible isolates) of 8,523 worldwide collected *H. influenzae* is available from Tristram *et al.* (2007). These data showed that virtually all *H. influenzae* were susceptible to fluoroquinolones, whereas susceptibility to aminopenicillins was observed in ca. 82% of the cases. Susceptibility to cephalosporins varied dramatically depending on the substance tested. Among the remaining non-β-lactam antibiotics tested, susceptibility to chloramphenicol was high at 98% whereas susceptibility to trimethoprim/sulphamethoxazole was seen in 78% of the isolates (Tristram *et al.*, 2007).

In veterinary medicine, bacteria belonging to the family *Pasteurellaceae* are currently not included in most national antimicrobial susceptibility monitoring and surveillance programs, such as the National Antimicrobial Resistance Monitoring System (NARMS) from the USA, the Norwegian monitoring program NORM-VET, or the Swedish Veterinary Antimicrobial Resistance Monitoring (SVARM) program. *A. pleuropneumoniae* from pigs had been included in the Danish Integrated Antimicrobial Resistance Monitoring and Research Programme (DANMAP) during the years 1997 and 1998 (http://www.danmap.org) and susceptibility data of *P. multocida* and *M. haemolytica* collected during 1996–2000 were included in the Dutch program Monitoring of Antimicrobial Resistance and Antimicrobial Usage in Animals in the Netherlands (MARAN) from 2002 (http://www.cidc-lelystad.wur.nl/UK/publications/other+publications/maran/default.htm). Neither the more recent DANMAP studies nor the latest MARAN studies from 2003 and 2004 monitored any *Pasteurellaceae* for their *in vitro* susceptibility. In contrast, the German GERM-Vet program included *P. multocida* from respiratory tract infections in swine in 2001 (Wallmann *et al.*, 2003), 2002/2003 (Wallmann *et al.*, 2004), and 2004–2006 (Kaspar *et al.*, 2007), *P. multocida* and *M. haemolytica* from respiratory tract infections of cattle in 2002/2003 (Wallmann *et al.*, 2004), and *P. multocida* from poultry with respiratory tract infections in 2004/2005 (Wallmann *et al.*, 2007). Moreover, *P. multocida* from dogs and cats suffering from either respiratory tract infections or infections of mouth, ear or skin were collected between 2004–2006 in the BfT-GermVet monitoring program and analysed for their *in vitro* susceptibility (Schwarz *et al.*, 2007).

Of the 442 porcine *P. multocida* isolates from 2002/2003, 27.8% were resistant to trimethoprim/sulphamethoxazole, followed by 5.7% being resistant to tetracyclines. All other resistances – if detected at all – were below the 1% level (Wallmann *et al.*, 2004). For the period

2004–2006, porcine *P. multocida* were differentiated into isolates from piglets (n = 229), weaners (n = 414), fattening pigs (n = 355), and breeding pigs (n = 113). Regardless of the production category, resistances to trimethoprim/sulphamethoxazole (5.1–8.0%) and tetracycline (4.1–8.3%) were observed most frequently while resistances to other antimicrobial agents were negligible (Kaspar *et al.*, 2007). Among the 28 *P. multocida* isolates from poultry, two (7.1%) were resistant to tetracyclines whereas a single isolate exhibited borderline resistance to enrofloxacin with an MIC of 2 μg/ml (Wallmann *et al.*, 2007). Of the 72 *P. multocida* isolates from respiratory tract infections of dogs and cats, solely sulphonamide resistance was seen in 31 (43%) isolates (Schwarz *et al.*, 2007). Sulphonamide resistance was also the only resistance property detected among 9 (45%) of the 20 canine/feline isolates from infections of mouth, ear and skin (Schwarz *et al.*, 2007). Among the 102 bovine *M. haemolytica* isolates from respiratory tract infections of cattle, 22 (21.6%) were resistant to tetracyclines and 7 (6.9%) resistant to trimethoprim/sulphamethoxazole while other resistances were detected – if at all – in up to five isolates (Wallmann *et al.*, 2004). These data confirmed that antimicrobial resistance is not very widespread among current porcine, bovine, avian, canine and feline *P. multocida* and bovine *M. haemolytica* isolates. In particular, resistance to newer agents such as ceftiofur, enrofloxacin, florfenicol, tilmicosin, or tulathromycin have rarely been detected.

Furthermore, a florfenicol-specific nationwide surveillance study on porcine and bovine respiratory tract pathogens has been conducted since the year 2000 in Germany. Among the 392 porcine *P. multocida*, 155 porcine *A. pleuropneumoniae*, 304 bovine *P. multocida*, and 297 bovine *M. haemolytica* isolates investigated between 2000 and 2005, not a single florfenicol-resistant isolate was detected and MIC_{50} and MIC_{90} data proved to be stable during the entire testing period (Priebe and Schwarz, 2003; Kehrenberg *et al.*, 2004; Schwarz, unpublished data). There are only few recently published data available for *A. pleuropneumoniae* (Gutiérrez-Martín *et al.*, 2006; Matter *et al.*, 2007), *Hist. somni* (Aarestrup *et al.*, 2004), and [*H.*] *parasuis* (Aarestrup *et al.*, 2004; Martín de la Fuente *et al.*, 2007). These studies, all following CLSI methodology, showed that *A. pleuropneumoniae* (Gutiérrez-Martín *et al.*, 2006) and [*H.*] *parasuis* (Martín de la Fuente *et al.*, 2007) from Spain showed distinctly higher MICs for most of the antimicrobial agents tested than the corresponding isolates from Switzerland (Matter *et al.*, 2007), Denmark (Aarestrup *et al.*, 2004), or UK (Martín de la Fuente *et al.*, 2007), respectively. The study on 80 bovine *Hist. somni* isolates from Denmark confirmed that these isolates were in general fully susceptible to the antimicrobial agents currently used for treatment of infections with these pathogens (Aarestrup *et al.*, 2004). Recently, a first study on antimicrobial susceptibility of *Avibacterium paragallinarum* strains has been published. Of the 18 strains collected between 1990 and 2003 in Taiwan, more than 75% were resistant to neomycin, streptomycin and erythromycin (Hsu *et al.*, 2007).

Antimicrobial resistance genes and resistance-mediating mutations among members of the family *Pasteurellaceae*

The following subchapters provide an update of the current status of knowledge with regard to resistance genes and resistance-mediating mutations known to occur in *Pasteurellaceae*. The focus will be on members of the genera *Pasteurella*, *Mannheimia*, *Aggregatibacter*, *Actinobacillus*, and *Haemophilus*, for which data on the genetic basis of resistance are available. As far as information is available, the association of the resistance genes with plasmids, transposons, or gene cassettes and their potential to spread across species and genus borders will be discussed.

Resistance to β-lactam antibiotics

In general, resistance of *Pasteurellaceae* to β-lactam antibiotics can be based on the production of a β-lactamase or the presence of penicillin-binding proteins (PBPs) with low affinity to β-lactams (Kehrenberg *et al.*, 2006b; Tristam *et al.*, 2007). Other mechanisms, such as reduced outer membrane permeability or multidrug efflux systems that can efficiently export β-lactams from the bacterial cell (Kehrenberg *et al.*, 2006b; Tristam *et al.*, 2007) have rarely – if at all – been demonstrated in *Pasteurellaceae*.

β-Lactam resistance mediated by β-lactamases

So far, four β-lactamase (*bla*) genes have been identified among *Pasteurellaceae*, bla_{ROB-1} (Livrelli *et al.*, 1988; Juteau and Levesque, 1990; Livrelli *et al.*, 1991; Azad *et al.*, 1992; San Millan *et al.*, 2007), bla_{TEM-1} (Naas *et al.*, 2001), bla_{TEM-15} (Pitout *et al.*, 2002), or bla_{PSE-1} (Wu *et al.*, 2003). According to the existing classification schemes of β-lactamases, the ROB-1 and TEM enzymes are assigned to the Ambler class A because of their structure and to the Bush class 2b on the basis of their substrate profile (Bush *et al.*, 1995). Members of this class can hydrolyse penicillins and first-generation cephalosporins, but are sensitive to inhibition by β-lactamase inhibitors such as clavulanic acid. The PSE-1 β-lactamase also belongs to the Ambler class A, but to the Bush class 2c. This enzyme, also known as CARB-2 β-lactamase, can hydrolyse carbenicillin and is also inactivated by clavulanic acid. While the ROB-1, TEM-1, and PSE-1 β-lactamases exhibit a narrow spectrum of activity, the TEM-15 β-lactamase, so far exclusively found in a limited number of *H. parainfluenzae* strains (Pitout *et al.*, 2002), represented an extended spectrum β-lactamase (ESBL) which also conferred resistance to cefotaxime, cefpodoxime and ceftriaxone (MIC >16 μg/ml).

Although initially identified in *H. influenzae* from a human infection (Medeiros *et al.*, 1986), ROB-1 β-lactamases have been detected in porcine *A. pleuropneumoniae* isolates (Juteau *et al.*, 1991; Matter *et al.*, 2007), porcine '*Actinobacillus porcitonsillarum*' isolates (Matter *et al.*, 2007), bovine and porcine *P. multocida*, *M. haemolytica* and [*P.*] *aerogenes* isolates (Philippon *et al.*, 1986; Livrelli *et al.*, 1988; Livrelli *et al.*, 1991), as well as porcine [*H.*] *parasuis* isolates (San Millan *et al.*, 2007). Hybridization studies and restriction analysis confirmed that the similar sized bla_{ROB-1}-carrying plasmids from *P. multocida*, *M. haemolytica*, and *H. influenzae* (Livrelli *et al.*, 1991) were also structurally closely related. The small bla_{ROB-1}-carrying plasmid pB1000 from [*H.*] *parasuis* was recently sequenced (San Millan *et al.*, 2007) and shown to be structurally related to the bla_{ROB-1}-carrying plasmid pAB2 from bovine *M. haemolytica* (Wood *et al.*, 1995) (Fig. 10.1) but also to the *tet*(B)-carrying tetracycline resistance plasmid pHS-Tet from [*H.*] *parasuis* (Lancashire *et al.*, 2005). Sequence comparisons confirmed that all so far sequenced bla_{ROB-1} genes from *M. haemolytica*, *A. pleuropneumoniae*, '*A. porcitonsillarum*', [*H.*] *parasuis*, and *H. influenzae* coded for an identical β-lactamase protein of 305 amino acids. Mutations in the bla_{ROB-1} gene which resulted in resistance to extended spectrum cephalosporins and β-lactamase inhibitors have been produced *in vitro* (Galan *et al.*, 2003). In contrast, only single reports described the detection of a TEM-1 β-lactamase in a *P. multocida* isolate of human origin (Naas *et al.*, 2001) and a PSE-1 β-lactamase in an avian *P. multocida* isolate (Wu *et al.*, 2003). The TEM-1 β-lactamase was confirmed by isoelectric focusing and sequence analysis of part of the bla_{TEM-1} gene, whereas the bla_{PSE-1} gene was completely sequenced. TEM-type β-lactamases have been identified in several *Haemophilus* spp. from humans, including *H. influenzae*, *H. parainfluenzae*, and [*H.*] *ducreyi* (McNicol and Ronald, 1984; Brunton *et al.*, 1986, MacLean *et al.*, 1992, Scriver *et al.*, 1994; Matic *et al.*, 2003; Tristram *et al.*, 2005). It should be noted that occasionally TEM-1 and TEM-15 type β-lactamases were detected in the same *H. parainfluenzae* isolate as confirmed by isoelectric focussing and sequence analysis (Pitout *et al.*, 2002). The results of survey studies on the dissemination of β-lactamases in *H. influenzae* (Molina *et al.*, 2003; Farrell *et al.*, 2005) suggested that there are so far not further identified β-lactamases in addition to the ROB-1 and TEM-1 enzymes.

β-Lactam resistance among *Pasteurellaceae* is often associated with small plasmids. These range in size between 4.1 and 4.4 kb in *P. multocida* (Livrelli *et al.*, 1988; Rosenau *et al.*, 1991; Naas *et al.*, 2001), between 4.2 and 5.2 kb in *M. haemolytica* (Craig *et al.*, 1989; Schwarz *et al.*, 1989a; Rossmanith *et al.*, 1991; Azad *et al.*, 1992; Chang *et al.*, 1992a; Murphy *et al.*, 1993), between 2.5 and 6.8 kb in *A. pleuropneumoniae* (Lalonde *et al.*, 1989; Ishii *et al.*, 1991; Juteau *et al.*, 1991; Chang *et al.*, 1992b, 2002), between 8.7 and 13.4 kb in '*A. porcitonsillarum*' (Matter *et al.*, 2007; Matter *et al.*, 2008) and between 4.4 and 10.5 kb in *H. influenzae* and [*H.*] *ducreyi* (McNicol and Ronald, 1984; Brunton *et al.*, 1986; Medeirros *et al.*, 1986; MacLean *et al.*, 1992). In *H. influenzae*, larger conjugative plasmids of approximately

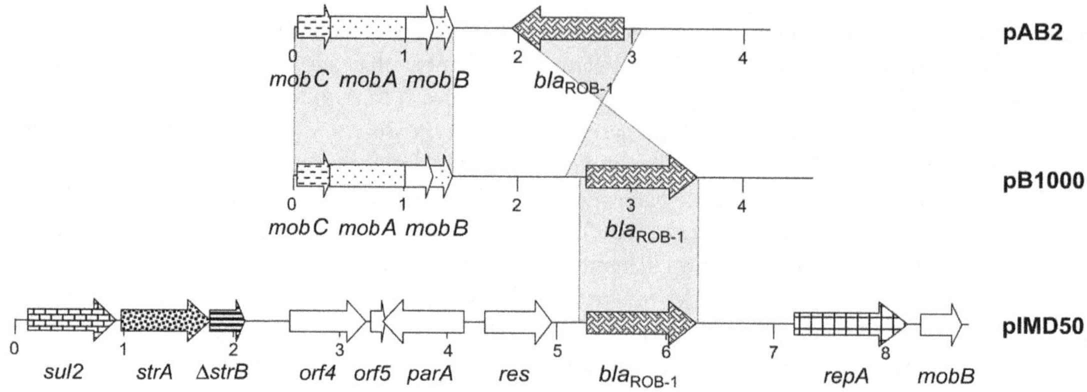

Figure 10.1 Structure and organization of the *bla*$_{ROB-1}$-carrying resistance plasmids pAB2 (accession no. Z21724) from *M. haemolytica*, pB1000 (accession no. DQ840517) from *[H.] parasuis*, and pIMD50 (accession no. AJ830711) from '*A. porcitonsillarum*'. It should be noted that another three pIMD50-related *bla*$_{ROB-1}$-carrying resistance plasmids from '*A. porcitonsillarum*' have been sequenced completely: pKMA5 (accession no. AM748705), pKM202 (accession no. AM748706), and pKMA1467 (accession no. AJ830712). The genes are shown as arrows with the arrowhead indicating the direction of transcription. A distance scale in kb is given below each map. The following genes are involved in antimicrobial resistance: *sul2* [sulphonamide resistance], *strA* and *strB* [streptomycin resistance], and *bla*$_{ROB-1}$ [β-lactam resistance]; plasmid replication: *rep*; mobilization functions: *mobA*, *mobB*, *mobC*; resolvase function: *res*; and DNA partition: *parA*. The prefix Δ indicates a truncated functionally inactive gene. The *orf4* and *orf5* of plasmid pIMD50 may code for proteins with unknown function.

40 kb, which occasionally mediate resistance to tetracycline, chloramphenicol or kanamycin, have also been detected (De Graaff *et al.* 1976; Elwell *et al.* 1975; Saunders *et al.* 1978; Laufs *et al.* 1981; Leaves *et al.* 2000). Maclean *et al.* (1992) reported that a considerable number of *[H.] ducreyi* strains from Thailand harboured co-resident *bla*$_{ROB-1}$- and *bla*$_{TEM-1}$-carrying plasmids. Most of the β-lactam resistance plasmids detected among *Pasteurellaceae* have been identified phenotypically by transformation or conjugation experiments.

β-Lactam resistance mediated by altered PBPs

There is also evidence that mutations in the genes coding for PBPs account for β-lactam resistance (Dabernat *et al.*, 2002; Tristram *et al.*, 2007). Virtually all work on altered PBPs among *Pasteurellaceae* refers to *H. influenzae*. This organism possesses eight PBPs, of which at least one, PBP3, is of relevance in terms of β-lactam resistance (Ubukata *et al.*, 2001; Dabernat *et al.*, 2002; Straker *et al.*, 2003; Hasegawa *et al.*, 2004; Fluit *et al.*, 2005). Analysis of the corresponding gene, *ftsI*, and comparisons of the deduced amino acid sequences revealed the presence of at least 24 amino acid substitutions in the transpeptidase region of PBP3 (for a review please see Tristram *et al.*, 2007). Among them, certain substitutions, such as N526K, R517H, M377I, S357N, and D350N, were more frequently found than others in these β-lactamase-negative ampicillin-resistant (BLNAR) *H. influenzae* strains. Because of the wide range of amino acid substitutions and combinations of substitutions in PBP3, it proved difficult to draw conclusions on the effects of specific amino acid substitutions and the β-lactam resistance phenotype of the corresponding strain. According to CLSI criteria (CLSI, 2006), BLNAR strains of *H. influenzae* should be considered as resistant to amoxicillin-clavulanic acid, ampicillin-sulbactam, cefaclor, cefetamet, cefonicid, cefprozil, cefuroxime, and loracarbef regardless of the results of *in vitro* susceptibility testing.

There is also evidence for the presence of a combination of β-lactamase production and alterations in PBP3 among *H. influenzae* strains. Such strains differ phenotypically from the β-lactamase producing strains by their resistance to amoxicillin-clavulanic acid. Accordingly, such strains were referred to as β-lactamase-positive amoxicillin-clavulanic acid-resistant (BLPACR) *H. influenzae*. Comparisons between BLNAR and BLPACR strains, which had the same substitutions in PBP3, revealed that the BLPACR strains had distinctly higher MICs of ampicillin whereas the MICs of amoxicillin-clavulanic acid and certain cephalosporins were virtually the same (Dabernat *et al.*, 2002; Turnidge *et al.*, 2006).

β-Lactam resistance mediated by active efflux

The acrAB system, which is widespread in Gram-negative bacteria, is known to export a variety of substances including penicillins from the bacterial cell. Expression of the *acrAB* operon is regulated by the repressor AcrR. Studies of the corresponding *acrR* gene in high-level BLNAR strains of *H. influenzae* (MICs of ampicillin: 4–16 μg/ml) revealed the presence of single nucleotide insertions which generated preliminary stop codons via frame-shift mutations (Kaczmarek *et al.*, 2004). In contrast, no alterations were seen in the *acrR* gene of low-level BLNAR strains or ampicillin-susceptible strains (Kaczmarek *et al.*, 2004). Transformation of the mutant *acrR* into *H. influenzae* Rd resulted in a statistically significant increase in the MIC value of ampicillin and thus confirmed that de-repression of the AcrAB efflux pump contributes to ampicillin resistance in these strains (Kaczmarek *et al.*, 2004).

Resistance to tetracyclines

Tetracycline resistance is a highly heterogeneous resistance property with more than 40 different resistance genes known to date (http://faculty.washington.edu/marilynr/). In bacteria of the genera *Pasteurella, Mannheimia, Actinobacillus,* and *Haemophilus,* at least seven different tetracycline resistance genes (*tet* genes), representing two different resistance mechanisms have been detected.

Tetracycline resistance mediated by specific exporters

Among the *tet* genes coding for membrane-associated proteins of the Major Facilitator Superfamily which specifically export tetracyclines from the bacterial cell, the genes *tet*(H), *tet*(B), *tet*(G), *tet*(L), and *tet*(K) have been identified in bacteria of the abovementioned four genera. No *tet* genes have been detected among other members of the *Pasteurellaceae* so far.

Studies from the late 1970s and 1980s reported transferable tetracycline resistance among *P. multocida* and *M. haemolytica* isolates of animal origin (Berman and Hirsh, 1978; Hirsh *et al.*, 1981; Hirsh *et al.*, 1985; Hirsh *et al.*, 1989), although the type of *tet* gene involved was not determined in any of these studies. In 1993, Hansen and co-workers identified a novel type of *tet* gene, designated *tet*(H) (Hansen *et al.*, 1993). The *tet*(H) gene was first detected on plasmid pVM111 which originated from a *P. multocida* isolate obtained from a turkey in the late 1970s in the USA (Berman and Hirsh, 1978). In a subsequent study, Hansen *et al.* (1996) screened tetracycline resistant *P. multocida* and *M. haemolytica* from infections of cattle and pigs in North America for the presence of the gene *tet*(H) and other *tet* genes. They found the *tet*(H) gene to be the predominant *tet* gene among the isolates studied. Since the *tet*(H) gene was located either in the chromosomal DNA or on plasmids, they speculated about the involvement of a transposable element in the spread of *tet*(H) (Hansen *et al.*, 1996). The corresponding transposon, Tn*5706*, was identified in 1998 on plasmid pPMT1 from a bovine *P. multocida* isolate (Kehrenberg *et al.*, 1998). Tn*5706* is a small, non-conjugative composite transposon of 4378 bp and represents the first and so far only known resistance mediating transposon identified among members of the genus *Pasteurella* (Hunt *et al.*, 2000). The *tetR-tet*(H) gene region in Tn*5706* is bracketed by inverted copies of the two closely related insertion sequences, IS*1596* and IS*1597* (Kehrenberg *et al.*, 1998) (Fig. 10.2). Truncated Tn*5706* elements in which these insertion sequences were deleted in part or completely have been found on small plasmids between 4.4–5.5 kb in isolates of *M. haemolytica, P. multocida* and [*P.*] *aerogenes* (Kehrenberg and Schwarz, 2000; Kehrenberg al., 2001). All so far sequenced *tet*(H) genes code for membrane associated efflux proteins of 400 amino acids, except that found on plasmid pPAT1. Due to a recombination between the terminal part of the *tet*(H) gene and the adjacent IS*1597* sequence (Kehrenberg and Schwarz, 2000), this gene codes only for a protein of 392 amino acids. Judging from the level of tetracycline resistance mediated by plasmid pPAT1, this terminal dele-

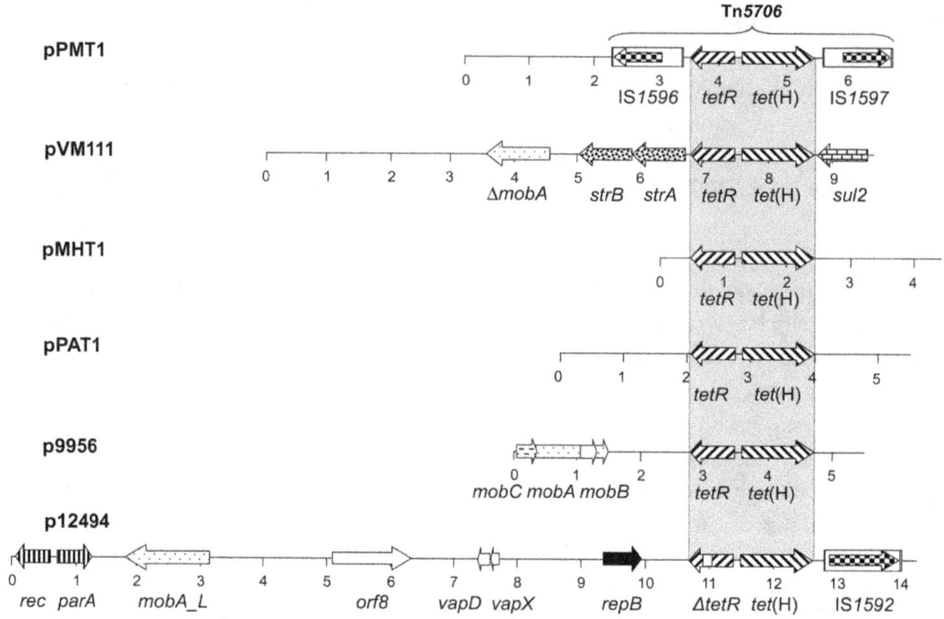

Figure 10.2 Structure and organization of the *tet*(H) carrying plasmids pPMT1 (accession no. Y15510) and pVM111 (accession nos. AJ514834 and U00792) both from *P. multocida*, pMHT1 (accession no. Y16103) from *M. haemolytica*, pPAT1 (accession no. AJ245947) from *[P.] aerogenes* (accession no. Z21724), and p9956 (accession no. AY362554) and p12494 (accession no. DQ517426), both from *A. pleuropneumoniae*. The genes are shown as arrows with the arrowhead indicating the direction of transcription. A distance scale in kb is given below each map. The following genes are involved in antimicrobial resistance: *tetR-tet*(H) [tetracycline resistance], *sul2* [sulphonamide resistance], as well as *strA* and *strB* [streptomycin resistance]; plasmid replication: *repB*; mobilization functions: *mobA*, *mobB*, *mobC*, *mobA_L*; recombination functions: *rec*; DNA partition: *par*; unknown function: *orf8*; virulence associated proteins: *vapD, vapX*. The Δ symbol indicates a truncated gene. The white boxes indicate the limits of the insertion sequences IS*1592*, IS*1596*, and IS*1597*; the arrows within these boxes indicate the reading frames of the corresponding transposase genes.

tion, however, has no impact on the activity of the respective TetH efflux protein (Kehrenberg and Schwarz, 2000). The *tetR-tet*(H) gene region of the *P. multocida* plasmid pVM111 was bracketed by the sulphonamide resistance gene *sul2* and the streptomycin resistance genes *strA/strB* and thus represented part of a novel resistance gene cluster (Kehrenberg *et al.*, 2003) (Fig. 10.2). Recent studies showed that the *tet*(H) gene is occasionally present in tetracycline resistant *A. pleuropneumoniae* (Blanco *et al.*, 2006; Blanco *et al.*, 2007). Two *tet*(H)-carrying plasmids, p9956 and p12494, of 5674 bp and 14,393 bp, respectively, have been isolated from porcine *A. pleuropneumoniae* and sequenced completely (Blanco *et al.*, 2007). Structural analysis showed that they differed distinctly from one another and from the *tet*(H)-carrying plasmids previously found in *Pasteurella* spp. and *Mannheimia* spp. (Fig. 10.2). In plasmid p12494, a 100-bp insertion in the *tetR* gene was detected which resulted in the loss of 60 amino acids at the C-terminus. MIC determination of the corresponding *A. pleuropneumoniae* strain in the presence and absence of tetracycline suggested that this *tet*(H) gene was constitutively expressed (Blanco *et al.*, 2007). The *tet*(H) gene was recently detected in two porcine *A. pleuropneumoniae* strains from Switzerland (Matter *et al.*, 2007), but also in bacteria outside of the family *Pasteurellaceae*, namely in *Moraxella* spp. and *Acinetobacter radioresistens* (Miranda *et al.*, 2003), both obtained from salmon farms in Chile. The *A. radioresistens* isolate also harboured the insertion sequence IS*1599* (Miranda *et al.*, 2003) which is closely related to the Tn*5706* associated insertion sequences IS*1596* and IS*1597*.

The gene *tet*(B) is located on the non-conjugative transposon Tn*10* (Chalmers *et al.*, 2000; Lawley *et al.*, 2000) and represents the most widely spread *tet* gene among *Enterobacteriaceae* (Chopra and Roberts, 2001). The *tet*(B) gene proved to be the dominant *tet* gene among porcine [*P.*] *aerogenes* isolates (Kehrenberg and Schwarz, 2001a), whereas it has been detected only in a single bovine *M. haemolytica* isolate from France (Chaslus-Dancla *et al.*, 1995) and in two porcine *P. multocida* isolates from the USA and Germany (Kehrenberg *et al.*, 2001a). In all cases but one, the *tet*(B) gene was located in one or two copies on the chromosome. Hybridization studies using *Sfu*I-digested whole cellular DNA of *tet*(B)-carrying *P. multocida* and [*P.*] *aerogenes* isolates suggested complete copies of Tn*10* in the majority of the isolates investigated (Kehrenberg *et al.*, 2001a; Kehrenberg and Schwarz, 2001a). In the only plasmid-borne case, the *tet*(B) gene proved to be part of a largely truncated Tn*10* element (Kehrenberg and Schwarz, 2001a). Studies on tetracycline resistant *A. pleuropneumoniae* from pigs (Wasteson *et al.*, 1996; Ouellet *et al.*, 2004; Blanco *et al.*, 2006) and *Agg. actinomycetemcomitans* from humans (Roe *et al.*, 1995) also revealed the presence of the gene *tet*(B). In porcine *A. pleuropneumoniae* from Spain and Switzerland, the *tet*(B) gene was the predominant *tet* gene (Blanco *et al.*, 2006; Matter *et al.*, 2007). This gene was also detected in a single field isolate of '*A. porcitonsillarum*' and in the '*A. porcitonsillarum*' reference strain CCUG46996 (Matter *et al.*, 2007). Among the Spanish *A. pleuropneumoniae* isolates, the *tet*(B) gene was found mainly on small plasmids, which were indistinguishable by their *Hin*dIII and *Dra*I restriction patterns (Blanco *et al.*, 2006). One such plasmid, p11745, of 5486 bp was sequenced completely. Surprisingly, neither the *tetR* repressor gene nor other parts of Tn*10* were detectable (Blanco *et al.*, 2006) (Fig. 10.3). A *tet*(B) gene, also without an accompanying *tetR* repressor gene, was identified on the mobilizable 5147-bp plasmid pHS-Tet from [*H.*] *parasuis* (Lancashire *et al.*, 2005) (Fig. 10.3). Tet(B)-mediated tetracycline resistance has also been found in other *Haemophilus* spp. from humans, including *Haemophilus parainfluenzae* (Levy *et al.*, 1984), *H. influenzae* and [*H.*] *ducreyi* (Marshall *et al.*, 1984).

The gene *tet*(G) was initially detected on a plasmid from *Vibrio anguillarum* (Zhao and Aoki, 1992). Since then, it has been detected in *Pseudomonas* spp., isolated from leaves of tetracycline-treated apple trees (Schnabel and Jones, 1999), and to be part of the multiresistance gene cluster present in *Salmonella* Typhimurium DT104 (Briggs and Fratamico, 1999) and other *Salmonella* serovars. It has also been found on the chromosome of six epidemiologically related *M. haemolytica* isolates from cattle (Kehrenberg *et al.*, 2001a) and on plasmid pJR1 from avian *P. multocida* (Wu *et al.*, 2003). Surprisingly, the *tet*(G) structural gene in plasmid pJR1 was found in the absence of a corresponding *tetR* repressor gene which is considered to be essential for the tetracycline-inducible expression of *tet*(G). Whether the lack of the repressor gene has an impact on the expression of *tet*(G)-mediated tetracycline resistance is unknown. It should be noted that plasmid pJR1 has not been transferred into susceptible recipient strains for phenotypic confirmation of the activity of the resistance genes found on this plasmid (Wu *et al.*, 2003).

The *tet*(L) gene, which is commonly found in Gram-positive cocci and *Bacillus* spp., was detected on small plasmids of *M. haemolytica* and *Mannheimia glucosida*, but also in the chromosomal DNA of single *M. haemolytica* and *P. multocida* strains, all originating from cattle in Belgium (Kehrenberg *et al.*, 2005a). One such plasmid of 5317 bp from *M. haemolytica*, designated pCCK3259, was sequenced completely (Fig. 10.3). Besides the *tet*(L) gene, it contained only the *mobABC* operon responsible for mobilization of the plasmid (Kehrenberg *et al.*, 2005a). In Gram-positive bacteria, the *tet*(L) gene is inducibly expressed via translational attenuation (Schwarz *et al.*, 1992). The corresponding regulatory region, however, was absent in plasmid pCCK3259 whereas all elements required for a constitutive expression of the *tet*(L) gene, including a suitable promoter with a −35 and a −10 region as well as a ribosome binding site, were detected upstream of the *tet*(L) gene (Kehrenberg *et al.*, 2005a). Two different sized *tet*(L)-carrying plasmids of 5.6 and >12 kb have also been identified in porcine *A. pleuropneumoniae* (Blanco *et al.*, 2006). The smaller one of these plasmids, p9555, of 5672 bp was sequenced completely and revealed striking structural similarities to plasmid pCCK3259 from *M. haemolytica* (Blanco

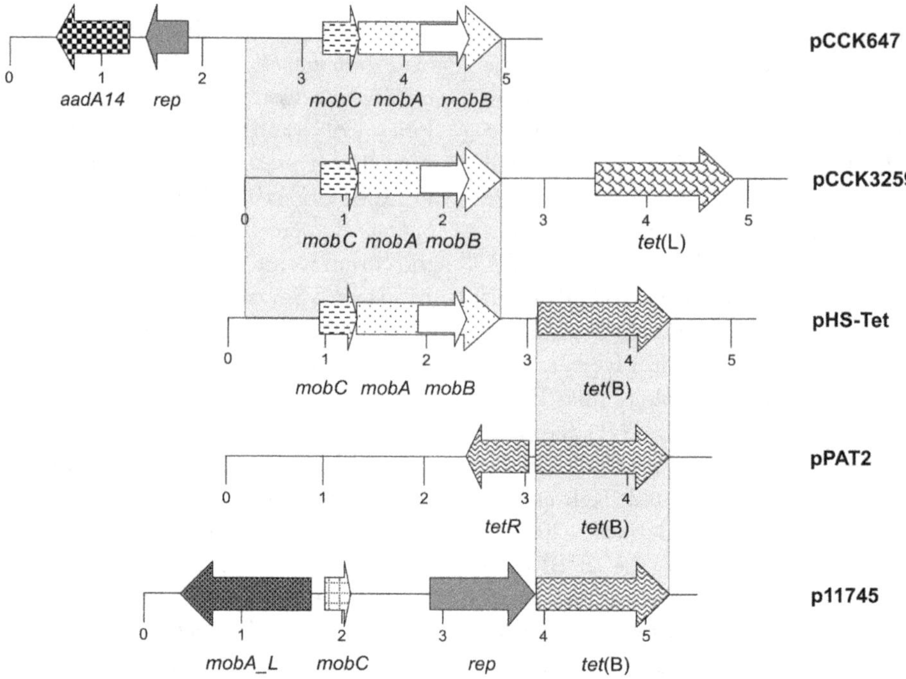

Figure 10.3 Structure and organization of the *tet*(B)-carrying tetracycline resistance plasmids pHS-Tet (accession no. AY862435) from *[H.] parasuis*, pPAT2 (accession no. AJ278685) from *[P.] aerogenes*, and p11745 (accession no. DQ176855) from *A. pleuropneumoniae*, the *tet*(L)-carrying tetracycline resistance plasmid pCCK3259 (accession no. AJ966516) from *M. haemolytica*, and the *aadA14*-carrying streptomycin/spectinomycin resistance plasmid pCCK447 (accession no. AJ884726) from *P. multocida*. The genes are shown as arrows with the arrowhead indicating the direction of transcription. A distance scale in kb is given below each map. The following genes are involved in antimicrobial resistance: *tetR*-*tet*(B), *tet*(B), and *tet*(L) [tetracycline resistance], and *aadA14* [streptomycin/spectinomycin resistance]; plasmid replication: *rep*; and mobilization functions: *mobA*, *mobB*, *mobC*. Gray shaded areas indicate the regions common to plasmids pCCK647, pCCK3259, and pHS-Tet or to plasmids pHS-Tet, pPAT2 and p11745.

et al., 2006). Again, the regulatory region of the *tet*(L) gene was also missing in plasmid p9555. While plasmid p9555 could be transferred into *E. coli* where it expressed tetracycline resistance (Blanco *et al.*, 2006), plasmid pCCK3259 did not replicate in *E. coli* but in a *P. multocida* host (Kehrenberg *et al.*, 2005a).

The *tet*(K) gene, which is commonly found on small plasmids in human and animal staphylococci, was identified by PCR in *[Haemophilus] aphrophilus* (Pang *et al.*, 1994) – meanwhile reclassified as *Aggregatibacter aphrophilus* (Nørskov-Lauritsen and Kilian, 2006) The tetracycline resistance gene *tet*(D) (Aoki and Takahashi, 1987) was detected in isolates of the fish pathogen *[Pasteurella] piscicida*. In 1995, this organism was reclassified as *Photobacterium damselae* subsp. *piscicida* (Gauthier *et al.*, 1995) and does not belong to the family *Pasteurellaceae* anymore.

Tetracycline resistance mediated by ribosome-protective proteins

Another two tetracycline resistance genes, *tet*(M) and *tet*(O), both coding for ribosome protective proteins, have been identified in *P. multocida*, *[H.] ducreyi*, or *A. pleuropneumoniae*. The gene *tet*(M) is associated with the conjugative transposon Tn*916* (Flannagan *et al.*, 1994). It is considered to be the most widespread *tet* gene among Gram-positive and Gram-negative bacteria (Chopra and Roberts, 2001). It has been detected by hybridization in the chromosomal DNA of two bovine *P. multocida* isolates (Chaslus-Dancla *et al.*, 1995; Hansen *et al.*, 1996), but also on a conjugative plasmid in *[H.] ducreyi* of human origin (Marshall *et al.*, 1984). The gene *tet*(O) previously identified in *Campylobacter* spp. and streptococci, was detected in the chromosomal DNA of a porcine *A. pleuropneumoniae* isolate (Ouellet *et al.*, 2004). More recently, the *tet*(O) gene was also detected in eight porcine *A. pleuropneumoniae* isolates from Spain (Blanco *et al.*, 2006). In these isolates, the *tet*(O) gene was located on small plasmids of ca. 6 kb. Sequencing of a 2489-bp region, which included the complete *tet*(O) gene, of one of these plasmids revealed 99% sequence identity to the *tet*(O) gene of *Campylobacter jejuni* (Blanco *et al.*, 2006).

Resistance to aminoglycosides and aminocyclitols

Resistance to aminoglycoside and aminocyclitol antibiotics is most frequently mediated by enzymes that inactivate the drugs by adenylation, acetylation, or phosphorylation. Moreover, mutations in chromosomal genes have also been identified to mediate resistance to selected members of these classes of antimicrobial agents (Schwarz *et al.*, 2006).

Resistance to streptomycin and/or spectinomycin by enzymatic inactivation

The first aminoglycoside resistance genes detected in *Pasteurella* and *Mannheimia* were those mediating streptomycin resistance. In 1978, Berman and Hirsh published a report on plasmids coding for streptomycin resistance along with sulphonamide resistance or sulphonamide and tetracycline resistance in *P. multocida* from turkeys (Berman and Hirsh, 1978). Streptomycin resistance is commonly associated with small non-conjugative plasmids of less than 15 kb in *P. multocida* (Berman and Hirsh, 1978; Silver *et al.*, 1979; Hirsh *et al.*, 1981; Hirsh *et al.*, 1985; Hirsh *et al.*, 1989; Schwarz *et al.*, 1989b; Yamamoto *et al.*, 1990; Cote *et al.*, 1991), [*P.*] *aerogenes* (Kehrenberg and Schwarz, 2001a), *M. haemolytica* (Zimmerman and Hirsh 1980; Chang *et al.*, 1992a), *Mannheimia varigena* (Kehrenberg and Schwarz 2002), '*A. porcitonsillarum*' (Matter *et al.*, 2007), *A. pleuropneumoniae* (Gilbride *et al.*, 1989; Willson *et al.*, 1989; Ishii *et al.*, 1990; Ishii *et al.*, 1991; Kiuchi *et al.*, 1992; Chang *et al.*, 2002), [*H.*] *ducreyi* (Dixon *et al.*, 1994), and *Av. paragallinarum* (Hsu *et al.*, 2007) (Fig. 10.4). In one case, streptomycin resistance was mediated by a conjugative multiresistance plasmid of approximately 113 kb in an avian *P. multocida* isolate (Hirsh *et al.*, 1989). Many of these plasmids carry additional resistance genes such as the sulphonamide resistance gene *sul2*, the kanamycin/neomycin resistance gene *aphA1*, the chloramphenicol resistance gene *catA3* (Fig. 10.4) and/or the ampicillin resistance gene $bla_{\mathrm{ROB-1}}$ (Fig. 10.1).

The *strA* gene is the predominant streptomycin resistance gene in bacteria of the genera *Pasteurella, Mannheimia,* and *Actinobacillus.* It codes for an aminoglycoside-3″-phosphotransferase of 269 amino acids and has been found together with the gene *strB*, which codes for an aminoglycoside-6-phosphotransferase of 278 amino acids. Both genes are part of transposon Tn5393 from *Erwinia amylovora* (Chiou and Jones, 1993). In the streptomycin resistant *Pasteurella, Mannheimia* and *Actinobacillus* isolates, *strA* is usually complete whereas various truncated *strB* genes have been identified (Kehrenberg and Schwarz 2001; Kehrenberg and Schwarz, 2002; Ito *et al.*, 2004). Strains that carry a functionally active *strA* gene but a largely truncated *strB* gene have been shown to be highly resistant to streptomycin (Kehrenberg and Schwarz, 2001b), suggesting that *strA* is the relevant gene for the expression of streptomycin resistance. Further support for this hypothesis comes from the observation that plasmid pSTOJO1, which carries an intact *strB* gene, but a functionally inactive *strA* gene that is disrupted by the integration of a *dfrA14* gene cassette, does not express streptomycin resistance (Ojo *et al.*, 2002). Studies on the prevalence and distribution of the *strA-strB* genes showed that the *strA* gene – in combination with a complete or a truncated copy of *strB* – occurs on plasmids or in the chromosomal DNA of a wide range of commensal and pathogenic bacteria from humans, animals, and plants (Sundin, 2000, 2002; Sundin and Bender, 1996).

An *aadA1* gene coding for an aminoglycoside-3″-adenyltransferase that mediates resistance to the aminoglycoside streptomycin and the aminocyclitol spectinomycin has been detected on the 5.2-kb plasmid pJR2 from avian *P. multocida* (Wu *et al.*, 2003). The *aadA1* gene is part of a gene cassette which is inserted into a relic of a class 1 integron in pJR2. The *intI1* gene coding for the integrase in the 5′ conserved segment of this integron is truncated, however without affecting the promoter, and the *sul1* gene in the

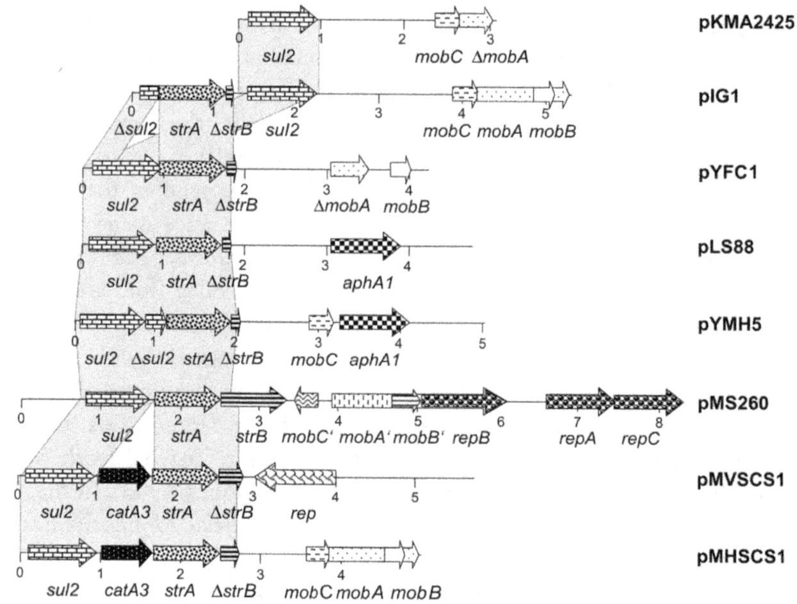

Figure 10.4 Structure and organization of selected *sul2*-based (multi-)resistance plasmids pKMA2425 (accession no. AJ830714) from *A. pleuropneumoniae*, pIG1 (accession no. U57647) from *P. multocida*, pYFC1 (accession no. M83717) from *M. haemolytica*, pLS88 (accession no. L23118) from *H. ducreyi*, pYMH5 (accession no. EF015636) from *Av. paragallinarum*, pMS260 (accession no. AB109805) from *A. pleuropneumoniae*, pMVSCS1 (accession no. AJ319822) from *M. varigena*, and pMHSCS1 (accession no. AJ249249) from *Mannheimia* unnamed taxon 10. The map of another *sul2*-based multiresistance plasmid, pIMD50 (accession no. AJ830711) from '*A. porcitonsillarum*', is displayed in Figure 1. The genes are shown as arrows with the arrowhead indicating the direction of transcription. A distance scale in kb is given below each map. The following genes are involved in antimicrobial resistance *sul2* [sulphonamide resistance], *strA* and *strB* [streptomycin resistance], *catA3* [chloramphenicol resistance], and *aphA1* [kanamycin/neomycin resistance]; plasmid replication: *rep*, *repA*, *repB*, *repC*; mobilization functions: *mobA*, *mobB*, *mobC*, *mobA'*, *mobB'*, *mobC'*. The prefix Δ indicates a truncated functionally inactive gene. Gray shaded areas indicate the regions common to different plasmids.

3′ conserved fragment is missing completely. Attempts to identify the *aadA* genes in bovine strains of *P. multocida* and *M. haemolytica* from Germany that are highly resistant to streptomycin and spectinomycin using PCR were unsuccessful (Schwarz *et al.*, 2004b). In all these isolates, streptomycin resistance was based on a *strA* gene whereas a spectinomycin resistance gene could not be identified.

A novel streptomycin-spectinomycin resistance gene, designated *aadA14,* was identified on a small 5.2 kb plasmid from a bovine *P. multocida* capsular type F isolate from Belgium (Kehrenberg *et al.*, 2005b) (Fig. 10.3). The corresponding isolate was obtained from case of fatal peritonitis in a calf (Catry *et al.*, 2005). The AadA14 adenyltransferase protein of 261 amino acids exhibited only 51.4–56.0% identity to the so far known AadA proteins and hence proved to be only distantly related to AadA proteins previously found in other bacteria.

Resistance to streptomycin or spectinomycin by mutations

Mutations in genes coding for specific ribosomal proteins can also confer streptomycin resistance. Comparative analysis of the chromosomal genes coding for the 124 amino acid ribosomal protein S12 from a high-level streptomycin resistant and a streptomycin susceptible *H. influenzae* revealed two amino acid substitutions in the respective proteins (Stuy and Walter, 1992).

Ribosomal mutations conferring spectinomycin resistance have been described in a variety of bacteria (Makosky and Dahlberg, 1987; De Stasio *et al.*, 1989; Brink *et al.*, 1994; Galimand *et al.*, 2000; O'Connor and Dahlberg, 2002). All these mutations were present in a specific region of helix 34 in 16S rRNA. This region encompassed the cross-linked positions 1063–1066 and 1090–1093, which are known to be involved in the binding of spectinomycin to the ribosome. Moreover, the *rpsE* gene coding for the ribosomal protein S5 is also relevant for the drug binding, and mutations in *rpsE* have been described to affect spectinomycin binding (Funatsu *et al.*, 1971; Davies *et al.*, 1998). The analysis of high-level spectinomycin-resistant *P. multocida* isolates (MICs of spectinomycin ≥ 4096 μg/ml), in which no enzymatic inactivation of the drug could be detected, revealed four different types of mutations: (a) a C1192G transversion in 16S rRNA in all six or (b) in five of the six rRNA operons, (c) the aforementioned transversion in only two of the six operons accompanied by a 3-bp deletion in *rpsE* that resulted in a change of the amino acids 32-SF-33 to 32-I, and (d) a 3-bp deletion in *rpsE* that resulted in the loss of 23-K without additional rRNA mutations (Kehrenberg and Schwarz 2007). Molecular modelling suggested that both types of mutations in the S5 protein have a negative impact on spectinomycin binding to the ribosome.

Resistance to kanamycin and neomycin by enzymatic inactivation

The first report of transferable kanamycin resistance among *Pasteurellaceae* was published in 1975 and referred to *H. influenzae* (Van *et al.*, 1975). Kanamycin/neomycin resistance has been associated with the gene *aphA1*, also known as *aph(3')-Ia*, which codes for an aminoglycoside-3′-phosphotransferase that mediates resistance to kanamycin and neomycin. This gene has been identified on transposon Tn*903* (Oka *et al.*, 1981). Later on, it was detected together with the streptomycin resistance genes *strA-strB* and the sulphonamide resistance gene *sul2* on the broad-host range plasmid pLS88 from [*H.*] *ducreyi* (Dixon *et al.*, 1994). A pLS88-related plasmid has recently also been identified in the *Av. paragallinarum* strain A14 (Hsu *et al.*, 2007) (Fig. 10.4). Further studies on kanamycin/neomycin-resistant [*P.*] *aerogenes*, *P. multocida*, and *M. glucosida* isolates identified the *aphA1* gene mostly in the chromosomal DNA of the respective strains (Kehrenberg and Schwarz, 2005b). In a single *P. multocida* isolate, this gene was found on the 5955-bp plasmid pCCK3152 along with complete *strA* and *sul2* genes and a truncated *strB* gene (Kehrenberg and Schwarz, 2005b).

A second kanamycin/neomycin resistance gene, *aphA3* [also known as *aph(3')-III*], was detected on the 5.1-kb plasmid pCCK411 in single strains of *P. multocida* and [*P.*] *aerogenes* (Kehrenberg and Schwarz, 2005b). In addition to the *aphA3* gene, this plasmid also carried a *mobABC* operon for mobilization. A not further specified aminoglycoside-3′-phosphotransferase gene mediating kanamycin resistance was also found together with a $bla_{\text{ROB-1}}$ gene on the 6-kb plasmid pTMY2 from *A. pleuropneumoniae* (Chang *et al.*, 2002).

Resistance to gentamicin by enzymatic inactivation

Although *P. multocida*, *A. pleuropneumoniae*, and '*A. porcitonsillarum*' strains with MICs of gentamicin of ≥ 32 μg/ml have been detected, all attempts to detect specific resistance genes, such as *aadB* [*ant(2")-Ia*], *aacC2* [*aac(3)-IIc*], or *aacC4* [*aac(3)-IVa*], have failed so far (Matter *et al.*, 2007; Kehrenberg and Schwarz, unpublished data). Moreover, attempts to transfer gentamicin resistance from *P. multocida* donors were also unsuccessful (Kehrenberg and Schwarz, unpublished data).

Resistance to phenicols

Resistance to non-fluorinated phenicols may be due to enzymatic inactivation by chloramphenicol acetyltransferases, whereas resistance to fluorinated and/or non-fluorinated phenicols can be mediated by phenicol-specific exporters (Schwarz *et al.*, 2004a, 2006). Other mechanisms, such as permeability barriers have rarely been detected among *Pasteurellaceae*.

Resistance to chloramphenicol by enzymatic inactivation

Resistance to chloramphenicol is most frequently due to enzymatic inactivation of the drug by chloramphenicol acetyltransferases (CATs). Two different types of CATs, A and B, specified by a number of different *catA* and *catB* genes, are currently known (Schwarz *et al.*, 2004a). The *catA* and *catB* genes are often located on plasmids, transposons or gene cassettes. Plasmids mediating chloramphenicol resistance have been identified in porcine *P. multocida* isolates (Yamamoto *et al.*, 1990), bovine *P. multocida* and *M. haemolytica* isolates (Vassort-Bruneau *et al.*, 1996), porcine [*P.*] *aerogenes* (Kehrenberg and Schwarz, 2001b), bovine *Mannheimia* spp. (Kehrenberg and Schwarz, 2001b; Kehrenberg and Schwarz, 2002), porcine *A. pleuropneumoniae* (Gilbride *et al.*, 1989; Lalonde *et al.*, 1989; Kawahara *et al.*, 1990; Ishii *et al.*, 1993), but also human *H. influenzae* (Roberts *et al.*, 1980). As previously seen with other resistance plasmids, those mediating chloramphenicol resistance were commonly less than 10 kb in size and occasionally carried additional resistance genes. Initial molecular studies on chloramphenicol resistance among *P. multocida* and *M. haemolytica* isolates included the detection of the three most frequently occurring *cat* genes among Gram-negative bacteria, *catA1*, *catA2*, and *catA3* (formerly known as *cat*I, *cat*II, and *cat*III), by specific PCR assays (Vassort-Bruneau *et al.*, 1996). The *catA3* gene was detected on small plasmids of 5.1 kb whereas the *catA1* gene was located on plasmids of either 17.1 or 5.5 kb (Vassort-Bruneau *et al.*, 1996). More recently, *catA3* genes have been identified as parts of chromosomal or plasmid-borne resistance gene clusters in porcine [*P.*] *aerogenes* and bovine *Mannheimia* isolates (Kehrenberg and Schwarz, 2001b; Kehrenberg and Schwarz, 2002). In these gene clusters, the *catA3* gene was always inserted between the sulphonamide resistance gene *sul2* and the streptomycin resistance gene *strA* (Kehrenberg and Schwarz, 2001b; Kehrenberg and Schwarz, 2002) (Fig. 10.4).

The gene *catA1* was first identified on transposon Tn*9* (Alton and Vapnek, 1979) and the *catA3* gene on plasmid R387 from *Shigella flexneri* and *E. coli* (Murray *et al.*, 1988). Similar or identical *catA1* and *catA3* genes have been detected in a wide variety of bacteria (Schwarz *et al.*, 2004a). Although the *catA2* gene is commonly found in *Haemophilus* spp. (Powell and Livermore, 1988; Murray *et al.*, 1990), this gene has not yet been detected in chloramphenicol resistant *Pasteurella*, *Mannheimia*, or *Actinobacillus*. However, a *catB2* gene, which codes for a different type of chloramphenicol acetyltransferase than the aforementioned *catA* genes (Schwarz *et al.*, 2004a), was found on plasmid pJR1 from avian *P. multocida* (Wu *et al.*, 2003). The *catB2* gene is part of a gene cassette and thus needs the integron-associated promoter for its expression. In plasmid pJR1, the *catB2* cassette was located outside of an integron structure. Gene cassettes located at secondary sites outside of integrons may be expressed if a suitable promoter is available (Ojo *et al.*, 2002). However, such a promoter was not identified in pJR1 (Wu *et al.*, 2003).

Resistance to chloramphenicol and florfenicol by active efflux

So far, resistance to florfenicol, a structural analogue of thiamphenicol, has rarely been detected among *Pasteurellaceae*. A gene designated *pp-flo* which codes for a chloramphenicol/florfenicol exporter was identified in the fish pathogen *P. damselae* subsp. *piscicida* (formerly known as *Pasteurella piscicida*) (Kim and Aoki, 1996b). The gene *pp-flo* is closely related to the gene *floR* which is part of the *Salmonella* Typhimurium DT104 multiresistance gene cluster (Briggs and Fratamico, 1999) and also has been found on plasmids or in the chromosome of various Gram-negative bacteria (Schwarz *et al.*, 2004a). Monitoring studies to specifically determine the MIC values of florfenicol among bovine and porcine respiratory tract pathogens showed that all *P. multocida*, *M. haemolytica*, and *A. pleuropneumoniae* isolates collected between 1994 and 2005 were susceptible to florfenicol (Hörmansdorfer and Bauer, 1996, 1998; Priebe and Schwarz, 2003; Kehrenberg *et al.*, 2004; Schwarz, unpublished data). Most recent data from the monitoring programs GE*RM*-Vet and BfT-GermVet conducted in Germany also confirmed that all *P. multocida* from pigs, poultry, dogs and cats were susceptible to florfenicol (Kaspar *et al.*, 2007; Schwarz *et al.*, 2007; Wallmann *et al.*, 2007).

The first report of a florfenicol resistant bovine *P. multocida* isolate was published in 2005 (Kehrenberg and Schwarz, 2005a). This isolate of capsular type A originated from a calf that had died from pneumonia in the UK. Florfenicol resistance was based on the presence of the gene *floR* which codes for an exporter of the Major Facilitator family that mediates the efflux of phenicols from the bacterial cell. This gene was located on the 10,874-bp plasmid pCCK381 (Fig. 10.5).

Interestingly, a florfenicol-resistant *Salmonella enterica* subsp. *enterica* serovar Dublin isolate was isolated from the same calf which carried a plasmid indistinguishable from pCCK381. This observation and the fact, that plasmid pCCK381 was able to replicate and express florfenicol resistance in different *E. coli* hosts (Kehrenberg and Schwarz, 2005a), strongly suggested that the plasmid pCCK381 can be exchanged between bacteria of different genera. Sequence analysis confirmed that this plasmid consisted of three segments which showed extended similarity to plasmids pDN1 from *Dichelobacter nodosus* (Whittle *et al.*, 2000), pMBSF1 from *E. coli* (Blickwede and Schwarz, 2004), and pRSV1 from *Vibrio salmonicida* (Sørum *et al.*, 1992) (Fig. 10.5). All these plasmids have been found either in bacteria such as *E. coli* from cattle and pigs, which have previously been shown to carry the *floR* gene, or in bacteria which cause diseases in fish and ruminants, such as coldwater vibriosis (*V. salmonicida*) or infectious pododermatitis (*D. nodosus*), for the control of which florfenicol is approved in several non-European Union countries. Although it is not possible to determine in retrospect where and when plasmid pCCK381 has evolved, the structural analysis suggested that this plasmid is most likely the result of interplasmid recombination processes (Kehrenberg and Schwarz, 2005a).

In 2006, a florfenicol resistance plasmid was identified in a bovine strain of [*Pasteurella*] *trehalosi* – meanwhile reclassified as *Bibersteinia trehalosi* (Blackall *et al.*, 2007) – which originated from a calf which had died from a respiratory tract infection in France (Kehrenberg *et al.*,

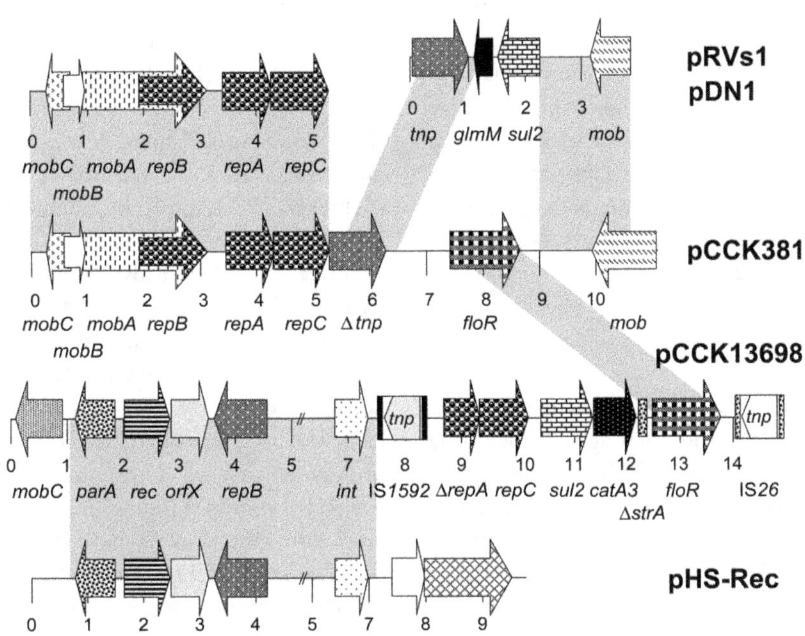

Figure 10.5 Structure and organization of the florfenicol resistance plasmids pCCK381 (accession no. AJ871969) from *P. multocida*, and pCCK13698 (accession no. AM183225) from *B. trehalosi*, in comparison to in part related plasmids, such as pDN1 (accession no. Y19120) from *Dichelobacter nodosus*, pRVs1 (accession no. AJ289135) from *Vibrio salmonicida*, and pHS-Rec (accession no. AY862436) from [*H.*] *parasuis*. The genes are shown as arrows with the arrowhead indicating the direction of transcription. A distance scale in kb is given below each map. The following genes are involved in antimicrobial resistance *sul2* [sulphonamide resistance], *strA* [streptomycin resistance], *catA3* [chloramphenicol resistance], and *floR* [chloramphenicol/florfenicol resistance]; plasmid replication: *rep*, *repA*, *repB*, *repC*; mobilization functions: *mobA*, *mobB*, *mobC*, *mob*; transposition functions: *tnp*; recombinase or integrase functions: *rec*, *int*; DNA partition: *parA*; and unknown functions: *orfX*, *orf1*, *orf2*. The prefix Δ indicates a truncated functionally inactive gene. The boxes in the map of pCCK13698 indicate the insertion sequences IS*1592* and IS*26* with the arrows within these boxes marking the transposase genes. Gray shaded areas indicate the regions common to different plasmids.

2006a). Again, the gene *floR* was responsible for florfenicol resistance and this time, the gene was located on the 14,969-bp plasmid pCCK13698. This plasmid differed structurally from plasmid pCCK381 and carried additional resistance genes such as the sulphonamide resistance gene *sul2*, the chloramphenicol acetyltransferase gene *catA3* and a truncated streptomycin resistance gene *strA* (Kehrenberg *et al.*, 2006a) (Fig. 10.5). Moreover, this plasmid shared almost 6.8 kb with the [*H.*] *parasuis* plasmid pHS-Rec (Lancashire *et al.*, 2005), and harboured two insertion sequences, IS*1592* and IS*26* (Kehrenberg *et al.*, 2006a). The structural analysis of plasmid pCCK13698 revealed that it was almost exclusively composed of segments previously associated with other plasmids, transposons, or insertion sequences and that interplasmid recombination processes might have played a role in the formation of this plasmid. While most of these plasmid and transposon segments have already been found in members of the family *Pasteurellaceae*, the occurrence of IS*26* – an insertion sequence widely distributed among *Enterobacteriaceae* – in the genus *Bibersteinia*, was a novel observation that supported the assumption of a gene flow between *Pasteurellaceae* and *Enterobacteriaceae* (Kehrenberg *et al.*, 2006a).

Resistance to chloramphenicol by impaired influx

Several porins of the outer membrane can act as entry for certain antimicrobial agents, including chloramphenicol. A permeability barrier based on the down-regulation of a 40-kDa outer membrane protein has been reported to account for chloramphenicol resistance in *H. influenzae* (Burns *et al.*, 1985).

Resistance to inhibitors of folic acid metabolism

Sulphonamides and trimethoprim are competitive inhibitors of different enzymatic steps in the folic acid metabolism. Sulphonamides represent structural analogues of *p*-aminobenzoic acid and inhibit the enzyme dihydropteroate synthase (DHPS), which – in the initial step of folic acid metabolism – catalyses the synthesis of dihydropteroic acid from dihydropteridin and *p*-aminobenzoic acid. Trimethoprim inhibits the enzyme dihydrofolate reductase (DHFR), which – in a later step of folic acid metabolism – reduces dihydrofolic acid to tetrahydrofolic acid. Resistance to both drugs is commonly mediated by replacement of susceptible DHPS or DHFRs by enzymes with reduced affinity to sulphonamides or trimethoprim, respectively. Moreover, overproduction of susceptible targets or mutations in chromosomal dihydropteroate synthase or dihydrofolate reductase genes, which alter substrate specificity, may also cause resistance (Schwarz *et al.*, 2006).

Resistance to sulphonamides by altered dihydropteroate synthases

Sulphonamide resistance among *Pasteurella*, *Mannheimia*, *Actinobacillus*, and *Haemophilus* isolates is commonly mediated by a type 2 dihydropteroate synthase (DHPS) with reduced affinity to sulphonamides. The corresponding gene, *sul2*, is frequently found on small plasmids in *P. multocida* (Hirsh *et al.*, 1981; Yamamoto *et al.*, 1990; Cote *et al.*, 1991; Wright *et al.*, 1997), *M. haemolytica* (Chang *et al.*, 1992a; Briggs *et al.*, 1998), *Mannheimia* spp. (Kehrenberg and Schwarz, 2001b; Kehrenberg and Schwarz, 2002), *A. pleuropneumoniae* (Willson *et al.*, 1989; Ishii *et al.*, 1990; Kawahara *et al.*, 1990; Ishii *et al.*, 1993; Chang *et al.*, 2002; Ito *et al.*, 2004; Matter *et al.*, 2007), '*A. porcitonsillarum*' (Matter *et al.*, 2007), *H. influenzae* (Enne *et al.*, 2002), and [*H.*] *ducreyi* (Dixon *et al.*, 1994).

The *sul2*-encoded DHPS proteins commonly consist of 271 amino acids. However, several variants ranging in size between 263 and 289 amino acids have also been reported (Dixon *et al.*, 1994; Kehrenberg and Schwarz, 2001b; Kehrenberg *et al.*, 2003; Kehrenberg and Schwarz, 2005c). Sequence analysis showed that single bp insertions downstream of codon 225 in the DHPS of the [*H.*] *ducreyi* plasmid pLS88 (Dixon *et al.*, 1994) resulted in a shortened C-terminus that differs from all so far known DHPS variants. A mutation in the translational stop codon of the DHPS from plasmids pYFC1 of *M. haemolytica* (Chang *et al.*, 1992a) and pTYM1 of *A. pleuropneumoniae* (Chang *et al.*, 2002) led to an extension of 12 amino acids at the C-terminus. In plasmid pCCK154 from *P. multocida*, the loss of a single 'A' at position 793 within the *sul2* gene caused a frame-shift mutation which led to the substitution of six codons and extended the reading frame by 18 codons (Kehrenberg and Schwarz 2005c). Finally, a recombination in the 3′ end of the *sul2* reading frame changed the final three codons and extended the *sul2* reading frame by one codon in plasmid pVM111 from *P. multocida* (Kehrenberg *et al.*, 2003). In addition to their location on small plasmids, *sul2* genes have also been detected on conjugative plasmids, such as pGS05 (Radström and Swedberg, 1988), or non-conjugative broad host range plasmids, such as RSF1010 (Scholz *et al.*, 1989) and pLS88 (Dixon *et al.*, 1994). Plasmids carrying *sul2* have also been identified in bacteria other than *Pasteurellaceae*, e.g. *E. coli* (Scholz *et al.*, 1989) or *P. damselae* subsp. *piscicida* (Kim and Aoki, 1996a).

Various studies revealed that the *sul2* gene is often linked to the *strA-strB* genes (Chang *et al.*, 1992a; Dixon *et al.*, 1994; Wright *et al.*, 1997; Kehrenberg and Schwarz 2001). PCR assays were developed to confirm the linkage of *sul2* and *strA* in both orientations (Kehrenberg and Schwarz 2001). In some cases, the *strA* gene followed by a truncated Δ*strB* gene was detected upstream of *sul2* (Wright *et al.*, 1997; Kehrenberg and Schwarz 2001). However in most of the *P. multocida*, [*P.*] *aerogenes* and *Mannheimia* isolates studied, the genes were found in the orientation *sul2–strA* whereas in one case a truncated Δ*strA* was found upstream of *sul2* (Chang *et al.*, 1992a; Dixon *et al.*, 1994; Kehrenberg and Schwarz 2001). Detailed studies of the non-coding spacer between *sul2* and *strA* revealed different lengths (Kehrenberg and Schwarz 2001; Ito *et al.*, 2004) and also showed that this region might represent a hot spot for recombination events. A *catA3* gene coding for chloramphenicol resistance was found to be inserted between *sul2* and *strA* via illegitimate recombination. Such *sul2–catA3–strA* clusters have been found on various plasmids as well as in the chromosomal DNA of [*P.*] *aerogenes*, *M. haemolytica*, *M. varigena* and *Mannheimia* taxon 10 isolates (Kehrenberg and Schwarz, 2001b; Kehrenberg and Schwarz, 2002). In plasmid pVM111 from an avian *P. multocida* isolate (Berman and Hirsh, 1978; Hansen *et al.*, 1993), a Tn*5706*-like *tetR-tet*(H) segment responsible for tetracycline resistance was also found to be inserted between *sul2* and *strA* via illegitimate recombination (Kehrenberg *et al.*, 2003). The resulting *sul2–tetR–tet*(H)–*strA* cluster, however, has not yet been detected on plasmids other than pVM111 or in the chromosomal DNA of *Pasteurella*, *Mannheimia*, or *Actinobacillus* isolates.

Recently, several small plasmids which carry only the sulphonamide resistance gene *sul2*, but not any other resistance genes, have been sequenced completely. These include the 3156-bp plasmid pKM2425 from *A. pleuropneumoniae* (Fig. 10.4), the 4065-bp plasmid pARD3079 (accession no. AM748707) from *A. pleuropneumoniae*, and the 4556-bp plasmid pKMA757 (accession no. AJ830713) from '*A. porcitonsillarum*' (Matter *et al.*, 2008).

Other *sul* genes, such as *sul1* or *sul3*, which also code for dihydropteroate synthetases with reduced affinity to sulphonamides, have not yet been detected among *Pasteurellaceae* (Matter *et al.*, 2007; Kehrenberg and Schwarz, unpublished data).

Resistance to sulphonamides by folP *mutations*

In high-level sulphonamide-resistant *H. influenzae*, which did not carry *sul1* or *sul2*, a 15-bp insertion into the chromosomal gene *folP* encoding a dihydropteroate synthase was detected. This insertion was shown to increase the MIC of sulphamethoxazole from 32 to 1024 μg/ml (Enne *et al.*, 2002). Such mutations have not yet been described to occur in other members of the *Pasteurellaceae*.

Resistance to trimethoprim by altered dihydrofolate reductases

A novel trimethoprim resistance gene was detected on the 11-kb plasmid pCCK154 from bovine *P. multocida* (Kehrenberg and Schwarz, 2005c). This plasmid was transferable into *E. coli* where it replicated and expressed high level resistance to sulphonamides and trimethoprim. Sequence analysis identified the gene *sul2* for sulphonamide resistance and a novel gene, designated *dfrA20*, for trimethoprim resistance. The *dfrA20* gene codes for a trimethoprim resistant dihydrofolate reductase of 169 amino acids, which is only distantly related to the dihydrofolate reductases of Gram-negative bacteria, but upon cluster analysis appears to be related to those found in the Gram-positive genera *Staphylococcus, Bacillus* and *Listeria* (Kehrenberg and Schwarz, 2005c).

Earlier studies on trimethoprim-resistant bovine *M. haemolytica* isolates from France revealed that trimethoprim resistance was not associated with plasmids and also was not transferable by conjugation. Hybridization experiments with gene probes specific for the genes *dfrA1–dfrA5* did not yield positive results (Escande *et al.*, 1991). Similar negative PCR results have been obtained recently for *A. pleuropneumoniae* and '*A. porcitonsillarum*' strains, in which none of 27 different *dfrA* or *dfrB* genes – including *dfrA20* – could be detected (Matter *et al.*, 2007). This observation led to the assumption that either other so far unknown trimethoprim resistance genes or other mechanisms are responsible for trimethoprim resistance in these isolates.

Resistance to trimethoprim by folH *mutations*

Overproduction of a trimethoprim-susceptible dihydrofolate reductase, encoded by the gene *folH*, was considered to be responsible for trimethoprim resistance in *H. influenzae* (De Groot *et al.*, 1988). A subsequent study by De Groot and co-workers (De Groot *et al.*, 1991) revealed that the over-expressed chromosomal dihydrofolate reductases of two *H. influenzae* strains also exhibited a 100- to 300-fold reduction in their inhibition by trimethoprim. This observation suggested that the over-expression of a chromosomal dihydrofolate reductase in combination with structural alterations that rendered the enzymes less susceptible to inhibition by trimethoprim accounted for trimethoprim resistance in these strains. Molecular analysis also showed that trimethoprim resistance might be due to alterations in the promoter region, but also alterations in the 5′ end and the central region of the *folH* gene which correspond to the trimethoprim binding domains in the dihydrofolate reductase protein (De Groot *et al.*, 1996). Moreover, these studies also suggested that amino acid exchanges in the C-terminus of the protein might result in trimethoprim resistance possibly due to changes in secondary structure and a consequential reduction of the trimethoprim binding affinity (De Groot *et al.*, 1996). Structural alterations in the gene coding for a chromosomal dihydrofolate reductase or in its promoter have not yet been described in other members of the *Pasteurellaceae*.

Resistance to macrolides

Many Gram-negative bacteria are believed to be innately resistant to macrolides due to permeability barriers or multidrug efflux pumps. However, chemical modification of the ribosomal target site by rRNA methylases or mutations in ribosomal proteins have also been described (Schwarz *et al.*, 2006)

Macrolide resistance due to rRNA methylases

Studies on the presence of macrolide resistance genes in bacteria of the genus *Actinobacillus* led to the identification of the rRNA methylase genes *erm*(A) and *erm*(C) in *A. pleuropneumoniae* (Wasteson *et al.*, 1996). Mating experiments showed that these genes were transferred into *Moraxella catarrhalis* and/or *Enterococcus faecalis* (Wasteson *et al.*, 1996). Similar studies with *Agg. actinomycetemcomitans* revealed the presence of *erm*(A), *erm*(B), *erm*(C), *erm*(F), and/or *erm*(Q) genes. All of these genes – except *erm*(A) – could be transferred alone or in various combinations into either *E. faecalis* or *H. influenzae* (Roe *et al.*, 1996). PCR directed analysis of bovine *P. multocida* and *M. haemolytica* which exhibited MICs of erythromycin of ≥16 µg/ml did not detect any of the three genes *erm*(A), *erm*(B) or *erm*(C) (Kehrenberg and Schwarz, unpublished data). Similar results were obtained by Matter and co-workers for *A. pleuropneumoniae* and '*A. porcitonsillarum*' strains with MICs of tilmicosin of ≥32 µg/ml, in which none of the genes *erm*(A), *erm*(B) or *erm*(C) could be detected by PCR (Matter *et al.*, 2007).

Macrolide resistance due to mutations in ribosomal proteins and rRNA

Analysis of macrolide-resistant *H. influenzae* isolates from humans revealed mutations in the genes coding for ribosomal proteins L4 and/or L22 and in the 23S rRNA (Peric *et al.*, 2003). The rRNA mutations mostly consisted of single bp exchanges at positions A2058G, G2160U, or C2164G while in one case also a GGA2160–2162UAU mutation was observed. Mutations in ribosomal proteins L4 or L22 resulted in either single amino acid exchanges, insertions of 2–6 amino acids, or deletions of 1–4 amino acids. Occasionally, combinations of 23S rRNA mutations with either mutations in ribosomal proteins L4 and/or L22, but also combinations of mutations in both ribosomal proteins were observed (Peric *et al.*, 2003).

Macrolide resistance due to active efflux

H. influenzae is considered to be intrinsically resistant to macrolides, lincosamides, streptogramin B antibiotics, and ketolides based on the activity of an efflux pump similar to AcrAB from *E. coli* (Sanchez *et al.*, 1997a; Sanchez *et al.*, 1997b; Peric *et al.*, 2003; Peric *et al.*, 2004; Bogdanovich *et al.*, 2006). The clinical significance of efflux-mediated macrolide resistance in *H. influenzae* is controversely discussed since the MICs conferred by this AcrAB homologue are comparatively low (Bogdanovich *et al.*, 2006). However, it was shown that high-level macrolide and ketolides-resistant *H. influenzae* isolates – in addition to efflux – usually exhibited mutations in 23S rRNA or ribosomal proteins L4 or L22 (Bogdanovich *et al.*, 2006). Moreover, L22 mutations failed to increase macrolide MICs in the absence of an efflux mechanism (Peric *et al.*, 2004). These observations suggest that active efflux is one component in a complex system, which contributes to macrolide resistance in *H. influenzae*.

Resistance to quinolones and fluoroquinolones

Quinolones are broad-spectrum antimicrobial agents which inhibit bacterial DNA gyrase and topoisomerase IV. Resistance is commonly due to mutational alterations in the genes coding for the different subunits of both enzymes, but also due to active efflux or protection of the enzymes by Qnr proteins (Schwarz *et al.*, 2006).

(Fluoro)Quinolone resistance due to target site mutations

Very little is known about (fluoro)quinolone resistance in *Pasteurella, Mannheimia, Actinobacillus,* or *Aggregatibacter*. Studies on *Pasteurella* spp. from humans and animals (Goldstein *et al.*, 2002; Kaspar *et al.* 2007; Schwarz *et al.*, 2007; Wallmann *et al.*, 2007), *A. pleuropneumoniae* (Matter *et al.*, 2007), and *Agg. actinomycetemcomitans* (Müller *et al.*, 2002) identified virtually all strains to be highly susceptible to the fluoroquinolones tested. A single report described the analysis of the quinolone resistance determining region (QRDR) of the genes *gyrA* and *parC* in *P. multocida* isolates which exhibited different levels of resistance to nalidixic acid (Cardenas *et al.*, 2001). A Ser83Ile alteration in GyrA was detected in an isolate that had an MIC to nalidixic acid of 256 µg/ml whereas Asp87Gly alterations were detected in isolates with MICs of 4 and 12 µg/ml (Cardenas *et al.*, 2001). None of these isolates exhibited resistance to fluoroquinolones.

In contrast, fluoroquinolone resistance has been detected occasionally in *H. influenzae*. Analysis of the quinolone determining regions in the target genes *gyrA*, *gyrB*, *parC* and/or *parE* revealed a wide range of mutations in resistant clinical isolates and laboratory-selected mutants (Georgiu *et al.*, 1996; Biedenbach and Jones, 2000; Davies *et al.*, 2000; Biedenbach and Jones, 2003; Elliott *et al.*, 2003; Clark *et al.*, 2004; Li *et al.*, 2004; Perez-Vasquez *et al.*, 2004; Rodriguez-Martinez *et al.*, 2006).

(Fluoro)Quinolone resistance due to other mechanisms

Although the AcrAB efflux pump has been shown in *Enterobacteriaceae* to be able to export also fluoroquinolones from the cell, the role of the AcrAB homologue of *H. influenzae* in (fluoro)quinolone resistance has not yet been demonstrated. Moreover, no *qnr* genes have been demonstrated in members of the *Pasteurellaceae*.

Conclusion

The present update on the molecular basis of antimicrobial resistance among members of the family *Pasteurellaceae* shows that an increasing number of resistance genes and resistance mediating mutations (Table 10.1) have been identified in pathogenic and commensal strains of various species and genera of the family *Pasteurellaceae* during recent years. Many of these resistance genes are located on mobile genetic elements such as plasmids or transposons and thus can be transferred horizontally between members of the same or even different bacterial species and genera. Several resistance plasmids isolated from

Table 10.1 Antimicrobial resistance genes and mutations identified in *Pasteurellaceae*

Antimicrobial agent(s)	Resistance gene or resistance mediating mutation(s)	Detected in						
		Pasteurella	*Mannheimia*	*Haemophilus*	*Actinobacillus*	*Avibacterium*	*Bibersteinia*	*Aggregatibacter*
Penicillins	*bla*$_{ROB-1}$	+	+	+	+	–	–	–
	bla$_{TEM-1}$	+	–	+	–	–	–	–
	bla$_{TEM-15}$	–	–	+	–	–	–	–
	bla$_{PSE-1}$	+	–	–	–	–	–	–
	Mutation in *ftsI* coding for penicillin binding protein 3	–	–	+	–	–	–	–
	Mutation in *acrR*	–	–	+	–	–	–	–
Tetracycline	*tet*(H)	+	+	–	+	–	–	–
	tet(B)	+	–	+	+	–	–	–
	tet(G)	+	+	–	–	–	–	–
	tet(L)	+	+	–	+	–	–	–
	tet(K)	–	–	–	–	–	–	+
	tet(M)	+	–	+	+	–	–	–
	tet(O)	–	–	–	+	–	–	–
Streptomycin	*strA*	+	+	+	+	+	+	–
	strB	+	+	+	+	+	–	–
Strepto-/spectinomycin	*aadA1*	+	–	–	–	–	–	–
	aadA14	+	–	–	–	–	–	–
Spectinomycin	Mutation in 16S rRNA	+	–	–	–	–	–	–
	Mutation in *rpsE* coding for ribosomal protein S5	+	–	–	–	–	–	–

Kanamycin/neomycin	*aphA1*	+	+	+	–	+	–	–
	aphA3	+	–	–	–	–	–	–
Chloramphenicol	*catA1*	+	–	–	–	–	–	–
	catA2	–	–	+	–	–	–	–
	catA3	+	+	–	–	–	+	–
	catB2	+	–	–	–	–	–	–
Chloramphenicol/ florfenicol	*floR*	+	–	–	–	–	+	–
Sulphonamides	*sul2*	+	+	+	+	+	+	–
	mutation in *folP*	–	–	+	–	–	–	–
Trimethoprim	*dfrA20*	+	–	–	–	–	–	–
	Mutation in *folH*	–	–	+	–	–	–	–
Macrolides	*erm*(A)	–	–	–	–	–	–	+
	erm(B)	–	–	–	–	–	–	+
	erm(C)	–	–	–	–	–	–	+
	erm(F)	–	–	–	–	–	–	+
	erm(Q)			–	–	–	–	+
	Mutation in *rplD* coding for ribosomal protein L4	–	–	+	–	–	–	–
	Mutation in *rplV* coding for ribosomal protein L22	–	–	+	–	–	–	–
(fluoro)quinolones	Mutation in *gyrA*	+	–	+	–	–	–	–
	Mutation in *gyrB*	–	–	+	–	–	–	–
	Mutation in *parC*	–	–	+	–	–	–	–
	Mutation in *parE*	–	–	+	–	–	–	–

the aforementioned *Pasteurellaceae* genera have been shown to be able to replicate and express their resistance properties not only in members of other genera of the *Pasteurellaceae*, but also in more distantly related bacteria, such as *E. coli*. The complete sequencing of some of the resistance plasmids also provided deeper insight into their development. These structural analyses revealed that plasmids, such as pCCK381 from *P. multocida* (Kehrenberg and Schwarz, 2005a), pCCK13698 from *B. trehalosi* (Kehrenberg *et al.*, 2006a), pHS-Tet from [*H.*] *parasuis* (Lancashire *et al.*, 2005), p12494 from *A. pleuropneumoniae* (Blanco *et al.*, 2007), or pCCK3259 from *M. haemolytica* (Kehrenberg *et al.*, 2005a), might represent the results of interplasmid recombination events. Such plasmids usually carry 'foreign' resistance genes in a plasmid backbone well adapted to replication in *Pasteurellaceae* hosts. These observations underline the enormous flexibility of the *Pasteurellaceae* to acquire and efficiently use resistance genes from other bacteria. In several cases, multiresistance plasmids have been identified in selected members of the *Pasteurellaceae*, some of which carry a cluster of resistance genes. Such resistance gene clusters, as the *sul2-tetR-tet*(H)*-strA* cluster on plasmid pVM111 from *P. multocida* (Kehrenberg *et al.*, 2003) or the *sul2-catA3-strA* cluster on plasmid pMVSCS1 from *M. varigena* (Kehrenberg and Schwarz, 2002), allow for the co-selection and the persistence of resistance genes even in the absence of a direct selective pressure. In addition, several novel resistance genes identified in *Pasteurellaceae*, such as the trimethoprim resistance gene *dfrA20* (Kehrenberg and Schwarz, 2005c) or the streptomycin/spectinomycin resistance gene *aadA14* (Kehrenberg *et al.*, 2005b), have so far not been detected in members of other bacterial families. Nevertheless, the analysis of the resistance genes present in *Pasteurellaceae* strongly suggests that there is a resistance gene flux between the *Pasteurellaceae* and other Gram-negative, but also Gram-positive bacteria.

Even though the current susceptibility status of *Pasteurellaceae* from infections in humans and animals looks rather favourable, especially for the newer drugs approved for use in human and veterinary medicine (Kaspar *et al.*, 2007; Schwarz *et al.*, 2007; Tristram *et al.*, 2007; Wallmann *et al.*, 2007), prescription and administration of antimicrobial agents should be undertaken with discretion supported by an accurate diagnosis, a careful choice of the respective antimicrobial agent(s), and the most appropriate dosing regimen to retain the efficacy of the antimicrobial drugs (Kehrenberg *et al.*, 2006b). Moreover, continuous monitoring of the antimicrobial susceptibility of *Pasteurellaceae* is required to early recognize changes in the susceptibility status and to analyse the respective strains for newly acquired/developed resistance genes and resistance-mediating mutations.

References

Aarestrup, F.M., Seyfarth, A.M., and Angen, Ø. (2004). Antimicrobial susceptibility of *Haemophilus parasuis* and *Histophilus somni* from pigs and cattle in Denmark. Vet. Microbiol. *101*, 143–146.

Alton, N.K., and Vapnek, D. (1979). Nucleotide sequence analysis of the chloramphenicol resistance transposon Tn*9*. Nature *282*, 864–869.

Aoki, T., and Takahashi, A. (1987). Class D tetracycline resistance determinants of R plasmids from the fish pathogens *Aeromonas hydrophila*, *Edwardsiella tarda*, and *Pasteurella piscicida*. Antimicrob. Agents Chemother. *31*, 1278–1280.

Azad, A.K., Coote, J.G., and Parton, R. (1992). Distinct plasmid profiles of *Pasteurella haemolytica* serotypes and the characterization and amplification in *E. coli* of ampicillin-resistance plasmids encoding ROB1-β-lactamase. J. Gen. Microbiol. *138*, 1185–1196.

Berman, S.M., and Hirsh, D.C. (1978). Partial characterization of R-plasmids from *Pasteurella multocida* isolated from turkeys. Antimicrob. Agents Chemother. *14*, 348–352.

Biedenbach, D.J., and Jones, R.N. (2000). Fluoroquinolone-resistant *Haemophilus influenzae*: frequency of occurrence and analysis of confirmed strains in the SENTRY antimicrobial surveillance program (North and Latin America). Diagn. Microbiol. Infect. Dis. *36*, 255–259.

Biedenbach, D.J., and Jones, R.N. (2003). Five-year analysis of *Haemophilus influenzae* isolates with reduced susceptibility to fluoroquinolones: prevalence results from the SENTRY antimicrobial surveillance program. Diagn. Microbiol. Infect. Dis. *46*, 55–61.

Blackall, P.J., Bojesen, A.M., Christensen, H., and Bisgaard M. (2007). Reclassification of [*Pasteurella*] *trehalosi* as *Bibersteinia trehalosi* gen. nov., comb. nov. Int. J. Syst. Evol. Microbiol. *57*, 666–674.

Blanco, M., Gutierrez-Martin, C.B., Rodriguez-Ferri, E.F., Roberts, M.C., and Navas, J. (2006). Distribution of tetracycline resistance genes in *Actinobacillus pleuropneumoniae* isolates from Spain. Antimicrob. Agents Chemother. *50*, 702–708.

Blanco, M., Kadlec, K., Gutierrez-Martin, C.B., Martin de la Fuente, A.J., Schwarz, S., and Navas, J. (2007). Nucleotide sequence and transfer properties of two novel types of *Actinobacillus pleuropneumoniae* plasmids carrying the tetracycline resistance gene *tet*(H). J. Antimicrob. Chemother. *60*, 864–867.

Blickwede, M., and Schwarz, S. (2004). Molecular analysis of florfenicol-resistant *Escherichia coli* from pigs. J. Antimicrob. Chemother. *53*, 58–64.

Bogdanovich, T., Bozdogan, B., and Appelbaum, P.C. (2006). Effect of efflux on telithromycin and macrolide susceptibility in *Haemophilus influenzae*. Antimicrob. Agents Chemother. *50*, 893–898.

Briggs, C.E., and Fratamico, P.M. (1999). Molecular characterization of an antibiotic resistance gene cluster of *Salmonella* Typhimurium DT104. Antimicrob. Agents Chemother. *43*, 846–849.

Briggs, R.E., Frank, G.H., Purdy, C.D., Zehr, E.S., and Loan, R.W. (1998). Rapid spread of a unique strain of *Pasteurella haemolytica* serotype 1 among transported calves. Am. J. Vet. Res. *59*, 401–405.

Brink, M.F., Brink, G., Verbeet, M.P., and de Boer, H.A. (1994). Spectinomycin interacts specifically with the residues G_{1064} and C_{1192} in 16S rRNA, thereby potentially freezing this molecule into an inactive conformation. Nucleic Acids Res. *22*, 325–331.

Brunton, J., Meier, M., Erhman, N., Clare, D., and Almawy R. (1986). Origin of small β-lactamase-specifying plasmids in *Haemophilus* species and *Neisseria gonorrhoeae*. J. Bacteriol. *168*, 374–379.

Burns, J.L., Mendelman, P.M., Levy, J., Stull, T.L., and Smith, A.L. (1985). A permeability barrier as a mechanism of chloramphenicol resistance in *Haemophilus influenzae*. Antimicrob. Agents Chemother. *27*, 46–54.

Bush, K., Jacoby, G.A., and Medeiros, A.A. (1995). A functional classification scheme for β-lactamases and its correlation with molecular structure. Antimicrob. Agents Chemother. *39*, 1211–1233.

Bywater R, Silley P, and Simjee S. (2006). Antimicrobial breakpoints – definitions and conflicting requirements. Vet Microbiol. *118*, 158–159.

Cardenas, M., Barbe, J., Llagostera, M., Miro, E., Navarro, F., Mirelis, B., Prats, G., and Badiola, I. (2001). Quinolone resistance-determining regions of *gyrA* and *parC* in *Pasteurella multocida* strains with different levels of nalidixic acid resistance. Antimicrob. Agents Chemother. *45*, 990–991.

Castanheira, M., Gales, A.C., Pignatari, A.C., Jones, R.N., and Sader, H.S. (2006). Changing antimicrobial susceptibility patterns among *Streptococcus pneumoniae* and *Haemophilus influenzae* from Brazil: Report from the SENTRY Antimicrobial Surveillance Program (1998–2004). Microb. Drug Resist. *12*, 91–98.

Catry, B., Chiers, K., Schwarz, S., Kehrenberg, C., Decostere, A., and de Kruif, A. (2005). A case of fatal peritonitis in calves caused by *Pasteurella multocida* capsular type F. J. Clin. Microbiol. *43*, 1480–1483.

Chalmers, S., Sewitz, R., Lipkow, K., and Crellin, P. (2000). Complete nucleotide sequence of Tn*10*. J. Bacteriol. *182*, 2970–2972.

Chang, C.-F., Yeh, T.-M., Chou, C.-C., Chang, Y.-F., and Chiang, T.-S. (2002). Antimicrobial susceptibility and plasmid analysis of *Actinobacillus pleuropneumoniae* isolated in Taiwan. Vet. Microbiol. *84*, 169–177.

Chang, Y.F., Ma, D.P., Bai, H.Q., Young, R., Struck, D.K., Shin, S.J., and Lein, D.H. (1992a). Characterization of plasmids with antimicrobial resistance genes in *Pasteurella haemolytica* A1. DNA Seq. *3*, 89–97.

Chang, Y.F., Shi, J., Shin, S.J., and Lein, D.H. (1992b). Sequence analysis of the ROB-1 β-lactamase gene from *Actinobacillus pleuropneumoniae*. Vet. Microbiol. *32*, 319–325.

Chaslus-Dancla, E., Lesage-Descauses, M.-C., Leroy-Sétrin, S., Martel, J-L., and Lafont, J.-P. (1995). Tetracycline resistance determinants, TetB and TetM, detected in *Pasteurella haemolytica* and *Pasteurella multocida* from bovine herds. J. Antimicrob. Chemother. *36*, 815–819.

Chiou, C.-S., and Jones, A.L. (1993). Nucleotide sequence analysis of a transposon (Tn*5393*) carrying streptomycin resistance genes in *Erwinia amylovora* and other Gram-negative bacteria. J. Bacteriol. *175*, 732–740.

Chopra, I., and Roberts, M.C. (2001). Tetracycline antibiotics: Mode of action, applications, molecular biology and epidemiology of bacterial resistance. Microbiol. Mol. Bio. Rev. *65*, 232–260.

Clark, C., Kosowska, K., Bozdogan, B., Credito, K., Dewasse, B., McGhee, P., Jacobs, M.R., and Appelbaum, P.C. (2004). *In vitro* selection of resistance in *Haemophilus influenzae* by 4 quinolones and 5 β-lactams. Diagn. Microbiol. Infect. Dis. *49*, 31–36.

Clinical and Laboratory Standards Institute. (2006). Performance standards for antimicrobial susceptibility testing; 16th informational supplement. M100-S16. CLSI, Wayne, PA.

Clinical and Laboratory Standards Institute. (2008). Performance standards for antimicrobial disk and dilution susceptibility tests for bacteria isolated from animals; approved standard – third edition. M31-A3. CLSI, Wayne, PA.

Cote, S., Harel, J., Higgins, R., and Jacques, M. (1991). Resistance to antimicrobial agents and prevalence of R plasmids in *Pasteurella multocida* from swine. Am. J. Vet. Res. *52*, 1653–1657.

Craig, F.F., Coote, J.G., Parton, R., Freer, J.H., and Gilmour, N.J. (1989). A plasmid which can be transferred between *Escherichia coli* and *Pasteurella haemolytica* by electroporation and conjugation. J. Gen. Microbiol. *135*, 2885–2890.

Dabernat, H., Delmas, C., Seguy, M., Pelissier, R., Faucon, G., Bennamani, S., and Pasquier, C. (2002). Diversity of β-lactam resistance conferring amino acid substitutions in penicillin-binding protein 3 of *Haemophilus influenzae*. Antimicrob. Agents Chemother. *46*, 2208–2218.

Daoud, Z., Cocozaki, A., and Hakime, N. (2006). Antimicrobial susceptibility patterns of *Haemophilus influenzae* and *Streptococcus pneumoniae* isolates in a Beirut general university hospital between 2000 and 2004. Clin. Microbiol. Infect. *12*, 86–90.

Davies, C., Bussiere, D.E., Golden, B.L., Porter, S.J., Ramakrishnan, V., and White, S.W. (1998). Ribosomal proteins S5 and L6: High resolution crystal structures and roles in protein biosynthesis and antibiotic resistance. J. Mol. Biol. *279*, 873–888.

Davies, T.A., Kelly, L.M., Hoellman, D.B., Ednie, L.M., Clark, C.L., Bajaksouzian, S., Jacobs, M.R., and Appelbaum, P.C. (2000). Activities and postantibiotic effects of gemifloxacin compared to those of 11 other agents against *Haemophilus influenzae* and *Moraxella catarrhalis*. Antimicrob. Agents Chemother. *44*, 633–639.

De Graaff, J., Elwell, L.P., and Falkow, S. (1976). Molecular nature of two β-lactamase-specifying plasmids isolated from *Haemophilus influenzae* type Bacteriol. *126*, 439–446.

De Groot, R., Campos, J., Moseley, S.L., and Smith, A.L. (1988). Molecular cloning and mechanism of trimethoprim resistance in *Haemophilus influenzae*. Antimicrob. Agents Chemother. *32*, 477–484.

De Groot, R., Chaffin, D.O., Kuehn, M., and Smith, A.L. (1991). Trimethoprim resistance in *Haemophilus influenzae* is due to altered dihydrofolate reductase(s). Biochem. J. *274*, 657–662.

De Groot, R., Sluijter, M., de Bruyn, A., Campos, J., Goessens, W.H., Smith, A.L., and Hermans, P.W. (1996). Genetic characterization of trimethoprim resistance in *Haemophilus influenzae*. Antimicrob. Agents Chemother. *40*, 2131–2136.

De Stasio, E.A., Moazed, D., Noller, H.F., and Dahlberg, A.E. (1989). Mutations in 16S ribosomal RNA disrupt antibiotic-RNA interactions. EMBO J. *8*, 1213–1216.

Dixon, L.G., Albritton, W.L., and Willson, P.J. (1994). An analysis of the complete nucleotide sequence of the *Haemophilus ducreyi* broad-host-range plasmid pLS88. Plasmid *32*, 228–232.

Elliott, E., Oosthuizen, D., Johnson, M.M., and Piddock, L.J.V. (2003). Fluoroquinolone resistance in *Haemophilus influenzae*. J. Antimicrob. Chemother. *52*, 734–735.

Elwell, L.P., De Graaff, J., Seibert, D., and Falkow, S. (1975). Plasmid-linked ampicillin resistance in *Haemophilus influenzae* type b. Infect. Immun. *12*, 404–410.

Enne, V.I., King, A., Livermore, D.M., and Hall, L.M. (2002). Sulfonamide resistance in *Haemophilus influenzae* mediated by acquisition of *sul2* or a short insertion in chromosomal *folP*. Antimicrob. Agents Chemother. *46*, 1934–1939.

Escande, F., Gerbaud, G., Martel, J.-L., and Courvalin, P. (1991). Resistance to trimethoprim and 2, 4-diamino-6, 7-diisopropyl-pteridine (O/129) in *Pasteurella haemolytica*. Vet. Microbiol. 26, 107–114.

Farrell, D.J., Morrissey, I., Bakker, S., Buckridge, S., and Felmingham, D. (2005). Global distribution of TEM-1 and ROB-1 β-lactamases in *Haemophilus influenzae*. J. Antimicrob. Chemother. *56*, 773–776.

Fedler, K.A., Biedenbach, D.J., and Jones, R.N. (2006). Assessment of pathogen frequency and resistance patterns among pediatric patient isolates: report from the 2004 SENTRY Antimicrobial Surveillance Program on 3 continents. Diagn. Microbiol. Infect. Dis. *56*, 427–436.

Felmingham, D., White, A.R., Jacobs, M.R., Appelbaum, P.C., Poupard, J., Miller, L.A., and Gruneberg, R.N. (2005). The Alexander Project: the benefits from a decade of surveillance. J. Antimicrob. Chemother. *56*, Suppl. 2:ii3-ii21.

Flannagan, S.E., Zitzow, L.A., Su, Y.A., and Clewell, D.B. (1994). Nucleotide sequence of the 18-kb conjugative transposon Tn*916* from *Enterococcus faecalis*. Plasmid *32*, 350–354.

Fluit, A.C., Florijn, A., Verhoef, J., and Milatovic, D. (2005). Susceptibility of European β-lactamase-positive and -negative *Haemophilus influenzae* isolates from the periods 1997/1998 and 2002/2003. J. Antimicrob. Chemother. *56*, 133–138.

Fukutomi, Y., Takayanagi, M., Kusaba, K., Nagasawa, Z., Ohkuma, M., Aoki, Y., and Nagayama, A. (2004). The isolation frequency and antimicrobial susceptibility of *Haemophilus influenzae* isolated in Saga University Hospital. Jpn. J. Antibiot. *57*, 187–195.

Funatsu, G., Schiltz, E., and Wittmann, H.G. (1971). Ribosomal proteins: XXVII. Localization of the amino acid exchanges in protein S5 from two *Escherichia coli* mutants resistant to spectinomycin. Mol. Gen. Genet. *114*, 106–111.

Galan, J.C., Morosini, M.I., Baquero, M.R., Reig, M., and Baquero, F. (2003). *Haemophilus influenzae* bla_{ROB-1} mutations in hypermutagenic Δ*ampC Escherichia coli* conferring resistance to cefotaxime and β-lactamase inhibitors and increased susceptibility to cefachlor. Antimicrob. Agents Chemother. *47*, 2551–2557.

Galimand, M., Gerbaud, G., and Courvalin, P. (2000). Spectinomycin resistance in *Neisseria* spp. due to mutations in 16S rRNA. Antimicrob. Agents Chemother. *44*, 1365–1366.

Gauthier, G., Lafay, B., Ruimy, R., Breittmayer, V., Nicolas, J.L., Gauthier, M., and Christen, R. (1995). Small subunit rRNA sequences and whole DNA relatedness concur for the reassignment of *Pasteurella piscicida* (Snieszko *et al.*) Janssen and Surgalla to the genus *Photobacterium* as *Photobacterium damsela* subsp. *piscicida* comb. nov. Int. J. Syst. Bacteriol. *45*, 139–144.

Georgiu, M., Munoz, R., Roman, F., Canton, R., Gomez-Lus, R., Campos, J. and de la Campa, A.G. (1996). Ciprofloxacin-resistant *Haemophilus influenzae* strains possess mutations in analogous positions of GyrA and ParC. Antimicrob. Agents Chemother. *40*, 1741–1744.

Gilbride, K.A., Rosendal, S., and Brunton, J.L. (1989). Plasmid mediated antimicrobial resistance in Ontario isolates of *Actinobacillus* (*Haemophilus*) *pleuropneumoniae*. Can. J. Vet. Res. *53*, 38–42.

Goldstein, E.J.C., Citron, D.M., Merriam, C.V., Warren, Y.A., Tyrrell, K.L., and Fernandez, H.T. (2002). *In vitro* activities of garenoxacin (BMS-28756) against 170 clinical isolates of nine *Pasteurella* species. Antimicrob. Agents Chemother. *46*, 3068–3070.

Gutiérrez-Martín, C.B., Garcíá del Blanco, N., Blanco, M., Navas, J., and Rodríguez-Ferri, E.F. (2006). Changes in antimicrobial susceptibility of *Actinobacillus pleuropneumoniae* isolated from pigs in Spain during the last decade. Vet. Microbiol. *115*, 218–222.

Hansen, L.M., Blanchard, P.C., and Hirsh, D.C. (1996). Distribution of *tet*(H) among *Pasteurella* isolates from the United States and Canada. Antimicrob. Agents Chemother. *40*, 1558–1560.

Hansen, L.M., McMurray, L.M., Levy, S.B., and Hirsh, D.C. (1993). A new tetracycline resistance determinant, TetH, from *Pasteurella multocida* specifying active efflux of tetracycline. Antimicrob. Agents Chemother. *37*, 2699–2705.

Hasegawa, K., Chiba, N., Kobayashi, R., Murayama, S.Y., Iwata, S., Sunakawa, K., and Ubukata, K. (2004). Rapidly increasing prevalence of β-lactamase-nonproducing, ampicillin-resistant *Haemophilus influenzae* type b in patients with meningitis. Antimicrob. Agents Chemother. *48*, 1509–1514.

Hirsh, DC., Martin, D.L., and Rhoades, K.R. (1981). Conjugal transfer of an R-plasmid in *Pasteurella multocida*. Antimicrob. Agents Chemother. *20*, 415–417.

Hirsh, D.C., Martin, D.L., and Rhoades, K.R. (1985). Resistance plasmids of *Pasteurella multocida* isolated from turkeys. Am. J. Vet. Res. *46*, 1490–1493.

Hirsh, D.C., Hansen, L.M., Dorfman, L.C., Snipes, K.P., Carpenter, T.E., Hird, D.W., and McCapes, R.H. (1989). Resistance to antimicrobial agents and prevalence of R plasmids in *Pasteurella multocida* from turkeys. Antimicrob. Agents Chemother. *33*, 670–673.

Hörmansdorfer, S., and Bauer J. (1996). Zur Resistenzsituation boviner Pasteurellen [Resistance pattern of bovine pasteurellae]. Berl. Münch. Tierärztl. Wschr. *109*, 168–171.

Hörmansdorfer, S., and Bauer, J. (1998). Zur Resistenz boviner und porciner Pasteurellen gegenüber Florfenicol und anderen Antibiotika [Resistance of bovine and porcine pasteurellae against florfenicol and other antibiotics]. Berl. Münch. Tierärztl. Wschr. *111*, 422–426.

Hsu, Y.M., Shieh, H.K., Chen, W.H., Sun, T.Y., and Shiang, J.-H. (2007). Antimicrobal susceptibility, plasmid profiles, and haemocin activities of *Avibacterium paragallinarum* strains. Vet. Microbiol. *124*, 209–218.

Hunt, M.L., Adler, B., and Townsend, K.M., (2000). The molecular biology of *Pasteurella multocida*. Vet. Microbiol. *72*, 3–25.

Ishii, H., Fukuyasu, T., Iyobe, S., and Hashimoto, H. (1993). Characterization of newly isolated plasmids from *Actinobacillus pleuropneumoniae*. Am. J. Vet. Res. *54*, 701–708.

Ishii, H., Hayashi, F., Iyobe, S., and Hashimoto, H. (1991). Characterization and classification of *Actinobacillus* (*Haemophilus*) *pleuropneumoniae* plasmids. Am. J. Vet. Res. 52, 1816–1820.

Ishii, H., Nakasone, Y., Shigehara, S., Honma, K., Araki, Y., Iyobe, S., and Hashimoto, H. (1990). Drug-susceptibility and isolation of a plasmid in *Haemophilus* (*Actinobacillus*) *pleuropneumoniae*. Jpn. J. Vet. Sci. 52, 1–9.

Ito, H., Ishii, H., and Akiba, M. (2004). Analysis of the complete nucleotide sequence of an *Actinobacillus pleuropneumoniae* streptomycin-sulfonamide resistance plasmid, pMS260. Plasmid *51*, 41–47.

Jacobs, M.R., Bajaksouzian, S., Windau, A., Appelbaum, P.C., Lin, G., Felmingham, D., Dencer, C., Koeth, L., Singer, M.E., and Good, C.E. (2002). Effects of various test media on the activities of 21 antimicrobial agents against *Haemophilus influenzae*. J. Clin. Microbiol. *40*, 3269–3276.

Jacobs, M.R., Bajaksouzian, S., Windau, A., Good, C.E., Lin, G., Pankuch, G.A., and Appelbaum, P.C. (2004). Susceptibility of *Streptococcus pneumoniae*, *Haemophilus influenzae*, and *Moraxella catarrhalis* to 17 oral antimicrobial agents based on pharmacodynamic parameters: 1998–2001 U S Surveillance Study. Clin. Lab. Med. *24*, 503–530.

Jacobs, M.R., Felmingham, D., Appelbaum, P.C., and Gruneberg, R.N. (2003). The Alexander Project 1998–2000: susceptibility of pathogens isolated from community-acquired respiratory tract infection to commonly used antimicrobial agents. J. Antimicrob. Chemother. 52, 229–246.

Jansen, W.T., Verel, A., Beitsma, M., Verhoef, J., and Milatovic, D. (2006). Longitudinal European surveillance study of antibiotic resistance of *Haemophilus influenzae*. J. Antimicrob. Chemother. 58, 873–877.

Jones, R.N., Mendes, C., Turner, P.J., and Masterton, R. (2005). An overview of the Meropenem Yearly Susceptibility Test Information Collection (MYSTIC) Program: 1997–2004. Diagn. Microbiol. Infect. Dis. 53, 247–256.

Juteau J.-M., and Levesque, R.C. (1990). Sequence analysis and evolutionary perspectives of ROB1-β-lactamase. Antimicrob. Agents Chemother. *34*, 1354–1359.

Juteau, J.-M., Sirois, M., Medeiros, A.A., and Levesque, R.C. (1991). Molecular distribution of ROB1-β-lactamase in *Actinobacillus pleuropneumoniae*. Antimicrob. Agents Chemother. *35*, 1397–1402.

Kaczmarek, F.S., Gootz, T.D., Dib-Hajj, F., Shang, W., Hallowell, S., and Cronan, M. (2004). Genetic and molecular characterization of β-lactamase-negative ampicillin-resistant *Haemophilus influenzae* with unusually high resistance to ampicillin. Antimicrob. Agents Chemother. *48*, 1630–1639.

Kaspar, H., Schröer, U., and Wallmann, J. (2007). Quantitative resistance level (MIC) of *Pasteurella multocida* isolated from pigs between 2004 and 2006: National Resistance Monitoring by the BVL. Berl. Münch. Tierärztl. Wochenschr. *120*, 442–451.

Kawahara, K., Kawase, H., Nakai, T., Kume, K., and Danbara, H. (1990). Drug resistance plasmids of *Actinobacillus pleuropneumoniae* serotype 2 strains isolated from swine. Kitasato Arch. Exp. Med. *63*, 131–136.

Kehrenberg, C., Catry, B., Haesebrouck, F., de Kruif, A., and Schwarz, S. (2005a). *tet*(L)-mediated tetracycline resistance in bovine *Mannheimia* and *Pasteurella* isolates. J. Antimicrob. Chemother. 56, 403–406.

Kehrenberg, C., Catry, B., Haesebrouck, F., de Kruif, A., and Schwarz, S. (2005b). A novel spectinomycin/streptomycin resistance gene, *aadA14*, from *Pasteurella multocida*. Antimicrob. Agents Chemother. *49*, 3046–3049.

Kehrenberg, C., Meunier, D., Targant, H., Cloeckaert, A., Schwarz, S., and Madec, J.-Y. (2006a). Plasmid-mediated florfenicol resistance in *Pasteurella trehalosi*. J. Antimicrob. Chemother. 58, 13–17.

Kehrenberg, C., Mumme, J., Wallmann, J., Verspohl, J., Tegeler, R., Kühn, T., and Schwarz, S. (2004).

Monitoring of florfenicol susceptibility among bovine and porcine respiratory tract pathogens collected in Germany during the years 2002 and 2003. J. Antimicrob. Chemother. *54*, 572–574.

Kehrenberg, C., Salmon, S.A., Watts, J.L., and Schwarz, S. (2001a). Tetracycline resistance genes in isolates of *Pasteurella multocida*, *Mannheimia haemolytica*, *Mannheimia glucosida*, and *Mannheimia varigena* from bovine and swine respiratory disease: intergeneric spread of plasmid pMHT1. J. Antimicrob. Chemother. *48*, 631–640.

Kehrenberg, C., Schulze-Tanzil, G., Martel, J.-L., Chaslus-Dancla, E., and Schwarz, S. (2001b). Antimicrobial resistance in *Pasteurella* and *Mannheimia*: epidemiology and genetic basis. Vet. Res. *32*, 323–339.

Kehrenberg, C., and Schwarz, S. (2000). Identification of a truncated, but functionally active *tet*(H) tetracycline resistance gene in *Pasteurella aerogenes* and *Pasteurella multocida*. FEMS Microbiol. Lett. *188*, 191–195.

Kehrenberg, C., and Schwarz, S. (2001a). Molecular analysis of tetracycline resistance in *Pasteurella aerogenes*. Antimicrob. Agents Chemother. *45*, 2885–2890.

Kehrenberg, C., and Schwarz, S. (2001b). Occurrence and linkage of genes coding for resistance to sulfonamides, streptomycin and chloramphenicol in bacteria of the genera *Pasteurella* and *Mannheimia*. FEMS Microbiol. Lett. *205*, 283–290.

Kehrenberg, C., and Schwarz, S. (2002). Nucleotide sequence and organisation of plasmid pMVSCS1 from *Mannheimia varigena*: Identification of a multiresistance gene cluster. J. Antimicrob. Chemother. *49*, 383–386.

Kehrenberg, C., and Schwarz, S. (2005a). Plasmid-borne florfenicol resistance in *Pasteurella multocida*. J. Antimicrob. Chemother. *55*, 773–775.

Kehrenberg, C., and Schwarz, S. (2005b). Molecular basis of resistance to kanamycin and neomycin in *Pasteurella* and *Mannheimia* isolates of animal origin. Abstracts of the ASM Conference on *Pasteurellaceae* 2005, American Society for Microbiology, Poster A47, p. 55.

Kehrenberg, C., and Schwarz, S. (2005c). Identification of *dfrA20*, a novel trimethoprim resistance gene from *Pasteurella multocida*. Antimicrob. Agents Chemother. *49*, 414–417.

Kehrenberg, C., and Schwarz, S. (2007). Mutations in 16S rRNA and ribosomal protein S5 associated with high-level spectinomycin resistance in *Pasteurella multocida*. Antimicrob. Agents Chemother. *51*, 2244–2246.

Kehrenberg, C., Tham, N.T.T., and Schwarz, S. (2003). New plasmid-borne antibiotic resistance gene cluster in *Pasteurella multocida*. Antimicrob. Agents Chemother. *47*, 2978–2980.

Kehrenberg, C., Walker, R.D., Wu, C.C., and Schwarz, S. (2006b). Antimicrobial resistance in members of the family *Pasteurellaceae*. In: Aarestrup, F.M., ed., Antimicrobial resistance in bacteria of animal origin. Washington, DC: American Society for Microbiology, ASM Press, 167–186.

Kehrenberg, C., Werckenthin, C., and Schwarz, S. (1998). Tn*5706*, a transposon-like element from *Pasteurella multocida* mediating tetracycline resistance. Antimicrob. Agents Chemother. *42*, 2116–2118.

Kim, E.H., and Aoki, T. (1996a). Sulfonamide resistance gene in a transferable R plasmid of *Pasteurella piscicida*. Microbiol. Immunol. *40*, 397–399.

Kim, E.H., and Aoki, T. (1996b). Sequence analysis of the florfenicol resistance gene encoded in the transferable R-plasmid of a fish pathogen, *Pasteurella piscicida*. Microbiol. Immunol. *40*, 665–669.

Kiuchi, A., Hara, M., and Tabuchi, K. (1992). Drug resistant plasmid of *Actinobacillus pleuropneumoniae* isolated from swine pleuropneumonia in Thailand. Kansenshogaku Zasshi. *66*, 1243–1247.

Lalonde, G., Miller, J.F., Tompkins, L.S., and O'Hanley, P. (1989). Transformation of *Actinobacillus pleuropneumoniae* and analysis of R factors by electroporation. Am. J. Vet. Res. *50*, 1957–1960.

Lancashire, J.F., Terry, T.D., Blackall, P.J., and Jennings, M.P. (2005). Plasmid-mediated tetracycline resistance in *Haemophilus parasuis*. Antimicrob. Agents Chemother. *49*, 1927–1931.

Laufs, R., Riess, F.C., Jahn, G., Fock, R., and Kaulfers, P.M. (1981). Origin of *Haemophilus influenzae* R factors. J. Bacteriol. *147*, 563–568.

Lawley, T.D., Burland, V.D., and Taylor, D.E. (2000). Analysis of the complete nucleotide sequence of the tetracycline resistance transposon Tn*10*. Plasmid *43*, 235–239.

Leaves, N.I., Dimopoulou, I., Hayes, I., Kerridge, S., Falla, T., Secka, O., Adegbola, R.A., Slack, M.P., Peto, T.E., and Crook, D.W. (2000). Epidemiological studies of large resistance plasmids in *Haemophilus*. J. Antimicrob. Chemother. *45*, 599–604.

Levy, S.B., Buu-Hoi, A., and Marshall, B. (1984). Transposon Tn*10*-like tetracycline resistance determinants in *Haemophilus parainfluenzae*. J. Bacteriol. *160*, 87–94.

Li, X., Mariano, N., Rahal, J.J., Urban, C.M., and Drlica, K. (2004). Quinolone-resistant *Haemophilus influenzae* in a long-term-care facility: nucleotide sequence characterization of alterations in the genes encoding DNA gyrase and DNA topoisomerase IV. Antimicrob. Agents Chemother. *48*, 3570–3572.

Livrelli, V.O., Darfeuille-Richaud, A., Rich, C., Joly, B.H., and Martel, J.-L. (1988). Genetic determinant of the ROB-1 β-lactamase in bovine and porcine *Pasteurella* strains. Antimicrob. Agents Chemother. *32*, 1282–1284.

Livrelli, V.O., Peduzzi, J., and Joly, B. (1991). Sequence and molecular characterization of the ROB1-β-lactamase gene from *Pasteurella haemolytica*. Antimicrob. Agents Chemother. *35*, 242–251.

MacLean, I.W., Slaney, L., Juteau, J.M., Levesque, R.C., Albritton, W.L., and Ronald, A.R. (1992). Identification of a ROB-1 β-lactamase in *Haemophilus ducreyi*. Antimicrob. Agents Chemother. *36*, 467–469.

Makosky, P.C., and Dahlberg, A.E. (1987). Spectinomycin resistance at site 1192 in 16S ribosomal RNA of *E. coli*: an analysis of three mutants. Biochimie *69*, 885–889.

Marchese, A., and Schito, G.C. (2007). Recent findings from multinational resistance surveys: are we

'PROTEKTed' from resistance? Int. J. Antimicrob. Agents. *29*, Suppl. 1:S2–5.

Marshall, B., Roberts, M., Smith, A., and Levy, S.B. (1984). Homogeneity of transferable tetracycline resistance determinants in *Haemophilus species*. J. Infect. Dis. *149*, 1028–1029.

Martín de la Fuente, A.J., Tucker, A.W., Navas, J., Blanco, M., Morris, S.J., and Gutiérrez-Martín, C.B. (2007). Antimicrobial susceptibility patterns of *Haemophilus parasuis* from pigs in the United Kingdom and Spain. Vet. Microbiol. *120*, 184–191.

Matic, V., Bozdogan, B., Jacobs, M.R., Ubukata, K., and Appelbaum, P.C. (2003). Contribution of β-lactamase and PBP amino acid substitutions to amoxicillin/clavulanate resistance in β-lactamase-positive, amoxicillin/clavulanate-resistant *Haemophilus influenzae*. J. Antimicrob. Chemother. *52*, 1018–1021.

Matter, D., Rossano, A., Limat, S., Vorlet-Fawer, L., Brodar, I., and Perreten, V. (2007). Antimicrobial resistance profile of *Actinobacillus pleuropneumoniae* and *Actinobacillus porcitonsillarum*. Vet. Microbiol. *122*, 146–156.

Matter, D., Rossano, A., Sieber, S., and Perreten, V. (2008). Small multidrug resistance plasmids in *Actinobacillus porcitonsillarum*. Plasmid *59*, 144–152.

McNicol, P.J., and Ronald, A.R. (1984). The plasmids of *Haemophilus ducreyi*. J. Antimicrob. Chemother. *14*, 561–573.

Medeiros, A.A., Levesque, R.C., and Jacoby, G.A. (1986). An animal source for the ROB1-β-lactamase of *Haemophilus influenzae* type b. Antimicrob. Agents Chemother. *29*, 212–215.

Miranda, C.D., Kehrenberg, C., Ulep, C., Schwarz, S., and Roberts, M.C. (2003). Diversity of tetracycline resistance genes from bacteria isolated from Chilean salmon farms. Antimicrob. Agents Chemother. *47*, 883–888.

Molina, J.M., Cordoba, J., Esteban, R., Lainez, B., Monsoliu, A., Gregori, V., Hernandez, A., Diosdado, N., and Gobernado, M. (2002). Study of the β-lactam resistance of *Haemophilus influenzae* conferred by the bla_{ROB-1} gene. Rev. Esp. Quimioter. *15*, 148–151.

Müller, H.P., Holderrieth, S., Burkhardt, U., and Höffler, U. (2002). In vitro antimicrobial susceptibility of oral strains of *Actinobacillus actinomycetemcomitans* to seven antibiotics. J. Clin. Peridontol. *29*, 736–742.

Murphy, G.L., Robinson, L.C., and Burrows, G.E. (1993). Restriction endonuclease analysis and ribotyping differentiate *Pasteurella haemolytica* serotype A1 isolates from cattle within a feedlot. J. Clin. Microbiol. *31*, 2303–2308.

Murray, I.A., Hawkins, J., Keyte, J.W., and Shaw, W.V. (1988). Nucleotide sequence analysis and overexpression of the gene encoding a type III chloramphenicol acetyltransferase. Biochem. J. *252*, 173–179.

Murray, I.A., Martinez-Suarez, J.V., Close, T.J., and Shaw, W.V. (1990). Nucleotide sequences of genes encoding the type II chloramphenicol acetyltransferases of *Escherichia coli* and *Haemophilus influenzae*, which are sensitive to inhibition by thiol-reactive reagents. Biochem. J. *272*, 505–510.

Naas, T., Benaoudia, F., Lebrun, L., and Nordmann, P. (2001). Molecular identification of TEM-1 β-lactamase in a *Pasteurella multocida* isolate of human origin. Eur. J. Clin. Microbiol. Infect. Dis. *20*, 210–213.

Nørskov-Lauritsen, N., and Kilian, M. (2006). Reclassification of *Actinobacillus actinomycetemcomitans*, *Haemophilus aphrophilus*, *Haemophilus paraphrophilus* and *Haemophilus segnis* as *Aggregatibacter actinomycetemcomitans* gen. nov., comb. nov., *Aggregatibacter aphrophilus* comb. nov. and *Aggregatibacter segnis* comb. nov., and emended description of *Aggregatibacter aphrophilus* to include V factor-dependent and V factor-independent isolates. Int. J. Syst. Evol. Microbiol. *56*, 2135–2146.

O'Connor, M., and Dahlberg, A.E. (2002). Isolation of spectinomycin resistance mutations in the 16S rRNA of *Salmonella enterica* serovar Typhimurium and expression in *Escherichia coli* and *Salmonella*. Curr. Microbiol. *45*, 429–433.

Ojo, K.K., Kehrenberg, C., Odelola, H.A., and Schwarz, S. (2002). Identification of a complete *dfrA14* gene cassette integrated at a secondary site in a resistance plasmid of uropathogenic *Escherichia coli* from Nigeria. Antimicrob. Agents Chemother. *46*, 2054–2055.

Oka, A., Sugisaki, H., and Takanami, M. (1981). Nucleotide sequence of the kanamycin resistance transposon Tn*903*. J. Mol. Biol. *147*, 217–226.

Ouellet, V., Forest, A., Nadeau, M., and Sirois, M. (2004). Characterization of tetracycline resistance determinants in *Actinobacillus pleuropneumoniae*. Abstracts of the 104th ASM General Meeting, *A-113*, 22.

Pang, Y., Bosch, T., and Roberts, M.C. (1994). Single polymerase chain reaction for the detection of the tetracycline resistance determinants TetK and TetL. Molec. Cell. Probes *8*, 417–422.

Perez-Vazquez, M., Roman, F., Aracil, B., Canton, R., and Campos, J. (2004). Laboratory detection of *Haemophilus influenzae* with decreased susceptibility to nalidixic acid, ciprofloxacin, levofloxacin, and moxifloxacin due to *gyrA* and *parC* mutations. J. Clin. Microbiol. *42*, 1185–1191.

Peric, M., Bozdogan, B., Galderisi, C., Krissinger, D., Rager, T., and Appelbaum, P.C. (2004). Inability of L22 ribosomal protein alteration to increase macrolide MICs in the absence of an efflux mechanism in *Haemophilus influenzae*. J. Antimicrob. Chemother. *54*, 393–400.

Peric, M., Bozdogan, B., Jacobs, M.R., and Appelbaum, P.C. (2003). Effects of an efflux mechanism and ribosomal mutations on macrolide susceptibility of *Haemophilus influenzae*. Antimicrob. Agents Chemother. *47*, 1017–1022.

Pitout, M., MacDonald, K., Musgrave, H., Lindique, C., Forward, K., Hiltz, M., and Davidson, R. (2002). Characterization of extended spectrum β-lactamase (ESBL) activity in *Haemophilus influenzae*, abstr. C2–645, p. 96. Abstr. 42nd Intersci. Conf. Antimicrob. Chemother., San Diego, CA, 2002. American Society for Microbiology, Washington, DC.

Philippon, A., Joly, B., Reynaud, D., Paul, G., Martel, J.-L., Sirot, D., Cluzel, R., and Nevot, P. (1986). Characterization of a β-lactamase from *Pasteurella multocida*. Ann. Inst. Pasteur Microbiol. *137A*, 153–158.

Powell, M., and Livermore, D.M. (1988). Mechanisms of chloramphenicol resistance in *Haemophilus influenzae* in the United Kingdom. J. Med. Microbiol. *27*, 89–93.

Priebe, S., and Schwarz, S. (2003). *In vitro* activities of florfenicol against bovine and porcine respiratory tract pathogens. Antimicrob. Agents Chemother. *47*, 2703–2705.

Qin, L., Watanabe, H., Asoh, N., Watanabe, K., Oishi, K., Mizota, T., and Nagatake, T. (2007). Short report: antimicrobial susceptibility and genetic characteristics of *Haemophilus influenzae* isolated from patients with respiratory tract infections between 1987 and 2000, including β-lactamase-negative ampicillin-resistant strains. Epidemiol. Infect. *135*, 665–668.

Radström, P., and Swedberg, G. (1988). RSF1010 and a conjugative plasmid contain *sul*II, one of two known genes for plasmid-borne sulfonamide resistance dihydropteroate synthase. Antimicrob. Agents Chemother. *32*, 1684–1692.

Roberts, M.C., Swenson, C.D., Owens, L.M., and Smith, A.L. (1980). Characterization of chloramphenicol-resistant *Haemophilus influenzae*. Antimicrob. Agents Chemother. *18*, 610–615.

Rodríguez-Martínez, J.M., López, L., García, I., and Pascual, A. (2006). Characterization of a clinical isolate of *Haemophilus influenzae* with a high level of fluoroquinolone resistance. *J. Antimicrob. Chemother. 57*, 577–578.

Roe, D.E., Braham, P.H., Weinberg, A., and Roberts, M.C. (1995). Characterization of tetracycline resistance in *Actinobacillus actinomycetemcomitans*. Oral Microbiol. Immunol. *10*, 227–232.

Roe, D.E., Weinberg, A., and Roberts, M.C. (1996). Mobile rRNA methylase genes coding for erythromycin resistance in *Actinobacillus actinomycetemcomitans*. J. Antimicrob. Chemother. *37*, 457–464.

Rosenau, A., Labigne, A., Escande, F., Courcoux, P., and Philippon, A. (1991). Plasmid-mediated ROB1-β-lactamase in *Pasteurella multocida* from human specimen. Antimicrob. Agents Chemother. *35*, 2419–2422.

Rossmanith, S.E.R., Wilt, G.R., and Wu, G. (1991). Characterization and comparison of antimicrobial susceptibilities and outer membrane protein and plasmid DNA profiles of *Pasteurella haemolytica* and certain other members of the genus *Pasteurella*. Am. J. Vet. Res. *52*, 2016–2022.

Sanchez, L., Leranoz, S., Puig, M., Loren, J.G., Nikaido, H., and Vinas, M. (1997a). Molecular basis of antimicrobial resistance in non-typeable *Haemophilus influenzae*. Microbiologia *13*, 309–314.

Sanchez, L., Pan, W., Vinas, M., and Nikaido, H. (1997b). The *acrAB* homolog of *Haemophilus influenzae* codes for a functional multidrug efflux pump. J. Bacteriol. *179*, 6855–6857.

San Millan, A., Escudero, J.A., Catalan, A., Nieto, S., Farelo, F., Gilbert, M., Moreno, M.A., Dominguez, L., and Gonzalez-Zorn, B. (2007). β-Lactam resistance in *Haemophilus parasuis* is mediated by plasmid pB1000 bearing bla_{ROB-1}. Antimicrob. Agents Chemother. *51*, 2260–2264.

Saunders, J.R., Elwell, L.P., Falkow, S., Sykes, R.B., and Richmond, M.H. (1978). β-Lactamases and R-plasmids of *Haemophilus influenzae*. Scand. J. Infect. Dis. Suppl. *1978*, 16–22.

Schnabel, E.L., and Jones, A.L. (1999). Distribution of tetracycline resistance genes and transposons among phylloplane bacteria in Michigan apple orchards. Appl. Environ. Microbiol. *65*, 4898–4907.

Scholz, P., Haring, V., Wittmann-Liebold, B., Ashman, K., Bagdasarian, M., and Scherzinger, E. (1989). Complete nucleotide sequence and gene organization of the broad-host-range plasmid RSF1010. Gene *75*, 271–288.

Schwarz, S., Alešík, E., Grobbel, M., Lübke-Becker, A., Werckenthin, C., Wieler, L.H., and Wallmann, J. (2007). Antimicrobial susceptibility of *Pasteurella multocida* and *Bordetella bronchiseptic* from dogs and cats as determined in the BfT-GermVet monitoring program 2004–2006. Berl. Münch. Tierärztl. Wochenschr. *120*, 423–430.

Schwarz, S., Böttner, A., Goosens, L., Hafez, H.M., Hartmann, K., Kaske, M., Kehrenberg, C., Kietzmann, M., Klarmann, D., Klein, G., Krabisch, P., Luhofer, G., Richter, A., Schulz, B., Sigge, C., Waldmann, K.H., Wallmann, J., and Werckenthin, C. (2008). A proposal of clinical breakpoints for amoxicillin applicable to porcine respiratory tract pathogens. Vet. Microbiol. *126*, 178–188.

Schwarz, S., Cardoso, M., and Wegener, H.C. (1992). Nucleotide sequence and phylogeny of the *tet*(L) tetracycline resistance determinant encoded by plasmid pSTE1 from *Staphylococcus hyicus*. Antimicrob. Agents Chemother. 36, 580–588.

Schwarz, S., Cloeckaert, A., and Roberts, M.C. (2006). Mechanisms and spread of bacterial resistance to antimicrobial agents. *In* F.M. Aarestrup (ed.) Antimicrobial resistance in bacteria of animal origin, ASM Press, Washington, DC, USA, 73–98.

Schwarz, S., Kehrenberg, C., Doublet, B., and Cloeckaert, A. (2004a). Molecular basis of bacterial resistance to chloramphenicol and florfenicol. FEMS Microbiol. Rev. *28*, 519–542.

Schwarz, S., Kehrenberg, C., Salmon, S.A., and Watts, J.L. (2004b). *In vitro* activities of spectinomycin and comparator agents against *Pasteurella multocida* and *Mannheimia haemolytica* from respiratory tract infections of cattle. J. Antimicrob. Chemother. *53*, 379–382.

Schwarz, S., Spies, U., Reitz, B., Seyfert, H.-M., Lämmler, C., and Blobel, H. (1989a). Detection and interspecies-transformation of a β-lactamase-encoding plasmid from *Pasteurella haemolytica*. Zentralbl. Bakteriol. Mikrobiol. Hyg. A *270*, 462–469.

Schwarz, S., Spies, U., Schäfer, F., and Blobel, H. (1989b). Isolation and interspecies-transfer of a plasmid from *Pasteurella multocida* encoding streptomycin resistance. Med. Microbiol. Immunol. *178*, 121–125.

Scriver, S.R., Walmsley, S.L., Kau, C.L., Hoban, D.J., Brunton, J., McGeer, A., Moore, T.C., Witwicki, E. Canadian Haemophilus Study Group, and Low. D.E. (1994). Determination of antimicrobial susceptibilities of Canadian isolates of *Haemophilus influenzae* and characterization of their β-lactamases. Antimicrob. Agents Chemother. *38*, 1678–1680.

Sener, B., Tunckanat, F., Ulusoy, S., Tunger, A., Soyletir, G., Mulazimoglu, L., Gurler, N., Oksuz, L., Koksal, I., Aydin, K., Yalcin, A.N., Ogunc, D., Acar, A., and Sievers, J. (2007). A survey of antibiotic resistance in *Streptococcus pneumoniae* and *Haemophilus influenzae* in Turkey, 2004–2005. J. Antimicrob. Chemother. *60*, 587–593.

Silver, R.P., Leming, B., Garon, C.F., and Hjerpe, C.A. (1979). R-plasmids in *Pasteurella multocida*. Plasmid 2, 493–497.

Sørum, H., Roberts, M.C., and Crosa, J.H. (1992). Identification and cloning of a tetracycline resistance gene from the fish pathogen *Vibrio salmonicida*. Antimicrob. Agents Chemother. *36*, 611–615.

Straker, K., Wootton, M., Simm, A.M., Bennett, P.M., MacGowan, A.P., and Walsh, T.R. (2003). Cefuroxime resistance in non-β-lactamase *Haemophilus influenzae* is linked to mutations in *ftsI*. J. Antimicrob. Chemother. *51*, 523–530.

Stuy, J.H., and Walter, R.B. (1992). Cloning, characterization, and DNA base sequence of the high-level streptomycin resistance gene *strA1* of *Haemophilus influenzae* Rd. J. Bacteriol. *174*, 5604–5608.

Sundin, G.W. (2000). Examination of basepair variants of the *strA-strB* streptomycin resistance genes from bacterial pathogens of humans, animals and plants. J. Antimicrob. Chemother. *46*, 848–849.

Sundin, G.W. (2002). Distinct recent lineages of the *strA-strB* streptomycin-resistance genes in clinical and environmental bacteria. Curr. Microbiol. *45*, 63–69.

Sundin, G.W., and Bender, C.L. (1996). Dissemination of the *strA-strB* streptomycin-resistance genes among commensal and pathogenic bacteria from humans, animals, and plants. Mol. Ecol. *5*, 133–143.

Tristram, S.G., Hawes, R., and Souprounov, J. (2005). Variation in selected regions of bla_{TEM} genes and promoters in *Haemophilus influenzae*. J. Antimicrob. Chemother. *56*, 481–484.

Tristram, S., Jacobs, M.R., and Appelbaum, P.C. (2007). Antimicrobial resistance in *Haemophilus influenzae*. Clin. Microbiol. Rev. *20*, 368–389.

Turnidge, J., Kahlmeter, G., and Kronvall, G. (2006). Statistical characterisation of bacterial wild-type MIC value distributions and the determination of epidemiological cut-off values. Clin. Microbiol. Infect. *12*, 418–425.

Ubukata, K., Shibasaki, Y., Yamamoto, K., Chiba, N., Hasegawa, K., Takeuchi, Y., Sunakawa, K., Inoue, M., and Konno, M. (2001). Association of amino acid substitutions in penicillin-binding protein 3 with β-lactam resistance in β-lactamase-negative ampicillin-resistant *Haemophilus influenzae*. Antimicrob. Agents Chemother. *45*, 1693–1699.

Van, A.D., Goldstein, F., Acar, J.F., and Bouanchaud, D.H. (1975). A transferable kanamycin resistance plasmid isolated from *Haemophilus influenzae*. Ann. Microbiol. (Paris). *126*, 397–399.

Vassort-Bruneau, C., Lesage-Decauses, M.-C., Martel, J.-L., Lafont, J.-P., and Chaslus-Dancla, E. (1996). CAT III chloramphenicol resistance in *Pasteurella haemolytica* und *Pasteurella multocida* isolated from calves. J. Antimicrob. Chemother. *38*, 205–213.

Wallmann, J., Kaspar, H., and Kroker, R. (2004). The prevalence of antimicrobial susceptibility of veterinary pathogens isolated from cattle and pigs: national antibiotic resistance monitoring 2002/2003 of the BVL. Berl. Münch. Tierärztl. Wochenschr. 117, 480–492.

Wallmann, J., Schröer, U., and, Kaspar, H. (2007). Quantitative resistance level (MIC) of bacterial pathogens (*Escherichia coli, Pasteurella multocida, Pseudomonas aeruginosa, Salmonella* sp., *Staphylococcus aureus*) isolated from chickens and turkeys: N ational Resistance Monitoring by the BVL 2004/2005. Berl. Münch. Tierärztl. Wochenschr. *120*, 452–463.

Wallmann, J., Schröter, K., Wieler, L.H., and Kroker, R. (2003). National antibiotic resistance monitoring in veterinary pathogens from sick food-producing animals: the German programme and results from the 2001 pilot study. Int. J. Antimicrob. Agents. *22*, 420–428.

Wasteson, Y., Roe, D.E., Falk, K., and Roberts, M.C. (1996). Characterization of tetracycline and erythromycin resistance in *Actinobacillus pleuropneumoniae*. Vet. Microbiol. *48*, 41–50.

Whittle, G., Katz, M.E., Clayton, E.H., and Cheetham, B.F. (2000). Identification and characterization of a native *Dichelobacter nodosus* plasmid, pDN1. Plasmid 43, 230–234.

Willson, P.J., Deneer, H.G., Potter, A., and Albritton, W. (1989.) Characterization of a streptomycin-sulfonamide resistance plasmid from *Actinobacillus pleuropneumoniae*. Antimicrob. Agents Chemother. 33, 235–238.

Wood, A.R., Lainson, F.A., Wright, F., Baird, G.D., and Donachie, W. (1995). A native plasmid of *Pasteurella haemolytica* serotype A1: DNA sequence analysis and investigation of its potential as a vector. Res. Vet. Sci. *58*, 163–168.

Wright, C.L., Strugnell, R.A., and Hodgson, A.L.M. (1997). Characterization of a *Pasteurella multocida* plasmid and its use to express recombinant proteins in *P. multocida*. Plasmid *37*, 65–79.

Wu, J.-R., Shieh, H.K., Shien, J.-H., Gong, S.-R., and Chang, P.-C. (2003). Molecular characterization of plasmids with antimicrobial resistant genes in avian isolates of *Pasteurella multocida*. Avian Dis. *47*, 1384–1392.

Yamamoto, J., Sakano, T., and Shimizu, M. (1990). Drug resistance and R plasmids in *Pasteurella multocida* isolates from swine. Microbiol. Immunol. *34*, 715–721.

Zhao, J. and Aoki, T. (1992). Nucleotide sequence analysis of the class G tetracycline resistance determinant from *Vibrio anguillarum*. Microbiol. Immunol. *36*, 1051–1060.

Zimmerman, M.L., and Hirsh, D.C. (1980). Demonstration of an R-plasmid in a strain of *Pasteurella haemolytica* isolated from feedlot cattle. Am. J. Vet. Res. *41*, 166–169.

Pasteurellaceae – the View from the Diagnostic Laboratory

11

Pat Blackall and Niels Nørskov-Lauritsen

Abstract

This chapter provides a centralized, organized overview of the isolation and identification of those members of the *Pasteurellaceae* that are commonly encountered in the clinical and veterinary diagnostic laboratories. The chapter covers both traditional methodologies as well as the modern molecular methods. Where possible, alternative methods are critically reviewed and recommendations on which methods are relevant for particular applications are provided. The chapter does not provide detailed methodologies but does provide identification tables that laboratories using the phenotypic identification approach should find useful.

Introduction

The family *Pasteurellaceae* presents a significant challenge to diagnostic microbiologists working in both clinical and veterinary laboratories. Members of the family are significant pathogens of both humans and animals. Ironically, while the family bears the name of one of the founders of the science of microbiology – Louis Pasteur – there is a relative paucity of centralized information available on this family in the standard text books widely used in both clinical and veterinary diagnostic laboratories. As an example, the latest *Manual of Clinical Microbiology* (Murray *et al.*, 2007) contains four chapters on the family *Enterobacteriaceae* while the family *Pasteurellaceae* is featured in one chapter (on the genus *Haemophilus*) and partly in another chapter (which covers *Pasteurella* and *Actinobacillus* as well as other fastidious or rarely encountered Gram-negative bacteria). Veterinary diagnostic microbiologists do have the advantage of specialized texts such as *Diseases of Swine* (Straw *et al.*, 1999), *Diseases of Poultry* (Saif *et al.*, 2003) and *A Laboratory Manual for the Isolation and Identification of Avian Pathogens* (Swayne *et al.*, 1998).

This chapter will attempt to address the relative deficiency of centralized information and will provide a critical review of the current status of methods for the isolation and identification of the family *Pasteurellaceae*. As molecular methods are now playing an increasing role in diagnostic laboratories, there will be an emphasis on these molecular methods – both in use directly on tissues or on cultures.

The strong host-association demonstrated by most members of the family *Pasteurellaceae* allows most of this chapter to be split into clinical and veterinary sections. This division is purely intended as a means of logical presentation and is not meant to highlight any distinction between the clinical and veterinary fields. Indeed, as a general observation, the isolation and identification of the family *Pasteurellaceae* is an area where clinical and veterinary microbiologists have a lot to learn from each other. There are common issues that impact on the work of both clinical and veterinary diagnostic laboratories and these issues are handled in shared sections.

Growth conditions and testing procedures

The *Pasteurellaceae* are obligate parasites adapted to living on mucosal surfaces of man and animals. They are propagated *in vitro* on rich media supplemented with serum, blood or yeast extract. A widely employed medium for isolation of *Pasteurellaceae* is the so-called chocolate agar plate, in which sheep or horse blood is added to the basic medium at a temperature of approximately 70°C. This medium is easily prepared and supports good growth of presently described *Pasteurellaceae*, although further additives are used in special circumstances such as the isolation of [*Haemophilus*] *ducreyi*.

Members of the genus *Aggregatibacter*, *Avibacterium paragallinarum and Histophilus somni* are capnophilic, and primary isolation may require the presence of 5–10% CO_2. Other species of the family may benefit from elevated levels of CO_2. The use of 5% carbon dioxide ensures a more uniform colony development of *Avibacterium gallinarum* (Blackall *et al.*, 2005). A general recommendation is that isolation plates for avian respiratory disease investigations should be incubated in 5–10% carbon dioxide and those of porcine respiratory disease investigations in air.

Two deficient metabolic pathways, giving rise to the dependence on specific growth factors, are of value in the differential identification of *Pasteurellaceae*. These factors are traditionally referred to as X and V factor, respectively. The X factor is of value in the identification of isolates within the genus *Haemophilus*, whereas V factor-dependent species are found in several genera parasitizing both man and animals.

Dependency on V factor is conveniently demonstrated by the satellitic growth of the investigated strain around a streak or a colony of a 'feeder' strain supplying the critical factor in excess (Grassberger, 1897). It is common to use a laboratory strain of *Staphylococcus aureus* as the feeder strain in clinical laboratories and a strain of *Staphylococcus epidermidis* or *Staphylococcus hyicus* in veterinary laboratories (Blackall and Matsumoto, 2003; Taylor, 1999). V factor dependent bacteria will grow to sizeable colonies immediately adjacent to the feeder strain, and demonstrate a clear gradient in growth with decreasing colonies 0.5–2 cm from the feeder strain. An alternative method is to use a paper disk impregnated with nicotinamide adenine dinucleotide (β-NAD) (Fig. 11.1A). The demonstration of satellitism is a robust method but is dependent on the absence of V factor in the medium. Cultivation of *Pasteurellaceae* is usually done on complex media, and if β-NAD or certain precursors are introduced in the formulations, a strain will not reveal its dependence on V factor. Furthermore, if the test of satellitism is done on blood agar plates, haemolytic strains will usually grow to sizeable colonies over the entire plate (Fig. 11.1B). The colonies may be larger in the vicinity of the β-NAD disc or the feeder strain, but sufficient V factor can be obtained from the lysis of erythrocytes to sustain significant growth. The V factor dependence of such strains is demonstrated in the absence of blood (Fig. 11.1A) or by use of autoclaved media, where β-NAD and precursors have been completely degraded.

The biochemical mechanism of V factor dependence is well known (Gerlach and Reidl, 2006). All members of the family *Pasteurellaceae* are incapable of synthesizing β-NAD *de novo* and acquire this essential nutrient from their environment either as β-NAD or a limited number of precursors. Extracellular pyridine nucleotides are degraded to nicotinamide riboside (NR) and internalized, and NR is converted to β-NAD via

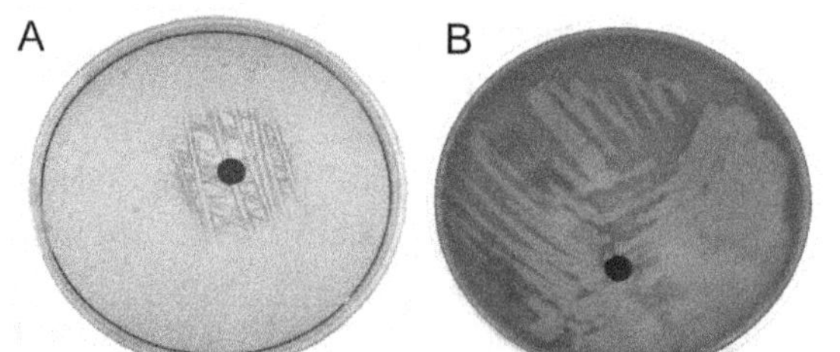

Figure 11.1 *H. pittmaniae* cultured for 24 h on different agars in the presence of a β-NAD-containing disc and photographed by transillumination. (A) Todd-Hewitt agar. Growth is apparent only around the β-NAD disc. (B) horse blood agar. Growth of haemolytic colonies is visible over the entire plate. A colour version of this figure is located in the plate section at the back of the book.

nicotinamide mononucleotide (NMN). Some species are capable of synthesizing NMN from nicotinamide and phosphoribosyl pyrophosphate in a reaction catalysed by nicotinamide phosphoribosyltransferase. Ample nicotinamide is present in complex media, and such isolates are therefore independent on exogenously supplied V factor (β-NAD, NMN or NR). Growth media can be prepared in which no nicotinamide is present, and in such circumstances, all *Pasteurellaceae* will exhibit satellitic growth around a feeder strain supplying V factor (Niven and O'Reilly, 1990).

The current definition of V factor dependence is therefore synonymous with the absence of a single functional enzyme (nicotinamide phosphoribosyltransferase), which may even be dispensable. With *Aggregatibacter aphrophilus*, some isolates are capable of synthesizing nicotinamide mononucleotide while others lack a functional nicotinamide phosphoribosyltransferase. V factor-dependent strains carry a *nad*V pseudogene, and the V factor-dependent phenotype is naturally reversible in strains competent for transformation (Nørskov-Lauritsen and Kilian, 2006). V factor-dependent and -independent biovars have also been documented for *Actinobacillus pleuropneumoniae* (Pohl *et al.*, 1983), *Haemophilus parainfluenzae* (Gromkova and Koornhof, 1990) and *Av. paragallinarum* (Mouahid *et al.*, 1992). Dependence on V factor is therefore not a character of paramount importance in classification of *Pasteurellaceae*, and occasional strains may give aberrant results in the routine diagnostic work. Nevertheless, V factor dependence is a highly valuable test in the initial characterization of presumptive members of *Pasteurellaceae* in the clinical and veterinary microbiological laboratory.

The haemin biosynthetic pathway is common to animals, plants and bacteria, irrespective of whether the final end product is cytochrome, haemoglobin or chlorophyll. The formation of protoporphyrin is carried out in successive enzymatic steps beginning with the condensation of four molecules of δ-aminolaevulinic acid and protohaem is finally formed from protoporphyrin by chelation of ferrous iron (White and Granick, 1963). The dependence on X factor is correlated to the inability to synthesize protoporphyrin (Biberstein *et al.*, 1963). Detection of porphyrins synthesized from δ-aminolaevulinic acid has become the standard method of investigating X factor dependence, whereas the use of paper discs impregnated with X factor on agar media may carry a high rate of misinterpretations (Kilian, 1974; Kilian, 2005).

In general, the details of media and reagents needed for the conventional identification of *Pasteurellaceae* are provided in a range of the standard text books found in diagnostic laboratories such as *Cowan and Steel's Manual for the Identification of Medical Bacteria* (Barrow and Feltham, 1993), *Manual of Clinical Microbiology* (Murray *et al.*, 2007) and *A Laboratory Manual for the Isolation and Identification of Avian Pathogens* (Swayne *et al.*, 1998). An excellent source for specialized methods is the information provided by Kilian and Frederiksen (1981). Specialized methods associated with the identification of veterinary pathogens are briefly mentioned in the section dealing specifically with these pathogens.

Taxonomic comments – clinical pathogens

Two species included in the Approved Lists of Bacterial Names (http://www.bacterio.cict.fr/), *Haemophilus aegyptius* and [*Haemophilus*] *paraphrohaemolyticus*, are difficult to circumscribe and may not deserve species rank. *H. aegyptius* was conceived as a species distinct from *Haemophilus influenzae* with a particular propensity to cause conjunctivitis (Pittman and Davis, 1950). It was believed to be the bacterium that Robert Koch noticed by microscopic examinations of purulent matter from cases of eye inflammation in Egypt in 1883, and later propagated *in vitro* by the American ophthalmologist John Weeks (the Koch-Weeks bacterium). However, the existence of the species *H. aegyptius* is controversial. The species cannot be validated by DNA-based methods (Casin *et al.*, 1986; Hedegaard *et al.*, 2001), and phenotypic traits allowing the unambiguous identification of *H. aegyptius* are not available. If cultivation of purulent discharge from cases of conjunctivitis yields haemophilic bacteria that comply by phenotype with *H. influenzae* biotype III, they should be reported as such until taxonomic questions regarding *H. aegyptius* have been resolved.

[*H.*] *paraphrohaemolyticus* was described by Zinnemann *et al.* (1971) as a species distinct from [*Haemophilus*] *parahaemolyticus* and discernible from the latter by the improved growth rate in the presence of CO_2. The type strain of [*H.*] *paraphrohaemolyticus* differs from the type strain of [*H.*] *parahaemolyticus* by the presence of β-galactosidase activity (ONPG test) and the absence of IgA1 protease activity (Kilian, 2005), and relatively large sequence differences between the two strains are observed with both 16S rRNA and housekeeping gene sequences (Nørskov-Lauritsen *et al.*, 2005). However, dependence on CO_2 is obsolete as a taxonomic criterion, and the other phenotypic traits (β-galactosidase and IgA1 protease) have not been validated on a larger collection of strains. The differentiation between [*H.*] *parahaemolyticus* and [*H.*] *paraphrohaemolyticus* is uncertain, and the latter species designation could be disregarded until taxonomic questions regarding [*H.*] *paraphrohaemolyticus* have been resolved.

Isolation methods – clinical pathogens

Cultivation from normally sterile body sites

Automated blood culture systems support the growth of *Haemophilus* spp. (Jamal *et al.*, 2006; Ziegler *et al.*, 1998). Samples, other than blood, obtained from normally sterile areas of the body are inoculated on different media and incubated in different atmospheres. A chocolate agar plate is routinely included and incubated in the presence of 5% CO_2 for 2 days or longer. Members of the *Pasteurellaceae* will usually form visible colonies under these conditions within 24 h of incubation.

H. influenzae of capsular serotype b (Hib) was a common cause of meningitis in small children before the introduction of vaccination, and continues to be so in many parts of the world. It is estimated that more than 300,000 children 0–4 years old annually will contract Hib meningitis on a world-wide basis, and that more than 90,000 will succumb to the infection (Peltola, 2000). The introduction of a polysaccharide and later a conjugate vaccine has had a dramatic impact on the incidence of Hib meningitis. In countries with effectual childhood vaccination programs, *H. influenzae* is rarely cultivated from the spinal fluid, and non-capsulate isolates or serotypes other than Hib are now the most commonly encountered serotypes (Bruun *et al.*, 2004; Cardines *et al.*, 2007). *Agg. aphrophilus* is not infrequently isolated from cases of brain abscesses (Carpenter *et al.*, 2007; Duel *et al.*, 1991; King and Tatum, 1962). If a neurosurgical department is being serviced by the laboratory, *Agg. aphrophilus* may be the most common species of *Pasteurellaceae* isolated from the central nervous system. Cases of *Actinobacillus ureae*, *H. parainfluenzae* and *Pasteurella multocida* meningitis are occasionally reported (Black *et al.*, 1988; De Castro *et al.*, 2007; Green *et al.*, 2002), usually from immunocompromised patients.

Pasteurellaceae species are sometimes isolated from blood in conjunction with particular infections such as endocarditis, meningitis or severe pneumonia. In a retrospective analysis of blood cultures obtained from four centres, *Haemophilus* spp. were cultivated in 26 of 59,203 positive blood cultures (Petti *et al.*, 2006). The isolation of *Aggregatibacter* spp. from blood is less frequent, but the occasional detection of these species may be clinically important due to association with infective endocarditis (Bangsborg *et al.*, 1988; Khairat, 1940; Paturel *et al.*, 2004). Cases of *P. multocida* bacteraemia are repeatedly reported (Breen *et al.*, 2000; Tseng *et al.*, 2001), whereas the isolation of other *Pasteurellaceae* from bloodstream infections is a rare event.

Haemophilus spp. are sometimes isolated from suppurative lesions in unusual anatomical sites, fluids from serosal cavities and gall bladder, gut content, pelvic inflammatory disease and the lumen of appendices removed at operations (Christensen *et al.*, 1990; Frankard *et al.*, 2004; Heinonen and Miettinen, 1994; Rogers *et al.*, 1960). Various *Haemophilus* spp. may be encountered at these sites. A large number of 16S rRNA gene sequences (from uncultured bacteria from normal and diseased human distal oesophagus) with similarity to [*H.*] *parahaemolyticus* have been deposited with the public databases (for example, nucleotide sequence accession number DQ633367).

Recovery of *Pasteurellaceae* from human urine specimens receives little attention in the clinical laboratory. Urine is usually not inoculated on media supporting the growth of fastidious organisms, and incubation may be limited to ambient air for 24 h. The X and V factor-independent [*Pasteurella*] *bettyae*, which demonstrates trophism for Bartholin's glands of the vagina, is sometimes cultured in significant numbers from human urine samples (unpublished observation). Serendipitous growth of *Haemophilus* species on blood agar is occasionally observed from human urine, if satellitism can be observed around colonies of other bacteria present in the urine. A few studies have systematically addressed the prevalence of *Haemophilus* spp. in human urine samples. Approximately 0.7% of urine sample from children with symptoms of urinary tract infection yielded significant growth of *Haemophilus* when cultivated on chocolate agar (Hansson *et al.*, 2007). An even distribution of *H. influenzae* and *H. parainfluenzae* was observed; however, *H. influenzae* was predominantly identified in girls and *H. parainfluenzae* in boys.

Non-selective isolation of *Pasteurellaceae* from non-sterile body sites

Isolation of *H. influenzae* from the respiratory tract, the middle ear and the conjunctiva is by far the most common finding of *Pasteurellaceae* in the clinical laboratory. Non-capsulated (non-typeable) *H. influenzae* is, together with *Streptococcus pneumoniae*, the most important bacterial pathogen of acute otitis media (Eskola and Kilpi, 2000), conjunctivitis (Buznach *et al.*, 2005) and exacerbations in chronic obstructive pulmonary disease (Sethi *et al.*, 2002). However, *H. influenzae* is part of the indigenous commensal flora of the pharynx, particularly in childhood, where point prevalence studies have shown carriage rates of 25–84% (Kilian, 2005). Processing of respiratory samples should therefore not aim at ultimate sensitivity, but rather employ semiquantitation of *H. influenzae*. If microscopic examination of lower respiratory tract specimens is employed (Fig. 11.2), the number of leukocytes per low-power field may correlate with inflammatory response; the inter-observer variability is, however, substantial (Al Balooshi *et al.*, 2003). Although not justified by rigorous scientific evidence, a commonly applied criterion is the requirement for a 'dominant' growth of *H. influenzae*, interpreted as amounts equal to, or not grossly exceeded by, other bacteria cultured in the sample. Growth on blood agar with a *S. aureus* streak is not a dependable method for quantitation of *H. influenzae* in lower respiratory tract specimens. The presence of indigenous flora may result in the modest, satellitic growth of *H. influenzae* being missed. In contrast, use of chocolate agar may result in clearly visible growth of *H. influenzae* in the same sample.

It should be emphasized that the criteria outlined above are not universally accepted, and that many laboratories use a selective isolation procedure in the processing of respiratory specimens to increase the recovery rate of *H. influenzae* (see below).

Pasteurellaceae other than *Haemophilus* isolated from the human respiratory tract include *A. ureae* and *A. hominis*, which are rarely reported. Infections with *A. hominis* appear to be strongly associated with immunosuppressed patients although the identification of the bacterium requires extended phenotypic testing and underreporting may occur (Friis-Moller *et al.*, 2001).

H. parainfluenzae is sometimes isolated from superficial wounds on the hands which may reflect a post-traumatic nursing of the wound by the mouth or the tongue. Quite a different situation is the introduction of oral flora into bruised tissue as a consequence of animal bites, which can result in severe infections necessitating hospitalization, surgical debridement and intravenous antibiotic therapy. *P. multocida* is by far the most commonly encountered animal *Pasteurella* spp. isolated from humans, while *Pasteurella canis*, *Pasteurella dagmatis* and *Pasteurella stomatis* are also frequently reported (Bisgaard and Falsen, 1986; Donnio *et al.*, 2004; Holst *et al.*, 1992). While clearly also reflecting pathogenic potential, the types of *Pasteurella* species associated with human infections also reflect man's preferred choice of pets. Humans cohabiting with dogs and cats will frequently be exposed to the oral flora of their pets in the form of licks or bites. Identification of *P. multocida* is a common event in the clinical laboratory. Identification of *Pasteurellaceae* from bites of other animals can pose significant challenge to clinical microbiologists. A strategy may

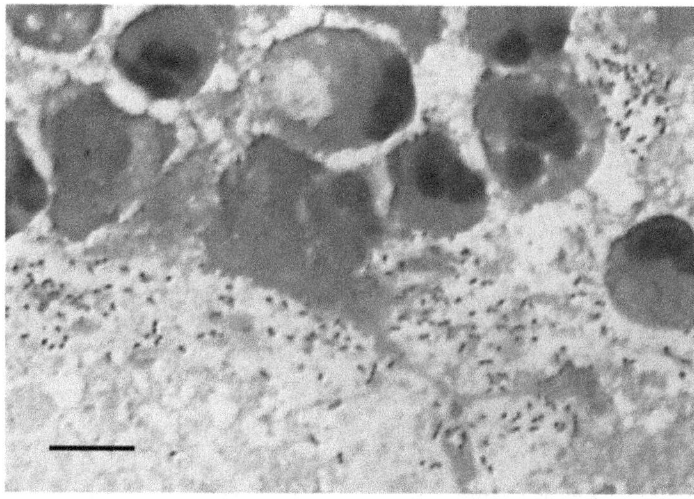

Figure 11.2 Microscopic examination of sputum from a patient with bronchiectasis and exacerbated expectoration. Neutrophil granulocytes are seen adjacent to clusters of Gram-negative coccobacilli. Original magnification ×1000, bar represents 10 μm. Number of squamous epithelial cells and leukocytes per low power field were 0 and > 25, respectively. The sample yielded *H. influenzae* in pure culture after incubation on growth media. A colour version of this figure is located in the plate section at the back of the book.

be to assign the unknown bacteria as a putative *Pasteurellaceae* on the basis of key characteristics (see later section on identification), and report it as such. The bacterium may then be referred to a veterinary institution with special interest in this field, or a more precise identification can be obtained by 16S rRNA gene sequencing.

Selective isolation of *Pasteurellaceae*

In routine clinical microbiological procedures, the use of antibiotics for selective isolation of *Pasteurellaceae* is limited to a few situations. Chancroid is caused by [*H.*] *ducreyi* and is characterized by ano-genital ulceration and lymphangitis (Albritton, 1989). It is a disease of resource poor settings and may be considered a tropical sexually transmitted disease. Material for culture is obtained from the undermined edge of the ano-genital ulcer after the removal of superficial pus with a cotton-tipped swab. It is recommended to directly inoculate the material onto culture medium and carry out the incubation at 33°C in a humidified atmosphere with 5% CO_2 for a minimum of 48–72 h (Lewis and Ison, 2006; Trees and Morse, 1995). The viability of [*H.*] *ducreyi* is lost after 24 h in transport media at room temperature but may be preserved for several days when kept at 4°C (Dangor *et al.*, 1993). To suppress the resident flora of skin, the selective media for isolation of [*H.*] *ducreyi* usually incorporate the glycopeptidic antibiotic vancomycin at concentrations of 3 mg/l. Different strains of [*H.*] *ducreyi* may grow preferentially on different culture media, and the use of more than one type of media is useful. Medium additives include haemoglobin, foetal calf serum or chocolatized horse blood and chemically defined growth promoting substances such as IsoVitaleX™.

The currently most widely employed agent for the selective isolation of *Pasteurellaceae* in epidemiological studies is the antibacterial glycopeptide bacitracin. Effective against many Gram-positive and some Gram-negative bacteria, bacitracin is used in concentrations of 75–300 μg/ml, sometimes supplemented with vancomycin (Ederer and Schurr, 1971; Slots, 1982). As mentioned above, some laboratories aim at a high isolation rate of *Haemophilus* species when processing samples from the respiratory tract. Application of a bacitracin disc (160 μg) on a chocolate blood agar may perform as well as bacitracin-incorporated chocolate blood agar (80 mg/l) (Nye *et al.*, 2001).

Conventional identification – genus level

Table 11.1 shows the differentiation, by phenotypic characteristics, of the currently recognized genera within the family *Pasteurellaceae*.

Table 11.1 Key characters for differentiation of genera within the family *Pasteurellaceae*.

	Genus												
Characteristic	1	2	3	4	5	6	7	8	9	10	11	12	13
Catalase	V	V	V	V	-	-	+	V	-	V	+	V	V
Symbiotic growth (β-NAD requirement)	–[a*]	–[b]	+ [V]	–	–	–	–	–	–	+[c]	–	–	V
Porphyrin	+	+	– [+]	+	+	+	+	+	–	+	nd	+	+
Methyl red	–	–	nd	–	–	nd	w	+	nd	–	nd	–	nd
Voges Proskauer	–	–	nd	–	V	+	–	–	–	–	nd	–	nd
Urease	+	–[d]	+ [V]	–	–	–	–	–	–	–	+	–	–
Indole	–	+	V	–	–	–	–	–	+	–	–	–	–
Phosphatase	+	+	+	+	–	+	+	+/w	+	+[e]	V	+	+
Pigment	–	–	nd	–	–	nd	–	–	+	V	nd	V	V
Acid from													
(-)-D-arabinose	–	V	nd	V	(+)	nd	(+)	–	nd	V	nd	–	–
Dulcitol	–	V	–	–	+	–	–	–	nd	–	–	–	nd
(-)-D-Fructose	+	+	– [V]	+	+	nd	+	+	–	+	–	+	+
(+)-D-Galactose	V	+	+	+	nd	nd	+	+	nd	V	–	–	V
Maltose	+	–	+	V	+	–	V	V	–	V	–	+	+

Characteristic	Genus 1	2	3	4	5	6	7	8	9	10	11	12	13
(-)-D-mannitol	+[f]	V	–	+	–	–	+	–	nd	V	–	+	V
(+)-D-mannose	V	+	– [+]	–	+	–	+	+	nd	+	–	+	V
Sucrose	+	+	– [V]	+	V	–	+	+	–	+	–	+	V
(+)-D-Trehalose	V	V	–	–	–	–	V	–	–	V	–	+	V
Glycosides	V	–	–	V	+	V	–	–	nd	–	nd	V	nd
ONPG	V	V	– [V]	V	+	+	+	V	nd	V	V	–	V
α-Glucosidase	V	+	–	–	–	nd	+	–	nd	+	nd	V	V
β-Glucuronidase	–	–	–	–	+	nd	–	–	nd	–	nd	–	V

Genera: 1, *Actinobacillus sensu stricto*; 2, *Pasteurella sensu stricto*; 3, *Haemophilus sensu stricto* (*includes H. influenzae, H. haemolyticus* and *H. aegypticus* – results for *H. parainfluenzae* and *H. pittmaniae* are given in [ ]); 4, *Mannheimia*; 5, *Lonepinella*; 6, *Phocoenobacter*; 7, *Gallibacterium*; 8, *Volucribacter*; 9, *Histophilus*; 10, *Avibacterium*, 11, *Nicoletella*, 12, *Bibersteinia*; 13, *Aggregatibacter*. Data based on Angen *et al.* (1999, 2003), Bisgaard and Mutters (1986), Blackall *et al.* (2005, 2007), Christensen and Bisgaard (2003, 2004), Christensen *et al.* (2003a, b), Christensen *et al.* (2004a, b), Kuhnert *et al.* (2004), Mutters *et al.* (1985a) and Nørskov-Lauritsen and Kilian (2006). Characters are scored as: +, 90% or more of the strains positive within 1–2 days; (+), 90% or more of the strains positive within 3–14 days; –, less than 10% of the strains are positive within 14 days; V, 11–89% of the strains are positive; w, weak positive. nd, no data available.

*Discrepant results are indicated by: [a]*Actinobacillus pleuropneumoniae* biovar 1 positive; [b]*Pasteurella multocida* might be positive; [c]*Avibacterium gallinarum* negative; some isolates of *Avibacterium paragallinarum* also negative (biovar 2); [d]*Pasteurella dagmatis* positive; [e]*Avibacterium paragallinarum* biovar 1 might be negative; [f]*Actinobacillus suis* negative

Conventional Identification – clinical pathogens

Once presumptive *H. influenzae* colonies are cultured from normally sterile body sites or found in significant amounts from specimens of the respiratory tract, a further characterization is carried out. Satellitic growth should be visible on blood agar, or, if suppressed by the mucosal commensal flora, it can subsequently be addressed by subculture. Various conclusive identification schemes are employed, depending on local preferences. Some laboratories use minimal criteria of limited specificity if significant amounts of haemophilic bacteria are cultured from the respiratory tract, eyes or ears. In this abbreviated approach, it is concluded that Gram-negative rods or coccobacilli with characteristic colonial morphology and satellitic growth are *H. influenzae*. Due to the preponderant presence of *H. influenzae* in these samples, this abbreviated identification scheme may arrive at the correct classification in the large majority of cases. In a retrospective characterization of more than 400 preserved bacteria reported as *H. influenzae* according to these minimal standards, the identification was confirmed for approximately 99% (N. Nørskov-Lauritsen, unpublished). If minimal criteria are employed in the identification of *H. influenzae* it is important to restrict the use of this approach to samples with dominant growth (as defined above) from particular anatomical sites (sputum and swabs from the ear or the eye). If an abbreviated identification scheme is used with samples obtained from wounds or urine, mis-identifications are likely due to the higher prevalence of other species, particularly *H. parainfluenzae*, at these sites.

Demonstration of an inability to synthesize porphyrins from δ-aminolaevulinic acid in a V factor-dependent isolate can be used as a confirmatory test for the identification of *H. influenzae* (Biberstein *et al.*, 1963; Kilian, 1974). Few other Gram-negative rods will demonstrate dependence on both X and V factor.

A recent characterization of a large number of presumptive *H. influenzae* strains isolated from the respiratory tract of patients with chronic obstructive pulmonary disease (COPD) could not confirm the identification for 102 of 258 strains (Murphy *et al.*, 2007). These strains clustered with the type strain of *H. haemolyticus* by 16S rRNA gene sequence, although many of the strains were non-haemolytic. In contrast to *H. influenzae*, the atypical strains were not associated with exacerbations in the COPD (Murphy *et al.*, 2007). This study gives rise to concern that bacteria of low pathogenicity may be wrongly reported as *H. influenzae* leading to a false interpretation of the patient's disease and unnecessary treatments with antibiotics. The misidentification is likely to be of minor importance in the routine microbiological service. In the retrospective characterization of preserved *H. influenzae* isolates (more than 200 were from the respiratory tract), only two strains could be classified as variant strains in accordance with the description of Murphy *et al.* (2007) (N. Nørskov-Lauritsen, unpublished). This striking difference may be related to the selected patient material used in the COPD study. Furthermore, the isolates from the COPD patients were collected in order to monitor the acquisition of new strains by a typing scheme, meaning that commensal isolates may have been included in the collection studied.

Other variant strains not formally named have been reported. Cryptic genospecies biotype IV is isolated from the genito-urinary tract and from neonatal infections (Quentin *et al.*, 1996), and cannot be distinguished from *H. influenzae* by phenotypic means. Fermentation of mannose or sucrose, production of H_2S (as evaluated by the lead acetate paper test), or production of gas from glucose should raise the suspicion of a variant strain not belonging to *H. influenzae sensu stricto* (Nørskov-Lauritsen *et al.*, 2008). The final identification can require genetic methods such as 16S rRNA gene sequencing.

To identify *Pasteurellaceae* other than *H. influenzae*, a more extended phenotypic characterization is carried out. Selected phenotypic traits are given in Table 11.2. The table is a restricted list of useful characters according to the personal experience of one of the authors (N.N.L.). Authoritative textbooks such as *Bergey's Manual on Systematic Bacteriology* should be consulted for a comprehensive phenotypic description. Tests for catalase and oxidase are of limited value for differentiation between human isolates of *Pasteurellaceae*. Weak or delayed reactions are

Table 11.2 Selected characters for differentiation of *Pasteurellaceae* species isolated from man

Characteristic	*H. influenzae*	*H. haemolyticus*	*H. parainfluenzae*	*H. pittmaniae*	*H. parahaemolyticus*
Porphyrin synthesis	–	–	+	+	+
Symbiotic growth (β-NAD requirement)	+	+	+	+	+
Haemolysis	–	+	V	+	+
Oxidase	+	+	+	+	+
Catalase	+	+	V	V	V
Indole	V	V	V	–	–
Urease	V	+	V	–	+
Ornithine decarboxylase	V	–	V	–	–
B-galactosidase (ONPG)	–	–	V	+	V
D-Glucose, gas production	–	V	V	V	V
Acid from					
(+)-D-Glucose	+	+	+	+	+
(–)-D-Fructose	–	–	+	+	+
Sucrose	–	–	+	+	+
Lactose	–	–	–	–	–
(–)-D-Ribose	+	+	–	–	–
(+)-D-Xylose	+	V	–	–	–
(+)D-Mannose	–	–	+	+	–
(+)-D-Galactose	+	+	+	+	V
Maltose	+	+	+	+	+

Data based on Friis-Moeller *et al.* (2001), Kilian and Frederiksen (1981), Kilian (2005), Nørskov-Lauritsen *et al.* (2005) and Nørskov-Lauritsen and Kilian (2006). Characters are scored as: +, 90% or more of the strains positive; –, less than 10% of the strains are positive; V, 11–89% of the strains are positive; w, weak positive. nd, no data available.

common, and there is a degree of subjectivity in deciding whether a particular species is considered to be positive, delayed, weak, variable or negative with these tests. This does not negate the existence of differential reactivity between some species of *Pasteurellaceae* or between this family and related taxa. A valuable example is the differentiation between *Agg. actinomycetemcomitans* and *Agg. aphrophilus*, where decomposition of hydrogen peroxide is usually observed with the former, but not with the latter. Another example is the initial differentiation between *Pasteurellaceae* and *Enterobacteriaceae*, where a positive result with the oxidase test is linked with the former family. Consensus interpretations of the results of the oxidase and catalase tests are included in Table 11.2.

Species included in the genus *Haemophilus* are dependent on V and/or X factor, *Aggregatibacter* species may be dependent on V factor, whereas *A. hominis*, *A. ureae* and [*P.*] *bettyae* show no growth factor requirement. Haemolysis is observed with strains of *H. haemolyticus*, [*H.*] *parahaemolyticus*, *H. pittmaniae* and some *H.*

[*H.*] *ducreyi*	*Agg. actinomycetemcomitans*	*Agg. aphrophilus*	*Agg. segnis*	*A. hominus*	*A. ureae*	[*P.*] *bettyae*	*P. multocida*
–	+	+	+	+	+	+	+
-	-	V	+	-	-	-	-
V	–	–	–	–	–	–	–
+	V	V	–	+	+	V	+
–	+	–	V	+	V	–	+
–	–	–	–	–	–	+	+
–	–	–	–	+	+	–	–
–	–	–	–	–	–	–	+
–	–	+	V	+	–	–	–
–	V	V	–	–	–	V	–
w	+	+	w	+	+	+	+
–	+	+	+	nd	nd	–	+
–	–	+	w	+	+	–	+
–	–	+	–	+	–	–	–
–	nd	V	–	+	nd	nd	+
–	V	–	–	+	–	–	+
–	V	+	w	V	V	V	+
–	V	V	w	+	–	–	+
–	+	+	w	+	+	–	–

parainfluenzae strains of biotype II and IV. *H. haemolyticus* strains are unable to synthesize porphyrins from δ-aminolaevulinic acid (X factor dependent) and [*H.*] *parahaemolyticus* strains do not ferment mannose. For practical purposes, [*H.*] *parahaemolyticus* may be conceived as a 'haemolytic *H. parainfluenzae* biotype III' and *H. pittmaniae* as a 'haemolytic *H. parainfluenzae* biotype V'.

The three tests used for biotyping of *H. influenzae* and *H. parainfluenzae* are production of indole, urease and ornithine decarboxylase (ODC) (Table 11.3). Strains being negative in the three tests could be assigned to *H. influenzae* biotype VIII or *H. parainfluenzae* biotype V, respectively. If a human X factor-independent isolate of *Pasteurellaceae* is negative for indole, urease and ODC, identification as *H. pittmaniae* or a species of the genus *Aggregatibacter* should be considered in addition to *H. parainfluenzae* biotype V. Haemolysis will suggest a classification as *H. pittmaniae*, whereas V factor independence will be observed with both *Agg. actinomycetemcomitans* and isolates of *Agg. aphrophilus* carrying a functional nicotinamide phosphoribosyltransferase. These species are separated by testing for catalase, β-galactosidase, and lactose fermentation (Table 11.2). If the specimen is obtained from the genitourinary tract, the rare finding of [*P.*] *bettyae* could be considered. The significant

Table 11.3 Key to the differentiation of the biotypes of *Haemphilus influenzae* and *Haemophilus parainfluenzae* as defined by Kilian (2005)

	H. influenzae biotypes									*H. parainfluenzae* biotypes							
Characteristic	I	II	III	IV	V	VI	VII	VIII		I	II	III	IV	V	VI	VII	VIII
Indole	+	+	–	–	+	–	+	–		–	–	–	+	–	+	+	+
Urease	+	+	+	+	–	–	–	–		–	+	+	+	–	–	+	–
Ornithine decarboxylase	+	–	–	+	+	+	–	–		+	+	–	+	–	+	–	–

growth of this organism *in vitro* contrasts with the stunted growth of *Aggregatibacter* spp. and should suggest the need for further phenotypic characterization.

V factor-dependent, X factor-independent isolates that are negative for indole, urease and ODC may be classified as *Agg. aphrophilus* (without a functional nicotinamide phosphoribosyltransferase), *Aggregatibacter segnis* or *H. parainfluenzae* biotype V. *Agg. aphrophilus* is capable of fermenting lactose. *Agg. segnis* typically shows sluggish growth and may take 48 h to reach a diameter of 0.5 mm on chocolate agar plates. Fermentation of sugars is protracted, and reactions are negative or weakly positive. Fermentation of sucrose is usually stronger than fermentation of glucose. Only quantitative differences in the amount of acid produced from carbohydrates may discriminate *Agg. segnis* from strains of *H. parainfluenzae*, biotype V.

Indole, urease, ODC and ONPG are also useful tests for the discrimination of *A. hominis*, *A. ureae*, *P. multocida* and [*P.*] *bettyae* (Table 11.2).

Commercial kit identification – clinical pathogens

The ability of a number of commercial kits to identify selected species of *Haemophilus* has been reported (Barbe *et al.*, 1994; Doern and Chapin, 1984; Janda *et al.*, 1987, 1989; Palladino *et al.*, 1990). The primary aim of the kits was to identify *H. influenzae* and *H. parainfluenzae* and they performed reasonably well when challenged with these species. However, if the aim of the characterization is to confirm or reject the identification of *H. influenzae*, porphyrin-based testing methods are more reliable than biochemical based identification schemes (Munson *et al.*, 2002).

A *Neisseria–Haemophilus* card for the VITEK® 2 was recently released by BioMerieux. It contains 30 biochemical tests and features a database comprising the majority of the organisms handled in this section with the exception of *A. hominis*, [*H.*] *ducreyi*, *H. pittmaniae*, [*P.*] *bettyae* and *P. multocida*. Information on the performance of the kit is, as yet, only scarce (Valenza *et al.*, 2007).

DNA-based detection and identification – clinical pathogens

PCR-based detection of *Pasteurellaceae* in clinical material has mainly focused on the rapid diagnosis of life-threatening infections, particularly *H. influenzae* meningitis, and improved detection of organisms which are difficult to culture, particularly [*H.*] *ducreyi*. A number of PCR assays have been described for the amplification of [*H.*] *ducreyi* DNA in swabs from genital ulcers (Gu *et al.*, 1998; Tekle-Michael *et al.*, 2001; Totten *et al.*, 2000). With respect to *H. influenzae* meningitis, the development of the PCR technology coincided with the implementation of the Hib vaccine that resulted in a dramatic decrease in the incidence of the disease in many countries. This may have diminished the demand for an assay targeting *H. influenzae*. Most experience has probably been obtained with the PCR assay of van Ketel *et al.* (1990), which included two primer sets, targeting the outer membrane protein P6 gene and the capsule export protein *bexA*, respectively. The P6 gene PCR was not specific for *H. influenzae* but reacted with DNA from *H. haemolyticus* and some strains of *H. parainfluenzae*. The BexA protein is necessary for capsule polysaccharide export and specific for capsulate strains. The PCR assay was reported to react with target DNA from 15 capsulate strains of all serotypes (van Ketel *et al.*, 1990). A failure to detect *bexA* in *H. influenzae* serotypes e and f, due to sequence variation within probe binding sites, was subsequently reported (Sam and Smith, 2005). The sequence diversity of *bexA* has recently been shown to reflect phylogenetic relations as revealed by the MLST scheme (Meats *et al.*, 2003), making *bexA* a difficult target for detection of capsulate phylogenetic group II strains (serotypes e and f, plus a few serotype a and b strains) (Zhou *et al.*, 2007).

To truly benefit from an advanced identification by PCR, the analysis must be fully automated. The processing of the sample must be initiated in the laboratory immediately after receipt. If the samples are processed only at predetermined times in order to save labour, much of the potential time savings associated with the molecular methods are lost. As fully automated robots for DNA isolation, PCR amplification and probe hybridization become available and affordable, the acute processing of a cerebrospinal fluid sample from suspected cases of meningitis will likely include a PCR assay. An additional benefit is the identification of the pathogen in cases where prior antibiotic therapy has made culturing impossible. The significance and clinical implications of the detection of bacterial DNA will need to be evaluated and defined. Presence of DNA from commensal organisms in a sample is not necessarily synonymous with bacterial infection at a particular anatomical site, but may convey clinically relevant information.

In parts of the world where laboratory facilities for immediate culturing of cerebrospinal fluid samples are not available, PCR assays may be advantageous for improving microbiological surveillance of bacterial meningitis (Chanteau *et al.*, 2006).

Commercial multiplex PCR detection kits to identify bacteria and fungi in whole blood for early sepsis diagnosis are becoming available. *H. influenzae* DNA is not detected by the LightCycler® SeptiFast kit from Roche Diagnostics, but is targeted in the VYOO™ multiplex PCR kit from SIRS-lab.

Isolation methods – veterinary pathogens

Non-selective isolation

For those organisms that do not show an *in vitro* requirement for nicotinamide adenine dinucleotide, the generally accepted universal isolation medium is blood agar. Typically, 5% sheep blood agar provides a good robust isolation medium. Specific examples of the use of 5% sheep blood agar as an isolation medium include *P. multocida* from pigs (White *et al.*, 1993), *Mannheimia* spp from cattle (Blackall *et al.*, 2001), *Actinobacillus* species from rabbits (Moyaert *et al.*, 2008) and *Hist. somni* from cattle (Gagea *et al.*, 2006). Some authors have reported use of 10% sheep blood agar as being generally suitable (Mohan *et al.*, 1994). Horse blood agar has also been used – with an example being *A. suis* from pigs (Wilson and McOrist, 2000).

When normally sterile tissues, e.g. lung, liver and heart, are being cultured, the lack of selective capacity of blood agar is not a significant issue. Indeed, even in some situations of a high level of background normal flora, blood agar has proven useful, e.g. for the isolation of *M. haemolytica* from bovine tonsils (Frank *et al.*, 1992) and *Pasteurellaceae* from bison tonsils (Ward *et al.*, 1999). The production of zones of haemolysis by some organisms e.g. *Bibersteinia trehalosi*, *Gallibacterium anatis* biovar *haemolytica* and *M. haemolytica* allows the recognition of these haemolytic organisms on blood agar, even in the face of background flora.

Other non-selective media have been used and/or recommended for non-growth factor dependent *Pasteurellaceae* including chocolate agar for *Pasteurella* species (Mohan *et al.*, 1994) and dextrose starch agar for *P. multocida* (Rimler *et al.*, 1998). Dextrose starch agar has been used in fowl cholera cases as it allows recognition of iridescent colonies, a characteristic associated with virulent strains of *P. multocida* (Glisson *et al.*, 2003). The iridescence is best seen at 18–24 hours of incubation with a stereo-microscope using obliquely transmitted light (Glisson *et al.*, 2003).

A range of media have been used for the isolation of the growth-factor dependent *Pasteurellaceae*. These growth factor dependent organisms include such well known pathogens as *A. pleuropneumoniae*, *Av. paragallinarum* and [*H.*] *parasuis* as well as such newly recognized organisms such as '*Actinobacillus porcitonsillarum*'. For the purpose of this chapter, the growth factor-dependent organisms of the family *Pasteurellaceae* will be termed haemophili.

Despite the range of media available, the use of 5% sheep blood agar with a nurse strain has been recommended for both pig and poultry pathogens such as *A. pleuropneumoniae* (Taylor, 1999) [*H.*] *parasuis* (Rapp-Gabrielson, 1999) and *Av. paragallinarum* (Blackall and Matsumoto, 2003). While some authors suggest the use of *S. aureus* as the nurse strain (Segales *et al.*, 1997), the general recommendation for both avian and porcine haemophili is to use a strain of *S. epidermidis* or *S. hyicus* (Blackall and Matsumoto, 2003; Taylor, 1999). The advantage of the use of blood agar plus a nurse colony is that the typical satellitic growth of the haemophili allows a ready recognition of the presence of these organisms.

A range of other complete media where satellitism is not a feature have been used as primary isolation media for the haemophili. These media include complete media such as chocolate agar for [*H.*] *parasuis* (Olvera *et al.*, 2007), PPLO (pleuropneumonia like organisms) agar for *A. pleuropneumoniae* (Kristensen *et al.*, 2004) and CLBA (Colombia blood agar base with 7% lysed horse blood) for *Av. paragallinarum* (Terzolo *et al.*, 1993).

Selective isolation

There have been a number of selective media developed and/or used for the non-growth factor dependent *Pasteurellaceae*. Selective media for the isolation of *P. multocida* from cattle nasopharynx (Catry *et al.*, 2006), chicken alimentary tract (Lee *et al.*, 2000), dog and cat oral cavity (Knight *et al.*, 1983; Muhairwa *et al.*, 2001), human pharynx and oral cavity (Avril *et al.*, 1990; Knight *et al.*, 1983), pig nostrils (Chanter *et al.*, 1989; de Jong and Borst, 1985; Lariviere *et al.*, 1993; Muhairwa *et al.*, 2001; Smith and Baskerville, 1983), poultry laryngotracheal and cloacal swabs (Muhairwa *et al.*, 2001) rabbit nostrils (Garlinghouse *et al.*, 1981) and wild bird internal organs and pond water (Moore *et al.*, 1994) have been described. There are a number of issues that influence the decision on whether to adopt a selective medium. A key issue is the source material and thus the competing flora. Some of the media used in the above studies have been developed specifically for use with samples obtained from the upper respiratory tract e.g. the medium of Knight *et al.* (1983) was developed for the isolation of *P. multocida* from oral swabs of humans, dogs and cats. In contrast, the medium developed by Lee *et al.* (2000) was designed to isolate *P. multocida* from the chicken alimentary tract while the selective broth of Moore *et al.* (1994) combined with plating onto blood agar has proven effective at isolating *P. multocida* from pond water. Another key issue to consider is that selective media may be more effective for some clones within a species population than other clones or alternatively some species more than other species within a genus. This variation within a species presumably explains why Smith and Baskerville (1983) caution that their medium – developed for *P. multocida* of capsule types A and D from pig nostrils did not perform well with a group B strain from haemorrhagic septicaemia or a capsule type A strain from a rabbit. A similar explanation of variation within a species would appear to explain why the selective medium for *Hist. somni* developed by Slee and Stephens (1985) proved suitable for use with Australian cattle but was found to be too inhibitory when used to grow *Hist. somni* isolated in Canada (Kwiecien and Little, 1989). At the species level, Muhairwa *et al.* (2001) have reported that while their selective medium is suitable for isolating *P. multocida* from free ranging chickens and their animal contacts, blood agar appears more suitable for other species of *Pasteurella*. The inhibitory nature of selective media is both a positive and a negative. As an example, the medium of Knight *et al.* (1983) was subsequently found to be more effective for the isolation of *P. multocida* from the pig nasal cavity if tellurite was deleted (Lariviere *et al.*, 1993).

To date, there has been no direct comparison of these various selective media for *P. multocida*. Hence, the decision on which – if any – of these media to use is very much based on the host animal, the nature of the samples being examined (in terms of their competing flora) and the selective pressure of the medium. Overall, the literature would suggest that when sampling the upper respiratory tract of animals, the relatively simple approach of Catry *et al.* (2006) appears suitable. When attempting isolation from the alimentary tract, the medium of Lee *et al.* (2000) appears the best option. For environmental samples such as pond water, the use of the selective broth of Moore *et al.* (1994) with plating onto non-selective blood agar would appear suitable.

A traditional selective approach for *P. multocida* has been the use of mouse inoculation. A full description of the method is provided by Muhairwa *et al.* (2001). The method selects only those clones of *P. multocida* that are pathogenic for mice (Muhairwa *et al.*, 2001). The practical difficulties of maintaining a mouse colony and the pressures to reduce animal experimentation mean that this approach holds little attraction for diagnostic laboratories.

When attempting isolation of haemophili from mixed infections or from sites with a high level of competing organisms, selective media or enrichment broths need to be considered. Several selective media for *A. pleuropneumoniae* have been reported (Gilbride and Rosendal, 1983; Jacobsen and Nielsen, 1995; Moller *et al.*, 1993; Wongnarkpet *et al.*, 1999). The medium developed by Jacobsen and Nielsen (1995) involves preparation of a meat infusion and is not likely to suit the majority of diagnostic laboratories. The only selective medium reported for *Av. paragallinarum* is the medium of Terzolo *et al.* (1993).

Somewhat more complex approaches involving dilution as well as selective agents have been reported for the porcine haemophili. Pijoan *et al.* (1983) developed a technique which combined both dilution and selection for the isolation of haemophili from pig lungs. Møller *et al.* (1993) have reported on the development of an optimized method for the isolation of haemophili from the upper respiratory tract (nasal cavity, tonsils) and lungs of pigs. Nasal swabs are simply directly rolled onto the selective medium and the sample further spread with a bent glass rod (called a hockey stick or Driglaski spatula). When sampling tonsils, Møller *et al.* (1993) combined the use of a selective medium with a dilution approach. The two selective media used by Møller *et al.* (1993) were chocolate agar with 300 μg/ml bacitracin and a 10% blood agar plate containing 300 μg/ml bacitracin and 100 μg/ml β-NAD. While no formal comparison of the two media were performed, Møller *et al.* (1993) note that the selective chocolate agar medium appears to yield more isolates of *A. pleuropneumoniae*.

An alternative for the isolation of *A. pleuropneumoniae* from the tonsils of pigs is immuno-magnetic separation (IMS). To date, IMS methods for *A. pleuropneumoniae* serovars 1 and 2 have been described (Angen *et al.*, 2001; Gagné *et al.*, 1998). IMS offers a high specificity and sensitivity – outperforming both PCR and culture when applied to the tonsils of pigs that had been exposed to an aerosol of *A. pleuropneumoniae* serovar 2 (Angen *et al.*, 2001). While limited to those laboratories with access to specific antisera, IMS represents a valuable tool for the isolation of *A. pleuropneumoniae* – particularly from tissues with a high background flora and/or a low level of *A. pleuropneumoniae* cells.

Overview of isolation methods

The optimal approach for the isolation of the non-growth factor dependent *Pasteurellaceae* is the use of a traditional non-selective medium such as Colombia base agar with 5% sheep blood agar. In circumstances where the samples are likely to contain high levels of competing flora, the use of a selective or enrichment medium (as outlined above) should be considered. Even when using a selective or enrichment medium, the sample should also be plated onto blood agar.

The most appropriate medium for the isolation of haemophili from normally sterile sites is 5% sheep blood agar with a nurse colony of *S. epidermidis* or *S. hyicus*. In situations where a complex respiratory disease with multiple agents is possible, the use of the selective broth method of Pijoan *et al.* (1983) for lung samples is an option to be considered. When attempting to isolate haemophilic organisms from heavily contaminated sites such as the nasal cavity or the tonsils, the selective agar and dilution methodology of Møller *et al.* (1993) should be considered. For those laboratories with access to high titre, specific antisera, IMS is a very practical alternative to the selective agar and dilution techniques.

Conventional identification – veterinary pathogens

A bewildering range of methods for the performance of the phenotypic tests necessary for the conventional identification of the veterinary *Pasteurellaceae* have been described in the literature. Essentially, provided that a method has adequate control strains included (preferably known reference strains from a reputable source) and those control strains give the expected reactions, a diagnostic laboratory can choose any of this vast range of methods. The following text provides comments and recommendations on phenotypic testing methods based on the personal experience and networks of one of the authors (PJB).

There are general precautions that need to be observed. The form of a carbohydrate needs to be carefully observed as reactions can vary. As an example, *B. trehalosi* cannot produce acid from (−)-D-arabinose and (+)-L-arabinose (Blackall *et al.*, 2007) while *Avibacterium volantium* is negative for (+)-L-arabinose fermentation and variable for (−)-D-arabinose (Blackall *et al.*, 2005). Care must also be taken with the quality of the reagents – as an example some forms of (+)-D-galactose can be contaminated with a sufficient level of (+)-D-glucose to give a false positive acid production due to the glucose (Blackall, unpublished data). Where possible, serum should not be used in fermentation media unless necessary as some sera have a capacity to cleave complex carbohydrates such as maltose – producing sufficient (+)-D-glucose to allow a false positive acid production from the liberated glucose. Additionally, some of the reagents/kits that are widely used for some of the key basic tests need to be used with caution and an awareness of their possible limitations. As an example, while no longer relevant in terms of the kits now available, it was shown over 20 years ago that commercial oxidase test systems do not necessarily perform well with *P. multocida* (Blackall, 1982).

An important issue, more widely appreciated than in the past, is that a growth factor requirement (i.e. a dependency upon the presence of NADH in laboratory media) is not a unique marker of the genus *Haemophilus*. β-NAD-dependent isolates of species such as *P. multocida* have been reported (Krause *et al.*, 1987). Conversely, β-NAD-independent forms of species commonly accepted as showing satellitism, e.g. *Av. paragallinarum*, have been reported (Garcia *et al.*, 2004; Mouahid *et al.*, 1992). Hence, diagnostic microbiologists must be aware of the fact that non-satellitic bacteria should be considered as suspect *Av. paragallinarum* or *A. pleuropneumoniae* if they show the typical biochemical properties of the species and have been obtained from either chickens or pigs (respectively) showing upper respiratory disease.

An alternative to the conventional approach outlined by Kilian and Frederiksen (1981) is the use of Diatabs® Diagnostic Tablets and test racks (Rosco, Denmark) which have been shown to produce equivalent results to the conventional media for the genus *Mannheimia* provided a heavy inoculum is used (Angen *et al.*, 2002). The tests that were validated by Angen *et al.* (2002) were carbohydrate fermentation tests as well as tests for ornithine decarboxylase, urease, β-glucosidase, α-fucosidase, β-galactosidase and β-xylosidase.

For *P. multocida* and related species, a micro-plate fermentation system has been reported (Blackall *et al.*, 1995). The plates can be prepared ahead of time and stored at −20°C, making the method more attractive for diagnostic laboratories (P. Blackall, unpublished). While suitable for *P. multocida*, this method is not suitable for *Mannheimia* or *Bibersteinia* (P. Blackall, unpublished).

For the haemophilic organisms of the *Pasteurellaceae*, there is a need for a more specialized media. The first requirement for the performance of conventional phenotypic identification of haemophilic organisms is a robust medium that supports good growth. As a general comment, *A. pleuropneumoniae* is an organism that requires little more than the addition of a suitable level of β-NAD or NADH (the reduced form of β-NAD). However, for *[H.] parasuis* and *Av. paragallinarum*, simple addition of β-NAD is generally not sufficient for vigorous growth and, hence, a reliable phenotypic test. A range of media have been used and proven suitable – chocolate agar and PPLO agar being common media. In the laboratory of one of the authors (PJB), a medium known originally as TM/SN has proven an excellent all-round medium for the growth of haemophilic organisms. The original formulation of this medium used a basal medium known as Test Medium Agar (Rimler, 1979) which prior to pouring was supplemented with 5% oleic-albumin complex, 1% chicken serum and 0.0025% NADH. In recent times, this medium has been simplified by using a commercially prepared blood agar base–BBL™ Blood Agar Base (Becton Dickinson, Sparks, MD, USA) and horse serum in place of chicken serum – and is now termed BA/SN (Turni and Blackall, 2007b).

The agar fermentation test (Blackall, 1983) offers a very practical means of performing carbohydrate fermentation tests for *A. pleuropneumoniae, Av. paragallinarum* and *[H.] parasuis*. Conventional broth based fermentation tests for the avian and porcine haemophili are also possible (Blackall, 1983).

While not commonly performed in the veterinary diagnostic laboratory, growth factor testing can be problematic. The selection of a test medium for the performance of growth factor tests using commercial discs has been shown to have a major influence (Blackall and Farrah, 1985). Use of a medium such as Isosensitest agar or Heart Infusion agar with *Av. paragallinarum* resulted in a majority of isolates either giving no result or a false result of a requirement for both X and V factor (Blackall and Farrah, 1985). If growth factor testing is to be performed, then a near complete medium that lacks only V factor (e.g. BA/S which is BA/SN without any NADH) should be used.

As noted earlier, Table 11.1 shows the differentiation of the currently recognized genera within the family *Pasteurellaceae*.

Tables 11.4–11.10 are identification tables assembled from the literature for the major organisms of interest. As there are a significant number of un-named species and taxa, these identification tables have been limited to the formally named species. Each table provides references to the major publications on which the tables are based. Where possible, a uniform set of properties has been used in each table. Where necessary, key tests that are relevant for one set of organisms are specifically included for those organisms. As with all identification tables, the tables are based on the available information and reflect the 10% rule i.e. a negative reaction means that less than 10% of strains are positive. This means that the tables should be used with a full awareness of possible variations. As an example, Table 11.8 shows *P. multocida* as being positive for indole production, acid from mannitol and the presence ornithine decarboxylase. While this is correct for the vast majority of *P. multocida* strains, Christensen *et al.* (2004a) have reported that strains of *P. multocida* that are negative for all three traits do occur – particularly associated with bovine pneumonia. Similarly, Table 11.8 shows that *P. multocida* does not show symbiotic growth. However, porcine growth-factor dependent *P. multocida* strains have been reported (Krause *et al.*, 1987).

Those diagnostic laboratories considering the use of conventional identification for the family *Pasteurellaceae* need to balance the resources available to them and the realities of achievable levels of diagnostic throughput. An example of this balancing is provided by the avian haemophili. Very basic laboratories with limited resources will clearly not be able to perform all of the tests shown in Table 11.4. However, a sliding scale of identification approaches that reflect the capacities of the particular diagnostic laboratory is possible. For laboratories with limited facilities, sufficient evidence for a diagnosis of infectious coryza would be the isolation of a catalase-negative, Gram-negative organism that exhibits satellitic growth and fails to grow in air, together with a history of a rapidly spreading acute coryza in a flock. Laboratories with a great capacity should determine growth factor requirements and the ability to ferment glucose, galactose, and trehalose. At this level, *Av. paragallinarum* can be differentiated from the other members of the genus *Avibacterium*. For laboratories with extensive resources, a complete phenotypic identification based on the properties shown in Table 11.4 is possible.

Commercial kit identification – veterinary pathogens

As a general comment, the use of commercial identification kits with members of the *Pasteurellaceae* is not a widely adopted practice. As well, the published literature on the use of commercial kits with veterinary *Pasteurellaceae* is relatively old – meaning that the kits are sometimes no longer relevant and that the publications often did not reflect our current understanding of the taxonomy and diversity of the family.

A clear example of the usefulness of a commercial kit is the ID Test HN 20 Rapid Kit (Nissui Pharmaceutical Company, Tokyo). This kit has been shown to be capable of effective

Table 11.4 Key characters for the differentiation of the species within the genera *Avibacterium* and *Gallibacterium*

Characteristic	*Av. gallinarum*	*Av. endocarditidis*	*Av. paragallinarum*	*Av. volantium*	*Av. avium*	*G. anatis*	
						Biovar anatis	Biovar haemolytica
Haemolysis (ovine blood)	–	–	–	–	–	–	+
CO_2 improves growth	+	–	+	–	–	–	–
Symbiotic growth (β-NAD requirement)	–	–	V	+	+	–	–
Catalase	+	+	–	+	+	+	+
Urease	–	–	–	–	–	–	–
Indole	–	–	–	–	–	–	–
ODC	–	–	–	V	–	–	–
Acid from							
(+)-L-Arabinose	–	–	–	–	–	–	–
(+)-D-arabinose	V	–	–	+	–	–	(+)
(+)-D-galactose	+	+	–	+	+	+	+
Lactose	V	–	–	V	–	V	V
Maltose	+	+	V	+	–	–	V
(–)-D-Mannitol	–	+	+	+	–	+	+
(+)-D-Mannose	+	+	+	+	+	+	+
(-)-D-Sorbitol	–	+	V	V	–	V	V
Trehalose	+	+	–	+	+	+	V
ONPG	V	+	–	+	–	+	+
α-Fucosidase	–	+	–	–	–	–	–
α-Galactosidase	–	+	–	–	–	–	–
α-Glucosidase	+	+	–	+	+	+	+
β-Glucosidase	–	+	–	–	–	–	–
β-Xylosidase	-	+	-	–	–	–	–
Host	Birds	Chickens	Chickens	Birds	Birds	Birds, cattle	Birds, cattle

Data based on Bisgaard *et al.* (2007), Blackall and Reid (1982), Blackall *et al.* (1989, 2005), Christensen *et al.* (2003a) and Mutters *et al.* (1985b). +, ≥ 90% positive; (+), > 90% positive or weak positive; –, ≤ 10% positive; V, 11–89% positive.

identification of [*Pasteurella*] *pneumotropica* and, indeed, is a recommended tool for the confirmation of this organism (when supported by a slide agglutination test) (Hayashimoto *et al.*, 2005).

The API ZYM system has been shown to be useful for the identification of '*Haemophilus somnus*' and '*Histophilus ovis*' although the system could not separate these two taxa (Cousins and Lloyd, 1988; Groom *et al.*, 1986). The subsequent recognition that these two taxa are part of the same species, now called *Hist. somni* (Angen *et al.*, 2003), provides additional information not fully appreciated at the time of these evaluations.

The promising evaluation of the RapID NH system in 1993 for the identification of taxa then called '*Haemophilus somnus*', *P. multocida*, '*P. haemolytica*' and *A. pleuropneumoniae* (Salmon *et al.*, 1993) has to be interpreted more carefully

Table 11.5 Key characters for the differentiation of the species within the genera *Mannheimia*, *Bibersteinia* and *Histophilus*

Characteristic	*M. haemolytica*	*M. glucosida*	*M. granulomatis*	*M. ruminalis*	*M. varigena*	*B. trehalosi*	*Hist. somni*
Haemolysis (ovine blood)	(+)	(+)	V	–	V	V	V
CO_2 improves growth	–	–	–	–	–	–	+
Symbiotic growth (β-NAD requirement)	–	–	–	–	–	–	–
Catalase	+	+	+	+	+	–	–
Urease	–	–	–	–	–	–	–
Indole	–	–	–	–	V	–	+
ODC	–	V	–	–	V	–	–
Acid from							
(+)-L-arabinose	–	V	–	–	+	–	–
(–)-D-arabinose	V	V	V	V	V	–	nd
(+)-D-galactose	+	+	+	+	+	–	–
Lactose	V	+	(+)	V	V	–	–
Maltose	+	+	V	V	V	+	–
(–)-D-mannitol	+	+	+	+	+	+	–
(+)-D-mannose	–	–	–	–	–	+	V
(–)-D-sorbitol	+	+	+	V	–	+	–
Trehalose	–	–	–	–	–	+	–
ONPG	V	+	V	+	V	–	+
α-Fucosidase	+	V	–	–	V	–	nd
β-Glucosidase	–	+	+	–	V	V	nd
β-Xylosidase	V	V	V	–	V	–	nd
Host	Ruminants	Sheep	Cattle, hares	Cattle, sheep	Cattle, pigs	Ruminants	Cattle, sheep

Data based on Angen *et al.* (1997, 2003) and Blackall *et al.* (2002, 2007). +, ≥ 90% positive; (+), > 90% positive or weak positive; –, ≤ 10% positive; V, 11–89% positive, nd, no data available.

with our current knowledge of the taxonomy of these organisms.

The study of Lindberg *et al.* (1998) on bacteria isolated in pig bite wounds of humans makes the key point that all identification systems – conventional or commercial – need to be used with an awareness of the abilities and limitations of the system. Lindberg *et al.* (1998) conclude that, for the family *Pasteurellaceae*, commercial systems only work satisfactorily in the hands of experienced microbiologists.

Overall, the view of one of the authors (PJB) is that commercial kits are not a preferred option for diagnostic laboratories. Where possible, diagnostic laboratories should attempt some form of traditional phenotypic identification or move to the genotypic approach. The approach suggested by Angen *et al.* (2002) is an exception and is a good example of the valid use of a commercial kit.

DNA-based detection and identification – veterinary pathogens

DNA based methods for detection and identification directly on animal tissues or on isolation plates and enrichment broths have become a realistic option for veterinary diagnostic labora-

Table 11.6 Key characters for the differentiation of the growth factor independent species within *Actinobacillus* associated with animals

Characteristic	*A. lignieresii*	*A. arthritidis*	*A. equuli* subsp. *equuli*	*A. equuli* subsp. *haemolyticus*	*A. suis*	*A. capsulatus*
Haemolysis (ovine blood)	–	–	–	(+)	+	–
CO_2 improves growth	–	–	–	–	–	–
Symbiotic growth (β-NAD requirement)	–	–	–	–	–	–
Catalase	V	+	+	+	+	d
Urease	+	+	+	+	+	+
Indole	–	–	–	–	–	–
ODC	–	–	–	–	–	–
Acid from						
(+)-L-Arabinose	–	V	V	–	+	+
(+)-D-Arabinose	+/(+)	V	–	–	–	–
(+)-D-Galactose	+	+/(+)	+	+	+	+
Lactose	(+)	(+)	+	+	+	+
Maltose	+	+	+	+	+	+
(–)-D-mannitol	+	+	+	V	–	+
(–)-D-mannose	+	+	+	+	+	+
(–)-D-sorbitol	–	+/(+)	–	–	–	+
Trehalose	–	–	+	+	+	+
ONPG	+	+	+	+	+	+
α-Galactosidase	–	+	+	+	+	+
α-Glucosidase	–	–	V	V	+	V
β-Glucosidase	–	–	–	+	+	+
β-Xylosidase	–	–	+	V	+	V
Host	Cattle, sheep	Horses	Horses, pigs	Horses	Pigs	Hares, rabbits

Data based on Blackall *et al.* (1997), Christensen *et al.* (2002), Christensen and Bisgaard (2004) and Kilian and Frederiksen (1981). +, ≥ 90% positive; (+), > 90% positive or weak positive; –, ≤ 10% positive; V, 11–89% positive.

tories. An extensive literature on PCR tests for the detection and/or identification of veterinary *Pasteurellaceae* exists.

The decision on whether to adopt a molecular diagnostic test requires careful consideration and comparison of the various advantages and disadvantages of the proposed test and the more traditional phenotypic based methodology. This comparison is influenced by the nature of the organism, the capacities of the diagnostic laboratory and the nature of the diagnostic samples available to the laboratory. The published literature on the PCR test for *Av. paragallinarum* is a good example of the possible scenarios that influence this decision of whether to adopt a molecular diagnostic test.

Av. paragallinarum is a fragile organism that requires considerable effort from the diagnostic laboratory in terms of specialized media for a full phenotypic identification (see previous section). The first, and so far only, PCR for *Av. paragallinarum* was reported by Chen *et al.* (1996) with the work being performed in a very experienced and well equipped laboratory in Australia. This initial work reported that the PCR showed 100% correlation with culture when used on nasal

Table 11.7 Key characters for the differentiation of the growth factor dependent species of *Actinobacillus* and *Haemophilus* found in pigs

Characteristic	*A. pleuropneumoniae*	'*A. porcitonsillarum*'	[*A.*] *minor*	[*A.*] *porcinus*	[*A.*] *indolicus*	[*H.*] *parasuis*
Haemolysis (ovine blood)	(+)	(+)	–	–	–	–
CO_2 improves growth	–	–	–	–	–	–
Symbiotic growth (β-NAD requirement)	V	+	+	+	+	+
Catalase	V	–	–	–	+	+
Urease	+	+	+	–	–	–
Indole	–	–	–	–	+	–
ODC	–	–	–	–	–	–
Acid from						
(+)-L-arabinose	–	–	–	–	–	–
(+)-D-galactose	+	+	V		+	+
Lactose	–	nd	+	V	V	–
Maltose	+	nd	+	V	+	+
(–)-D-Mannitol	+	–	–	V	–	–
(+)-D-Mannose	+	nd	+	V	+	+
(–)-D-Sorbitol	–	–	–	V	–	–
Trehalose	–	nd	V	V	V	–
ONPG	+	+	+	+	+	+
α-Fucosidase	–	nd	–	V	V	+

Data based on Gottschalk *et al.* (2003), Kilian and Frederiksen (1981), Møller and Kilian (1990) Møller *et al.* (1996) and Tonpitak *et al.* (2007). +, ≥ 90% positive; (+), > 90% positive or weak positive; –, ≤ 10% positive; V, 11–89% positive, nd. no data available.

swabs obtained from artificially infected chickens (Chen *et al.*, 1996). The test was then transferred to a laboratory in China and used on both artificial infection samples and typical field samples that suffered all the typical problems of slow transport and poor temperature control (Chen *et al.*, 1998a). When used on field samples, the PCR gave significantly more positives (15/39) than traditional culture (8/39) with two of the six PCR confirmed outbreaks not being detected by culture (Chen *et al.*, 1998a). Further work in China showed that the majority of nasal swabs from artificially infected chickens were still positive by PCR after 6 months' storage at either 4°C or –20°C (Chen *et al.*, 1998b). In contrast, swabs were negative by culture after 3 days – regardless of storage temperature (Chen *et al.*, 1998b). Overall, these studies emphasize that, in situations where samples may not be optimal, where transport of samples are often delayed and where laboratories lack the resources to have expensive, complex media routinely available, PCR can be a very attractive option.

In a similar vein, Turni and Blackall (2007a) have compared traditional culture with the PCR of Oliveira *et al.* (2001) for the detection of [*H.*] *parasuis* in artificially infected pigs. Under optimal conditions (good quality samples, rapid laboratory processing, a suitable growth medium and experienced laboratory diagnosticians), traditional culture was significantly better than PCR in one challenge trial and as good as PCR in the other trial. However, Turni and Blackall (2007a) concluded that PCR should be considered when (a) sampling conditions are suboptimal, (b) transport delays may reduce the viability of any [*H.*] *parasuis* present and (c) when the laboratory lacks expertise and/or suitable media.

Table 11.8 Key characters for the differentiation of the species of *Pasteurella sensu stricto* associated with animal infections

Characteristic	*P. multocida*	*P. canis*	*P. dagmatis*	*P. stomatis*	[*P.*] *langaaensis*
Haemolysis (ovine blood)	–	–	–	–	–
CO_2 improves growth	–	–	–	–	–
Symbiotic growth (β-NAD requirement)	–	–	–	–	–
Catalase	+	+	+	+	+
Urease	–	–	+	–	–
Indole	+	+	+	+	–
ODC	+	+	–	–	–
Acid from					
(+)-L-Arabinose	V	–	–	–	–
(+)-D-Galactose	+	+	+	+	+
Lactose	–	–	nd	–	+
Maltose	–	–	+	–	–
(–)-D-Mannitol	+	–	–	–	+
(+)-D-Mannose	+	+	+	+	+
(–)-D-Sorbitol	V	–	–	–	–
Trehalose	V	V	+	+	–
Host	Most mammals, birds	Dogs	Dogs, cats	Cats, dogs	Chickens

Data based on Christensen *et al.* (2004a) and Mutters *et al.* (1985a). +, ≥ 90% positive; (+), > 90% positive or weak positive; –, ≤ 10% positive; V, 1–89% positive; nd, no data available.

To date, around a dozen different PCR tests for the detection and/or identification of *A. pleuropneumoniae* have been reported. Fittipaldi *et al.* (2003) have evaluated eight of these tests. Following the initial evaluation using pure cultures and spiked pig tonsils, two tests – the *omlA* PCR (Savoye *et al.*, 2000) and a commercial PCR kit (the Adiavet App PCR) were evaluated with field samples. Direct PCR testing of tonsils, by both methods, gave more positives than culture. Both PCR tests gave even more positives if they were applied to culture plates – with both PCR tests giving more positives than an IMS method (Fittipaldi *et al.*, 2003). Hence, based on current knowledge, the *omlA* PCR (Savoye *et al.*, 2000) and the Adiavet App PCR would appear to be the most suitable assays for the direct identification of *A. pleuropneumoniae* in sites such as the tonsils. In laboratories with good quality media and experienced bacteriologists, direct PCR should not be necessary on lung samples. The two duplex PCR assays described by Tonpitak *et al.* (2007) appear a good method for confirming the identification of '*A. porcitonsillarum*', an organism of uncertain pathogenicity and which is very difficult to separate from *A. pleuropneumoniae* on phenotypic testing.

A range of PCR assays for the detection and/or identification of *P. multocida* have been described. For those laboratories considering a species specific conventional PCR, the assays of Townsend *et al.* (1998) and Miflin and Blackall (2001) are suitable assays. Neither assay has been validated for direct use on animal tissues. Hence, for laboratories considering direct tissue assays, the 5′ *Taq* nuclease assay of Corney *et al.* (2007) is the better option – provided the technology for such an assay is available. The haemorrhagic septicaemia specific PCR of Townsend *et al.* (1998) is an important diagnostic tool for the identification of the forms of *P. multocida* associated with this major disease. Atrophic rhinitis is another

Table 11.9 Key characters for the differentiation of incorrectly assigned species of *Pasteurellaceae* associated with animal infections

Characteristic	*[A.] rossii*	*[A.] seminis*	*[P.] aerogenes*	*[P.] caballi*	*[P.] mairii*	*[Pasteurella] testudinis*	*[P.] pneumotropica*
Haemolysis (ovine blood)	V	–	–	–	V	+	–
CO_2 improves growth	–	+	–	–	–	–	–
Symbiotic growth (β-NAD requirement)	–	–	–	–	–	–	–
Catalase	V	+	+	–	+	+	+
Urease	+	–	+	–	+	+	V
Indole	V	–	–	–	–	+	+
ODC	–	V	V	+	+	–	V
Acid from							
(+)-L-arabinose	+	V	V	–	+	V	V
(+)-D-arabinose	V	–	+	–	–	nd	nd
(+)-D-galactose	+	V	+	+	+	nd	+
Lactose	+	–	+	+	–	V	–
Maltose	V	V	+	V	–	V	+
(–)-D-mannitol	+	V	V	+	+	V	V
(+)-D-mannose	V	–	+	+	+	nd	+
(–)-D-sorbitol	V	–	–	–	+	V	–
Trehalose	–	–	–	–	–	V	V
ONPG	+	–	+	+	+	V	V
α-Fucosidase	–	nd	–	–	–	nd	nd
α-Galactosidase	V	nd	–	–	–	nd	nd
α-Glucosidase	–	nd	–	–	–	nd	nd
β-Glucosidase	V	+	V	–	+	nd	nd
β-Xylosidase	V	+	V	–	V	nd	nd
Host	Pigs	Sheep	Pigs	Pigs, Horses	Pigs	Tortoises	Laboratory animals

Data based on Blackall *et al.* (1997), Christensen *et al.* (2006; 2005), Kilian and Frederiksen (1981), Sneath and Stevens (1990) and Snipes and Biberstein (1982). +, ≥90% positive; (+), >90% positive or weak positive; –, ≤ 10% positive; V, 11–89% positive; nd, no data available.

important disease associated with *P. multocida*. A number of PCR assays specifically for toxigenic *P. multocida* have been reported – varying from conventional PCR to nested PCR to a multiplex PCR (Choi and Chae, 2001; Kamp *et al.*, 1996; Lichtensteiger *et al.*, 1996; Nagai *et al.*, 1994; Register and DeJong, 2006). There has been no formal direct comparison of all of these assays so laboratories considering these assays need to consult the original publications as part of the decision making process. The multiplex PCR of Register and Dejong (2006) is an attractive option but it currently lacks validation with clinical material.

While single agent PCR assays have been developed for [*Actinobacillus*] *seminis* (Appuhamy *et al.*, 1998) and *Hist. somni* (Angen *et al.*, 1998), the multiplex PCR for both agents as well as *Brucella ovis* reported by Saunders *et al.* (2007) appears highly suitable for diagnostic laboratories.

Two PCR assays for [*H.*] *parasuis* have been reported. The assay originally developed by Oliveira *et al.* (2001) is now known to give false-

Table 11.10 Key characters for the differentiation of species in the genus *Volucribacter*.

Characteristic	*V. psittacicida*	*V. amazonae*
Haemolysis (ovine blood)	–	–
CO_2 improves growth	–	–
Symbiotic growth (β-NAD requirement)	–	–
Catalase	V	+
Urease	–	–
Indole	–	–
ODC	–	–
Acid from		
(+)-L-Arabinose	–	V
(+)-D-Arabinose	–	V
(–)-L-Fucose	–	+
(+)-D-Galactose	+/(+)	+/(+)
meso-Inositol	–	+
Lactose	–	V
Maltose	–	+
(–)-D-Mannitol	–	–
(+)-D-Mannose	+	+
(–)-D-Sorbitol	–	–
Trehalose	–	–
ONPG	–	+
α-Fucosidase	–	–
α-Galactosidase	–	V
α-Glucosidase	–	–
β-Glucosidase		
β-Xylosidase	+	+
Host	Chickens, parrots, budgerigars	Parrots, parakeets

Data based on Christensen *et al.* (2004b). +, ≥ 90% positive; (+), > 90% positive or weak positive; –, ≤ 10% positive; V, 1–89% positive.

positive reactions when tested with pure cultures of [*Actinobacillus*] *indolicus*, [*Actinobacillus*] *minor* and [*Actinobacillus*] *porcinus* (Angen *et al.*, 2007). The [*H.*] *parasuis* PCR reported by Angen *et al.* (2007) overcomes these non-specificities, although both assays seem to function acceptably when used on samples from systemic locations. As the Oliveira *et al.* (2001) PCR is more sensitive than the Angen *et al.* (2007) PCR, laboratories considering the adoption of a [*H.*] *parasuis* PCR will need to select the assay that best meets their particular requirement for specificity and sensitivity.

The [*P.*] *pneumotropica* PCR of Nozu *et al.* (1999) gave a slightly better ability to correctly identify field isolates of this species than the earlier assay of Wang *et al.* (1996) and appears to be the best option for this species.

As noted earlier, the PCR for *Av. paragallinarum* has proven useful in laboratories in both the developed and the developing world and is a suitable tool for direct detection of *Av. paragallinarum* in nasal swabs (Chen *et al.*, 1996, 1998a,b). The assay does not function well with infraorbital sinus swabs collected at necropsy (P. Blackall, unpublished) and is best applied directly to nasal mucus obtained by squeezing the sinus of the live chicken.

An alternative approach to molecular identification of the *Pasteurellaceae* has been reported by Catry *et al.* (2004). This approach, called tDNA-PCR, is based on the DNA fragment patterns generated by consensus primers complementary to the conserved edges of the tRNA genes (Catry *et al.*, 2004). The reproducibility of the method and the ability to exchange tDNA-PCR fingerprints has been shown in other studies on other bacterial genera (Baele *et al.*, 2001; Baele *et al.*, 2002). The evaluation of the tDNA-PCR performed by Catry *et al.* (2004) was undertaken before a full understanding of the true phylogenetic relationships in the family was available. Thus, of the six '*Pasteurella*' species examined by Catry *et al.* (2004) and found to produce specific patterns, only two are truly members of the genus *Pasteurella*. Within the genus *Mannheimia*, tDNA-PCR generated species-specific patterns for *M. granulomatis* and *M. varigena* (Catry *et al.*, 2004). The other three species within *Mannheimia* (*M. haemolytica*, *M. glucosida* and *M. ruminalis*) formed a cluster of related patterns that could not be easily separated (Catry *et al.*, 2004). Overall, tDNA-PCR appears to have some potential for the identification of *Pasteurellaceae* but further studies performed with a full appreciation of our current understanding of the phylogeny of the family are required.

Overall, a range of DNA-based assays (including DNA sequencing – reviewed below) are now available for the veterinary *Pasteurellaceae*. Laboratories need to critically evaluate their particular needs and the established characteristics of the various possible assays. Current technology means that live cultures are still necessary for many species to allow serovar determination and antibiotic resistance tests to be performed.

Identification by DNA sequencing

Amplification of target genes by the PCR, followed by sequencing of the product (amplicon) and comparison with deposited sequences in the public databases, is a powerful technique to identify bacteria. The 16S rRNA gene is by far the most commonly chosen target. The gene is ubiquitously present and sufficiently conserved to make possible the application of PCR primers capable of hybridizing with a very broad range of bacterial 16S rRNA genes. Yet, the diversity of the gene is sufficient to assign an unknown strain to genus and commonly also to species on the basis of a ca 500-nucleotide sequence arising from a single sequencing reaction. It may be difficult or impossible to discern closely related species as *A. lignieresii* and *A. pleuropneumoniae* (Nørskov-Lauritsen *et al.*, 2004).

Other target genes which have been used with *Pasteurellaceae* include the translation initiation factor 2 gene *inf*B (Hedegaard *et al.*, 2001), the RNA polymerase beta subunit gene *rpo*B (Korczak *et al.*, 2004), the recombination/repair protein gene *recN* (Kuhnert and Korczak, 2006) and the *sodA* gene (Gautier *et al.*, 2005). Housekeeping gene sequence comparison is important in taxonomic and phylogenetic studies (see Chapters 1 and 2), but of minor value in the diagnostic laboratory due to a higher rate of recombination compared with 16S rRNA gene, and the frequent incidence of amplification failures due to mutation events involving the primer annealing sites.

Bacterial DNA for PCR can be isolated and purified by various methods, but a simple boiling of cells in water for 10 min is usually sufficient with *Pasteurellaceae* and other Gram-negative bacteria (Miflin and Blackall, 2001; Sacchi *et al.*, 2005). DNA sequences are usually 'BLASTed' (Basic Local Alignment Search Tool) against the depositories at the public databases for example at NCBI (http://www.ncbi.nlm.nih.gov/BLAST/). The program will find related sequences in the database and list them, with alignments, according to the level of identity. Target sequences can be traced by their accession number and a link to an eventual publication is given. Problems may arise with target sequences based on uncultivated bacteria and incorrectly identified strains. At the Ribosomal Database Project (http://rdp.cme.msu.edu/), it is possible to limit the search to type strains, as well as sequences from uncultured bacteria can be filtered. The last option can be helpful, but restricting the search to type strains does not solve the problems with mislabelled strains, especially as these databases may not be efficiently curated. As an example, by September 2007, the accession number M75073 still refers to the 16S rRNA sequence of the type strain of [*H.*] *parahaemolyticus*, although it clearly is not based on the correct strain (Hedegaard *et al.*, 2001; Nørskov-Lauritsen *et al.*, 2005; Olsen *et al.*, 2005).

Primers widely used with *Pasteurellaceae* and other proteobacteria have been described (Dewhirst *et al.*, 1989; Kuhnert *et al.*, 2002; Paster and Dewhirst, 1988). In the laboratory of one of the authors (NNL), a single set of primers (B5 and B4) described by Ley *et al.* (1998) are routinely used for identification of unusual bacteria deemed clinically important, when traditional phenotypic testing is inconclusive. This primer set have worked well with a broad range of Gram-positive and Gram-negative bacteria and can also be used to detect low amounts of bacterial DNA in clinical samples (Voldstedlund *et al.*, 2003). If both detection and identification are done in the same laboratory, it may be advisable to use different sets of primers for the different purposes. Primers can then be designed to assure that the amplicons of propagated material cannot serve as template in the PCR-reactions used with the broad detection methods applied to the clinical samples.

Commercial, automated DNA sequencing systems for identification of a range of pathogens are also now available e.g. the MicroSeq® Microbial Identification System (Applied Biosystems) or the SmartGene IDNS™ platform (Simmon *et al.*, 2006). The latter has specifically been adapted for *Pasteurellaceae* by including 16S rRNA genes as well as *rpoB* (Korczak *et al.*, 2004).

References

Al Balooshi, N., Jamsheer, A., and Botta, G.A. (2003). Impact of introducing quality control/quality assurance (QC/QA) guidelines in respiratory specimen processing. Clin. Microbiol. Infec. *9*, 810–815.

Albritton, W.L. (1989). Biology of *Haemophilus ducreyi*. Microbiol. Rev. *53*, 377–389.

Angen, O., Aalbaek, B., Falsen, E., Olsen, J.E., and Bisgaard, M. (1997). Relationships among strains classified with the ruminant *Pasteurella haemolytica*-complex using quantitative evaluation of phenotypic data. Zentralbl. Bakteriol. *285*, 459–479.

Angen, O., Ahrens, P., and Tegtmeier, C. (1998). Development of a PCR test for identification of *Haemophilus somnus* in pure and mixed cultures. Vet. Microbiol. *63*, 39–48.

Angen, O., Mutters, R., Caugant, D.A., Olsen, J.E., and Bisgaard, M. (1999). Taxonomic relationships of the *[Pasteurella] haemolytica* complex as evaluated by DNA–DNA hybridizations and 16S rRNA sequencing with the proposal of *Mannheimia haemolytica* gen. nov., comb. nov., *Mannheimia granulomatis* comb. nov., *Mannheimia glucosida* sp. nov., *Mannheimia ruminalis* sp. nov. and *Mannheimia varigena* sp. nov. Int. J. Syst. Bacteriol. *49*, 67–86.

Angen, O., Heegaard, P.M., Lavritsen, D.T., and Sorensen, V. (2001). Isolation of *Actinobacillus pleuropneumoniae* serotype 2 by immunomagnetic separation. Vet. Microbiol. *79*, 19–29.

Angen, O., Ahrens, P., and Bisgaard, M. (2002). Phenotypic and genotypic characterization of *Mannheimia (Pasteurella) haemolytica*-like strains isolated from diseased animals in Denmark. Vet. Microbiol. *84*, 103–114.

Angen, O., Ahrens, P., Kuhnert, P., Christensen, H., and Mutters, R. (2003). Proposal of *Histophilus somni* gen. nov., sp. nov. for the three species incertae sedis *'Haemophilus somnus', 'Haemophilus agni'* and *'Histophilus ovis'*. International J. Syst. Evol. Microbiol. *53*, 1449–1456.

Angen, O., Oliveira, S., Ahrens, P., Svensmark, B., and Leser, T.D. (2007). Development of an improved species specific PCR test for detection of *Haemophilus parasuis*. Vet. Microbiol. *119*, 266–276.

Appuhamy, S., Low, J.C., Parton, R., and Coote, J.G. (1998). Specific PCR primers from the 16S-23S rRNA spacer region for the rapid detection and identification of *Actinobacillus seminis*. J. Appl. Microbiol. *85*, 941–948.

Avril, J.L., Donnio, P.Y., and Pouedras, P. (1990). Selective medium for *Pasteurella multocida* and its use to detect oropharyngeal carriage in pig breeders. J. Clin. Microbiol. *28*, 1438–1440.

Baele, M., Storms, V., Haesebrouck, F., Devriese, L.A., Gillis, M., Verschraegen, G., de Baere, T., and Vaneechoutte, M. (2001). Application and evaluation of the interlaboratory reproducibility of tRNA intergenic length polymorphism analysis (tDNA-PCR) for identification of *Streptococcus* species. J. Clin. Microbiol. *39*, 1436–1442.

Baele, M., Vaneechoutte, M., Verhelst, R., Vancanneyt, M., Devriese, L.A., and Haesebrouck, F. (2002). Identification of *Lactobacillus* species using tDNA-PCR. J. of Microbiol. Methods *50*, 263–271.

Bangsborg, J.M., Tvede, M., and Skinhoj, P. (1988). *Haemophilus segnis* endocarditis. J. Infect. *16*, 81–85.

Barbe, G., Babolat, M., Boeufgras, J.M., Monget, D., and Freney, J. (1994). Evaluation of API NH, a new 2-hour system for identification of *Neisseria* and *Haemophilus* species and *Moraxella catarrhalis* in a routine clinical laboratory. J. Clin. Microbiol. *32*, 187–189.

Barrow, G.I., and Feltham, R.K.A., eds. (1993). Cowan and Steel's Manual for the Identification of Medical Bacteria, 3rd edn (Cambridge, Cambridge University Press).

Biberstein, E.L., Mini, P.D., and Gills, M.G. (1963). Action of *Haemophilus* cultures on {delta}-aminolevulinic acid. J. Bacteriol. *86*, 814–819.

Bisgaard, M., and Falsen, E. (1986). Reinvestigation and reclassification of a collection of 56 human isolates of *Pasteurellaceae*. Acta Path. Micro. Im. B *94*, 215–222.

Bisgaard, M., and Mutters, R. (1986). Characterization of some previously unclassified *"Pasteurella"* spp. obtained from the oral cavity of dogs and cats and description of a new species tentatively classified with the family *Pasteurellaceae* Pohl 1981 and provisionally called taxon 16. Acta Path. Micro. Im. B *94*, 177–814.

Bisgaard, M., Christensen, J.P., Bojesen, A.M., and Christensen, H. (2007). *Avibacterium endocarditidis* sp. nov., isolated from valvular endocarditis in chickens. Int. J. Syst. Evol. Microbiol. *57*, 1729–1734.

Black, C.T., Kupferschmid, J.P., West, K.W., and Grosfeld, J.L. (1988). *Haemophilus parainfluenzae* infections in children, with the report of a unique case. Rev. Infect. Dis. *10*, 342–346.

Blackall, P.J. (1982). An evaluation of two commercial systems for the detection of oxidase activity in the genus *Pasteurella*. Antonie van Leeuwenhoek, International J. Gen. Mol. Microbiol. *48*, 461–464.

Blackall, P.J. (1983). An evaluation of methods for the detection of carbohydrate fermentation patterns in avian *Haemophilus* species. J. Microbiol. Met. *1*, 275–281.

Blackall, P.J., and Farrah, J.G. (1985). An evaluation of commercial discs for the determination of the growth factor requirements of the avian haemophili. Vet. Microbiol. *10*, 125–131.

Blackall, P.J., and Matsumoto, M. (2003). Infectious coryza. In Diseases of Poultry, Y.M. Saif, H.J. Barnes, J.R. Glisson, A.M. Fadly, L.R. McDougald, and D. Swayne, eds. (Ames, Iowa, Iowa State University Press), pp. 691–703.

Blackall, P.J., and Reid, G.G. (1982). Further characterization of *Haemophilus paragallinarum* and *Haemophilus avium*. Vet. Microbiol. *7*, 359–367.

Blackall, P.J., Eaves, L.E., and Rogers, D.G. (1989). Biotyping of *Haemophilus paragallinarum* by hemagglutination serotyping, and carbohydrate fermentation patterns, and antimicrobial drug resistance patterns. Avian Dis. *33*, 491–496.

Blackall, P.J., Pahoff, J.L., Marks, D., Fegan, N., and Morrow, C.J. (1995). Characterisation of *Pasteurella multocida* isolates from fowl cholera outbreaks on turkey farms. Aust. Vet. J. *72*, 135–138.

Blackall, P.J., Bisgaard, M., and McKenzie, R.A. (1997). Characterisation of Australian isolates of *Actinobacillus capsulatus, Actinobacillus equuli, Pasteurella caballi* and Bisgaard Taxa 9 and 11. Aust. Vet. J. *75*, 52–55.

Blackall, P.J., Angen, O., Fegan, N., Blackall, L.L., Mutters, R., and Bisgaard, M. (2001). Characterisation of a novel *Mannheimia* sp from Australian feedlot cattle. Aust. Vet. J. *79*, 634–639.

Blackall, P.J., Bisgaard, M., and Stephens, C.P. (2002). Phenotypic characterisation of Australian sheep and cattle isolates of *Mannheimia haemolytica, Mannheimia granulomatis* and *Mannheimia varigena*. Aust. Vet. J. *80*, 87–91.

Blackall, P.J., Christensen, H., Beckenham, T., Blackall, L.L., and Bisgaard, M. (2005). Reclassification of *Pasteurella gallinarum, [Haemophilus] paragallinarum, Pasteurella avium* and *Pasteurella volantium* as *Avibacterium gallinarum* gen. nov., comb. nov., *Avibacterium paragallinarum* comb. nov., *Avibacterium avium* comb. nov. and *Avibacterium volantium* comb. nov. Int. J Syst. Evol. Microbiol. *55*, 353–362.

Blackall, P.J., Bojesen, A.M., Christensen, H., and Bisgaard, M. (2007). Reclassification of *[Pasteurella] trehalosi* as *Bibersteinia trehalosi* gen. nov., comb. nov. Int. J. Syst. Evol. Microbiol. *67*, 666–674.

Breen, D., Schonell, M., Au, T., and Reiss-Levy, E. (2000). *Pasteurella multocida*: a case report of bacteremic pneumonia and 10-year laboratory review. Pathology *32*, 152–153.

Bruun, B., Gahrn-Hansen, B., Westh, H., and Kilian, M. (2004). Clonal relationship of recent invasive *Haemophilus influenzae* serotype f isolates from Denmark and the United States. J. Med. Microbiol. *53*, 1161–1165.

Buznach, N., Dagan, R., and Greenberg, D. (2005). Clinical and bacterial characteristics of acute bacterial conjunctivitis in children in the antibiotic resistance era. Pediatr. Infect. Dis. J. *24*, 823–828.

Cardines, R., Giufre, M., Mastrantonio, P., Degli Atti, M.L., and Cerquetti, M. (2007). Nontypeable *Haemophilus influenzae* meningitis in children: phenotypic and genotypic characterization of isolates. Pediatr. Infect. Dis. J. *26*, 577–582.

Carpenter, J., Stapleton, S., and Holliman, R. (2007). Retrospective analysis of 49 cases of brain abscess and review of the literature. Eur. J. Clin. Microbiol. *26*, 1–11.

Casin, I., Grimont, F., and Grimont, P.A. (1986). Deoxyribonucleic acid relatedness between *Haemophilus aegyptius* and *Haemophilus influenzae*. Ann. Inst. Pasteur Mic. *137B*, 155–163.

Catry, B., Baele, M., Opsomer, G., de Kruif, A., Decostere, A., and Haesebrouck, F. (2004). tRNA-intergenic spacer PCR for the identification of *Pasteurella* and *Mannheimia* spp. Vet. Microbiol. *98*, 251–260.

Catry, B., Decostere, A., Schwarz, S., Kehrenberg, C., de Kruif, A., and Haesebrouck, F. (2006). Detection of tetracycline-resistant and susceptible pasteurellaceae in the nasopharynx of loose group-housed calves. Vet. Res. Com. *30*, 707–715.

Chanteau, S., Sidikou, F., Djibo, S., Moussa, A., Mindadou, H., and Boisier, P. (2006). Scaling up of PCR-based surveillance of bacterial meningitis in the African meningitis belt: indisputable benefits of multiplex PCR assay in Niger. T. Roy. Soc. r Trop. Med. H. *100*, 677–680.

Chanter, N., Goodwin, R.F., and Rutter, J.M. (1989). Comparison of methods for the sampling and isolation of toxigenic *Pasteurella multocida* from the nasal cavity of pigs. Res. Vet. Sci. *47*, 355–358.

Chen, X., Miflin, J.K., Zhang, P., and Blackall, P.J. (1996). Development and application of DNA probes and PCR tests for *Haemophilus paragallinarum*. Avian Dis. *40*, 398–407.

Chen, X., Chen, Q., Zhang, P., Feng, W., and Blackall, P.J. (1998a). Evaluation of a PCR test for the detection of *Haemophilus paragallinarum* in China. Avian Pathol. *27*, 296–300.

Chen, X., Song, C., Gong, Y., and Blackall, P.J. (1998b). Further studies on the use of a polymerase chain reaction test for the diagnosis of infectious coryza. Avian Pathol. *27*, 618–624.

Choi, C., and Chae, C. (2001). Enhanced detection of toxigenic *Pasteurella multocida* directly from nasal swabs using a nested polymerase chain reaction. Vet. J. *162*, 255–258.

Christensen, H., and Bisgaard, M. (2003). The genus *Pasteurella*. In The Prokaryotes, M. Dworkin, ed. (New York, Springer).

Christensen, H., and Bisgaard, M. (2004). Revised definition of *Actinobacillus sensu stricto* isolated from animals: A review with special emphasis on diagnosis. Vet. Microbiol. *99*, 13–30.

Christensen, H., Bisgaard, M., Angen, O., and Olsen, J.E. (2002). Final classification of Bisgaard taxon 9 as *Actinobacillus arthritidis* sp. nov. and recognition of a novel genomospecies for equine strains of *Actinobacillus lignieresii*. Int. J. Syst. Evol. Microbiol. 52, 1239–1246.

Christensen, H., Bisgaard, M., Bojesen, A.M., Mutters, R., and Olsen, J.E. (2003a). Genetic relationships among avian isolates classified as *Pasteurella haemolytica*, '*Actinobacillus salpingitidis*' or *Pasteurella anatis* with proposal of *Gallibacterium anatis* gen. nov., comb. nov. and description of additional genomospecies within *Gallibacterium* gen. nov. Int. J. Syst. Evol. Microbiol. 53, 275–287.

Christensen, H., Foster, G., Christensen, J.P., Pennycott, T., Olsen, J.E., and Bisgaard, M. (2003b). Phylogenetic analysis by 16S rDNA sequence comparison of avian taxa of Bisgaard and characterization and description of two new taxa of *Pasteurellaceae*. J. Appl. Microbiol. 95, 354–363.

Christensen, H., Angen, O., Olsen, J.E., and Bisgaard, M. (2004a). Revised description and classification of atypical isolates of *Pasteurella multocida* from bovine lungs based on genotypic characterization to include variants previously classified as biovar 2 of *Pasteurella canis* and *Pasteurella avium*. Microbiology *150*, 1757–1767.

Christensen, H., Bisgaard, M., Aalbaek, B., and Olsen, J.E. (2004b). Reclassification of Bisgaard taxon 33, with proposal of *Volucribacter psittacicida* gen. nov., sp. nov. and *Volucribacter amazonae* sp. nov. as new members of the *Pasteurellaceae*. Int. J. Syst. Evol. Microbiol. 54, 813–818.

Christensen, H., Kuhnert, P., Bisgaard, M., Mutters, R., Dziva, F., and Olsen, J.E. (2005). Emended description of porcine *[Pasteurella] aerogenes*, *[Pasteurella] mairii* and *[Actinobacillus] rossii*. Int. J. Syst. Evol. Microbiol. 55, 209–223.

Christensen, H., Hommez, J., Olsen, J.E., and Bisgaard, M. (2006). *[Pasteurella] caballi* infection not limited to horses–a closer look at taxon 42 of Bisgaard. Lett. Appl. Microbiol. 43, 424–429.

Christensen, J.J., Kirkegaard, E., and Korner, B. (1990). *Haemophilus* isolated from unusual anatomical sites. Scand. J. Infect. Dis. 22, 437–444.

Corney, B.G., Diallo, I., Wright, L., Hewitson, G., De Jong, A., Burrell, P., Duffy, P., Stephens, C.P., Rodwell, B., Boyle, D.B., and Blackall, P.J. (2007). *Pasteurella multocida* detection by 5′ *Taq* nuclease assay: a new tool for use in diagnosing fowl cholera. J. Microbiol. Met. 69, 376–380.

Cousins, D.V., and Lloyd, J.M. (1988). Rapid identification of *Haemophilus somnus, Histophilus ovis* and *Actinobacilus seminis* using the API ZYM system. Vet. Microbiol. *17*, 75–81.

Dangor, Y., Radebe, F., and Ballard, R.C. (1993). Transport media for *Haemophilus ducreyi*. Sex. Transm. Dis. *20*, 5–9.

De Castro, N., Pavie, J., Lagrange-Xelot, M., Bouvry, D., Delisle, F., Parrot, A., and Molina, J.M. (2007). Severe *Actinobacillus ureae* meningitis in an immunocompromised patient: Report of one case and review of the literature. Scand. J. Infect. Dis. *39*, 1076–1079.

de Jong, M.F., and Borst, G.H. (1985). Selective medium for the isolation of *P multocida* and *B bronchiseptica*. Vet.Rec. *116*, 167.

Dewhirst, F.E., Paster, B.J., and Bright, P.L. (1989). *Chromobacterium, Eikenella, Kingella, Neisseria, Simonsiella*, and *Vitreoscilla* species comprise a major branch of the beta group *Proteobacteria* by 16S ribosomal ribonucleic acid sequence comparison: transfer of *Eikenella* and *Simonsiella* to the family *Neisseriaceae* (emend.). Int. J. Syst. Bacteriol. *39*, 258–266.

Doern, G.V., and Chapin, K.C. (1984). Laboratory identification of *Haemophilus influenzae*: effects of basal media on the results of the satellitism test and evaluation of the RapID NH system. J. Clin. Microbiol. *20*, 599–601.

Donnio, P.-Y., Lerestif-Gautier, A.-L., and Avril, J.-L. (2004). Characterization of *Pasteurella* spp. Strains Isolated from Human Infections. J. Comp. Pathol. *130*, 137–142.

Duel, P., Siboni, K., and Jensen, T.G. (1991). Intracranial abscesses in Odense Hospital. Survey of bacteriology, epidemiology, and treatment with antibiotics, 1963–1989. Dan. Med. Bull. *38*, 407–410.

Ederer, G.M., and Schurr, M.L. (1971). Optimal bacitracin concentration for selective isolation medium for *Haemophilus*. Am. J. Med. Technol. *37*, 304–305.

Eskola, J., and Kilpi, T. (2000). Potential of bacterial vaccines in the prevention of acute otitis media. Pediatr. Inf. Dis. J. *19*, S72–78.

Fittipaldi, N., Broes, A., Harel, J., Kobisch, M., and Gottschalk, M. (2003). Evaluation and Field Validation of PCR Tests for Detection of *Actinobacillus pleuropneumoniae* in Subclinically Infected Pigs. J. Clin. Microbiol. *41*, 5085–5093.

Frank, G.H., Briggs, R.E., and Debey, B.M. (1992). Bovine tonsils as reservoirs for *Pasteurella haemolytica*: colonisation, immune response, and infection of the nasopharynx. Paper presented at: Pasteurellosis in production animals (Bali, Indonesia, Australian Centre for International Agricultural Research).

Frankard, J., Rodriguez-Villalobos, H., Struelens, M.J., and Jacobs, F. (2004). *Haemophilus parainfluenzae*: an underdiagnosed pathogen of biliary tract infections? Eur. J. Clin. Microbiol. *23*, 46–48.

Friis-Moller, A., Christensen, J.J., Fussing, V., Hesselbjerg, A., Christiansen, J., and Bruun, B. (2001). Clinical Significance and Taxonomy of *Actinobacillus hominis*. J. Clin. Microbiol. *39*, 930–935.

Gagea, M.I., Bateman, K.G., van Dreumel, T., McEwen, B.J., Carman, S., Archambault, M., Shanahan, R.A., and Caswell, J.L. (2006). Diseases and pathogens associated with mortality in Ontario beef feedlots. J. Vet. Diagn. Invest. *18*, 18–28.

Gagné, A., Lacouture, S., Broes, A., D'Allaire, S., and Gottschalk, M. (1998). Development of an immunomagnetic method for selective isolation of *Actinobacillus pleuropneumoniae* serotype 1 from tonsils. J. Clin. Microbiol. *36*, 251–254.

Garcia, A.J., Angulo, E., Blackall, P.J., and Ortiz, A.M. (2004). The presence of nicotinamide adenine dinucleotide-independent *Haemophilus paragallinarum* in Mexico. Avian Dis. *48*, 425–429.

Garlinghouse, L.E., Jr., DiGiacomo, R.F., Van Hoosier, G.L., Jr., and Condon, J. (1981). Selective media for *Pasteurella multocida* and *Bordetella bronchiseptica*. Lab. Anim. Sci. *31*, 39–42.

Gautier, A.-L., Dubois, D., Escande, F., Avril, J.-L., Trieu-Cuot, P., and Gaillot, O. (2005). Rapid and accurate identification of human isolates of *Pasteurella* and related species by sequencing the *sodA* gene. J. Clin. Microbiol. *43*, 2307–2314.

Gerlach, G., and Reidl, J. (2006). NAD^+ utilization in *Pasteurellaceae*: simplification of a complex pathway. J. Bacteriol. *188*, 6719–6727.

Gilbride, K.A., and Rosendal, S. (1983). Evaluation of a selective medium for isolation of *Haemophilus pleuropneumoniae*. Can. J. Comp. Med. *47*, 445–450.

Glisson, J.R., Hofacre, C.L., and Christensen, J.P. (2003). Fowl cholera. In Diseases of Poultry, Y.M. Saif, H.J. Barnes, J.R. Glisson, A.M. Fadly, L.R. McDougald, and D.A. Swayne, eds. (Ames, Iowa State University Press), pp. 658–676.

Gottschalk, M., Broes, A., Mittal, K.R., Kobisch, M., Kuhnert, P., Lebrun, A., and Frey, J. (2003). Non-pathogenic *Actinobacillus* isolates antigenically and biochemically similar to *Actinobacillus pleuropneumoniae*: a novel species? Vet. Microbiol. *92*, 87–101.

Grassberger, R. (1897). Beiträge zür Bakteriologie der influenza. Zentralbalt. Hyg. Infectionskr. *25*, 453–475.

Green, B.T., Ramsey, K.M., and Nolan, P.E. (2002). *Pasteurella multocida* meningitis: case report and review of the last 11 y. Scand. J. Inf. Dis. *34*, 213–217.

Gromkova, R., and Koornhof, H. (1990). Naturally occurring NAD-independent *Haemophilus parainfluenzae*. J. Gen. Microbiol. *136*, 1031–1035.

Groom, S.C., Hazlett, M.J., and Little, P.B. (1986). An evaluation of the API ZYM system as a means of identifying *Haemophilus somnus* and related taxa. Can. J. Vet. Res. *50*, 238–244.

Gu, X., Rossau, R., Jannes, G., Ballard, R., Laga, M., and Van Dyck, E. (1998). The *rrs* (16S)-*rrl* (23S) ribosomal intergenic spacer region as a target for the detection of *Haemophilus ducreyi* by a heminested-PCR assay. Microbiology *144*, 1013–1019.

Hansson, S., Svedhem, A., Wennerstrom, M., and Jodal, U. (2007). Urinary tract infection caused by *Haemophilus influenzae* and *Haemophilus parainfluenzae* in children. Pediatr. Nephrol. 22, 1321–1325.

Hayashimoto, N., Aiba, T., Itoh, K., Kato, M., Kawamoto, E., Kiyokawa, S., Morichika, Y., Muraguchi, T., Narita, T., Okajima, Y., *et al.* (2005). Identification procedure for *Pasteurella pneumotropica* in microbiologic monitoring of laboratory animals. Exp. Anim. *54*, 123–129.

Hedegaard, J., Okkels, H., Bruun, B., Kilian, M., Mortensen, K.K., and Nørskov-Lauritsen, N. (2001). Phylogeny of the genus *Haemophilus* as determined by comparison of partial *inf*B sequences. Microbiology *147*, 2599–2609.

Heinonen, P.K., and Miettinen, A. (1994). Laparoscopic study on the microbiology and severity of acute pelvic inflammatory disease. Eur. J. Obstet. Gyn. R.B. *57*, 85–89.

Holst, E., Rollof, J., Larsson, L., and Nielsen, J.P. (1992). Characterization and distribution of *Pasteurella* species recovered from infected humans. J. Clin. Microbiol. *30*, 2984–2987.

Jacobsen, M.J., and Nielsen, J.P. (1995). Development and evaluation of a selective and indicative medium for isolation of *Actinobacillus pleuropneumoniae* from tonsils. Vet. Microbiol. *47*, 191–197.

Jamal, W., Tamaray, G., Pazhoor, A., and Rotimi, V.O. (2006). Comparative evaluation of BacT/ALERT 3D and BACTEC systems for the recovery of pathogens causing bloodstream infections. Med. Prin. Pract. *15*, 223–227.

Janda, W.M., Malloy, P.J., and Schreckenberger, P.C. (1987). Clinical evaluation of the Vitek *Neisseria-Haemophilus* Identification card. J. Clin. Microbiol. *25*, 37–41.

Janda, W.M., Bradna, J.J., and Ruther, P. (1989). Identification of *Neisseria* spp., *Haemophilus* spp., and other fastidious gram-negative bacteria with the MicroScan *Haemophilus-Neisseria* identification panel. J. Clin. Microbiol. *27*, 869–873.

Kamp, E.M., Bokken, G.C., Vermeulen, T.M., de Jong, M.F., Buys, H.E., Reek, F.H., and Smits, M.A. (1996). A specific and sensitive PCR assay suitable for large-scale detection of toxigenic *Pasteurella multocida* in nasal and tonsillar swabs specimens of pigs. J. Vet. Diagn. Inv. *8*, 304–309.

Khairat, O. (1940). Endocarditis due to a new species of *Haemophilus*. J. Pathol. Bacteriol *50*, 497–505.

Kilian, M. (1974). A rapid method for the differentiation of *Haemophilus* strains. The porphyrin test. Acta Pathol. Microbiol. Im. B *82*, 835–842.

Kilian, M. (2005). Genus *Haemophilus* Winslow, Braodhurst, Buchanan, Rogers and Smith 1917, 561 AL. In Bergey's Manual of Systematic Bacteriology, D.J. Brenner, N.R. Kreig, J.T. Staley, and G.M. Garrity, eds. (New York, Springer-Verlag), pp. 883–904.

Kilian, M., and Frederiksen, W. (1981). Appendix – Identification tables for *Haemophilus-Pasteurella-Actinobacillus* group. In *Haemophilus, Pasteurella* and *Actinobacillus*, M. Kilian, W. Frederiksen, and E.L. Biberstein, eds. (London, Academic Press).

King, E.O., and Tatum, H.W. (1962). *Actinobacillus actinomycetemcomitans* and *Hemophilus aphrophilus*. J. Infect. Dis. *111*, 85–94.

Knight, D.P., Paine, J.E., and Speller, D.C. (1983). A selective medium for *Pasteurella multocida* and its use with animal and human specimens. J. Clin. Pathol. *36*, 591–594.

Korczak, B., Christensen, H., Emler, S., Frey, J., and Kuhnert, P. (2004). Phylogeny of the family *Pasteurellaceae* based on *rpoB* sequences. Int. J. Syst. Evol. Microbiol. *54*, 1393–1399.

Krause, T., Bertschinger, H.U., Corboz, L., and Mutters, R. (1987). V-factor dependent strains of *Pasteurella multocida* subsp. *multocida*. Zbl. Bakt. Hyg. A *266*, 255–260.

Kristensen, C., Angen, O., Andreasen, M., Takai, H., Nielsen, J.P., and Jorsal, S.E. (2004). Demonstration of airborne transmission of *Actinobacillus pleurop-*

neumoniae serotype 2 between simulated pig units located at close range. Vet. Microbiol. *98*, 243–249.

Kuhnert, P., and Korczak, B.M. (2006). Prediction of whole-genome DNA–DNA similarity, determination of G+C content and phylogenetic analysis within the family *Pasteurellaceae* by multilocus sequence analysis (MLSA). Microbiology *152*, 2537–2548.

Kuhnert, P., Frey, J., Lang, N.P., and Mayfield, L. (2002). Phylogenetic analysis of *Prevotella nigrescens, Prevotella intermedia* and *Porphyromonas gingivalis* clinical strains reveals a clear species clustering. Int. J. Syst. Evol. Microbiol. *52*, 1391–1395.

Kuhnert, P., Korczak, B., Falsen, E., Straub, R., Hoops, A., Boerlin, P., Frey, J., and Mutters, R. (2004). *Nicoletella semolina* gen. nov., sp. nov., a new member of *Pasteurellaceae* isolated from horses with airway disease. J. Clin. Microbiol. *42*, 5542–5548.

Kwiecien, J.M., and Little, P.B. (1989). Failure of a selective medium to isolate *Haemophilus somnus* strains. Aust. Vet. J. *66*, 159–160.

Lariviere, S., Leblanc, L., Mittal, K.R., and Martineau, G.-P. (1993). Comparison of isolation methods for the recovery of *Bordetella bronchiseptica* and *Pasteurella multocida* from the nasal cavities of piglets. J. Clin. Microbiol. *31*, 364–367.

Lee, C.W., Wilkie, I.W., Townsend, K.M., and Frost, A.J. (2000). The demonstration of *Pasteurella multocida* in the alimentary tract of chickens after experimental oral infection. Vet. Microbiol. *72*, 47–55.

Lewis, D.A., and Ison, C.A. (2006). Chancroid. Sex. Transm. Infect. *82 Suppl 4*, iv19–20.

Ley, B.E., Linton, C.J., Bennett, D.M., Jalal, H., Foot, A.B., and Millar, M.R. (1998). Detection of bacteraemia in patients with fever and neutropenia using 16S rRNA gene amplification by polymerase chain reaction. Eur. J. Clin. Microbiol. *17*, 247–253.

Lichtensteiger, C.A., Steenbergen, S.M., Lee, R.M., Polson, D.D., and Vimr, E.R. (1996). Direct PCR analysis for toxigenic *Pasteurella multocida*. J. Clin. Microbiol. *34*, 3035–3039.

Lindberg, J., Frederiksen, W., Gahrn-Hansen, B., and Bruun, B. (1998). Problems of identification in clinical microbiology exemplified by pig bite wound infections. Zentralbl. Bakteriol. *288*, 491–499.

Meats, E., Feil, E.J., Stringer, S., Cody, A.J., Goldstein, R., Kroll, J.S., Popovic, T., and Spratt, B.G. (2003). Characterization of Encapsulated and Noncapsulated *Haemophilus influenzae* and Determination of Phylogenetic Relationships by Multilocus Sequence Typing. J. Clin. Microbiol. *41*, 1623–1636.

Miflin, J.K., and Blackall, P.J. (2001). Development of a 23S rRNA-based PCR assay for the identification of *Pasteurella multocida*. Lett. Appl. Microbiol. *33*, 216–221.

Mohan, K., Sadza, M., Madsen, M., Hill, F.W.G., and Pawandiwa, A. (1994). Phenotypic characterization of Zimbabwean isolates of *Pasteurella multocida*. Vet. Microbiol. *38*, 351–357.

Moller, K., and Kilian, M. (1990). V factor-dependent members of the family *Pasteurellaceae* in the porcine upper respiratory tract. J. Clin. Microbiol. *28*, 2711–2716.

Moller, K., Andersen, L.V., Christensen, G., and Kilian, M. (1993). Optimalization of the detection of NAD dependent Pasteurellaceae from the respiratory tract of slaughterhouse pigs. Vet. Microbiol. *36*, 261–271.

Moller, K., Fussing, V., Grimont, P.A.D., Paster, B.J., Dewhirst, F.E., and Kilian, M. (1996). *Actinobacillus minor* sp. nov., *Actinobacillus porcinus* sp. nov., and *Actinobacillus indolicus* sp. nov., three new V factor-dependent species from the respiratory tract of pigs. Int. J. Syst. Bacteriol. *46*, 951–956.

Moore, M.K., Cicnjak-Chubbs, L., and Gates, R.J. (1994). A new selective enrichment procedure for isolating *Pasteurella multocida* from avian and environmental samples. Avian Dis. *38*, 317–324.

Mouahid, M., Bisgaard, M., Morley, A.J., Mutters, R., and Mannheim, W. (1992). Occurrence of V-factor (NAD) independent strains of *Haemophilus paragallinarum*. Vet. Microbiol. *31*, 363–368.

Moyaert, H., Decostere, A., Baele, M., Hermans, K., Tavernier, P., Chiers, K., and Haesebrouck, F. (2008). An unusual *Actinobacillus equuli* strain isolated from a rabbit with Tyzzer's disease. Vet. Microbiol. *In Press, Accepted Manuscript*.

Muhairwa, A.P., Mtambo, M.M.A., Christensen, J.P., and Bisgaard, M. (2001). Occurrence of *Pasteurella multocida* and related species in village free ranging chickens and their animal contacts in Tanzania. Veterinary Microbiology *78*, 139–153.

Munson, E., Pfaller, M., Koontz, F., and Doern, G. (2002). Comparison of porphyrin-based, growth factor-based, and biochemical-based testing methods for identification of *Haemophilus influenzae*. Eur. J. Clin. Microbiol. *21*, 196–203.

Murphy, T.F., Brauer, A.L., Sethi, S., Kilian, M., Cai, X., and Lesse, A.J. (2007). *Haemophilus haemolyticus*: a human respiratory tract commensal to be distinguished from *Haemophilus influenzae*. J. Inf. Dis. *195*, 81–89.

Murray, P.R., Baron, E.J., Jorgensen, J.H., Landry, M.L., and Pfaller, M.A. (2007). Manual of Clinical Microbiology, 9th edn (Washington, D.C, ASM Press).

Mutters, R., Ihm, P., Pohl, S., Frederiksen, W., and Mannheim, W. (1985a). Reclassification of the genus *Pasteurella* Trevisan 1887 on the basis of deoxyribonucliec acid homology, with proposals for the new species *Pasteurella dagmatis, Pasteurella canis, Pasteurella stomatis, Pasteurella anatis*, and *Pasteurella langaa*. Int. J. Syst. Bacteriol. *35*, 309–322.

Mutters, R., Piechulla, K., Hinz, K.-H., and Mannheim, W. (1985b). *Pasteurella avium* (Hinz and Kunjara 1977) comb. nov. and *Pasteurella volantium* sp. nov. Int. J. Syst. Bacteriol. *35*, 5–9.

Nagai, S., Someno, S., and Yagihashi, T. (1994). Differentiation of toxigenic from nontoxigenic isolates of *Pasteurella multocida* by PCR. J. Clin. Microbiol. *32*, 1004–1010.

Niven, D.F., and O'Reilly, T. (1990). Significance of V-factor dependency in the taxonomy of *Haemophilus* species and related organisms. Int. J. Syst. Bacteriol. *40*, 1–4.

Nørskov-Lauritsen, N., Christensen, H., Okkels, H., Kilian, M., and Bruun, B. (2004). Delineation of the genus *Actinobacillus* by comparison of partial *infB* sequences. Int. J. Syst. Evol. Microbiol. *54*, 635–644.

Nørskov-Lauritsen, N., Bruun, B., and Kilian, M. (2005). Multilocus sequence phylogenetic study of the genus *Haemophilus* with description of *Haemophilus pittmaniae* sp. nov. Int. J. Syst. Evol. Microbiol. *55*, 449–456.

Nørskov-Lauritsen, N., and Kilian, M. (2006). Reclassification of *Actinobacillus actinomycetemcomitans*, *Haemophilus aphrophilus*, *Haemophilus paraphrophilus* and *Haemophilus segnis* as *Aggregatibacter actinomycetemcomitans* gen. nov., comb. nov., *Aggregatibacter aphrophilus* comb. nov. and *Aggregatibacter segnis* comb. nov., and emended description of *Aggregatibacter aphrophilus* to include V factor-dependent and V factor-independent isolates. Int. J. Syst. Evol. Microbiol. *56*, 2135–2146.

Nørskov-Lauritsen, N., Overballe, M.D., and Kilian, M. (2008). Delineation of the species *Haemophilus influenzae* by phenotype, multi-locus sequence phylogeny, and detection of marker genes. Submitted.

Nozu, R., Goto, K., Ohashi, H., Takakura, A., and Itoh, T. (1999). Evaluation of PCR as a means of identification of *Pasteurella pneumotropica*. Exp. Anim. *48*, 51–54.

Nye, K.J., Fallon, D., Gee, B., Howe, S., Messer, S., Turner, T., and Warren, R.E. (2001). A comparison of the performance of bacitracin-incorporated chocolate blood agar with chocolate blood agar plus a bacitracin disk in the isolation of *Haemophilus influenzae* from sputum. J. Med. Microbiol. *50*, 472–475.

Oliveira, S., Galina, L., and Pijoan, C. (2001). Development of a PCR test to diagnose *Haemophilus parasuis* infections. J. Vet. Diagn. Invest. *13*, 495–501.

Olsen, I., Dewhirst, F.E., Paster, B.J., and Busse, H.-J. (2005). Family *Pasteurellaceae*. In Bergey's Manual of Systematic Bacteriology, D.J. Brenner, N.R. Kreig, and J.T. Staley, eds. (New York, Springer), pp. 851–856.

Olvera, A., Cerda-Cuellar, M., Nofrarias, M., Revilla, E., Segales, J., and Aragon, V. (2007). Dynamics of *Haemophilus parasuis* genotypes in a farm recovered from an outbreak of Glasser's Disease. Vet. Microbiol. *123*, 230–237.

Palladino, S., Leahy, B.J., and Newall, T.L. (1990). Comparison of the RIM-H rapid identification kit with conventional tests for the identification of *Haemophilus* spp. J. Clin. Microbiol. *28*, 1862–1863.

Paster, B.J., and Dewhirst, F.E. (1988). Phylogeny of campylobacters, wolinellas, *Bacteroides gracilis*, and *Bacteroides ureolyticus* by 16S ribosomal ribonucleic acid sequencing. Int. J. Syst. Bacteriol. *38*, 56–62.

Paturel, L., Casalta, J.P., Habib, G., Nezri, M., and Raoult, D. (2004). *Actinobacillus actinomycetemcomitans* endocarditis. Clin. Microbiol. Infect. *10*, 98–118.

Peltola, H. (2000). Worldwide *Haemophilus influenzae* Type b Disease at the Beginning of the 21st Century: Global Analysis of the Disease Burden 25 Years after the Use of the Polysaccharide Vaccine and a Decade after the Advent of Conjugates. Clin. Microbiol. Rev. *13*, 302–317.

Petti, C.A., Bhally, H.S., Weinstein, M.P., Joho, K., Wakefield, T., Reller, L.B., and Carroll, K.C. (2006). Utility of Extended Blood Culture Incubation for Isolation of *Haemophilus*, *Actinobacillus*, *Cardiobacterium*, *Eikenella*, and *Kingella* Organisms: a Retrospective Multicenter Evaluation. J. Clin. Microbiol. *44*, 257–259.

Pijoan, C., Morrison, R.B., and Hilley, H.D. (1983). Dilution technique for isolation of *Haemophilus* from swine lungs collected at slaughter. J. Clin. Microbiol. *18*, 143–145.

Pittman, M., and Davis, D.J. (1950). Identification of the Koch-Weeks bacillus (*Haemophilus aegyptius*). J. Bacteriol. *59*, 413–426.

Pohl, S., Bertschinger, H.U., Frederiksen, W., and Mannheim, W. (1983). Transfer of *Haemophilus pleuropneumoniae* and the *Pasteurella haemolytica*-like organism causing porcine necrotic pleuropneumonia to the genus *Actinobacillus* (*Actinobacillus pleuropneumoniae* comb. nov.) on the basis of phenotypic and deoxyribonucleic acid relatedness. Int. J. Syst. Bacteriol. *33*, 510–514.

Quentin, R., Ruimy, R., Rosenau, A., Musser, J., and Christen, R. (1996). Genetic identification of cryptic genospecies of *Haemophilus* causing urogenital and neonatal infections by PCR using specific primers targeting genes coding for 16S rRNA. J. Clin. Microbiol. *34*, 1380–1385.

Rapp-Gabrielson, V.J. (1999). *Haemophilus parasuis*. In Diseases of Swine, B. Straw, S. D'Allaire, W.L. Mengeling, and D.J. Taylor, eds. (Ames, Iowa State University Press), pp. 475–481.

Register, K.B., and DeJong, K.D. (2006). Analytical verification of a multiplex PCR for identification of *Bordetella bronchiseptica* and *Pasteurella multocida* from swine. Vet. Microbiol. *117*, 201–210.

Rimler, R.B. (1979). Studies of the pathogenic avian haemophili. Avian Dis. *23*, 1006–1018.

Rimler, R.B., Sandhu, T.S., and Glisson, J.R. (1998). Pasteurellosis, Infectious serositis, and Pseudotuberculosis. In A laboratory manual for the isolation and identification of avian pathogens, D.E. Swayne, ed. (Philadelphia, American Association of Avian Pathogens), pp. 17–25.

Rogers, K.B., Zinnemann, K., and Foster, W.P. (1960). The isolation and identification of *Haemophilus* spp, from unusual lesions in children. J. Clin. Pathol. *13*, 519–524.

Sacchi, C.T., Alber, D., Dull, P., Mothershed, E.A., Whitney, A.M., Barnett, G.A., Popovic, T., and Mayer, L.W. (2005). High Level of Sequence Diversity in the 16S rRNA Genes of *Haemophilus influenzae* Isolates Is Useful for Molecular Subtyping. J. Clin. Microbiol. *43*, 3734–3742.

Saif, Y.M., Barnes, H.J., Glisson, J.R., Fadly, A.M., McDougald, L.R., and Swayne, D.A. (2003). Diseases of Poultry, 11th edn (Ames, Iowa, Iowa State University Press).

Salmon, S.A., Watts, J.L., and Yancey, R.J., Jr. (1993). Evaluation of the RapID NH system for identification of *Haemophilus somnus*, *Pasteurella multocida*, *Pasteurella haemolytica*, and *Actinobacillus pleuropneumoniae* isolated from cattle and pigs with respiratory disease. J. Clin. Microbiol. *31*, 1362–1363.

Sam, I.-C., and Smith, M. (2005). Failure to detect capsule gene *bexA* in *Haemophilus influenzae* types e and f by real-time PCR due to sequence variation within probe binding sites. J. Med. Microbiol. *54*, 453–455.

Saunders, V.F., Reddacliff, L.A., Berg, T., and Hornitzky, M. (2007). Multiplex PCR for the detection of *Brucella ovis*, *Actinobacillus seminis* and *Histophilus somni* in ram semen. Aust. Vet. J. *85*, 72–77; quiz 85.

Savoye, C., Jobert, J.L., Berthelot-Herault, F., Keribin, A.M., Cariolet, R., Morvan, H., Madec, F., and Kobisch, M. (2000). A PCR assay used to study aerosol transmission of *Actinobacillus pleuropneumoniae* from samples of live pigs under experimental conditions. Vet. Microbiol. *73*, 337–347.

Segales, J., Domingo, M., Solano, G.I., and Pijoan, C. (1997). Immunohistochemical detection of *Haemophilus parasuis* serovar 5 in formalin-fixed, paraffin-embedded tissues of experimentally infected swine. J. Vet. Diagn. Invest. *9*, 237–243.

Sethi, S., Evans, N., Grant, B.J., and Murphy, T.F. (2002). New strains of bacteria and exacerbations of chronic obstructive pulmonary disease. New Engl. J. Med. *347*, 465–471.

Simmon, K.E., Croft, A.C., and Petti, C.A. (2006). Application of SmartGene IDNS software to partial 16S rRNA gene sequences for a diverse group of bacteria in a clinical laboratory. J. Clin. Microbiol. *44*, 4400-4406

Slee, K J., and Stephens, L.R. (1985). Selective medium for isolation of *Haemophilus somnus* from cattle and sheep. Vet. Rec. *116*, 215–217.

Slots, J. (1982). Selective medium for isolation of *Actinobacillus actinomycetemcomitans*. J. Clin. Microbiol *15*, 606–609.

Smith, I.M., and Baskerville, A.J. (1983). A selective medium for the isolation of *P. multocida* in nasal specimens from pigs. Brit. Vet. J. *139*, 476–486.

Sneath, P.H., and Stevens, M. (1990). *Actinobacillus rossii* sp. nov., *Actinobacillus seminis* sp. nov., nom. rev., *Pasteurella bettii* sp. nov., *Pasteurella lymphangitidis* sp. nov., *Pasteurella mairi* sp. nov., and *Pasteurella trehalosi* sp. nov. Int. J. Syst. Bacteriol. *40*, 148–153.

Snipes, K.P., and Biberstein, E.L. (1982). *Pasteurella testudinis* sp. nov.: a parasite of desert tortoises (*Gopherus agassizi*). Int. J. Syst. Bacteriol. *32*, 201–210.

Straw, B.E., D'Allaire, S., Mengeling, W.L., and Taylor, D.J. (1999). Diseases of Swine, 9th edn (Ames, Iowa, Iowa State University Press).

Swayne, D.A., Glisson, J.R., Jackwood, M.W., Pearson, J.E., and Reed, W.M. (1998). A Laboratory Manual for The Isolation and Identification of Avian Pathogens, 4th edn, American Association of Avian Pathologists).

Taylor, D.J. (1999). *Actinobacillus pleuropneumoniae*. In Diseases of Swine, B. Straw, S. D'Allaire, W.L. Mengeling, and D.J. Taylor, eds. (Ames, Iowa State University Press), pp. 343–354.

Tekle-Michael, T., Van Dyck, E., Abdellati, S., and Laga, M. (2001). Development of a heminested polymerase chain reaction assay for the detection of *Haemophilus ducreyi* in clinical specimens. Int. J. STD AIDS *12*, 797–803.

Terzolo, H.R., Paolicchi, F.A., Sandoval, V.E., Blackall, P.J., Yamaguchi, T., and Iritani, Y. (1993). Characterization of isolates of *Haemophilus paragallinarum* from Argentina. Avian Dis. *37*, 310–314.

Tonpitak, W., Rohdeb, J., and Gerlach, G.F. (2007). Prevalence of *'Actinobacillus porcitonsillarum'* in porcine tonsils and development of diagnostic duplex-PCR differentiating between *'Actinobacillus porcitonsillarum'* and *Actinobacillus pleuropneumoniae*. Vet. Microbiol. *122*, 157–165.

Totten, P.A., Kuypers, J.M., Chen, C.-Y., Alfa, M.J., Parsons, L.M., Dutro, S.M., Morse, S.A., and Kiviat, N.B. (2000). Etiology of genital ulcer disease in Dakar, Senegal, and comparison of PCR and serologic assays for detection of *Haemophilus ducreyi*. J. Clin. Microbiol. *38*, 268–273.

Townsend, K.M., Frost, A.J., Lee, C.W., Papadimitriou, J.M., and Dawkins, J.S. (1998). Development of PCR assays for species- and type-specific identification of *Pasteurella multocida* isolates. J. Clin. Microbiol. *36*, 1096–1100.

Trees, D., and Morse, S. (1995). Chancroid and *Haemophilus ducreyi*: an update. Clin. Microbiol. Rev. *8*, 357–375.

Tseng, H.K., Su, S.C., Liu, C.P., and Lee, C.M. (2001). *Pasteurella multocida* bacteremia due to non-bite animal exposure in cirrhotic patients: report of two cases. J. Microbiol. Immunol. Infect. *34*, 293–296.

Turni, C., and Blackall, P.J. (2007a). Comparison of sampling sites and detection methods for *Haemophilus parasuis*. Aust. Vet. J. *85*, 177–184.

Turni, C., and Blackall, P.J. (2007b). An evaluation of the *apx*IVA based PCR-REA method for differentiation of *Actinobacillus pleuropneumoniae*. Vet. Microbiol. *121*, 163–169.

Valenza, G., Ruoff, C., Vogel, U., Frosch, M., and Abele-Horn, M. (2007). Microbiological evaluation of the new VITEK 2 *Neisseria-Haemophilus* (NH) identification card. J. Clin. Microbiol. *45*, 3493–3497.

van Ketel, R.J., de Wever, B., and van Alphen, L. (1990). Detection of *Haemophilus influenzae* in cerebrospinal fluids by polymerase chain reaction DNA amplification. J. Med. Microbiol. *33*, 271–276.

Voldstedlund, M., Pedersen, L.N., Fuursted, K., and Nielsen, L.P. (2003). Different polymerase chain reaction-based analyses for culture-negative endocarditis caused by *Streptococcus pneumoniae*. Scand. J. Infect. Dis. *35*, 757–759.

Wang, R.F., Campbell, W., Cao, W.W., Summage, C., Steele, R.S., and Cerniglia, C.E. (1996). Detection of *Pasteurella pneumotropica* in laboratory mice and rats by polymerase chain reaction. Lab. Anim. Sci. *46*, 81–85.

Ward, A.C., Dyer, N.W., and Fenwick, B.W. (1999). *Pasteurellaceae* isolated from tonsillar samples of commercially-reared American bison (Bison bison). Can. J. Vet. Res. *63*, 161–165.

White, D.C., and Granick, S. (1963). Hemin biosynthesis in *Haemophilus*. J. Bacteriol. *85*, 842–850.

White, M.P., Mukkur, T.K.S., Cameron, R.D.A., and Love, R.J. (1993). Epidemiology of pasteurella pneumonia in pigs. In Pasteurellosis in Production Animals, B.E. Patten, T.L. Spencer, R.B. Johnson, D.

Hoffmann, and L. Lehane, eds. (Canberra, Australian Centre for International Agricultural Research), pp. 92–97.

Wilson, R.J., and McOrist, S. (2000). *Actinobacillus suis* infection in pigs in Australia. Aust. Vet. J. *78*, 317–319.

Wongnarkpet, S., Pfeiffer, D.U., Morris, R.S., and Fenwick, S.G. (1999). An on-farm study of the epidemiology of *Actinobacillus pleuropneumoniae* infection in pigs as part of a vaccine efficacy trial. Preventative Vet. Med. 39, 1–11.

Zhou, J., Law, D.K.S., Sill, M.L., and Tsang, R.S.W. (2007). Nucleotide sequence diversity of the *bexA* Gene in serotypeable *Haemophilus influenzae* strains recovered from invasive disease patients in Canada. J. Clin. Microbiol. *45*, 1996–1999.

Ziegler, R., Johnscher, I., Martus, P., Lenhardt, D., and Just, H.-M. (1998). Controlled clinical laboratory comparison of two supplemented aerobic and anaerobic media used in automated blood culture systems to detect bloodstream infections. J. Clin. Microbiol. *36*, 657–661.

Zinnemann, K., Rogers, K.B., Frazer, J., and Devaraj, S.K. (1971). A haemolytic V-dependent CO_2-preferring *Haemophilus* species (*Haemophilus paraphrohaemolyticus* nov. spec.). J. Med. Microbiol. *4*, 139–143.

Index

I

K

L

Q

R

S

Figure 1.3 Typical infection (infectious coryza) by primary pathogen (*Avibacterium paragallinarum*). With permission from and thanks to Professor Pat Blackall.

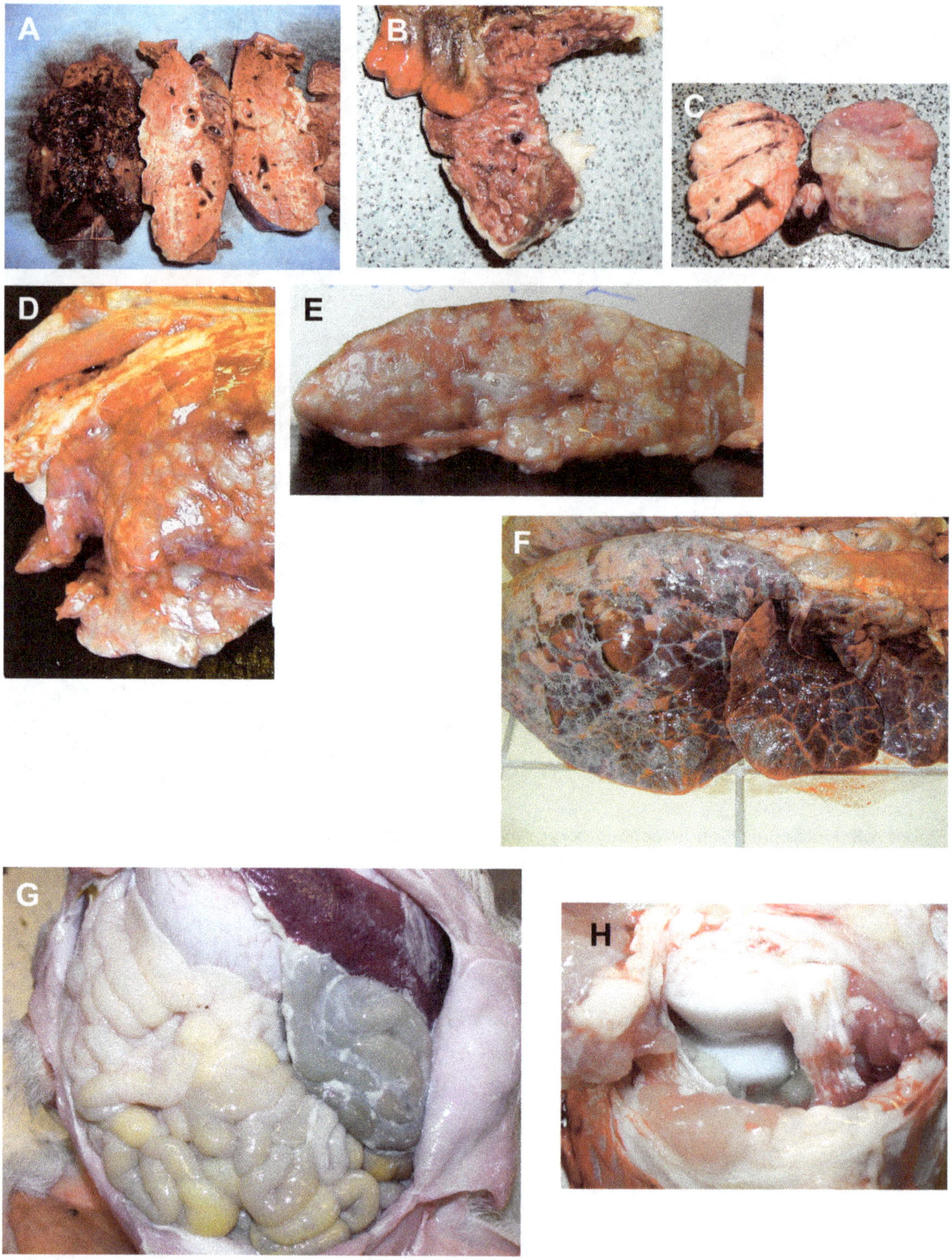

Figure 1.4 (A–E) Typical infection associated with the opportunistic pathogen *Pasteurella multocida*. (A) Pneumonia (fowl cholera) in turkey (cross-section of lung) and (B) pneumonia (fowl cholera) in chicken (cross-section of lung). (C) Normal chicken lung (normal left, affected right). (D and E) Necrotizing bronchopneumonia of pig (E cross-section of lung). (F) Typical bovine infection (acute haemorrhagic fibrinonecrotizing pneumonia) associated with the opportunistic pathogen *Mannheimia haemolytica*. (G and H) Typical infection of pig associated with the opportunistic pathogen [*Haemophilus*] *parasuis*. (G) Acute fibrinopurulent peritonitis and (H) acute fibrinous arthritis. With permission from and thanks to Professor M. Bisgaard, Professor H. E. Jensen and PhD student S. Pors.

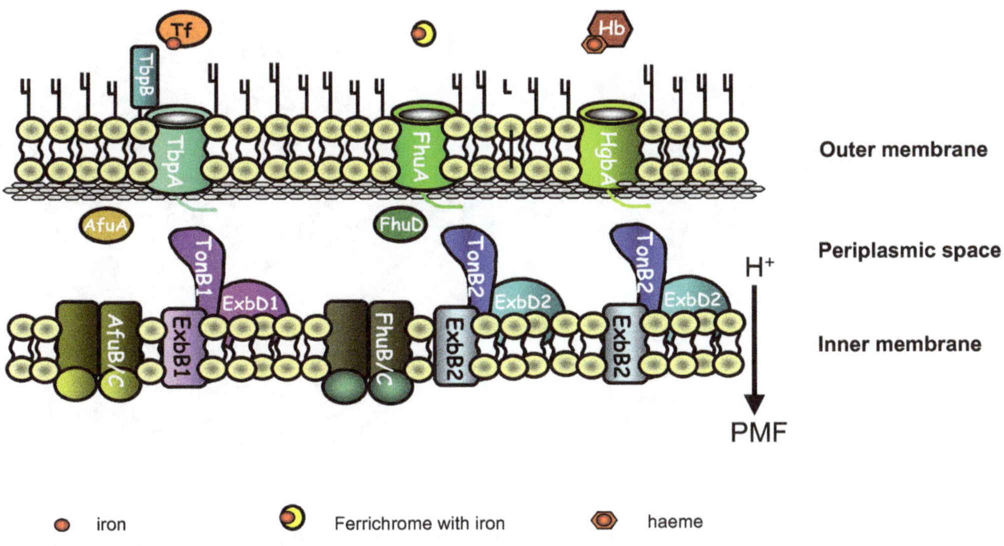

Figure 8.1 Representation of iron-uptake systems identified in *A. pleuropneumoniae.* The AfuABC proteins are homologous to the FbpABC proteins of *Neisseria*, responsible for transporting iron, from transferrin (Tf), into the cell. Tf is bound to TbpBA that is energized by the TonB1 system, one of two Ton systems that transduces energy from the proton motive force (PMF) at the inner membrane. Ferrichrome is taken up by FhuA, bound by FhuD and delivered to FhuBC for iron transport into the cell. Haemoglobin (Hb) interacts with HgbA, whereupon haem is removed and crosses the outer membrane. Ferrichrome and haem uptake are energized by the TonB2 system.

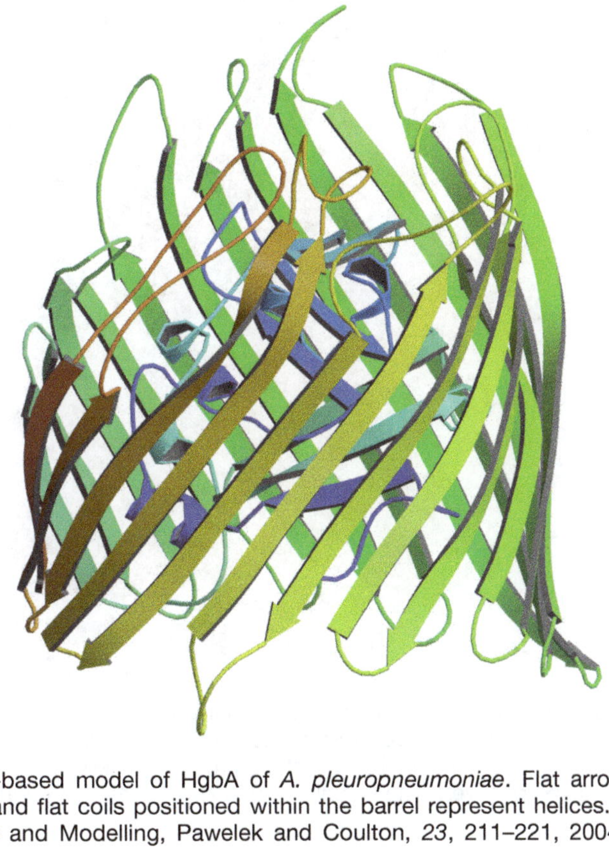

Figure 8.4 Homology-based model of HgbA of *A. pleuropneumoniae*. Flat arrows represent β-strands, lines represent loops and flat coils positioned within the barrel represent helices. Reprinted from Journal of Molecular Graphics and Modelling, Pawelek and Coulton, *23*, 211–221, 2004, with permission from Elsevier.

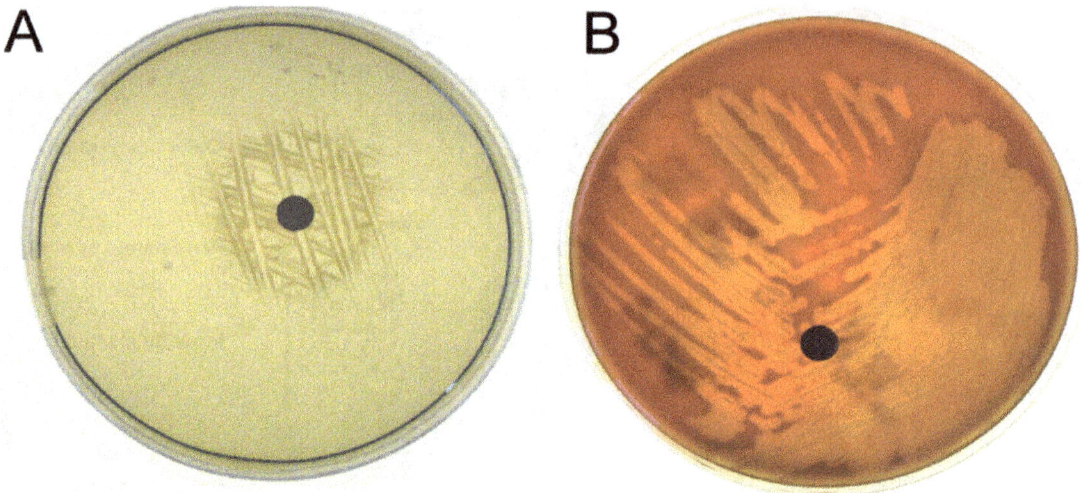

Figure 11.1 *H. pittmaniae* cultured for 24 h on different agars in the presence of a NAD-containing disc and photographed by transillumination. (A) Todd-Hewitt agar. Growth is apparent only around the NAD disc. (B) horse blood agar. Growth of haemolytic colonies is visible over the entire plate.

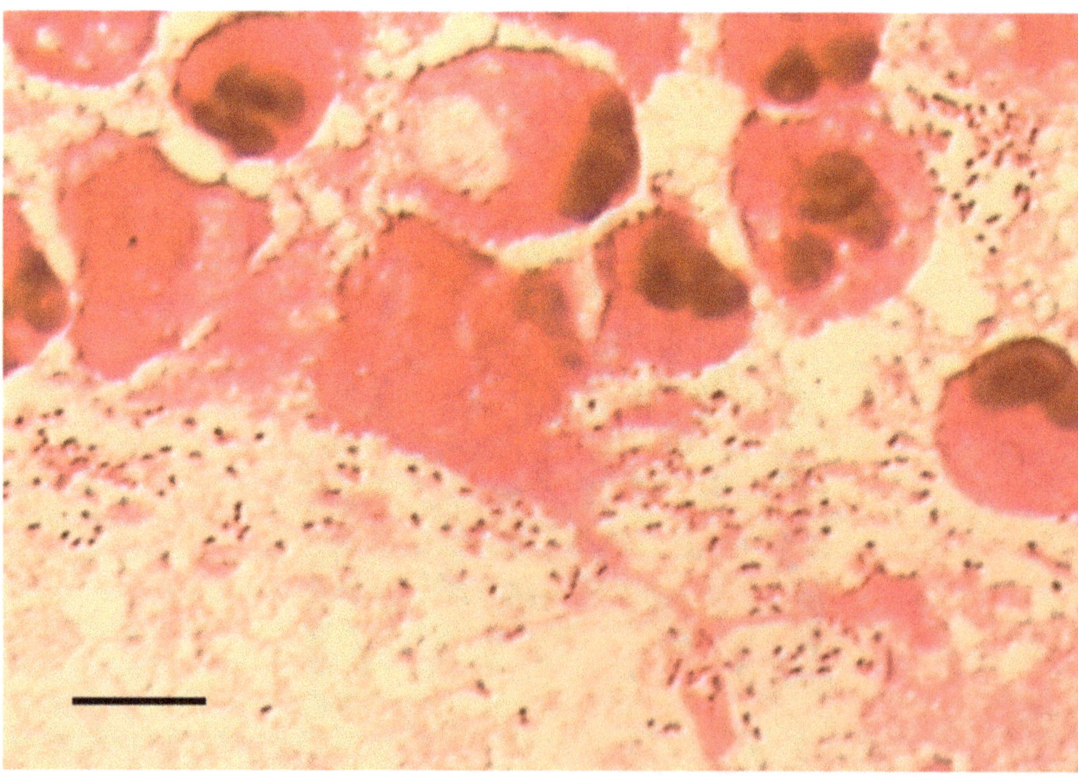

Figure 11.2 Microscopic examination of sputum from a patient with bronchiectasis and exacerbated expectoration. Neutrophil granulocytes are seen adjacent to clusters of Gram-negative coccobacilli. Original magnification ×1000, bar represents 10 μm. Number of squamous epithelial cells and leukocytes per low power field were 0 and > 25, respectively. The sample yielded *H. influenzae* in pure culture after incubation on growth media.

www.ingramcontent.com/pod-product-compliance
Lightning Source LLC
LaVergne TN
LVHW062227200926
848484LV00006B/1492